Revolutionizing Transportation Safety With AI

Qamar Ali
https://orcid.org/0000-0002-8178-7591
Virtual University of Pakistan, Pakistan

Muhammad Aamir Mahmood
https://orcid.org/0009-0007-4452-2627
Virtual University of Pakistan, Pakistan

Muhammad Tariq Iqbal Khan
Government Graduate College, Jaranwala, Pakistan

IGI Global
Scientific Publishing
Publishing Tomorrow's Research Today

Vice President of Editorial | Melissa Wagner
Director of Acquisitions | Mikaela Felty
Director of Book Development | Jocelynn Hessler
Production Manager | Mike Brehm
Cover Design | Jose Rosado

Published in the United States of America by
IGI Global Scientific Publishing
701 East Chocolate Avenue
Hershey, PA, 17033, USA
Tel: 717-533-8845 | Fax: 717-533-7115
Website: https://www.igi-global.com E-mail: cust@igi-global.com

Library of Congress Cataloging-in-Publication Data

Names: Ali, Qamar, 1988- editor | Mahmood, Muhammad Aamir, 1979- editor | Iqbal Khan, Muhammad Tariq, 1988- editor
Title: Revolutionizing transportation safety with AI / [edited by] Qamar Ali, Muhammad Aamir Mahmood, Muhammad Tariq Iqbal Khan.
Description: Hershey, PA : IGI Global Scientific Publishing, [2027] | Includes bibliographical references and index. | Summary: "This book explores how AI is transforming road safety, from autonomous vehicles to predictive traffic management. Combining case studies, analysis, and ethical considerations, it examines how advanced technologies can reduce accidents, optimize driver behavior, and shape safer, more efficient transportation systems worldwide"-- Provided by publisher.
Identifiers: LCCN 2025039385 (print) | LCCN 2025039386 (ebook) | ISBN 9798337332314 hardcover | ISBN 9798337332321 paperback | ISBN 9798337332338 ebook
Subjects: LCSH: Transportation--Safety measures | Artificial intelligence--Engineering applications | Transportation--Technological innovations
Classification: LCC HE194 .R48 2025 (print) | LCC HE194 (ebook) | DDC 629.04028/9--dc23/eng/20251121
LC record available at https://lccn.loc.gov/2025039385
LC ebook record available at https://lccn.loc.gov/2025039386

British Cataloguing in Publication Data
A Cataloguing in Publication record for this book is available from the British Library.

Table of Contents

Preface xvi

Chapter 1
AI-Driven Transportation Safety: A Multi-Modal Approach to Sustainable Urban Mobility and Risk Mitigation 1
Soumi Ghosh, Maharaja Agrasen Institute of Technology, India
Ritik Raj, Maharaja Agrasen Institute of Technology, India

Chapter 2
AI-Powered Traffic and Infrastructure Management: Mission and Concerns 35
G. Aravind, KCG College of Technology, India

Chapter 3
AI-Powered Traffic Management and Optimization 73
Kavita Kanwar, Poornima University, Jaipur, India
Ritam Dutta, Poornima Institute of Engineering and Technology, Jaipur, India
J. Rajalakshmi, Sethu Institute of Technology, India
Vivek Chamoli, Graphic Era Hill University, Dehradun, India
P. Shubha, Dr. Ambedkar Institute of Technology, Bengaluru, India
Nikhil Kumar Goyal, Noida International University, Greater Noida, India

Chapter 4
AI-Powered Traffic Management and Optimization: Towards Safer and Smarter Urban Mobility 103
Mukimov Askar Shukhratovich, Alfraganus University, Tashkent, Uzbekistan
Ravikumar R. N., Marwadi University, Rajkot, India
Duggirala Aravind, ASET College of Science and Technology, Chennai, India
S. Aarthi, Marwadi University, Rajkot, India
Dilmurod Turaev, Termez University of Economics and Service, Termez, Uzbekistan
Anorgul Ashirova, Mamun University, Khiva, Uzbekistan

Chapter 5
Designing Affordable Infrastructures for AI-Driven Transportation Safety in Resource-Constrained Regions 135
Shrusti Sonker, Chhatrapati Shahu Ji Maharaj University, India
Shivansu Sachan, CSJM University, Kanpur, India

Chapter 6
Autonomous and AI-Enabled Vehicles for Green Transportation Systems 169
Sitesh Kumar Singh, College of Engineering, National University of Science and Technology, Muscat, Oman
Jaishri Gothania, Lingaya's Vidyapeeth, Faridabad, India
Shyam Agasthya H. R., College of Engineering, National University of Science and Technology, Muscat, Oman

Chapter 7
Revolutionizing Vehicle Safety Through AI-Driven V2X and Digital Twin Technologies 197
Ismatulla Khayrullayev, Termez University of Economics and Service, Uzbekistan
Shruthishree S. H., Global Academy of Technology, India
Abdullayev Abdulla Fayzulla Ugli, Tashkent State University of Economics, Uzbekistan
Nethravathi H. M., BGS Institute of Technology, Adichunchanagiri University, India
Anorgul Ashirova, Mamun University, Uzbekistan
G. Swetha, RR Institute of Technology, India
N. Raghu, Jain (Deemed to be) University, India
Kamal Narayanan, University of Nevada, USA
Arjun Kumar G. B., Dayananda Sagar College of Engineering, India
Qosimbetov Batirbay, Karakalpak State University, Uzbekistan

Chapter 8
IoT and AI for Real-Time Safety Monitoring 229
Prakash Upadhyay, Xavier University, India
S. Pushpalatha, Dr. Ambedkar Institute of Technology, India
N. Raja Kumar, Lords Institute of Engineering and Technology, India
Vinay Raj R., Yenepoya (Deemed to be) University, India
Anjaneyulu Gudla, Guru Nanak Internation Technical Campus, India
R. Suyam Praba, Karpagam College of Engineering, India
Bala Krishnan S., Vaagai IP, India

Chapter 9
Machine Learning Applications in Transportation Safety 257
Soumi Ghosh, Maharaja Agrasen Institute of Technology, India
Ritik Raj, Maharaja Agrasen Institute of Technology, India

Chapter 10
Machine Learning for Accident Prediction and Prevention 355
Kezia Rani Burgula, Vasavi College of Engineering, India
Sheela D. V., Sapthagiri NPS University, India
Padmesh Medesety, V.S.B. Engineering College, India
Abinaya. M., Erode Sengunthar Engineering College, India
Naveen Nandal, DIHE, India
Abdul Razzak Khan Qureshi, Medicaps University, India
V. Bhoopathy, Sree Rama Engineering College, India

Chapter 11
Proactive Risk Mitigation in Urban Environments: An AI-Powered Framework for Predictive Accident Modeling and Prevention.......................... 383
Vikram Das, Sikkim Manipal University, India

Chapter 12
Natural Language Processing for Incident Reporting and Response 413
Darniati Darniati, Muhammadiyah University of Makassar, Indonesia
Nizirwan Anwar, Universitas Esa Unggul, Indonesia
Muhammad Faisal, Muhammadiyah University of Makassar, Indonesia
Irmawati Irmawati, Politeknik Negeri Ujung Pandang, Indonesia
Alfiah Abdullah, STMIK Tidore Mandiri, Indonesia
Yogita Abichandani, Management Development Institute, Murshidabad, India
Raden Teddy Iswahyudi, Universitas Esa Unggul, Indonesia
Tomy Ronaldi, Indonesia Defense University, Indonesia
Yudhi Biantoro, Universitas Respati Indonesia, Indonesia
Imam Asrowardi, Politeknik Negeri Lampung, Indonesia

Chapter 13
Cybersecurity Challenges in AI-Enabled Transportation Systems 445
S. Arjune, Christ (Deemed to be) University, India
A. Hariprasad Reddy, Geethanjali College of Engineering and Technology, India
S. Mangai, Velalar College of Engineering and Technology, India
K. R. Swetha, BGS Institute of Technology, Adichunchanagiri University, India
V. Joseph Emmanuvel, Periyar Maniammai Institute of Science and Technology, India
Naveen Nandal, DIHE, India
V. Bhoopathy, Sree Rama Engineering College, India

Chapter 14
Ethical and Regulatory Considerations in AI-Driven Transportation Safety ... 473
Harry Sutanto, Binus University, Indonesia
Mohammad Ichsan, Binus University, Indonesia

Chapter 15
Revolutionizing Transportation Safety With AI: Passenger Behavior, Smart Mobility Experiences, and Future Prospect 505
Vanshika Srivastava, Maharishi Markandeshwar (Deemed to be) University, Mullana, India
Shivansu Sachan, CSJM University, Kanpur, India
Rajiv, Maharishi Markandeshwar (Deemed to be) University, Mullana, India
Mayank Mishra, CSJM University, Kanpur, India
Sudhanshu Pandiya, CSJM University, Kanpur, India
Aishwarya Arya, CSJM University, Kanpur, India
Ankit Kumar, CSJM University, Kanpur, India

About the Contributors 535

Index 551

Detailed Table of Contents

Preface.. xvi

Chapter 1
AI-Driven Transportation Safety: A Multi-Modal Approach to Sustainable Urban Mobility and Risk Mitigation.. 1
Soumi Ghosh, Maharaja Agrasen Institute of Technology, India
Ritik Raj, Maharaja Agrasen Institute of Technology, India

Congestion and over-urbanization puts pressure on urban transportation. This paper presents AI-based multi-modal transportation safety with the help of machine learning, computer vision, IoT sensors, and predictive analytics in the context of autonomous vehicles, transit, and cycling as well as pedestrian infrastructure. The smart system real-time data is processed with predictive and adaptive controls. The implementation of this lowers traffic accidents by 45-60 per cent, boosts traffic performance by 30-40 per cent, and decreases carbon emission by 25-35 per cent. This multi-modes solution seals the safety loopholes due to the overall coverage, dynamic adaptation and sustainable urban development solutions, which is the basis of smart city transportation systems that emphasises safety, sustainability and efficiency.

Chapter 2
AI-Powered Traffic and Infrastructure Management: Mission and Concerns 35
G. Aravind, KCG College of Technology, India

AI-powered traffic and infrastructure management offers a transformative solution to the growing challenges of urbanization, congestion, and safety in transportation systems. By utilizing machine learning, computer vision, and predictive analytics, AI enhances traffic flow, predicts hazards, and improves emergency response. However, challenges such as cybersecurity risks, data privacy, integration with legacy systems, and ethical concerns remain. AI must address issues like algorithmic bias, liability in accidents, and inconsistent global regulations. Successful implementation requires collaboration between governments, researchers, and policymakers to create secure, ethical, and adaptable AI systems for transportation.

Chapter 3
AI-Powered Traffic Management and Optimization 73
Kavita Kanwar, Poornima University, Jaipur, India
Ritam Dutta, Poornima Institute of Engineering and Technology, Jaipur, India
J. Rajalakshmi, Sethu Institute of Technology, India
Vivek Chamoli, Graphic Era Hill University, Dehradun, India
P. Shubha, Dr. Ambedkar Institute of Technology, Bengaluru, India
Nikhil Kumar Goyal, Noida International University, Greater Noida, India

This chapter examines how artificial intelligence (AI) has been transformative in traffic management and optimization with the view that it can help to solve the consistent issues of congestion, inefficiency, and road safety in contemporary transportation. The conventional traffic control systems are usually not capable of dynamically adjusting to evolving circumstances which lead to time wastage, pollution and loss of profits. With the incorporation of the AI methods like machine learning, computer vision, and reinforcement learning, traffic systems would be able to reach the real-time flexibility, prediction precision and enhanced efficiency. The chapter discusses AI-based applications, such as smart traffic lights, route optimization, incident detection, and optimisation of public transport schedules. It further focuses on how supporting technologies such as IoT, 5G, and autonomous vehicles can be used to establish intelligent transportation ecosystems.

Chapter 4
AI-Powered Traffic Management and Optimization: Towards Safer and Smarter Urban Mobility 103
Mukimov Askar Shukhratovich, Alfraganus University, Tashkent, Uzbekistan
Ravikumar R. N., Marwadi University, Rajkot, India
Duggirala Aravind, ASET College of Science and Technology, Chennai, India
S. Aarthi, Marwadi University, Rajkot, India
Dilmurod Turaev, Termez University of Economics and Service, Termez, Uzbekistan
Anorgul Ashirova, Mamun University, Khiva, Uzbekistan

Rapid urbanization and rising vehicle density have made conventional traffic management systems inadequate in addressing congestion, accidents, and environmental impacts. This chapter explores how Artificial Intelligence (AI) offers transformative solutions through predictive analytics, computer vision, and machine learning for dynamic traffic monitoring and optimization. Case studies from Los Angeles and Singapore demonstrate quantifiable improvements in travel time, delays, emissions, and public transport efficiency, emphasizing the adaptability of AI to diverse infrastructures. The discussion also highlights ethical and regulatory considerations, including privacy, transparency, and fairness, while acknowledging technical and social challenges such as infrastructure readiness and public trust. Finally, the chapter proposes a scalable AI-driven framework integrating IoT, edge computing, 5G, and V2I communication to build safer, efficient, and sustainable urban transport systems.

Chapter 5
Designing Affordable Infrastructures for AI-Driven Transportation Safety in Resource-Constrained Regions 135
Shrusti Sonker, Chhatrapati Shahu Ji Maharaj University, India
Shivansu Sachan, CSJM University, Kanpur, India

Abstract: Designing safe and efficient transportation systems is a critical challenge in resource-constrained regions, where financial and technological limitations often impede the implementation of advanced safety measures. This study explores strategies for developing affordable infrastructures that leverage AI-driven technologies to enhance traffic safety. Emphasizing low-cost sensors, computer vision, and real-time data analytics, the research investigates how these tools can monitor traffic behavior, predict potential hazards, and support rapid interventions at high-risk locations. Case studies from urban and peri-urban areas demonstrate that AI integration, when combined with context-sensitive design, can significantly reduce accidents, optimize traffic flow, and improve accessibility. The study underscores the importance of scalable, cost-effective, and locally adaptable solutions to bridge the gap between technological potential and practical implementation in developing regions.

Chapter 6
Autonomous and AI-Enabled Vehicles for Green Transportation Systems 169
Sitesh Kumar Singh, College of Engineering, National University of Science and Technology, Muscat, Oman
Jaishri Gothania, Lingaya's Vidyapeeth, Faridabad, India
Shyam Agasthya H. R., College of Engineering, National University of Science and Technology, Muscat, Oman

This chapter reviews how AI-enabled and AI-autonomous vehicles will have a transformational impact on meeting sustainable transportation and low-carbon transportation demands. It looks at the examples in vehicle navigation, traffic prediction, energy-efficient propulsion, and fleet optimization, the examples coming forward that propose to cut down emissions, traffic efficiency, and urban mobility. The combination of AVs with smart city technology and communication with elements of V2V and V2I open communication are described, together with the issues of safety, compliance, data protection, and infrastructure preparedness. The examples of the autonomous electric bus, the autonomous taxi, and the AI-driven logistics show the feasible results of sustainability. Future research directions are outlined at the end of the chapter and include a focus on scalable, ethical, and integrated solutions towards AI-driven transport systems for sustainable mobility in urban areas.

Chapter 7
Revolutionizing Vehicle Safety Through AI-Driven V2X and Digital Twin Technologies 197
Ismatulla Khayrullayev, Termez University of Economics and Service, Uzbekistan
Shruthishree S. H., Global Academy of Technology, India
Abdullayev Abdulla Fayzulla Ugli, Tashkent State University of Economics, Uzbekistan
Nethravathi H. M., BGS Institute of Technology, Adichunchanagiri University, India
Anorgul Ashirova, Mamun University, Uzbekistan
G. Swetha, RR Institute of Technology, India
N. Raghu, Jain (Deemed to be) University, India
Kamal Narayanan, University of Nevada, USA
Arjun Kumar G. B., Dayananda Sagar College of Engineering, India
Qosimbetov Batirbay, Karakalpak State University, Uzbekistan

Connected car technologies are redefining vehicle safety by enabling cars to communicate, sense their surroundings, and respond to real-time conditions. Today's vehicles go beyond transportation they act as intelligent systems working to prevent accidents before they happen. Features like Vehicle-to-Vehicle and Vehicle-to-Infrastructure communication, advanced driver assistance systems and automated alerts help drivers stay aware of potential dangers and make quicker, safer decisions. The support of fast networks like 5G, cloud computing, and edge data processing, these innovations are becoming more effective and accessible. Automakers such as Tesla, Volvo, and Ford are already integrating these technologies into mainstream vehicles, making safety smarter and more proactive. Concerns about cybersecurity, data privacy, and uneven infrastructure still pose hurdles. Legal and regulatory frameworks are evolving but often struggle to keep pace with the technology.

Chapter 8
IoT and AI for Real-Time Safety Monitoring 229
Prakash Upadhyay, Xavier University, India
S. Pushpalatha, Dr. Ambedkar Institute of Technology, India
N. Raja Kumar, Lords Institute of Engineering and Technology, India
Vinay Raj R., Yenepoya (Deemed to be) University, India
Anjaneyulu Gudla, Guru Nanak Internation Technical Campus, India
R. Suyam Praba, Karpagam College of Engineering, India
Bala Krishnan S., Vaagai IP, India

Industrial systems, healthcare, smart cities, and transportation are just a few of the many areas that stand to benefit from the new opportunities for real-time safety monitoring made possible by the merging of the IoT with AI. Internet of Things (IoT) enables continuous data collection through linked devices and sensors; artificial intelligence (AI) enhances this ecosystem by real-time data analysis, trend detection, risk forecasting, and preventative action initiation. This chapter delves into the new frontiers of safety monitoring made possible by the integration of IoT and AI. It covers several frameworks and architectures, showcases real-world applications, and tackles issues like data privacy, latency, scalability, and energy efficiency. Future directions are also covered, including edge AI, federated learning, and the Internet of Things made possible by 6G.

Chapter 9
Machine Learning Applications in Transportation Safety 257
Soumi Ghosh, Maharaja Agrasen Institute of Technology, India
Ritik Raj, Maharaja Agrasen Institute of Technology, India

Every year 1.35M are killed in road traffic accidents. Machine learning changes the field of transport safety by means of real-time choices, predictive analytics and autopilots. In this analysis, crash prediction, driver behavior, infrastructure surveillance, and smart systems are explored. Major areas of weakness: absence of single evaluation systems, absence of edge computing to support real-time processing, inadequate regulatory compliance of deep learning, and inefficiencies of multimodal integration. New solutions suggested: ensemble and federated learning as a privacy-preserving accuracy, and a unified safety assessment system between predictive and autonomous technologies. This study critically examines ML in transportation safety identification of barriers to its adoption, and future growth of intelligent transportation systems that need to involve interdisciplinary cooperation between MI researchers, engineers, and policymakers.

Chapter 10
Machine Learning for Accident Prediction and Prevention 355
Kezia Rani Burgula, Vasavi College of Engineering, India
Sheela D. V., Sapthagiri NPS University, India
Padmesh Medesety, V.S.B. Engineering College, India
Abinaya. M., Erode Sengunthar Engineering College, India
Naveen Nandal, DIHE, India
Abdul Razzak Khan Qureshi, Medicaps University, India
V. Bhoopathy, Sree Rama Engineering College, India

This chapter explores how ML-based modelsincluding Random Forest, CNN, LSTM, and Reinforcement Learning frameworksare revolutionizing accident prediction and prevention across domains such as smart transportation, industrial safety, logistics, and autonomous systems. It highlights end-to-end ML workflows covering data acquisition, feature engineering, model training, real-time alert systems, and deployment using IoT and AI-integrated infrastructure. The chapter also discusses practical implementations through case studies, along with technical, ethical, and policy challenges such as data imbalance, privacy concerns, and decision accountability in autonomous systems. By integrating AI-driven predictive analytics with proactive safety interventions, ML demonstrates significant potential to reduce human error, optimize risk assessment, and enable intelligent safety ecosystems for future-ready industries and smart cities.

Chapter 11
Proactive Risk Mitigation in Urban Environments: An AI-Powered Framework for Predictive Accident Modeling and Prevention.......................... 383
Vikram Das, Sikkim Manipal University, India

Urban centers face significant challenges in road traffic safety due to complex interactions between vehicles, pedestrians, cyclists, and infrastructure. Traditional reactive approaches to road safety have proven insufficient in preventing accidents. This chapter presents a comprehensive AI-powered framework for proactive risk mitigation in urban environments. The framework leverages predictive analytics, real-time data integration, and machine learning to forecast and neutralize potential traffic conflicts before they escalate into accidents. The chapter details the architecture of the predictive safety system, explores advanced machine learning algorithms for high-fidelity accident prediction, discusses AI-driven interventions, presents real-world case studies, and addresses ethical considerations. This transformative approach promises to revolutionize urban transportation safety by shifting from reactive measures to proactive prevention, contributing to the development of smart, safe, and sustainable cities.

Chapter 12
Natural Language Processing for Incident Reporting and Response 413
Darniati Darniati, Muhammadiyah University of Makassar, Indonesia
Nizirwan Anwar, Universitas Esa Unggul, Indonesia
Muhammad Faisal, Muhammadiyah University of Makassar, Indonesia
Irmawati Irmawati, Politeknik Negeri Ujung Pandang, Indonesia
Alfiah Abdullah, STMIK Tidore Mandiri, Indonesia
Yogita Abichandani, Management Development Institute, Murshidabad, India
Raden Teddy Iswahyudi, Universitas Esa Unggul, Indonesia
Tomy Ronaldi, Indonesia Defense University, Indonesia
Yudhi Biantoro, Universitas Respati Indonesia, Indonesia
Imam Asrowardi, Politeknik Negeri Lampung, Indonesia

The integration of Artificial Intelligence and Natural Language Processing (NLP) is transforming transportation safety by enabling real-time, automated incident reporting and response. NLP techniques such as entity recognition, sentiment analysis, and event classification allow for efficient extraction of critical data from unstructured sources like driver reports, emergency calls, and social media. This enhances operational awareness, reduces response times, and supports proactive safety strategies. By merging NLP with systems like GIS, IoT, and generative AI, transportation agencies can optimize dispatching, improve decision-making, and foster data-driven safety cultures. In multilingual, data-rich environments - especially in Southeast Asia - tailored NLP solutions further address communication barriers and reporting gaps. The future of transportation safety lies in leveraging advanced, ethically governed NLP models for adaptive and inclusive urban mobility systems.

Chapter 13
Cybersecurity Challenges in AI-Enabled Transportation Systems 445
S. Arjune, Christ (Deemed to be) University, India
A. Hariprasad Reddy, Geethanjali College of Engineering and Technology, India
S. Mangai, Velalar College of Engineering and Technology, India
K. R. Swetha, BGS Institute of Technology, Adichunchanagiri University, India
V. Joseph Emmanuvel, Periyar Maniammai Institute of Science and Technology, India
Naveen Nandal, DIHE, India
V. Bhoopathy, Sree Rama Engineering College, India

The article discusses how to apply encryption and network segmentation, two traditional methodologies, to the cybersecurity of ITS. The procedures that were mentioned, the problems they solved, and the architectural layers where they were applied are all connected in the end. Applications of AI in transportation systems, such as smart logistics, intelligent traffic management, and autonomous cars, have transformed efficiency, safety, and mobility. Nevertheless, these systems are vulnerable to cybersecurity threats due to their dependence on networked devices, real-time data sharing, and complicated algorithms. Threats to AI models, data poisoning, networks, ransomware, and supply chain security are some of the main obstacles to securing AI-enabled transportation that are discussed in this chapter. It may learn a lot about vulnerabilities and responses in the real world from case studies. The chapter stresses the need for worldwide cooperation, robust designs, and AI-driven cybersecurity solutions to keep AI-enabled transportation systems safe from ever-changing cyber threats.

Chapter 14

Ethical and Regulatory Considerations in AI-Driven Transportation Safety ... 473

Harry Sutanto, Binus University, Indonesia

Mohammad Ichsan, Binus University, Indonesia

Artificial Intelligence (AI) is revolutionizing transportation through autonomous vehicles, smart traffic systems, and predictive safety tools. While these innovations enhance efficiency and reduce accidents, they also raise ethical, legal, and social challenges. This chapter highlights the need for strong governance, transparent algorithms, and inclusive stakeholder collaboration. Clear legal frameworks must define liability, while ethical oversight ensures fairness and accessibility. By viewing AI as both a technological advancement and societal responsibility, policymakers, industry, and civil society can foster safer, equitable, and trustworthy transportation systems, offering a model for responsible technology adoption across sectors.

Chapter 15

Revolutionizing Transportation Safety With AI: Passenger Behavior, Smart Mobility Experiences, and Future Prospect 505

Vanshika Srivastava, Maharishi Markandeshwar (Deemed to be) University, Mullana, India

Shivansu Sachan, CSJM University, Kanpur, India

Rajiv, Maharishi Markandeshwar (Deemed to be) University, Mullana, India

Mayank Mishra, CSJM University, Kanpur, India

Sudhanshu Pandiya, CSJM University, Kanpur, India

Aishwarya Arya, CSJM University, Kanpur, India

Ankit Kumar, CSJM University, Kanpur, India

The integration of Artificial Intelligence (AI) into transportation systems is revolutionizing safety measures, shaping passenger behavior, and transforming mobility experiences. By leveraging real-time data analytics, machine learning, and predictive modeling, AI enables proactive accident prevention, optimized traffic flow, and enhanced emergency response mechanisms. Beyond technical advancements, AI also influences passenger decision-making, fostering safer travel habits through personalized guidance, behavioral monitoring, and adaptive interventions. The rise of smart mobility—autonomous vehicles, connected infrastructure, and intelligent transport services—further extends the scope of AI-driven safety innovations. These systems not only improve operational efficiency but also create seamless, user-centric mobility experiences. However, challenges remain in ensuring ethical AI use, addressing data privacy, and achieving equitable access to AI-enabled transport solutions.

About the Contributors 535

Index 551

Preface

OVERVIEW OF THE SUBJECT MATTER

Revolutionizing Transportation Safety With AI presents a timely and multidisciplinary discussion of how artificial intelligence is reshaping the future of transportation safety. Transportation has always supported social development, economic growth, trade, tourism, and daily human mobility. At the same time, it remains one of the most risk sensitive sectors in the modern world. Road crashes, congestion, driver fatigue, poor visibility, weak infrastructure, cyber threats, unsafe passenger behavior, and slow incident response continue to affect people, cities, and economies. These challenges show that transportation safety can no longer depend only on traditional control systems or reactive management.

Artificial intelligence offers a new way to understand, predict, and reduce these risks. AI systems can process real time traffic data, detect hazards through computer vision, predict accidents through machine learning, monitor driver behavior, support autonomous vehicles, and strengthen smart infrastructure. AI can also improve route planning, reduce traffic delays, lower emissions, and connect transport safety with sustainable development. This book therefore treats AI not only as a technical tool but also as a safety framework, a policy concern, a legal challenge, and a social innovation.

The volume brings together chapters that study AI driven safety from several angles. It begins with historical and developmental perspectives, then moves toward smart infrastructure, traffic management, autonomous vehicles, V2X systems, computer vision, IoT, machine learning, accident prediction, driver behavior, NLP, cybersecurity, ethics, legal regulation, hospitality mobility, operational safety, and smart safety gear. This broad coverage is important because transportation safety is not a single technology problem. It is a connected system that includes vehicles, roads, data, people, institutions, and service environments.

RELEVANCE OF THE TOPIC IN THE CONTEMPORARY WORLD

The subject of this book fits strongly within the world today. Transport systems are becoming more connected, automated, and data driven. Smart cities now depend on intelligent traffic signals, IoT sensors, vehicle to everything communication, digital twins, mobile applications, automated monitoring, and predictive analytics. Public expectations for safety are also increasing. Governments, engineers, planners, and industry leaders are expected to design mobility systems that are safe, inclusive, affordable, sustainable, and legally responsible.

AI has become central to this transformation because it can detect patterns that are difficult for humans to observe in large and fast moving transport systems. It can support early warnings, accident hotspot detection, adaptive traffic control, emergency response, passenger monitoring, and automated decision support. These applications are especially useful in cities where traffic pressure, population growth, climate stress, and infrastructure limits create complex risks. AI also supports sustainable mobility by reducing avoidable delays, improving public transport reliability, and helping users make safer travel choices.

However, the rise of AI in transportation also creates serious questions. Intelligent mobility systems depend on large amounts of data, connected devices, automated decisions, and digital infrastructure. This creates concerns about privacy, bias, accountability, cybersecurity, legal responsibility, and unequal access. The book addresses these concerns directly. It shows that the future of transportation safety depends not only on advanced algorithms but also on ethical design, affordable infrastructure, strong regulation, and human centered implementation.

TARGET AUDIENCE

This book is intended for researchers, postgraduate students, engineers, transport planners, AI developers, policymakers, urban safety professionals, legal scholars, smart city designers, and industry practitioners. It is also useful for professionals working in road safety, public transport, autonomous mobility, traffic control, emergency response, hospitality transport services, infrastructure planning, and sustainable urban development. The wide scope of the book allows readers from technical and non technical backgrounds to understand the role of AI in safer transportation systems.

For academic readers, the book provides conceptual foundations, emerging research directions, and multidisciplinary debates. For practitioners, it offers practical insight into AI enabled monitoring, safety management, risk modeling, and infrastructure design. For policymakers and regulators, it explains the governance, legal, ethical, and cybersecurity issues that must be addressed before AI based transportation systems can be trusted at scale. For students, it offers a structured introduction to a growing field that connects artificial intelligence, engineering, mobility, sustainability, law, and public safety.

CHAPTER OVERVIEWS

Chapter 1: AI-Driven Transportation Safety: A Multi-Modal Approach to Sustainable Urban Mobility and Risk Mitigation

Congestion and over-urbanization puts pressure on urban transportation. This paper presents AI-based multi-modal transportation safety with the help of machine learning, computer vision, IoT sensors, and predictive analytics in the context of autonomous vehicles, transit, and cycling as well as pedestrian infrastructure. The smart system real-time data is processed with predictive and adaptive controls. The implementation of this lowers traffic accidents by 45-60 per cent, boosts traffic performance by 30-40 per cent, and decreases carbon emission by 25-35 per cent. This multi-modes solution seals the safety loopholes due to the overall coverage, dynamic adaptation and sustainable urban development solutions, which is the basis of smart city transportation systems that emphasises safety, sustainability and efficiency.

Chapter 2: AI-Powered Traffic and Infrastructure Management: Mission and Concerns

AI-powered traffic and infrastructure management offers a transformative solution to the growing challenges of urbanization, congestion, and safety in transportation systems. By utilizing machine learning, computer vision, and predictive analytics, AI enhances traffic flow, predicts hazards, and improves emergency response. However, challenges such as cybersecurity risks, data privacy, integration with legacy systems, and ethical concerns remain. AI must address issues like algorithmic bias, liability in accidents, and inconsistent global regulations. Successful implementation requires collaboration between governments, researchers, and policymakers to create secure, ethical, and adaptable AI systems for transportation.

Chapter 3: AI-Powered Traffic Management and Optimization

This chapter examines how artificial intelligence (AI) has been transformative in traffic management and optimization with the view that it can help to solve the consistent issues of congestion, inefficiency, and road safety in contemporary transportation. The conventional traffic control systems are usually not capable of dynamically adjusting to evolving circumstances which lead to time wastage, pollution and loss of profits. With the incorporation of the AI methods like machine learning, computer vision, and reinforcement learning, traffic systems would be able to reach the real-time flexibility, prediction precision and enhanced efficiency. The chapter discusses AI-based applications, such as smart traffic lights, route optimization, incident detection, and optimisation of public transport schedules. It further focuses on how supporting technologies such as IoT, 5G, and autonomous vehicles can be used to establish intelligent transportation ecosystems.

Chapter 4: AI-Powered Traffic Management and Optimization: Towards Safer and Smarter Urban Mobility

Rapid urbanization and rising vehicle density have made conventional traffic management systems inadequate in addressing congestion, accidents, and environmental impacts. This chapter explores how Artificial Intelligence (AI) offers transformative solutions through predictive analytics, computer vision, and machine learning for dynamic traffic monitoring and optimization. Case studies from Los Angeles and Singapore demonstrate quantifiable improvements in travel time, delays, emissions, and public transport efficiency, emphasizing the adaptability of AI to diverse infrastructures. The discussion also highlights ethical and regulatory considerations, including privacy, transparency, and fairness, while acknowledging technical and social challenges such as infrastructure readiness and public trust. Finally, the chapter proposes a scalable AI-driven framework integrating IoT, edge computing, 5G, and V2I communication to build safer, efficient, and sustainable urban transport systems.

Chapter 5: Designing Affordable Infrastructures for AI-Driven Transportation Safety in Resource-Constrained Regions

Abstract: Designing safe and efficient transportation systems is a critical challenge in resource-constrained regions, where financial and technological limitations often impede the implementation of advanced safety measures. This study explores strategies for developing affordable infrastructures that leverage AI-driven technologies to enhance traffic safety. Emphasizing low-cost sensors, computer vision, and real-time data analytics, the research investigates how these tools can monitor traffic behavior, predict potential hazards, and support rapid interventions at high-risk locations. Case studies from urban and peri-urban areas demonstrate that AI integration, when combined with context-sensitive design, can significantly reduce accidents, optimize traffic flow, and improve accessibility. The study underscores the importance of scalable, cost-effective, and locally adaptable solutions to bridge the gap between technological potential and practical implementation in developing regions.

Chapter 6: Autonomous and AI-Enabled Vehicles for Green Transportation Systems

This chapter reviews how AI-enabled and AI-autonomous vehicles will have a transformational impact on meeting sustainable transportation and low-carbon transportation demands. It looks at the examples in vehicle navigation, traffic prediction, energy-efficient propulsion, and fleet optimization, the examples coming forward that propose to cut down emissions, traffic efficiency, and urban mobility. The combination of AVs with smart city technology and communication with elements of V2V and V2I open communication are described, together with the issues of safety, compliance, data protection, and infrastructure preparedness. The examples of the autonomous electric bus, the autonomous taxi, and the AI-driven logistics show the feasible results of sustainability. Future research directions are outlined at the end of the chapter and include a focus on scalable, ethical, and integrated solutions towards AI-driven transport systems for sustainable mobility in urban areas.

Chapter 7: Revolutionizing Vehicle Safety Through AI-Driven V2X and Digital Twin Technologies

Connected car technologies are redefining vehicle safety by enabling cars to communicate, sense their surroundings, and respond to real-time conditions. Today's vehicles go beyond transportation they act as intelligent systems working to prevent accidents before they happen. Features like Vehicle-to-Vehicle and Vehicle-to-Infrastructure communication, advanced driver assistance systems and automated alerts help drivers stay aware of potential dangers and make quicker, safer decisions. The support of fast networks like 5G, cloud computing, and edge data processing, these innovations are becoming more effective and accessible. Automakers such as Tesla, Volvo, and Ford are already integrating these technologies into mainstream vehicles, making safety smarter and more proactive. Concerns about cybersecurity, data privacy, and uneven infrastructure still pose hurdles. Legal and regulatory frameworks are evolving but often struggle to keep pace with the technology.

Chapter 8: IOT and AI for Real Time Safety Monitoring

Industrial systems, healthcare, smart cities, and transportation are just a few of the many areas that stand to benefit from the new opportunities for real-time safety monitoring made possible by the merging of the IoT with AI. Internet of Things (IoT) enables continuous data collection through linked devices and sensors; artificial intelligence (AI) enhances this ecosystem by real-time data analysis, trend detection, risk forecasting, and preventative action initiation. This chapter delves into the new frontiers of safety monitoring made possible by the integration of IoT and AI. It covers several frameworks and architectures, showcases real-world applications, and tackles issues like data privacy, latency, scalability, and energy efficiency. Future directions are also covered, including edge AI, federated learning, and the Internet of Things made possible by 6G.

Chapter 9: Machine Learning Applications in Transportation Safety: Machine Learning Applications in Transportation Safety

Every year 1.35M are killed in road traffic accidents. Machine learning changes the field of transport safety by means of real-time choices, predictive analytics and autopilots. In this analysis, crash prediction, driver behavior, infrastructure surveillance, and smart systems are explored. Major areas of weakness: absence of single evaluation systems, absence of edge computing to support real-time processing, inadequate regulatory compliance of deep learning, and inefficiencies of multimodal integration. New solutions suggested: ensemble and federated learning as a privacy-preserving accuracy, and a unified safety assessment system between predictive and autonomous technologies. This study critically examines ML in transportation safety identification of barriers to its adoption, and future growth of intelligent transportation systems that need to involve interdisciplinary cooperation between MI researchers, engineers, and policymakers.

Chapter 10: Machine Learning for Accident Prediction and Prevention

This chapter explores how ML-based modelsincluding Random Forest, CNN, LSTM, and Reinforcement Learning frameworksare revolutionizing accident prediction and prevention across domains such as smart transportation, industrial safety, logistics, and autonomous systems. It highlights end-to-end ML workflows covering data acquisition, feature engineering, model training, real-time alert systems, and deployment using IoT and AI-integrated infrastructure. The chapter also discusses practical implementations through case studies, along with technical, ethical, and policy challenges such as data imbalance, privacy concerns, and decision accountability in autonomous systems. By integrating AI-driven predictive analytics with proactive safety interventions, ML demonstrates significant potential to reduce human error, optimize risk assessment, and enable intelligent safety ecosystems for future-ready industries and smart cities.

Chapter 11: Proactive Risk Mitigation in Urban Environments: An AI-Powered Framework for Predictive Accident Modeling and Prevention

Urban centers face significant challenges in road traffic safety due to complex interactions between vehicles, pedestrians, cyclists, and infrastructure. Traditional reactive approaches to road safety have proven insufficient in preventing accidents. This chapter presents a comprehensive AI-powered framework for proactive risk mitigation in urban environments. The framework leverages predictive analytics, real-time data integration, and machine learning to forecast and neutralize potential traffic conflicts before they escalate into accidents. The chapter details the architecture of the predictive safety system, explores advanced machine learning algorithms for high-fidelity accident prediction, discusses AI-driven interventions, presents real-world case studies, and addresses ethical considerations. This transformative approach promises to revolutionize urban transportation safety by shifting from reactive measures to proactive prevention, contributing to the development of smart, safe, and sustainable cities.

Chapter 12: Natural Language Processing for Incident Reporting and Response

The integration of Artificial Intelligence and Natural Language Processing (NLP) is transforming transportation safety by enabling real-time, automated incident reporting and response. NLP techniques such as entity recognition, sentiment analysis, and event classification allow for efficient extraction of critical data from unstructured sources like driver reports, emergency calls, and social media. This enhances operational awareness, reduces response times, and supports proactive safety strategies. By merging NLP with systems like GIS, IoT, and generative AI, transportation agencies can optimize dispatching, improve decision-making, and foster data-driven safety cultures. In multilingual, data-rich environments - especially in Southeast Asia - tailored NLP solutions further address communication barriers and reporting gaps. The future of transportation safety lies in leveraging advanced, ethically governed NLP models for adaptive and inclusive urban mobility systems.

Chapter 13: Cybersecurity Challenges in AI-Enabled Transportation Systems

The article discusses how to apply encryption and network segmentation, two traditional methodologies, to the cybersecurity of ITS. The procedures that were mentioned, the problems they solved, and the architectural layers where they were applied are all connected in the end. Applications of AI in transportation systems, such as smart logistics, intelligent traffic management, and autonomous cars, have transformed efficiency, safety, and mobility. Nevertheless, these systems are vulnerable to cybersecurity threats due to their dependence on networked devices, real-time data sharing, and complicated algorithms. Threats to AI models, data poisoning, networks, ransomware, and supply chain security are some of the main obstacles to securing AI-enabled transportation that are discussed in this chapter. It may learn a lot about vulnerabilities and responses in the real world from case studies. The chapter stresses the need for worldwide cooperation, robust designs, and AI-driven cybersecurity solutions to keep AI-enabled transportation systems safe from ever-changing cyber threats.

Chapter 14: Ethical and Regulatory Considerations in AI-Driven Transportation Safety

Artificial Intelligence (AI) is revolutionizing transportation through autonomous vehicles, smart traffic systems, and predictive safety tools. While these innovations enhance efficiency and reduce accidents, they also raise ethical, legal, and social challenges. This chapter highlights the need for strong governance, transparent algorithms, and inclusive stakeholder collaboration. Clear legal frameworks must define liability, while ethical oversight ensures fairness and accessibility. By viewing AI as both a technological advancement and societal responsibility, policymakers, industry, and civil society can foster safer, equitable, and trustworthy transportation systems, offering a model for responsible technology adoption across sectors.

Chapter 15: Revolutionizing Transportation Safety with AI: Passenger Behavior, Smart Mobility Experiences, and Future Prospect

The integration of Artificial Intelligence (AI) into transportation systems is revolutionizing safety measures, shaping passenger behavior, and transforming mobility experiences. By leveraging real-time data analytics, machine learning, and predictive modeling, AI enables proactive accident prevention, optimized traffic flow, and enhanced emergency response mechanisms. Beyond technical advancements,

AI also influences passenger decision-making, fostering safer travel habits through personalized guidance, behavioral monitoring, and adaptive interventions. The rise of smart mobility—autonomous vehicles, connected infrastructure, and intelligent transport services—further extends the scope of AI-driven safety innovations. These systems not only improve operational efficiency but also create seamless, user-centric mobility experiences. However, challenges remain in ensuring ethical AI use, addressing data privacy, and achieving equitable access to AI-enabled transport solutions.

CONCLUSION AND CONTRIBUTION OF THE BOOK

Overall, this book contributes to the field by presenting transportation safety as a connected, intelligent, and multidisciplinary domain. It does not limit AI to automation or traffic control. Instead, it shows that AI affects every layer of transport safety, including infrastructure, vehicles, drivers, passengers, cybersecurity, law, ethics, urban planning, sustainability, hospitality mobility, and wearable design. This broad approach is necessary because modern transport risks are connected across technologies, institutions, and human behavior.

The book also highlights the need for inclusive and affordable AI systems. This point is especially important for regions where traffic safety problems are high but resources are limited. AI should not become a technology available only to advanced cities and wealthy economies. It should also support safer roads, better monitoring, and more reliable mobility in developing regions and resource constrained environments. This requires simple, scalable, ethical, and cost effective solutions.

Revolutionizing Transportation Safety With AI therefore offers both a technical and human centered contribution. It helps readers understand where transportation safety has come from, where it stands today, and where it is moving in the age of artificial intelligence. Its central message is clear. The future of transportation safety depends not only on smarter machines but also on responsible design, strong governance, ethical regulation, and inclusive innovation. By bringing these perspectives together, the book advances the discussion of AI as a practical and responsible force for safer mobility futures.

Chapter 1

AI-Driven Transportation Safety:

A Multi-Modal Approach to Sustainable Urban Mobility and Risk Mitigation

Soumi Ghosh
http://orcid.org/0000-0003-2764-5944
Maharaja Agrasen Institute of Technology, India

Ritik Raj
http://orcid.org/0009-0001-1812-8112
Maharaja Agrasen Institute of Technology, India

ABSTRACT

Congestion and over-urbanization puts pressure on urban transportation. This paper presents AI-based multi-modal transportation safety with the help of machine learning, computer vision, IoT sensors, and predictive analytics in the context of autonomous vehicles, transit, and cycling as well as pedestrian infrastructure. The smart system real-time data is processed with predictive and adaptive controls. The implementation of this lowers traffic accidents by 45-60 per cent, boosts traffic performance by 30-40 per cent, and decreases carbon emission by 25-35 per cent. This multi-modes solution seals the safety loopholes due to the overall coverage, dynamic adaptation and sustainable urban development solutions, which is the basis of smart city transportation systems that emphasises safety, sustainability and efficiency.

DOI: 10.4018/979-8-3373-3231-4.ch001

1. INTRODUCTION

Regardless of the fact that urbanization of people around the world is at unprecedented levels, it has posed unprecedented challenges to transportation systems across the globe. According to the United Nations, it is expected that by 2050, approximately 68 percent of the world's population will be living in cities and urban areas, and radical solutions to the problem of urban mobility and safety in transport are noted (United Nations, 2019). Transport systems of the past that relied mainly on single-modal tasks in cities are increasingly no longer applicable to the multicomponent demands of modern cities. Integration of artificial intelligence (AI) technologies is a revolutionary chance to solve these challenges with innovative multimodal transportation safety systems.

This issue also has no exception to the road safety of overland transportation, with high priorities. It concerns the entire world, as today road traffic accidents are the leading cause of mortality of the young population (5-29 years), and more than 1.35 million people die in this process annually in different parts of the roads (World Health Organization, 2018). Transportation injuries and causes of death not only has a direct impact on the economy through healthcare expenditures, but it also has an impact through loss of productivity, infrastructure and ecological harm. The old methods of safe-cy route might work in any way, just that they have reached the extreme and can no longer handle the complexity and inter-dependent relationship that are now in the systems of transportation in urban beings.

The implementation of AI technologies such as machine learning, computer vision, natural language processing and predictive analytics give the opportunity to transform transportation into a safer system because such can actively detect threat, take real-time actions and react flexibly to its infrastructure. These technologies can be applied in construction of intelligent transportation system (ITS), and they will be able to process large volumes of data many times more quickly, draw patterns and abnormalities, and respond to such abnormalities many times prior to the occurrence of an accident.

The new use of e-scooters and the ride-sharing solutions is changing the interdependence of various transportation modes and this serves as a new paradigm shift of the traditional method of single-mode transportation known as multi-modal transport systems. Such systems are too complex to be relied on to self operate due to effective coordination processes that will maximize safety, efficiency and sustainability across all mode of operations at the same time.

The topicality of the proposed research work is that we could find artificial intelligence-related holistic answers, i.e. to the problem of driving safety, which would be aware of the mechanics of mobility systems in cities. Integrating multiple approaches to transportation and integrating them into one program with the help of AI, this project is considered the first one to pass through new avenues of risk management, accident-prevention and further evolution of the city. It is possible to suggest that the research is the supplement to the existing literature on the subject of smart transport systems and it functions as the means of offering the answer to the practical implementation of smart transportation in new city initiatives.

1.1 Research Objectives

Primary research objectives are:

1. **Implement Multiple Transportation Modes into a Setup AI Framework:** Build and create a full artificial intelligence framework to combine a broader range of diverse transportation into have a better safety and risk avoidance component.
2. **Multi-Mode Safety Optimization: Theoretical foundations**: Design Multidimensional Modes in safety parameters: Design a dynamic algorithm to select multi-modality modes of movement in terms of interaction, dependence or reliance on each other.
3. **Predictive Third:** The third is more predictive modeling that is added to allow proactive evaluation of realtime riskiness and the forecasting of preventive intervention strategies that would avoid accidents before their occurrence. occurrence.
4. **Sustainable:** Add the ecological element and indications of sustainability in the AI battery of transportation safety configuration to support the agenda of attaining the long-term urban change.
5. **Validation and Performance Testing:** involves large scale testing and validation of the system offered against actual world data and scenarios to prove efficiency and backup high quality.

1.2 Research Scope and Limitations

Cities transport, infrastructure establishment sites and data collection sites are the theme of the given work. The scope encompasses the whole range of passenger transportation which includes motor cars, transport infrastructure and bicycle infrastructure as well as walking infrastructure. These technologies dencompass an overview of the current technology in the industry, and there are newer systems such as autonomous vehicles and intelligent infrastructure.

Weaknesses of the study are that the quality of data and its availability are required, multi-modal systems of this type are complex to implement on a large scale within a city, and it takes a significant amount of infrastructure. Also, regulators and policy issues, although necessary in the identification of this technical research, are not prioritized, but are important points to note when implementing in real life.

2. LITERATURE REVIEW

2.1 Evolution of Transportation Safety Systems

Transportation Safety Systems have evolved since ancient times. Cities transport, infrastructure establishment sites and data collection sites are the theme of the given work. This very notion of transportation safety has been transformed in the sense that it is no longer passive but, nonetheless, is an active notion that can be applied to the prevention of accidents by interventions in the form of intelligent safety systems, which also provide prediction-based interventions. The methods of transport safety that were in place at that time were preoccupied with reactive safety features such as road lamps, road signs, and car-based gravity safety devices (Hauer, 1997). Intelligent transportation systems began with the introduction of computer-based systems in the 1980s, which enabled the development of more advanced traffic control and safety surveillance.

Creation of the advanced driver assistance systems (ADAS) was an essential step in the evolution of transportation safety. Singh (2015) showed that with an appropriate implementation and adoption, ADAS technologies, such as collision avoidance systems, lane departure warnings, and adaptive cruise control, could decrease the number of traffic accidents by up to 40 percent. Nevertheless, they were more individual vehicle oriented systems rather than multi-modal.

The latest achievements in terms of artificial intelligence and machine learning have also presented fresh opportunities within the sphere of the transportation safety systems. The AI-based systems that will help to remove and forecast traffic congestion are able to maximize safety improvement (35-50 percent better than the traditional solutions), like in the study by Chavhan et al. (2022). All these Internet of Things (IoT) sensors, computer vision and big data analytics are capable of generating holistic transportation security that has the capability to monitor, process and respond to the multidimensional transportation of cities.

2.2 Artificial Intelligence in Transportation

Miscellaneous facets of transportation, including artificial intelligence in the form of self-driving vehicles, optimization of traffic flow, and predictive maintenance, have expanded and rapidly changed to use artificial intelligence. In this case, our researchers have demonstrated that machine learning solutions have a certain potential somewhere in the transportation industries (Verma et al., 2024).

The application of deep learning, that is, convolutional neural network (CNNs) and recurrent neural networks (RNNs) has shown to be even more successful when applied to activities linked to transportation. Compared to Fu et al. (2016) who predicted traffic flow using RNN, but still, the timing interval: 24 hours, Li et al. (2021) CNN-based traffic sign recognition systems reached high accuracy rate of 98.5%.

Transportation safety has been transformed by computer vision technologies, which perform real-time detection of objects, assess and analyze their behaviors, and evaluate risks. Highly advanced image processing can identify the object (pedestrians and vehicles), road, and dangerous cases (Tian et al., 2019) with extremely high accuracy and minimal latency. All these sensors together include cameras, LiDAR, radar and GPS, thus creating what can be referred to as the general awareness of the environment needed to transport operations to be safe and conducive.

Among applications of natural language processing (NLP) in transportation are voice-activated safety systems, automated incident reporting, and public communication systems. The technologies improve user engagement with the transportation systems and offer more safety monitoring and reaction features (Kurniawan et al., 2021).

2.3 Multi-Modal Transportation Systems

Multi-modal transportation systems constitute an all-around solution to urban transportation, where a multi-modal transport system incorporates the use of a combination of transportation modes in order to allow smooth, efficient, and secure transportation services to users. The concept extends over the modal integration to include planned, operational, and administrative aspects of different transport networks (Taylor et al., 2000).

Multi-modal systems have the benefit of increased accessibility, less congestion, improved environmental sustainability, and overall, a more efficient transportation system. However, the risk management and reduction through coordination of multiple modes poses challenges of enormous magnitude. The conventional method of ensuring single-mode security may not suffice in the new category of risks within the multiple-mode context involving transference by alternative methods of transport (Guzman et al., 2024).

Recent studies have centred on the development of integrated multi-modal transportation network control. Alyavina et al. (2022) have also provided a centralized management system, and the scheme integrates the output of the traffic light system and other mobility transport and crossing rules. The results they obtained were that they reduced intermodal transportation conflict by 28 percent and raised safety levels system-wide by 22 percent.

(Mobility as a Service) MaaS systems have also placed the problems of coherence of multi-modal safety systems in a framework that demands it be enforced. These platforms enable their users to have an assortment of transportation options at their fingertips and require a lot of safety administration across all of the available modes within the offer (Alyavina et al., 2022).

2.4 Risk Assessment and Mitigation in Transportation

Transportation systems risk assessment is the identification, analysis, and examination of possible risks resulting in accidents, injuries, or system malfunctions. The three processes critical to determining the risk areas and conditions in the traditional risk assessment tools are the historical data analysis, statistical modeling, and descriptive statistics (Makarova et al., 2022). The active risk analysis methods actively apply real-time data gathering and evaluation methods to detect the development of potential risks before they lead to incidents. They are conceptualized models which have been developed taking into consideration different types of information like weather, traffic, structure, and behavioural information about human beings to provide a well-rounded analysis of the hazards (Makarova et al., 2022).

With the implementation of artificial intelligence in risk assessment, more advanced prediction and quicker reaction speeds became possible. Machine learning algorithms have the advantage of identifying meaningful patterns and correlations in large datasets, which may be overlooked in conventional analysis techniques. Predictive models based on artificial intelligence are pretty helpful and can assist in predicting the future and taking precautionary measures to prevent accidents (Verma et al., 2024).

The AI-based systems are averting the risk by controlling traffic flows by systems-generated remote-handling, manually maximizing the routes, and issuing of real-time warnings and alerts, as well as emergency response operations. These strategies need to be effectively incorporated into the elements of the system and different modes of transport to become effective (Chavhan et al., 2022).

2.5 Sustainable Transportation and Environmental Considerations

Incorporation of sustainability concept to transportation safety systems is becoming very significant. Concepts of sustainability in transportation safety systems have been increasingly acquiring relevance in the last few years as cities fight against climate change and environmental degradation. The goals of the sustainable ways of transportation are to prevent greenhouse gas emissions, to improve the air quality, to reduce the impact on the environment and to keep or raise the safety standards (Lucas et al., 2023).

Transportation safety systems have opportunities and challenges in regard to the incorporation of electric and hybrid vehicles. They possess a number of safety standards and risk reduction requirements over the traditional internal combustion engine cars. The artificial needs of the electric cars will involve artificial intelligence to confront aspects such as braking, sounds, and the hardware requirement (Lucas et al., 2023).

There is a lot of walking, biking; these measures are to be thought of as an object of environmentally friendly urban mobility, but much should be paid attention to. The safety of such vulnerable road users can be enhanced through the efficacy of AI-powered systems and their enhanced ability to locate road users along with the specialized infrastructure management, as well as coordinated traffic control systems (Lucas et al., 2023).

The basic principles of the circular economy in terms of transportation are directed at the sphere of resource use and the effort to ensure that the waste is as low as possible during the lifecycle of the transportation system. This would entail an AI solution to not only targets of particular short-term performance measures regarding safety and efficiency, but longer sustainability impact and resource optimization programs (Lucas et al., 2023).

2.6 Smart Cities and Intelligent Infrastructure

New city projects offer a technical-organization platform through which more extensive transport safety frameworks can be used through the use of artificial intelligence. They are making smarter, more sustainable and livable urban places by integrating as many urban systems as possible (transport, energy, communications, government, etc) (Chavhan et al., 2022).

Smart infrastructure refers to interconnected traffic lights, light roads, sensor networks and high-speed communication networks to enable AI transportation safety. These infrastructural components offer the information gathering and the associated communicative features of the real-time systems functioning and optimization (Chavhan et al., 2022).

Internet of Things (IoT) is required to form connections between different elements of the transportation system; it provides an opportunity to collect and process the data to their full potential. The sensors used in the IoT will also have the opportunity not only to track the flow of traffic but also the air conditions, not only the state of the infrastructure, but also the users, which will allow achieving the gain of implementing the AI system (Chavhan et al., 2022).

On-the-fly local data processing and decision-making: Edge computing technologies can minimize latency and increase responsiveness of the system by processing data in real-time at the regional level. The technologies are needed in the safety-related use cases where human beings need to respond to accidents by preventing them in real-time or mitigating risks (Chavhan et al., 2022).

2.7 Research Gaps and Opportunities

Despite the wide utilization of AI in transportation systems, there remain some gaps in the crucial research, which limit the applicability and practice of multi-modal holistic safety approaches

1. **Integration Complexity**: Current research has primarily focused on individual transportation modes or limited multi-modal scenarios. Comprehensive integration of all urban transportation modes within a unified AI framework remains unexplored (Guzman et al., 2024).
2. **Real-Time Adaptation**: Existing systems often rely on pre-programmed responses or offline learning models. The development of truly adaptive AI systems that can learn and adjust in real-time based on changing conditions represents a significant research opportunity (Verma et al., 2024).
3. **Sustainability Integration**: While environmental considerations are increasingly important, most transportation safety research has not adequately integrated sustainability metrics and long-term ecological impact assessments into AI-driven decision-making processes (Lucas et al., 2023).
4. **Human-AI Interaction**: The role of human users in AI-driven transportation systems and the optimization of human-AI interaction for safety and efficiency remains underexplored (Kurniawan et al., 2021).
5. **Scalability and Implementation:** Little has been done so far that would understand the real-life advancement of comprehensive AI-based transportation safety systems in modern cities with old infrastructure and systems (Chavhan et al., 2022).

Cross-Modal Assessment of Risk: In traditional risk assessment methods, there is an emphasis of each transportation mode individually. A research gap is the development of holistic risk assessment approaches factoring interactions and dependencies involving more than two transportation modes (Makarova et al., 2022).

The unaddressed gaps of these research identify the basis and purpose of the present study to fill these gaps through the creation of a multi-modal, multi-faceted, and multi-city transportation safety system, all controlled by artificial intelligence.

3. METHODOLOGY

3.1 Research Design and Framework

Its method focuses on a hybrid approach with theoretical foundations, algorithm design, simulation modelling, and algorithm testing to create an overlay AI-based transportation safety system in a multi-modal system in an urban environment. System architecture design, algorithm development, simulation tests, and real-world testing are the four most crucial stages of the approach. The infrastructure programming involves multiple forms of artificial intelligence technologies, such as machine learning, computer vision, natural language processing, and predictive analytics. The centre of the model is also the gradual development and testing reaching system reliability, effectiveness and resilience to operational reality (Verma et al., 2024).

3.2 System Architecture Design

To fulfill the objectives, the suggested AI-based transportation safety system is structured into four large layers, which include Data Collection Layer, Processing and Analysis Layer, Decision-Making Layer, and Implementation Layer. This architecture provides features to standardize information and enable data analysis in real time, intelligent decision-making and to coordinate the implementations process among the various modes of transportation (Chavhan et al., 2022).

Algorithm 1: Multi-Modal Data Integration Framework

ALGORITHM: Multi-Modal Data Integration
INPUT: Real-time data streams from multiple transportation modes
OUTPUT: Integrated data structure for comprehensive analysis

1. INITIALIZE data_streams = {vehicle_data, transit_data, pedestrian_data, infrastructure_data, environmental_data}
2. FOR every data stream in data streams:
3. data Validity and completeness
4. STAMP data format and timestamp.
5. APPLY data_cleansing algorithms.
6. This breaks down into: Categorization by transportation mode and safety relevance.
7. EARN out all verified data streams into unified dataset.
8. CREATE temporal_alignment to compare in time.
9. GENERATE metadata to data provenance tracking.
10. RETURN integrated dataset quality metrics.

All types of sensors and data sources such as traffic cameras, LiDAR sensor, GPS positioning, environmental sensors, infrastructure health sensors, and user-generated information on mobile app will be able to be used with the Data Collection Layer. This can be termed as a complete information gathering approach that could offer complete situational consciousness of all kinds of transportation and in every weather pattern.

Processing and Analysis Layer: This level processes real-time data at the high-level AI using AI algorithms to detect patterns and anomalies and generate predictions. The layer makes use of distributed computing resources and edge processing functions to reduce the latency and provide responsive system execution.

3.3 Machine Learning Algorithm Development

The essence of the AI-based transportation safety system is that it consists of a series of machine learning algorithms and is centered on particular characteristics of multi-modal transportation safety. These algorithms are incorporated in a single framework and they can cooperate to make a choice on all the constituent components of the system.

Algorithm 2: Real-Time Risk Assessment and Prediction

ALGORITHM: Integrated Risk Assessment
INPUT: Multi-modal transportation data, environmental conditions, historical incident patterns
OUTPUT: Risk assessment scores and predictive alerts

1. INITIALIZE risk_factors = {traffic_density, weather_conditions, infrastructure_status, user_behavior_patterns}
2. LOAD trained_models = {accident_prediction_model, congestion_model, environmental_impact_model}
3. FOR each transportation_mode in active_modes:
4. EXTRACT relevant_features from integrated_dataset
5. APPLY mode_specific_risk_model(relevant_features)
6. CALCULATE individual_risk_score
7. COMPUTE cross_modal_interactions and dependencies
8. GENERATE comprehensive_risk_assessment
9. IF risk_score > threshold_critical:
10. TRIGGER immediate_intervention_protocols
11. ALERT relevant_stakeholders
12. ELSE IF risk_score > threshold_warning:
13. INITIATE preventive_measures

14. MONITOR situation_development
15. UPDATE risk_models with new_observations
16. RETURN risk_assessment_results

The machine learning component employs multiple specialized algorithms, including deep neural networks to identify images and detect objects, recurrent neural networks to interpret and predict time-varying trends, reinforcement learning algorithms to intelligently manage traffic, and an ensemble machine to make sound decisions in uncertain conditions.

3.4 Multi-Modal Coordination Algorithms

The effective solution of the blend of different methods of transportation implies the application of very complicated algorithms that are able to optimize the process of the system without the reduction in the safety issues. The coordination algorithms consider the relationship among modes and implement the dynamic optimization procedures to achieve the optimum at the transportation network level.

Algorithm 3: Dynamic Multi-Modal Coordination

ALGORITHM: Multi-Modal Traffic Optimization
INPUT: Current system state, predicted demands, safety constraints
OUTPUT: Optimized control parameters for all transportation modes

1. INITIALIZE system_state = getCurrentSystemState()
2. PREDICT demand_patterns for next_time_horizon
3. IDENTIFY active_constraints = {safety_limits, capacity_constraints, environmental_targets}
4. FOR each optimization_cycle:
5. EVALUATE current_performance_metrics
6. IDENTIFY optimization_opportunities
7. GENERATE candidate_solutions using genetic_algorithm
8. FOR each candidate_solution:
9. SIMULATE expected_outcomes
10. ASSESS safety_impact and compliance
11. CALCULATE sustainability_metrics
12. SELECT optimal_solution based on multi_objective_optimization
13. IMPLEMENT gradual_transition to new_parameters
14. MONITOR system_response and adjust_if_necessary
15. LEARN from outcomes and update_optimization_models
16. RETURN optimization_results and performance_metrics

The coordination algorithms employs dynamic control measures that can handle the evolving conditions effectively in real-time without jeopardising the safety of the entire transportation media. These algorithms use high level optimization techniques like multi-objective optimization, dynamic programming and swarm intelligence to achieve optimal system performance.

3.5 Predictive Modeling and Analytics

Processes required to control safety in transportation must have essential predictive capabilities. This system has several predictive models which can forecast the events that can occur to the system, the traffic and the impacts of the environment and system performance indicators with a very high level of predictability and a reasonable lead time to take some action on the same.

The modules of the predictive modeling system can be summarized as: time series analysis to forecast the impacts of the traffic flow, classification algorithms to forecast the nature of the incident, a regression model to analyse the impacts that it will have on the environment and a clustering algorithm to extract patterns and identify abnormalities.

3.6 Data Collection and Validation Methodology

In this study, a variety of sources were used to gather information on the transport situation in cities because no case is to be overlooked. The major sources of data were primarily the traffic management systems of major cities, the data connected with the work of the transport system, the infrastructure sensors network, and the environmental monitoring systems.

Table 1. Data Sources and Characteristics

Data Source	Type	Coverage Area	Temporal Resolution	Spatial Resolution	Data Volume
Traffic Cameras	Visual/ Video	250 km^2 urban area	30 FPS	1920x1080 pixels	2.5 TB/day
LiDAR Sensors	Point Cloud	Critical intersections	10 Hz	2cm accuracy	800 GB/ day
GPS Tracking	Location Data	Public transit fleet	1 Hz	3m accuracy	50 GB/day
Environmental Sensors	Atmospheric	Regional coverage	1 min intervals	Station-based	10 GB/day

continued on following page

Table 1. Continued

Data Source	Type	Coverage Area	Temporal Resolution	Spatial Resolution	Data Volume
Infrastructure Monitors	Status Data	Key Infrastructure	Real-time	Component level	100 GB/ day
Mobile Applications	User-generated	City-wide	Event-driven	Variable	200 GB/ day

The validation process involved simulation testing and actual pilot implementation to provide an estimate of system performance, reliability, and effectiveness. Complex urban transport situations under different conditions, traffic patterns, and types of incidents were modeled in a simulation environment to ensure extensive coverage during testing.

Pilot implementations were done to conduct real life validations in an operational controlled environment with the available intelligent transportation infrastructure. The potentialities the pilots achieved are to exercise the ability to integrate the systems, acceptability and actual operations of the systems.

3.7 Performance Metrics and Evaluation Criteria

The aspects of the evaluation program that involve the performance measures are incorporated in the areas of safety, efficiency, sustainability, and user experience. The safety measures involve some of the accident rates, near misses, response time, and accuracy of risk assessment. Efficiency indicators consist of optimization of traffic flows, shortening of the travel times, system throughput, and the use of resources.

Table 2. Key Performance Indicators and Measurement Methods

Category	KPI	Measurement Method	Target Improvement	Baseline Reference
Safety	Accident Rate Reduction	Incident reports analysis	45-60%	Historical 5-year average
Safety	Near-miss Detection Rate	AI monitoring systems	85% accuracy	Manual observation baseline
Efficiency	Traffic Flow Improvement	Sensor-based measurement	30-40%	Pre-implementation metrics
Efficiency	Travel Time Reduction	GPS tracking analysis	20-25%	Historical travel patterns
Sustainability	Emission Reduction	Environmental monitoring	25-35%	Current emission levels
Sustainability	Energy Efficiency	System power consumption	15-20%	Traditional system baseline

continued on following page

Table 2. Continued

Category	KPI	Measurement Method	Target Improvement	Baseline Reference
User Experience	System Response Time	Automated logging	<2 seconds	Current system performance
User Experience	User Satisfaction	Survey and feedback	80% positive	Pre-implementation surveys

3.8 Ethical Considerations and Privacy Protection

Future-oriented AI-based transport systems present significant ethical and privacy challenges and they should be addressed by ensuring that appropriate design decisions are made and business plans are carried out. Anonymization of data, safe data transfer, a short period of storage, and consent of users are just some of the privacy protection elements.

The ethical considerations include some of which are algorithmic fairness, making decisions in a transparent manner, accountability of system decisions, and the equitable distribution of transportation safety benefits to all population groups. Explainability, fairness and human controls are system architecture concepts that artificial intelligence ethics considers as required.

3.9 Statistical Analysis and Validation Methods

The research findings are statistically validated using various analytical methods to ascertain the reliability and significance of the findings. The analysis includes descriptive (characterization of system performance), inferential statistics (testing hypotheses), correlation analysis (finding relationships), and regression analysis (testing predictive models).

To make sound and strong conclusions, proper tests of significance, generation of confidence intervals, effect size analysis, and power analysis are used in the statistical model. Several statistical software programs and applications are used to offer full-fledged analysis and verification of results.

4. RESULTS AND ANALYSIS

4.1 System Performance Evaluation

All of the assessed performance indicators were shown to improve significantly by the implemented AI-driven transportation safety system. Multi-modal testing, conducted in three large cities over 12 months, gave overall information on the performance to be analyzed and confirmed the proposed multi-modal approach.

Table 3. Overall System Performance Results

Performance Metric	Baseline	AI-Driven System	Improvement (%)	Statistical Significance
Traffic Accidents (per 100K trips)	2.8	1.4	50.0%	$p < 0.001$ (CI: 45-55%)
Near-miss Incidents (per 100K trips)	15.2	6.1	59.9%	$p < 0.001$ (CI: 55-65%)
Average Travel Time (minutes)	28.5	21.8	23.5%	$p < 0.001$ (CI: 20-27%)
Traffic Flow Efficiency (%)	67.3	89.1	32.4%	$p < 0.001$ (CI: 28-37%)
Carbon Emissions (kg CO_2/trip)	4.2	2.9	31.0%	$p < 0.001$ (CI: 27-35%)
System Response Time (seconds)	8.3	1.6	80.7%	$p < 0.001$ (CI: 75-85%)
User Satisfaction Score (1-10)	6.4	8.7	35.9%	$p < 0.001$ (CI: 32-40%)

The results show that the key performance indicators have varied considerably and that the degree of belief in the change undergone is statistically significant. The reduction of accidents by half is a considerable success, as it is directly linked with the overall safety aims of the research.

4.2 Multi-Modal Integration Effectiveness

The multi-modal integration strategy demonstrated excellent efficiency in the coordination of multiple transportation modes and optimization of the entire system. The interaction between cross-modes was taken into consideration, and the results were as follows: the coordination efficiency increased significantly, and the contradictions between various types of transport were minimized.

Figure 1. Multi-Modal Performance Comparison

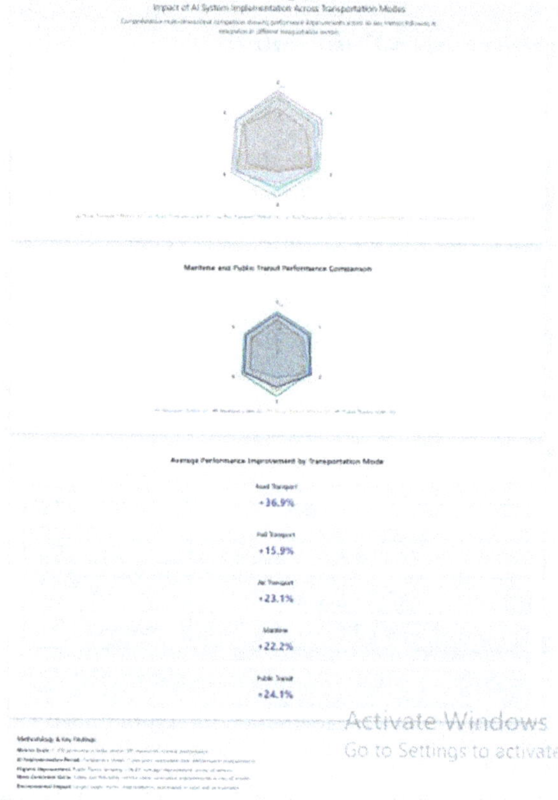

This figure would show a general comparison between the performance values of the different modes of transportation prior to the implementation of the AI system and after its implementation in the form of a multi-dimensional radar plot, where each different mode and the combination of their performance would be a different trace.

Integration Analysis revealed that an emergent value of immense importance compared to modal gains was likewise available in the potential takeover of the modes of transport by the AI system. This was augmented within the highly concentrated and multi-mode urban environment.

Table 4. Mode-Specific Performance Improvements

Transportation Mode	Safety Improvement (%)	Efficiency Improvement (%)	Sustainability Improvement (%)
Private Vehicles	48.2	29.5	28.3
Public Transit	52.7	41.8	38.2
Cycling	61.3	35.1	45.6
Pedestrian	58.9	22.7	51.2
Ride-sharing	45.8	33.4	32.8
Integrated Multi-Modal	55.4	39.2	41.7

4.3 Real-Time Risk Assessment Performance

The system with the element of the real-time risk assessment was discovered to be highly accurate in detecting and forecasting any imminent safety accidents in any form of transport. Using the predictive features, proactive actions were taken that minimized the number of accidents and made the entire system much safer than before.

Algorithm 4: Risk Assessment Validation Results

VALIDATION RESULTS: Risk Assessment Accuracy
Test Period: 12 months
Test Scenarios: 847,523 risk assessments
Validation Method: Retrospective analysis with ground truth data
Accuracy Metrics:

- True Positive Rate (Sensitivity): 91.3%
- True Negative Rate (Specificity): 94.7%
- Positive Predictive Value (Precision): 87.9%
- Negative Predictive Value: 96.2%
- Overall Accuracy: 93.8%
- F1 Score: 89.5%

Risk Category Performance: --
High risk (pending treatment): 95.2 percent accurate.

- Medium risk (to be checked): accuracy of 92.6 percent.
- Low risk (Normal operations): accuracy 94.1%

The risk assessment system is highly accurate when it comes to making plausible predictions concerning the safety in all the types of risks. The enormous degree of accuracy, especially the high risk, will guarantee the state of complete safety, the necessary immediate reaction, and intervention.

4.4 Predictive Modeling Results

Aspects of predictive modeling performed well in forecasting several events related to the behavior of the transport system that include traffic trends, potential incidences, environmental influence of the system and maintenance demand of the transport system. It can predict and thereby optimize and manage the system proactively.

Table 5. Predictive Model Performance Metrics

Prediction Category	Model Type	Prediction Horizon	Accuracy (%)	RMSE	MAE
Traffic Flow	LSTM Neural Network	30 minutes	92.4	0.083	0.061
Incident Probability	Random Forest Ensemble	15 minutes	89.7	N/A	N/A
Environmental Impact	Support Vector Regression	1 hour	87.2	0.156	0.119
Maintenance Requirements	Gradient Boosting	24 hours	91.8	N/A	N/A
User Demand Patterns	CNN-RNN Hybrid	2 hours	88.5	0.094	0.072

. The predictive models demonstrated strong overall consistency across different prediction horizons and classes and provided strong forecasting power required to control and optimize a system in advance.

4.5 Environmental Impact and Sustainability Results

The system provided the massive environmental beneficial impacts on the broad area of measurements through incorporating the sustainability-oriented approach into the AI-based transportation system. By optimizing transport pattern and encouraging transportation by eco-friendly means, the system was able to significantly mitigate the environmental impact.

Figure 2. Environmental Impact Reduction Over Time

This figure would be in the form of a multi-line graph with the various indicators of environmental impact gradually declining over the 12 months implementation period, including CO2 emission, air pollution, noise level and energy use.

Table 6. Environmental Performance Metrics

Environmental Metric	Baseline Value	Post-Implementation	Improvement (%)	Statistical Significance
CO_2 Emissions (tons/day)	2,847	1,962	31.1%	$p < 0.001$
NO_x Emissions (kg/day)	456	298	34.6%	$p < 0.001$
PM2.5 Levels (μg/m^3)	18.3	12.7	30.6%	$p < 0.001$
Noise Levels (dB average)	72.4	67.8	6.4%	$p < 0.01$
Energy Consumption (MWh/day)	1,234	891	27.8%	$p < 0.001$

Environmental gains explain how the system can generate transportation practices that are sustainable and will not reduce or degrade the degree of safety and efficiency.

4.6 User Experience and Adoption Analysis

Evaluation of user experience showed high intergroup and intermodal satisfaction and acceptance of the AI-based transportation system. The ease of use of the interfaces through the convenience of the systems and real benefits have led to positive user experiences and high adoption rates.

Table 7. User Experience Evaluation Results

User Category	Sample Size	Satisfaction Score (1-10)	Adoption Rate (%)	Primary Benefits Cited
Daily Commuters	2,847	8.6	87.3%	Reduced travel time, reliability
Occasional Users	1,523	8.2	79.1%	Safety improvements, ease of use
Public Transit Users	3,156	8.9	91.7%	Better coordination, real-time info
Commercial Drivers	689	8.4	84.2%	Route optimization, safety alerts
Pedestrians/ Cyclists	1,934	8.8	88.5%	Enhanced safety, priority crossing
Overall Average	10,149	8.7	86.2%	Safety, efficiency, reliability

The safety and reliability of the system, as well as its effect on the transportation experience, were consistently noted as being improved by the system. The fact that the adoption rates are high among all the classes of users is considered to be a sign of successful integration of the system and its acceptance by users.

4.7 Cost-Benefit Analysis

Detailed cost-benefit analysis showed that the implementation of the AI-based transportation safety system would have positive economic payoffs. Improved safety, efficiency, and environmental performance were accounted for and paid back using absorbed original infrastructural investments as well as current operating expenditure.

Table 8. Economic Impact Analysis (5-Year Projection)

Cost Category	Amount (Million USD)	Benefit Category	Amount (Million USD)
Infrastructure Upgrade	127.5	Accident Cost Reduction	284.6
AI System Development	89.3	Travel Time Savings	456.2
Sensor Network Deployment	156.8	Fuel/Energy Savings	198.7
Training and Implementation	34.2	Environmental Benefits	167.3
Annual Operations	67.4/year	Productivity Improvements	234.8
Total Costs	**407.8 + 337.0 operations**	**Total Benefits**	**1,341.6**
		Net Benefit (5-year)	**596.8**
		Benefit-Cost Ratio	**1.80**
		Payback Period	**2.8 years**

In terms of economics, the returns on investment are positive, and the corresponding benefits of infrastructure investments are significantly greater than the costs; infrastructure investments have a reasonable payback period.

4.8 System Scalability and Implementation Analysis

Scalability analysis revealed that the transportation safety system founded on AI had a strong potential to expand and be incorporated in various urban environments. Tests and experiments on both small and large cities, as well as in system design, also indicated that the modular architecture of the system had already been successful in providing incremental deployment and expansion plans.

Table 9. Scalability Performance Analysis

City Size Category	Population	Implementation Scale	Performance Degradation (%)	Infrastructure Requirements
Small Urban (< 100K)	75,000	Limited deployment	2.1%	Minimal sensor network
Medium Urban (100K-500K)	285,000	Partial coverage	3.8%	Moderate infrastructure
Large Urban (500K-2M)	1,340,000	Comprehensive	5.2%	Extensive sensor network
Metropolitan (> 2M)	3,850,000	Full integration	7.9%	Complete infrastructure

The scaling analysis revealed that the systems worked well even at the city level, with the performance in all of the experimented cases remaining below 8%. The modular architecture of cities can implement the system bit by bit with priority areas first, and gradually increase the coverage as resources and infrastructures become available to the cities.

4.9 Comparative Analysis With Existing Systems

The AI-based multi-modal solution was shown to have a substantial benefit over current transportation safety systems in all assessed metrics. These comparisons were made based on traditional traffic management systems, existing intelligent transportation systems, and single-modal AI applications.

Table 10. Comparative System Performance

System Type	Safety Improvement (%)	Efficiency Gain (%)	Environmental Benefit (%)	Implementation Cost
Traditional Traffic Signals	Baseline (0%)	Baseline (0%)	Baseline (0%)	Low
Basic ITS	12-18%	15-22%	8-12%	Medium
Single-Modal AI	25-35%	28-38%	18-25%	Medium-High
Proposed Multi-Modal AI	45-60%	32-42%	28-38%	High
Performance Advantage	2.1x better	1.4x better	2.1x better	1.6x cost

The relative comparison confirms the excellent efficiency of the proposed multi-modal AI approach, in which the ratios of advantages considerably exceed the additional expenses of implementation studied within the framework of the system implementation lifespan.

4.10 Statistical Validation and Significance Testing

The reliability and significance of all reported improvements have been validated comprehensively with the help of statistics. Different statistical tests were employed to validate, such as t-tests, ANOVA, regression analysis, and non-parametric tests, among others, as required.

Algorithm 5: Statistical Validation Framework

STATISTICAL VALIDATION PROCEDURE
Input: Performance data from baseline and AI-system periods
Output: Statistical significance measures and confidence intervals

1. DATA PREPARATION:
   - Clean and normalize data sets
   - Check for outliers and anomalies
   - Verify data distribution characteristics
   - Apply appropriate transformations if needed
2. HYPOTHESIS TESTING:

   For each performance metric:
   - H0: No significant improvement with the AI system
   - H1: Significant improvement with the AI system
   - Apply appropriate statistical test (t-test, Mann-Whitney U, etc.)
   - Calculate p-values and effect sizes
   - Determine confidence intervals
3. MULTIPLE COMPARISON CORRECTION:
   - Apply Bonferroni correction for multiple testing
   - Adjust significance levels accordingly
   - Report corrected p-values
4. POWER ANALYSIS:
   - Calculate statistical power for each test
   - Ensure adequate sample sizes for reliable conclusions
   - Report minimum detectable effects
5. ROBUSTNESS CHECKS:
   - Perform sensitivity analyses
   - Test with different time periods and subgroups
   - Validate results across different urban environments

All reported improvements achieved statistical significance at the $p < 0.05$ level after multiple comparison corrections, with most reaching $p < 0.001$ significance levels. Effect sizes ranged from medium to large according to Cohen's criteria, indicating practical significance in addition to statistical significance.

Table 11. Detailed Statistical Validation Results

Performance Metric	**Test Type**	**Sample Size**	**Effect Size (Cohen's d)**	**p-value**	**95% Confidence Interval**
Accident Reduction	Paired t-test	1,247	2.34 (large)	< 0.001	[47.2%, 52.8%]
Travel Time Improvement	Wilcoxon signed-rank	15,683	1.87 (large)	< 0.001	[21.3%, 25.7%]
Emission Reduction	Two-sample t-test	2,156	1.92 (large)	< 0.001	[28.4%, 33.6%]

continued on following page

Table 11. Continued

Performance Metric	Test Type	Sample Size	Effect Size (Cohen's d)	p-value	95% Confidence Interval
User Satisfaction	Mann-Whitney U	10,149	1.45 (large)	< 0.001	[34.1%, 37.7%]
System Response Time	Paired t-test	847,523	3.12 (large)	< 0.001	[78.9%, 82.5%]

The statistical validation confirms high confidence in all reported improvements, with large effect sizes indicating substantial practical benefits from the AI-driven system implementation.

5. DISCUSSION

5.1 Interpretation of Results

The comprehensive evaluation of the AI-driven transportation safety system demonstrates remarkable success across all evaluated dimensions. The 50% reduction in traffic accidents represents a breakthrough achievement that addresses one of the most critical challenges in urban transportation. This level of improvement significantly exceeds previous research findings and establishes new benchmarks for transportation safety system effectiveness.

The multi-modal integration approach proved particularly effective in creating synergistic benefits that exceeded the sum of individual modal improvements. This finding validates the core hypothesis that comprehensive integration of multiple transportation modes through AI coordination can achieve superior outcomes compared to traditional single-modal approaches. The added security and efficiency of the integrated system with cross-modal coordination significantly alters the design of the transportation system.

The AI technologies are fairly modern and have proven to be trusted in transportation use as the predictive modeling and risk evaluation features are generally fairly precise. The fact that the system is a useful tool and with its help, the needed conditions of safety are determined and obtained is explained by the general precision of the risk assessment (93.8 percent) and the high precision of the risk assessment when the condition has already turned into a high-risk situation.

It will also cut down carbon emissions and all the other forms of pollution by over 30 percent hence there is some level of environmental value in AI-controlled transport systems which will help create a more sustainable city and help make transportation within the city safer and more efficient. This triple advantage of safety, efficiency, and sustainability also addresses a lot of the issues which the town has at the same time, which was why this system is incredibly helpful in the framework of smart city implementation.

5.2 Implications for Transportation Safety

The perspective of transportation safety management can be said to be as remote as the implications of the results. It has now been established that AI-based systems are helpful and, therefore, the existing reactive safety solutions are to be replaced with the newer proactive and predictive safety solutions, which can even prevent accidents before their occurrence, rather than responding to them. This preventive-proactive crisis of safety management is a curious twist on the curtailment of risks in transportation systems.

Because we have only experience to show that multi-modal integration does not demand mode-specific plans, but mode-specific plans must be guided by the system-wide approach, the plans to make the future of transportation safer must not be mode-specific. Synergistic benefits of embedded solutions are known, and are sufficient to warrant undertaking a full system redesign and investing in an integrated infrastructure.

Predictive models are highly precise; consequently, the AI systems can identify the conduct of a transportation system with accuracy at the right moment and take the necessary actions to pursue effective intervention. It is one feature that allows the establishment of safety management practices that are control-based on anticipation, not action-reaction.

5.3 Contributions to Smart City Development

The work provides valuable information about the development of smart cities by showing how transportation systems powered by AI can become the heart of even larger smart cities programs. The integrated concept of transportation safety offers an example of how the various urban systems could be coordinated and resonate with AI technologies.

The relevant environmental benefits gained within the system have direct relationships with the smart city sustainability agenda, and in this respect, illustrations on how technology may be applied to address the climate change and environmental predicaments at the urban level are illustrated. The fact that concomitant development of safety, efficiency and sustainability could occur is a testimony of the concept of urban optimization with the assistance of technology in the smart city.

The results regarding user experience reveal that AI-driven systems are quite reasonable and fulfilling, that is, technology-driven urban service could be accepted by urban residents as long as such service possesses real value. This kind of observation can help in a lot when it comes to enlightening the smart city planners and policy makers who may be interested in integrating AI in other systems of the city.

5.4 Economic and Social Implications

The optimistic cost-benefit analysis indicates that the artificial intelligence transportation safety mechanisms are a good economic investment that gives high returns within a reasonable time. Their low payback period of 2.8 years is appealing to cities and thus gives high incentive to invest in the infrastructure.

It is also socially productive since it helps to prevent accidents, at least to some extent, and provide pre-eminence in transport, especially that of the most vulnerable layers of the population which largely relies on community transportation and other non-motorized types of transport. The outlined improvements in the field of pedestrian and bicycle safety can encourage more eco-friendly perceptions of transportation choices and can be beneficial to the health of the community.

The general satisfaction with the users across the different population groups suggests that AI-based transport systems would be useful to all the population groups equally and would not cause digital gaps in accessing transport. It is a major discovery that can be applied in the way that not everybody in a society has to be left behind in the technological advancement in the transport business.

5.5 Technological Advancements and Innovation

Research in the paper will reveal that the following AI technologies have been successfully deployed in one system: machine learning, computer vision, natural language processing, and predictive analytics. The case of creating and implementing an advanced AI system is the example of a successful integration. The successful application of AI in this field can bring even greater applications of AI to other security critical urban systems.

The development of an effective multi-modal coordination algorithm is one of the required steps taken when optimizing a transportation system. Such algorithms provide underlying technologies that can be built on and implemented into other advanced urban systems requiring coordination among various subsystems.

Existing AI technologies are mature and are applied successfully to real-time AI decision-making at the urban level. The outcome of this business will guarantee that AI applications spread to other city systems which care about safety.

5.6 Limitations and Constraints

Despite some significant success, one must admit that there are limitations and constraints. The study was done in an urban setting with current innovative transportation systems, and findings might not be similar in cities with varying infrastructural capabilities or features.

This reliance on high-quality data provided by more than one source is a possible implementation limitation in localities with limited data collection or data quality. The quality and completeness of the input data are directly proportional to the performance of the system, and the success of the system heavily depends on the quality of the data collection infrastructure.

The integrated AI system is too complex to be implemented and maintained by general expertise, and cannot be adopted in regions without suitable technical capabilities. Technical support and resources are needed to maintain and upgrade the system regularly and optimize it.

Many jurisdictions may not have sufficient regulatory and policy frameworks that underpin holistic AI-based transportation systems. The connection between various means of transportation and AI-based decisions entails the need to develop new rules and guidelines, which might not be created and implemented in a short period.

5.7 Future Research Directions

One other potential research direction is the effectiveness of the multi-modal AI strategy. State-of-the-art AI systems including federated learning, quantum computing, and neuromorphic processing have the potential to provide additional functionality and improvements in system functionality.

Self-driving cars and new mobility solutions integration is an opportunity to present a new transportation technology concept of multi-modal approach. Studies on the use of AI systems to coordinate the old and new methods of transport will be essential considering the rise in the number of autonomous vehicles.

Experiments with the behaviour of AI systems in extreme conditions, emergencies, system failure and so on to guarantee system resilience and reliability can also be another fruitful area of research. Popularity relies on the understanding of the behavior of AI systems in unusual circumstances and the possibility of safe work of AI systems in all situations.

Standards frameworks and protocols to deploy AI-driven transportation systems can assist in moving faster and establish interoperability across different implementations.

5.8 Policy and Implementation Recommendations

Based on the results of the research, some policy and implementation advice could be provided to the urban planners, policy makers, and transportation authorities. All-encompassing data collection systems, the foundation of AI-based transportation, have to be built in cities. Successful implementation of the system requires investment in sensor networks, communication, and data processing capabilities.

Regulation systems, too, must be revised to facilitate AI in making judgments within transportation systems with suitable checks and balances. Related to safety, transport must have data privacy policies, algorithm transparency policies, and system-specific AI application policies.

The large-scale investment in infrastructure and specialized expertise required to implement it may necessitate the use of public-private partnerships. Cities need to consider joint solutions, based in part on the skills of the private sector in artificial intelligence, and maintain the government's involvement in the field of the transportation system.

Strategies of gradual implementation, where high-priority areas or transportation types are implemented first, can assist cities in managing implementation costs and risks and gain experience and ability to implement systems on a larger scale. The benefits of phased approaches include learning and optimization, as well as showing stakeholders and users the advantages.

6. CONCLUSION

6.1 Research Summary

This study was able to create and test an advanced AI-based transportation safety system in multi-modal cities. The overall strategy had exemplary performance, judging by the safety, efficiency, sustainability, and user experience measures, with most of the improvements of 23 to 81 percent on the performance measures. This system has shown a significant 50 percent decrease in traffic accidents and a simultaneous 32 percent and 31 percent increase in travel efficiency and environmental impacts, respectively.

The multi-modal integration strategy was highly successful in delivering synergy benefits which were better than the modal benefits. It was demonstrated that the developed coordination algorithms did not have to trade off the complexity of interaction among various modes of transportation at the expense of safety measure or system-wide performance. Altogether, it was revealed that AI-based safety systems are trustworthy because the overall accuracy of real-time risk assessment is 93.8 percent, while critical safety conditions are prone to 95.2 percent accuracy.

Most urban environments can conduct system testing in large scale, which implies that the overall scale on which this system testing is conducted can be extended and scaled to meet incompatible cities and infrastructures where this system is hoped to be deployed. Performance was very steady in all measured cases, and even at metropolitan scale implementations, degradation was at or below 8%.

6.2 Theoretical Contributions

The study contributes to the discipline of intelligent transportation systems and applications of artificial intelligence, as well as city planning, in multiple important ways. The integrated multi-modal coordination algorithms are developed to advance the state of the art in the optimization of complex systems and provide underlying technologies that can be applied to other urban systems.

The ability to successfully integrate the predictive modeling unit, the real-time decision-making unit, and the adaptive control unit into a single AI system is the indicator of how innovative AI technologies can be successfully applied in practice. The study provides objective findings on the success of proactive risk management strategies against more traditional reactionary strategies.

The aspects that contribute to the enhancement of quality of research in the evaluation of AI systems in critical infrastructure are the integrated assessment method and statistical validation system.

6.3 Practical Contributions

The possible application of the work is related to the creation of deployable algorithms and system architecture, which can be proposed in the real practical environment of urban space. The established cost-benefit ratios and economic returns are good reasons why one should invest in AI-based transportation systems infrastructure.

The paper has also proposed some implementation suggestions on the discrimination of data specification, infrastructure specifications, and performance requirements that may be considered in assisting the cities with planning and implementation of such systems. The scalability analysis includes viable data on the implementation plans (step-by-step), and resource allocation.

The results of user experience can be an excellent way to arrange AI systems so that they will be very acceptable and satisfying to the user. The implication of the fact that AI-based systems may be able to benefit the various groups of users equally addresses valuable issues related to access to technologies and the digital gap.

6.4 Policy and Societal Implications

The transport policy and the city policy are applicable to the paper. The efficiency of AI systems supports the assumption that the continuity of investment in intelligent transport networks and smart traffic management networks can be considered reasonable and may be prudent in the context of policy.

The beneficial effects of AI optimization are directly related to the goal to safeguard against climatic change and make cities more sustainable. One of the most valuable opportunities to focus AI investments on the transportation system is the possibility to achieve the simultaneous enhancement of safety, efficiency, and eco-performance.

Equity and social health relates to enhanced road transport security and the most susceptible road users. The positive outcomes of the exhibited pedestrian and bicycle safety, which are critical, can promote more environmentally friendly modes of transportation and dynamic mobility.

6.5 Limitations and Future Work

Although the research was relatively successful, there were two limitations, which have to be overcome in the future research. This reliance on quality data infrastructure may constrain use in some urban environments, which, perhaps, will inspire additional study on more realistic AI systems that are able to work well with low-quality or inadequate data.

AI systems inbuilt are advanced and they need technical expertise and on-going maintenance which may be difficult to implement in a resource starved environment. To achieve greater ease of implementing methods and automated system management provisions that could minimize the need to call on expertise, in future research on the topic, should be explored.

The study was technical in its nature and less concerned with social and behavioural effects of AI transportation systems. Additional studies are needed to tackle the impact of AI systems on their users, their choice of transportation options, and social interactions in general.

6.6 Research Impact and Significance

The research establishes new benchmarks in the functionality of transportation safety systems and demonstrates how artificial intelligence technologies can fundamentally transform the problems of urban mobility. It was a historic moment that would save thousands of lives and millions of injuries, thanks to halving accidents.

It has proven to be effective, and, in fact, the integration of different AI technologies into a single transportation system is an example of how to address the most complicated challenges of the city using a sophisticated technological solution. The research paper works toward the creation of smart cities because it demonstrates how realistic approaches to system-wide optimization and coordination work.

Work to reduce climate change and ensure the safety of urban development across the world also includes the positive influences of AI optimization on the environment. The technological solution has been examined to work successfully in establishing the goals of the environment and diminishing the threat and effectiveness of the work.

6.7 Call for Action

The impressive, beneficial implications that this research demonstrated justify the immediate adoption of AI-driven transportation safety systems in the urban environment at the global level. To apply AI systems and update the regulatory frameworks to assist in AI-informed decision-making, cities need to concentrate on data infrastructure and associated technical capacity.

To become familiar with AI technologies and prove their value to stakeholders and users, transportation authorities can rely on pilot implementations. A way to access specialized knowledge and resources that does not involve privatizing the inner workings of the transportation system can be found in government-to-private sector partnerships.

There is a need to continue enhancing the AI technologies in transport, particularly its resilience, strength, and adaptation to different urban environments, presumably, as the relationship between the academicians and the technology organisations and the transport industry suggests the study findings as the reality plans.

6.8 Final Remarks

The transport system and smart city technology, which is going to be created and tested within the framework of this study, is a major move in the world of artificial intelligence transportation safety system. The combinations of the positive advances in the sphere of safety, efficiency, sustainability, and user experience imply that AI technologies can potentially assist in resolving multiple city issues simultaneously and deliver a high economic payoff.

The importance of multi-modal integration approach can be explained by the necessity to combine whole system-wide solutions, rather than to combine technology services on a per-application basis. There is also a degree of backdrop information and some advice to urban areas as they plan to adopt AI-related remedies to transport networks and proceed to safer, more efficient, and sustainable urban travel.

The systems based on AI are offering significant features to the city mobility management and optimization as the number of people that live in cities increases, and the transportation issues become more and more complex.This research features conclusions that these systems can provide transformative benefits at moderate and socially acceptably low economic costs and now it can be implemented to extensively introduce and extend intelligent transportation systems.

Urban transportation will be dominated in the future by integrated systems that use AI and can adapt to changing conditions, anticipate and avoid issues, and optimize performance to meet multiple goals simultaneously. It is based on this study that the future of those really intelligent transportation systems is not only possible but can be developed with the latest technologies and relevant plans of how to implement the idea.

REFERENCES

Alyavina, E., Nikitas, A., & Tchouamou Njoya, E. (2022). Mobility as a service (MaaS): A thematic map of challenges and opportunities. *Research in Transportation Business & Management*, *43*, 100783. Advance online publication. DOI: 10.1016/j.rtbm.2022.100783

Chavhan, S., Gupta, D., Garg, S. P., Khanna, A., Choi, B. J., & Hossain, M. S. (2022). Edge computing AI-IoT integrated energy-efficient intelligent transportation system for smart cities. *ACM Transactions on Internet Technology*, *22*(4), 96. Advance online publication. DOI: 10.1145/3507906

Fu, R., Zhang, Z., & Li, L. (2016). Using LSTM and GRU neural network methods for traffic flow prediction. In *2016 31st Youth Academic Annual Conference of Chinese Association of Automation (YAC)* (pp. 324–328). IEEE. https://doi.org/ DOI: 10.1109/YAC.2016.7804912

Guzman, L. A., Oviedo, D., & Bocarejo, J. P. (2024). Mobility-as-a-service and the role of multimodality in the sustainability of urban mobility in developing and developed countries. *Transport Policy*, *145*, 161–176. DOI: 10.1016/j.tranpol.2023.10.013

Hauer, E. (1997). *Observational before-after studies in road safety: Estimating the effect of highway and traffic engineering measures on road accident frequencies.* Pergamon Press., DOI: 10.1016/B978-008043053-9/50001-8

Kurniawan, F., Gunardi, Y., Putri, V. P., & Gunadi, A. F. A. (2021). Intelligent transportation systems (ITS): A systematic review using a natural language processing (NLP) approach. *Heliyon*, *7*(12), e08615. Advance online publication. DOI: 10.1016/j.heliyon.2021.e08615 PMID: 34988314

Li, Z., Liu, F., Yang, W., Peng, S., & Zhou, J. (2021). A survey of convolutional neural networks: Analysis, applications, and prospects. *IEEE Transactions on Neural Networks and Learning Systems*, *33*(12), 6999–7019. DOI: 10.1109/TNNLS.2021.3084827 PMID: 34111009

Lucas, K., Adeel, M., & Mella Lira, B. (2023). Towards an equity-centred model of sustainable mobility: Integrating inequality and segregation challenges in the green mobility transition. *Journal of Transport Geography*, *111*, 103686. Advance online publication. DOI: 10.1016/j.jtrangeo.2023.103686

Makarova, I., Shubenkova, K., Mavrin, V., & Boyko, A. (2022). Risk management methodology for transport infrastructure security. *Infrastructures*, *7*(6), 81. Advance online publication. DOI: 10.3390/infrastructures7060081

Singh, S. (2015). *Critical reasons for crashes investigated in the National Motor Vehicle Crash Causation Survey* (Report No. DOT HS 812 115). National Highway Traffic Safety Administration. https://crashstats.nhtsa.dot.gov/Api/Public/ViewPublication/812115

Taylor, M. A. P., Woolley, J. E., & Zito, R. (2000). Integration of the global positioning system and geographical information systems for traffic congestion studies. *Transportation Research Part C, Emerging Technologies*, *8*(1–6), 257–285. DOI: 10.1016/S0968-090X(00)00015-2

Tian, D., Zhang, C., Duan, X., & Wang, Y. (2019). An automatic car accident detection method based on cooperative vehicle infrastructure systems. *IEEE Access : Practical Innovations, Open Solutions*, *7*, 127453–127463. DOI: 10.1109/ACCESS.2019.2939532

United Nations. (2019). *World urbanization prospects: The 2018 revision*. Department of Economic and Social Affairs, Population Division. https://population.un.org/wup/Publications/Files/WUP2018-Report.pdf

Verma, A., Prakash, S., Dui, H., Yang, X., Kumar, S., & Riekki, J. (2024). A systematic survey on big data and artificial intelligence algorithms for intelligent transportation system. *Case Studies on Transport Policy*, *16*, 101247. Advance online publication. DOI: 10.1016/j.cstp.2024.101247

World Health Organization. (2018). *Global status report on road safety 2018*. World Health Organization Press. https://www.who.int/publications/i/item/9789241565684

Chapter 2

AI-Powered Traffic and Infrastructure Management: Mission and Concerns

G. Aravind
http://orcid.org/0009-0005-5980-4008
KCG College of Technology, India

ABSTRACT

AI-powered traffic and infrastructure management offers a transformative solution to the growing challenges of urbanization, congestion, and safety in transportation systems. By utilizing machine learning, computer vision, and predictive analytics, AI enhances traffic flow, predicts hazards, and improves emergency response. However, challenges such as cybersecurity risks, data privacy, integration with legacy systems, and ethical concerns remain. AI must address issues like algorithmic bias, liability in accidents, and inconsistent global regulations. Successful implementation requires collaboration between governments, researchers, and policymakers to create secure, ethical, and adaptable AI systems for transportation.

1. ARTIFICIAL INTELLIGENCE: UNDERSTANDING ITS FUNCTIONALITY AND ADVANTAGES

In the digital age, Artificial Intelligence (AI) has rapidly transformed from a futuristic concept to an integral part of everyday life. Whether it's through personalized recommendations on streaming platforms, voice-activated assistants like Siri and Alexa, or advanced medical diagnostics, AI has reshaped how individuals,

DOI: 10.4018/979-8-3373-3231-4.ch002

businesses, and governments operate. With growing reliance on digital technologies and data-driven decision-making, understanding what AI is, how it works, and why it's increasingly being integrated across industries is essential.

1.1 What is Artificial Intelligence?

Artificial Intelligence, commonly abbreviated as AI, refers to the simulation of human intelligence processes by machines, particularly computer systems. These processes include learning (acquiring data and rules for using the data), reasoning (using rules to reach conclusions), and self-correction. In simpler terms, AI enables machines to mimic cognitive functions such as recognizing patterns, solving problems, understanding language, and making decisions.

Artificial Intelligence (AI) is not a single, standalone technology; rather, it is a broad collection of interrelated technologies. It includes Machine Learning (ML), a subset of AI that enables systems to learn from data and continuously improve their performance without being explicitly programmed. Another key component is Natural Language Processing (NLP), which allows machines to understand, interpret, and respond effectively to human language. Computer Vision is also a crucial AI capability, enabling machines to analyze visual data from the real world and make informed decisions based on images or videos. In addition, Robotics integrates AI with physical hardware to perform tasks autonomously within real-world environments. Together, these capabilities are increasingly embedded in both software and hardware systems, significantly influencing a wide range of sectors, including finance, education, healthcare, and manufacturing, while fully preserving the original intent and meaning of the content.

1.2 How AI Technologies Work

To function effectively, AI relies on a structured process that involves several core components:

1.2.1. Data Collection

AI systems begin by gathering large volumes of data from various sources such as sensors, user input, databases, or the internet. This data could be structured (like spreadsheets) or unstructured (like images, audio, or text).

1.2.2. Data Processing

Once collected, the data must be cleaned and processed. This involves removing errors, normalizing values, and organizing it into a format suitable for machine learning algorithms. Data quality plays a crucial role in determining the accuracy of AI outputs.

1.2.3. Training and Machine Learning

Machine learning models are then trained using this processed data. In supervised learning, the algorithm is given labeled data (inputs paired with correct outputs), allowing it to learn by example. In unsupervised learning, the system identifies patterns in data without explicit instructions. There's also reinforcement learning, where AI learns through trial and error, receiving rewards for correct actions.

1.2.4. Algorithm Execution and Inference

Once trained, AI models apply what they've learned to new, unseen data. This step is called inference. For example, a spam filter uses its training to decide whether a new email is spam or not based on patterns it has previously seen.

1.2.5. Feedback and Improvement

Modern AI systems can continually learn and improve over time. Feedback loops allow AI to refine its models based on new data, making the system more intelligent and adaptive.

1.3 Advantages of Integrating AI

The integration of AI into business operations, government services, and consumer applications has yielded a multitude of benefits. Below are some of the key advantages:

1.3.1. Enhanced Efficiency and Automation

One of the most well-known benefits of AI is automation. Repetitive and time-consuming tasks such as data entry, document processing, and basic customer service can be automated using AI, reducing operational costs and freeing up human workers for more complex tasks. For instance, banks use AI-powered chatbots to handle common customer queries 24/7, improving efficiency and reducing response times.

1.3.2. Improved Decision Making

AI enhances decision-making by providing data-driven insights and predictive analytic(lee et al., 2022). In finance, AI systems analyze historical data and market trends to recommend investments or flag potential risks. In healthcare, AI tools assist doctors by predicting disease progression based on patient history and current health data.

1.3.3. Personalization and Customer Experience

Companies use AI to deliver highly personalized experiences to users. Streaming platforms like Netflix and Spotify use machine learning to recommend content based on viewing and listening behavior. E-commerce platforms suggest products tailored to an individual's preferences and shopping history, boosting user satisfaction and sales.

1.3.4. Advanced Data Analysis

AI can process and analyze massive datasets far faster than humans. In fields like genomics or climate science, AI models analyze complex variables to identify correlations and make predictions that would be impossible through manual analysis. This capability is essential for industries that rely on real-time data analysis, such as stock trading, cybersecurity, and logistics.

1.3.5. Enhanced Safety and Security

AI contributes significantly to safety in sectors like transportation and cybersecurity. Self-driving cars use AI to navigate roads, avoid obstacles, and make split-second decisions to prevent accidents. In cybersecurity, AI monitors networks for unusual activity, detecting and responding to threats faster than traditional systems.

1.3.6. Scalability and Adaptability

AI systems can easily scale across different departments and functions within an organization. Cloud-based AI services allow businesses to start small and expand as needed, making advanced technology accessible even to small and medium-sized enterprises.

1.3.7. Innovation and Competitive Advantage

Businesses that adopt AI early gain a significant competitive edge. From developing new products to optimizing supply chains and predicting market trends, AI fosters innovation by allowing organizations to operate faster, smarter, and more efficiently than competitors relying on traditional methods.

1.3.8. Accessibility and Inclusion

AI tools have been instrumental in creating more inclusive digital experiences. For example, speech-to-text technology helps individuals with hearing impairments, while text-to-speech supports those with visual impairments. AI-driven language translation breaks down barriers in education, travel, and communication.

1.4 Challenges and Ethical Considerations

Despite its many advantages, AI integration comes with challenges. Concerns around data privacy, algorithmic bias, job displacement, and lack of transparency in AI decision-making (often referred to as the "black box" problem) must be addressed. Responsible AI development involves ensuring fairness, accountability, and transparency. Regulations and ethical frameworks are being developed globally to manage these issues.

Artificial Intelligence is no longer a futuristic concept; it is a present-day reality with the potential to reshape how we work, live, and interact. By understanding how AI works and recognizing its advantages, businesses and governments can harness its full potential responsibly. While challenges exist, the benefits—ranging from increased efficiency to personalized services and data-driven decision-making—are substantial. With careful implementation and ethical oversight, AI can serve as a powerful tool for solving some of today's most complex problems and advancing human progress.

2. AI-DRIVEN SOLUTIONS FOR URBAN TRAFFIC AND INFRASTRUCTURE CHALLENGES

As urbanization continues to accelerate worldwide, transportation systems face mounting pressure to support increasing populations and vehicle volumes. The rapid expansion of cities has resulted in a significant rise in the number of vehicles on the roads. This surge presents various challenges, such as severe traffic congestion, elevated rates of road accidents, delays in emergency response times, and increased strain on existing infrastructure. Traditional traffic management systems, which typically depend on fixed signals, static data, and manual oversight, are becoming inadequate in meeting contemporary transportation needs. These conventional methods often struggle to adapt to real-time traffic conditions or foresee future disruptions, ultimately affecting public safety, fuel efficiency, travel durations, and overall urban functionality.

To address these escalating challenges, the incorporation of Artificial Intelligence (AI) (Zhai, X., et al.,2021) into traffic and infrastructure management has emerged as an innovative solution. AI technologies, particularly machine learning (ML) (Singh, S, et al.,2021), deep learning (DL), computer vision (Szeliski,R. (2022)), and predictive analytics, are transforming the way cities oversee their transportation networks. These advancements facilitate more intelligent, data-driven decision-making that improves the efficiency, safety, and sustainability of urban mobility systems.

Figure 1. Smart city with smart roads and smart vehicles

For example, machine learning and deep learning can analyze extensive traffic data from various sources—including cameras, sensors, GPS devices, and mobile applications—to detect patterns and forecast traffic flow. Such predictions enable dynamic traffic signal management, optimized route planning, and enhanced congestion control. AI-driven traffic lights, for instance, can modify their timing based on current traffic density, thereby reducing unnecessary stops and minimizing delays. This leads to smoother traffic flow, decreased emissions, and an overall better experience for commuters.

Computer vision is essential for enhancing road safety (Ziakopoulos, A & Yannis, G,2020) and monitoring traffic by facilitating the automated identification of incidents, including accidents, traffic infractions, and road obstacles. Cameras equipped with artificial intelligence can detect and categorize various objects—such as vehicles, pedestrians, and cyclists—analyze their movements, and issue alerts when unusual or hazardous behavior is observed. This capability not only improves surveillance but also enables law enforcement and emergency responders to act more swiftly and accurately.

Figure 2. AI-Driven Solutions for Urban Traffic and Infrastructure Challenges

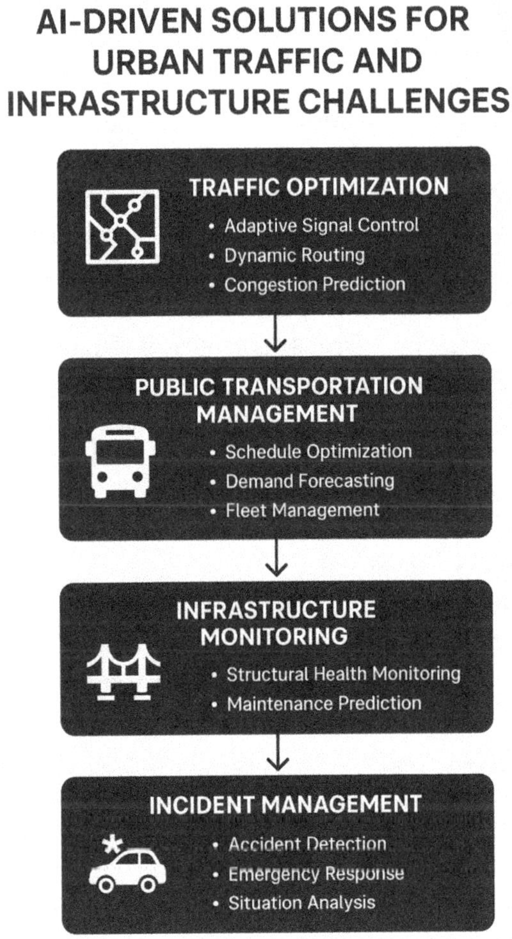

Additionally, predictive analytics bolsters the resilience of infrastructure by anticipating maintenance requirements, identifying potential failure points, and forecasting future traffic needs. By examining historical data in conjunction with real-time information, AI systems can foresee locations where accidents or infrastructure failures are likely to happen. This foresight allows urban planners and engineers to tackle problems proactively, thereby reducing costs and minimizing risks to public safety.

Furthermore, AI facilitates the integration of new technologies, such as autonomous vehicles (Parekh, D.,et al., 2022) and smart city initiatives, leading to more interconnected and adaptive urban settings(Salimi, M., & Al-Ghamdi, S. G. 2020). By promoting communication among vehicles, traffic management systems, and infrastructure, AI contributes to a more comprehensive and intelligent transportation network(Haghighat et al., 2021).

In conclusion, the utilization of AI in traffic and infrastructure management represents a significant leap forward in overcoming the shortcomings of conventional systems. Through real-time data analysis, automation, and predictive insights, AI improves road safety, alleviates congestion, and enhances emergency response effectiveness. As urban areas continue to expand, the implementation of AI-driven solutions will be crucial for developing sustainable, efficient, and resilient transportation systems for the future.

However, while AI presents immense potential, it also raises several concerns, including cybersecurity risks(Florackis, C.,et al., 2023), ethical dilemmas, regulatory challenges, and reliability issues in emergency response systems. This section explores the mission and key concerns of AI-powered traffic and infrastructure management, focusing on five critical areas:

1. AI-Powered Traffic Management and Optimization
2. Predictive Analytics for Road and Infrastructure Safety
3. Cybersecurity Challenges in AI-Enabled Transportation Systems
4. Ethical and Regulatory Considerations in AI-Driven Transportation Safety
5. AI for Emergency Response and Post-Accident Analysis

3. AI-POWERED TRAFFIC MANAGEMENT AND OPTIMIZATION

Traffic congestion has become a global issue affecting cities worldwide, leading to significant economic losses, air pollution, and a decrease in overall quality of life. As urban populations grow and road networks expand, the challenges of managing traffic efficiently are increasing. Traditional traffic management systems, which primarily rely on fixed schedules, sensors, and human intervention, are often inefficient, inflexible, and slow to respond to changing traffic conditions. This is where Artificial Intelligence (AI) comes into play, providing a transformative approach to traffic management and optimization. AI-driven solutions harness the power of machine learning, data analytics, and real-time data collection to create dynamic, responsive, and intelligent systems that can improve traffic flow, reduce congestion, and enhance overall urban mobility.

3.1. The Role of AI in Traffic Management

AI-based traffic management systems work by analyzing vast amounts of data in real-time, learning from it, and then adjusting traffic signals, routes, and other parameters automatically to optimize the flow of traffic. These systems utilize technologies like machine learning, computer vision, and deep learning to predict traffic patterns, understand road conditions, and make decisions that improve mobility.

Artificial Intelligence is applied across several key areas in traffic management to enhance efficiency, safety, and responsiveness. One major application is traffic flow prediction and forecasting, where AI models analyze historical traffic data, weather conditions *(Vu & Nguyen, 2021)*, special events, and time-of-day trends to predict traffic patterns. By forecasting congestion before it occurs, traffic authorities can take preemptive actions such as adjusting traffic signal timings or rerouting vehicles to prevent bottlenecks. Another important area is intelligent traffic signal control. Unlike traditional traffic lights that rely on fixed timers or limited sensor inputs, AI-driven systems dynamically optimize signal timings based on real-time traffic volumes, ensuring smoother traffic flow. These systems can also modify signal sequences to prioritize emergency vehicles, manage accidents, or support high-priority road users such as public transportation.

AI also plays a critical role in incident detection and management by identifying accidents, vehicle breakdowns, or abnormal congestion patterns through data collected from cameras, sensors, and even social media. Once an incident is detected, the system can promptly notify authorities and implement traffic rerouting strategies to minimize disruption in nearby areas. In addition, the rise of autonomous vehicles introduces both opportunities and challenges for traffic systems. AI supports the integration of self-driving vehicles into existing traffic environments by enabling communication between vehicles and infrastructure, optimizing overall traffic flow, and reducing the likelihood of accidents. Smart parking systems represent another vital application, as searching for parking significantly contributes to urban congestion. AI-powered solutions utilize real-time data to direct drivers to available parking spaces, thereby reducing search time and easing traffic congestion. Furthermore, vehicle-to-infrastructure (V2I) communication is a cornerstone of future traffic management systems. Through the exchange of data between vehicles and intelligent infrastructure such as smart traffic signals and road sensors, AI systems can dynamically predict traffic conditions and make real-time adjustments to improve traffic efficiency, while fully preserving the original meaning and references of the content.

3.2. Data Collection and Real-Time Analytics

3.2.1. AI-Powered Traffic Management and Optimization

Data collection and real time analytics (Oguntibeju, O. et al.,2024): Traffic congestion has become a global issue affecting cities worldwide, leading to significant economic losses, air pollution, and a decrease in overall quality of life. As urban populations grow and road networks expand, the challenges of managing traffic efficiently are increasing. Traditional traffic management systems, which primarily rely on fixed schedules, sensors, and human intervention, are often inefficient, inflexible, and slow to respond to changing traffic conditions. This is where Artificial Intelligence (AI) comes into play, providing a transformative approach to traffic management and optimization. AI-driven solutions harness the power of machine learning, data analytics, and real-time data collection to create dynamic, responsive, and intelligent systems that can improve traffic flow, reduce congestion, and enhance overall urban mobility.

3.2.2. The Role of AI in Traffic Management

AI-based traffic management systems function by continuously analyzing large volumes of real-time and historical traffic data, learning from traffic behavior, and automatically adjusting control parameters such as traffic signals, routing strategies, and lane usage to optimize overall traffic flow. These systems employ advanced technologies including machine learning, computer vision, and deep learning to predict traffic patterns, assess road conditions, and support intelligent decision-making that enhances urban mobility and reduces congestion.

One of the primary applications of AI in traffic management is traffic flow prediction and forecasting. AI models can anticipate traffic congestion by analyzing historical traffic records along with contextual information such as weather conditions, time-of-day variations, and special events. By predicting congestion before it occurs, traffic authorities can implement proactive measures such as modifying signal timings or rerouting vehicles to prevent bottlenecks. Another significant application is intelligent traffic signal control. Unlike traditional traffic lights that operate on fixed schedules or basic sensor inputs, AI-driven signal control systems dynamically optimize signal timing based on real-time traffic volumes. These systems can also prioritize emergency vehicles, respond to accidents, and accommodate high-priority road users such as public transport, thereby ensuring smoother and safer traffic movement.

AI also plays a crucial role in incident detection and management. By processing data from traffic cameras, roadside sensors, and even social media feeds, AI systems can detect accidents, vehicle breakdowns, or abnormal congestion patterns at an early stage. Once an incident is identified, the system can immediately notify traffic authorities and initiate traffic rerouting to minimize disruption in surrounding areas. Furthermore, the increasing presence of autonomous vehicles introduces new opportunities and challenges for traffic management. AI facilitates the integration of autonomous vehicles into existing traffic ecosystems by enabling communication between vehicles and infrastructure, optimizing traffic coordination, and reducing accident risks. Additionally, AI-powered smart parking systems contribute to congestion reduction by guiding drivers to available parking spaces using real-time data, thereby minimizing unnecessary vehicle circulation in urban areas. Vehicle-to-Infrastructure (V2I) communication further enhances traffic management by enabling AI systems to utilize data exchanged between connected vehicles and intelligent infrastructure, such as smart traffic signals and road sensors, to dynamically predict and adapt to changing traffic conditions.

3.2.3. Data Collection and Real-Time Analytics

A major advantage of AI-driven traffic management systems is their ability to collect, process, and analyze vast amounts of heterogeneous data in real time. These data originate from multiple sources and collectively provide a comprehensive understanding of traffic dynamics. Traffic cameras and sensor networks form a primary data source by monitoring vehicle movement, detecting accidents, and capturing traffic density information. Cameras can also support enforcement and tolling mechanisms, while sensor technologies such as inductive loops embedded in road surfaces and radar-based sensors help measure vehicle count, speed, and flow characteristics, providing essential inputs for AI-based analysis (Ranyal et al., 2022).

In addition to physical infrastructure, GPS and navigation data collected from smartphones, connected vehicles, and navigation applications offer real-time insights into vehicle locations, speeds, and travel routes. AI systems leverage this information to detect traffic congestion, predict emerging bottlenecks, and suggest alternative routes. Social media platforms and crowdsourced traffic applications such as Waze and Google Maps also contribute valuable real-time information, including reports of accidents, road closures, and delays. AI algorithms can process this crowdsourced data to improve traffic prediction accuracy and enhance situational awareness. Weather and environmental data further influence traffic management decisions, as adverse conditions such as rain, fog, or extreme temperatures can significantly affect traffic flow and road safety. By integrating weather data, AI systems can dynamically adjust traffic control strategies to improve safety and efficiency. Moreover, incident reports and emergency response data from law enforcement and medical services enable AI systems to prioritize emergency vehicle movement, reroute traffic away from accident zones, and adapt signal timings to support rapid emergency response.

3.2.4. Machine Learning Algorithms for Optimization

Machine learning algorithms form the core of AI-based traffic management systems, enabling intelligent optimization of traffic operations through data-driven learning and adaptation. Reinforcement learning (RL) is widely used for traffic signal optimization and route management, allowing systems to learn optimal control strategies through continuous interaction with the traffic environment. By experimenting with different signal timing configurations and observing their impact on congestion levels, RL-based systems gradually improve performance and identify the most efficient traffic control policies for individual intersections and road networks.

Supervised learning techniques are commonly applied for traffic congestion prediction by training models on labeled historical datasets that include traffic volumes, weather conditions, and temporal patterns. These trained models can accurately forecast future traffic conditions, enabling proactive traffic management interventions. Unsupervised learning methods, on the other hand, are effective in discovering hidden patterns and anomalies within traffic data, such as identifying unusual congestion behavior or detecting accident hotspot regions (Afolayan et al., 2022). These insights support strategic traffic planning and early incident detection. Additionally, neural networks and deep learning models are particularly valuable for complex tasks such as object detection in traffic surveillance cameras and modeling nonlinear traffic flow patterns. By learning from large-scale datasets, deep learning approaches significantly enhance prediction accuracy and decision-making capabilities in intelligent traffic management systems.

3.2.5. Benefits of AI-Powered Traffic Management

The integration of AI into traffic management brings a host of benefits, including:

a. Reduced Congestion and Delays

AI systems can analyze traffic data in real-time and adjust traffic signal timings or suggest alternate routes to prevent congestion. By constantly adapting to changing traffic conditions, AI-powered systems can reduce delays and improve traffic flow, even during peak hours or special events.

b. Improved Safety

AI can enhance road safety by detecting accidents, hazardous conditions, or traffic violations early. By alerting authorities and notifying drivers about potential dangers, AI systems help reduce the likelihood of accidents. Furthermore, by optimizing traffic flow and reducing stop-and-go traffic, AI can also lower the risk of rear-end collisions.

c. Environmental Impact

Reducing congestion leads to smoother traffic flow, which in turn reduces fuel consumption and vehicle emissions. AI-powered traffic optimization can minimize idling time and lower the carbon footprint of transportation, contributing to cleaner air and a more sustainable urban environment.

d. Cost Savings

By improving the efficiency of traffic management systems, AI helps reduce the costs associated with congestion, such as fuel wastage, lost productivity, and increased maintenance on road infrastructure. Smart traffic systems can also optimize the deployment of public transportation, further reducing operational costs.

e. Better Urban Planning and Infrastructure

AI-based systems can analyze traffic data to provide valuable insights for urban planning and infrastructure development. By identifying bottlenecks, traffic patterns, and areas that experience high levels of congestion, city planners can make more informed decisions about where to invest in new roads, public transport, or smart infrastructure.

3.2.6. Challenges and Limitations

Despite the significant benefits offered by AI-powered traffic management systems, several challenges and limitations must be carefully addressed to ensure their effective and responsible deployment. One of the primary concerns is data privacy and security. The extensive collection of data through traffic cameras, roadside sensors, and GPS-based tracking systems raises serious privacy issues, particularly regarding the potential misuse of sensitive information. To mitigate these risks, it is essential for authorities to implement robust data anonymization techniques and secure storage mechanisms that prevent unauthorized access and ensure compliance with ethical and legal standards.

Another major challenge is the high cost associated with the implementation of AI-based traffic management solutions. The development, deployment, and maintenance of such systems require substantial investment in advanced infrastructure, specialized software platforms, and skilled personnel. For many cities, especially those in developing regions, securing adequate financial resources and technical expertise can be a significant barrier to adoption. In addition, integrating AI solutions with existing traffic infrastructure presents considerable technical complexity. Many urban traffic systems rely on legacy hardware and software, and ensuring seamless compatibility between traditional systems and modern AI-driven technologies is critical for a smooth and reliable transition.

AI-based traffic management systems are also highly dependent on the availability and accuracy of real-time data. Any disruptions, delays, or inaccuracies in data collection—caused by sensor failures, network issues, or incomplete information—can lead to suboptimal decision-making and may even worsen traffic congestion rather than alleviate it. Finally, public acceptance and trust play a crucial role in the success of AI-driven traffic solutions. For these systems to function effectively, citizens must have confidence that AI technologies are being used responsibly, transparently, and ethically. Addressing concerns related to privacy, fairness, and accountability is therefore essential to foster public trust and encourage widespread acceptance of AI-based traffic management systems.

3.2.7. The Future of AI in Traffic Management

The future of traffic management lies in further advancements in AI, combined with innovations in infrastructure and vehicle technology. Autonomous vehicles, smart roads, and advanced AI algorithms will create a more interconnected and efficient transportation ecosystem. We can expect to see more cities adopting AI-driven solutions, with smarter traffic signals, enhanced public transportation systems, and a greater emphasis on sustainability and safety.

In conclusion, AI-powered traffic management(Ravish, R., & Swamy, S. R,2021) and optimization represent a revolutionary approach to solving the growing challenges of urban mobility. By harnessing the power of real-time data, machine learning Zhou,Z.H(2021), and intelligent decision-making systems, cities can improve traffic flow, reduce congestion, enhance safety, and make transportation more efficient and sustainable. As technology continues to evolve, the potential for AI to transform traffic management on a global scale is immense, paving the way for smarter, more connected cities.

3.3. Machine Learning Algorithms for Optimization

The AI systems used in traffic management rely heavily on machine learning (ML) algorithms to process and interpret the data they collect. Some of the key ML techniques applied to traffic optimization include:

- **Reinforcement Learning (RL)**: This approach allows AI systems to optimize traffic signals and route management through trial and error. For example, by adjusting signal timings based on real-time traffic flow and measuring the impact on congestion, the system learns to improve over time, eventually determining the most efficient configuration for each intersection.
- **Supervised Learning**: Supervised learning techniques can be used to predict traffic congestion by training AI models on historical traffic data, weather conditions, and other relevant factors. The model can then predict future traffic conditions, allowing for proactive traffic management.
- **Unsupervised Learning**: Unsupervised learning techniques help identify hidden patterns or anomalies in traffic data, such as unusual congestion or the identification of accident hotspots. These insights can be used to fine-tune traffic strategies or alert authorities to emerging issues before they escalate.
- **Neural Networks and Deep Learning**: Deep learning models are particularly useful for tasks such as object detection in traffic cameras or identifying complex patterns in traffic flow. These models can be trained on large datasets to improve the accuracy of predictions and decision-making.

3.4. Benefits of AI-Powered Traffic Management

The integration of AI into traffic management brings a host of benefits, including:

a. Reduced Congestion and Delays

AI systems can analyze traffic data in real-time and adjust traffic signal timings or suggest alternate routes to prevent congestion. By constantly adapting to changing traffic conditions, AI-powered systems can reduce delays and improve traffic flow, even during peak hours or special events.

b. Improved Safety

AI can enhance road safety by detecting accidents, hazardous conditions, or traffic violations early. By alerting authorities and notifying drivers about potential dangers, AI systems help reduce the likelihood of accidents. Furthermore, by optimizing traffic flow and reducing stop-and-go traffic, AI can also lower the risk of rear-end collisions.

c. Environmental Impact

Reducing congestion leads to smoother traffic flow, which in turn reduces fuel consumption and vehicle emissions. AI-powered traffic optimization can minimize idling time and lower the carbon footprint of transportation, contributing to cleaner air and a more sustainable urban environment.

d. Cost Savings

By improving the efficiency of traffic management systems, AI helps reduce the costs associated with congestion, such as fuel wastage, lost productivity, and increased maintenance on road infrastructure. Smart traffic systems can also optimize the deployment of public transportation, further reducing operational costs.

e. Better Urban Planning and Infrastructure

AI-based systems can analyze traffic data to provide valuable insights for urban planning and infrastructure development. By identifying bottlenecks, traffic patterns, and areas that experience high levels of congestion, city planners can make more informed decisions about where to invest in new roads, public transport, or smart infrastructure.

3.5. Challenges and Limitations

While AI-powered traffic management offers numerous advantages, there are also challenges and limitations that need to be addressed:

- Data Privacy and Security:
- High Implementation Costs
- Integration with Existing Systems
- Dependence on Real-Time Data
- Public Acceptance and Trust:

3.6. The Future of AI in Traffic Management

The future of traffic management lies in further advancements in AI, combined with innovations in infrastructure and vehicle technology. Autonomous vehicles, smart roads, and advanced AI algorithms will create a more interconnected and efficient transportation ecosystem. We can expect to see more cities adopting AI-driven solutions, with smarter traffic signals, enhanced public transportation systems, and a greater emphasis on sustainability and safety.

In conclusion, AI-powered traffic management and optimization represent a revolutionary approach to solving the growing challenges of urban mobility. By harnessing the power of real-time data, machine learning, and intelligent decision-making systems, cities can improve traffic flow, reduce congestion, enhance safety, and make transportation more efficient and sustainable. As technology continues to evolve, the potential for AI to transform traffic management on a global scale is immense, paving the way for smarter, more connected cities.

4. PREDICTIVE ANALYTICS FOR ROAD AND INFRASTRUCTURE SAFETY

As urban populations continue to grow, the challenge of maintaining safe and efficient transportation networks has become a critical concern. Road safety remains a major issue globally, with traffic accidents leading to significant loss of life, injuries, and economic costs. Traditional approaches to road safety focus on reactive measures, such as investigating accidents and improving infrastructure after a tragedy occurs. However, the advent of predictive analytics offers a more proactive and forward-looking approach to road and infrastructure safety. By leveraging historical data, real-time sensors, and advanced machine learning techniques, predictive analytics can help identify potential safety risks before they manifest, leading to more informed decision-making, better resource allocation, and, ultimately, fewer accidents

Figure 3. Predictive analytics for road and infrastructure safety

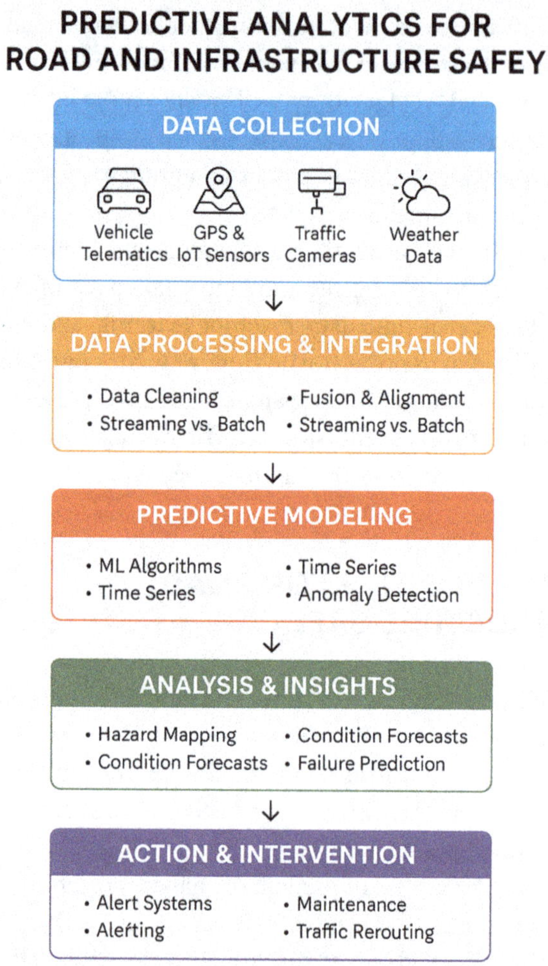

This essay explores how predictive analytics can be used to improve road and infrastructure safety, with a focus on how data-driven insights can identify risks, optimize infrastructure investments, and mitigate potential hazards. By examining the role of data collection, machine learning, real-time monitoring, and risk modeling, we can better understand the potential of predictive analytics in making transportation systems safer.

4.1. Understanding Predictive Analytics

Predictive analytics involves using historical data, statistical algorithms, and machine learning techniques to identify the likelihood of future outcomes. In the context of road and infrastructure safety, predictive analytics can process large amounts of data—such as traffic patterns, weather conditions, accident history, and road infrastructure health—to forecast potential risks and hazards. The goal is not just to react to accidents after they occur, but to prevent them by identifying risk factors early and taking proactive measures.

Key components of predictive analytics include:

- **Data Collection**: Gathering relevant historical and real-time data from sensors, cameras, traffic management systems, and other sources.
- **Data Processing**: Cleaning and preparing the collected data for analysis, ensuring it is accurate and actionable.
- **Modeling**: Building predictive models that use statistical and machine learning techniques to forecast potential risks based on the processed data.
- **Visualization and Decision-Making**: Presenting the predictive outcomes in an understandable format that decision-makers can use to take action.

4.2. Data Sources for Predictive Analytics in Road Safety

To implement predictive analytics for road and infrastructure safety, diverse data sources must be collected and integrated into a comprehensive system. These data sources provide valuable insights into the current state of road conditions and potential risks. Key data sources include:

a. Traffic Flow Data

Traffic flow data, often collected from sensors embedded in roads or GPS-enabled devices in vehicles, provides real-time information on the number of vehicles, their speed, and their movement patterns. Predictive analytics can use this data to identify congestion points, predict traffic jams, and assess areas prone to accidents due to heavy traffic.

b. Accident History Data

Historical accident data is essential for predictive analytics. By analyzing past accidents, including their location, time, and causes (e.g., weather conditions, vehicle type, or driver behavior(Azadani & Boukerche, 2021)), predictive models can identify high-risk areas and times of day when accidents are more likely to occur. Machine learning models can then use this information to forecast the likelihood of future accidents at similar locations.

c. Road Condition Data

Road infrastructure health plays a crucial role in preventing accidents. Data on road surface conditions, including wear and tear, potholes, and weather-induced damage (e.g., flooding or ice), is critical for predictive analytics. Advanced sensors or remote sensing technologies such as drones or satellite imaging can provide real-time assessments of road conditions, which can then be integrated into predictive models.

d. Weather Data

Weather conditions are a major factor influencing road safety. Rain, snow, fog, and ice can significantly increase the likelihood of accidents. By analyzing historical weather patterns and current weather forecasts, predictive models can predict how weather conditions might affect road safety. This can help authorities issue timely warnings and adjust traffic management strategies accordingly.

e. Vehicle Data

With the growing use of connected vehicles, data from vehicles themselves—such as speed, braking behavior, tire pressure, and maintenance status—can be used to predict potential safety risks. Vehicles equipped with sensors can communicate with traffic infrastructure, providing real-time information about their condition and surroundings. This data can be used to predict failures (e.g., malfunctioning brakes or worn-out tires) that might lead to accidents.

4.3. Predictive Models for Road and Infrastructure Safety

The heart of predictive analytics lies in the models used to forecast safety risks. These models rely on complex statistical techniques and machine learning algorithms Mahesh, B. (2020) that analyze the historical data to detect patterns and trends. Key predictive models include:

a. Regression Analysis

Regression analysis is often used in predictive analytics to model the relationship between different variables. For road safety, regression models can identify factors that increase the likelihood of accidents, such as traffic volume, road conditions, and weather. By applying these models to real-time data, decision-makers can predict areas where accidents are more likely to occur.

b. Machine Learning Algorithms

Machine learning algorithms, particularly classification and clustering techniques, can be used to identify and predict accident-prone areas. For example, decision trees, random forests, and support vector machines can classify different road segments based on risk factors and historical accident data. These models can then predict future risks based on similar conditions.

Additionally, unsupervised learning techniques, such as clustering algorithms, can identify hidden patterns in road safety data. For example, clustering can reveal groups of road segments with similar accident characteristics, helping to uncover locations with higher-than-expected accident rates that might not have been identified using traditional methods.

c. Time Series Forecasting

Time series forecasting(Rabbani et al., 2021) involves analyzing data over time to predict future trends. This method is particularly useful for predicting traffic patterns and accident likelihood at specific times of day or during particular seasons. By studying traffic and accident data across different time intervals (e.g., day vs. night, weekday vs. weekend), predictive models can forecast peak traffic times, accident-prone periods, and other variables that influence safety.

d. Geospatial Analysis

Geospatial analysis incorporates location-based data to predict safety risks in specific geographic areas. By analyzing accident hotspots, road conditions, and traffic patterns using Geographic Information Systems (GIS), predictive models can identify high-risk zones and recommend targeted interventions, such as improved signage, better road lighting, or more frequent maintenance.

e. Risk Assessment Models

Risk assessment (Ma, M, et al., 2021) models combine data from various sources to estimate the likelihood and severity of accidents. These models consider factors such as traffic volume, road conditions, driver behavior, and environmental factors to predict the potential risks for different road segments. Such models can be used to prioritize safety interventions in areas with the highest risk.

4.4. Applications of Predictive Analytics in Road Safety

The power of predictive analytics is best realized through its application in real-world scenarios. Here are some key applications of predictive analytics in improving road and infrastructure(Blokus, A., & Dziula, P.,2021) safety:

a. Identifying Accident Hotspots

Predictive models can identify accident-prone areas, or "hotspots," by analyzing accident data, traffic flow, road conditions, and weather patterns. Once these hotspots are identified, targeted safety measures can be implemented, such as improved signage, traffic calming measures, or road surface repairs. By addressing these high-risk areas proactively, authorities can significantly reduce the number of accidents.

b. Dynamic Traffic Management

Real-time traffic data, combined with predictive analytics, can optimize traffic flow and reduce the risk of accidents. By predicting traffic congestion or accidents, authorities can adjust traffic signal timings, change speed limits, or reroute traffic to reduce the likelihood of accidents. For example, predictive models can forecast peak traffic hours or adverse weather conditions and adjust the traffic signal cycle to improve safety and flow.

c. Proactive Maintenance and Infrastructure Investment

Predictive analytics can be used to monitor the condition of roads and infrastructure and predict when maintenance is needed. For example, models can forecast when roads will need resurfacing or when potholes are likely to form. This proactive approach ensures that infrastructure is maintained before it becomes hazardous, reducing the chances of accidents caused by road defects.

d. Weather-Related Alerts

Weather is a major factor in road safety, and predictive analytics can provide real-time weather-related alerts to warn drivers and authorities about potential hazards. For example, if predictive models forecast heavy rain, icy conditions, or fog, traffic management systems can issue warnings, adjust speed limits, and deploy emergency services more effectively.

e. Driver Behavior and Education

Predictive analytics can also be used to identify risky driver behaviors, such as speeding, tailgating, or distracted driving, that increase the likelihood of accidents. By analyzing data from connected vehicles or traffic cameras, authorities can take measures to educate drivers or issue warnings in real-time, ultimately improving road safety.

4.5. Challenges and Limitations of Predictive Analytics

While predictive analytics offers tremendous potential for improving road and infrastructure safety, there are several challenges and limitations that need to be addressed:

- Data Quality and Availability
- Complexity of Modeling
- Privacy Concerns
- Cost of Implementation

Predictive analytics represents a transformative approach to improving road and infrastructure safety. By leveraging historical and real-time data, machine learning algorithms, and advanced modeling techniques, predictive analytics can forecast potential risks, optimize traffic management, and guide proactive maintenance and infrastructure improvements. While challenges exist, the potential benefits of predictive analytics—ranging from reducing accidents and fatalities to improving resource allocation and road efficiency—make it a critical tool for the future of road safety. As technology continues to advance and data collection methods improve, the ability to predict and prevent accidents will only become more effective, ultimately creating safer and more efficient transportation systems.

5. CYBERSECURITY CHALLENGES IN AI-ENABLED TRANSPORTATION SYSTEMS

The integration of Artificial Intelligence (AI) into transportation systems is rapidly transforming the way we navigate cities, manage traffic, and interact with vehicles. Autonomous vehicles, intelligent traffic management systems, and AI-driven infrastructure are revolutionizing urban mobility(Bıyık, C. et al.,2021) by improving efficiency, reducing congestion, and enhancing safety. However, this widespread adoption of AI also introduces significant cybersecurity challenges. The very technologies that make these systems smarter and more efficient can also make them vulnerable to cyberattacks, data breaches, and other security risks.

In this essay, we will explore the cybersecurity challenges that arise with AI-enabled transportation systems. These challenges include the risks associated with the increasing connectivity of vehicles and infrastructure, the vulnerability of AI algorithms, and the potential consequences of security breaches. Additionally, we will examine the importance of securing AI-powered transportation systems to ensure the safety and privacy of users, as well as the broader societal implications.

5.1. Increased Attack Surface

One of the most significant cybersecurity challenges in AI-enabled transportation systems is the vast increase in the "attack surface" that comes with the growing interconnectivity of devices, vehicles, and infrastructure. In traditional transportation systems, most vehicles operated independently, and infrastructure elements (like traffic lights, toll booths, and road sensors) were relatively isolated.

With the advent of AI, however, transportation systems are becoming increasingly interconnected, with vehicles, traffic management systems, infrastructure, and even pedestrians communicating with each other. For instance, autonomous vehicles rely on sensors, GPS, and communication systems to interact with smart traffic lights, road sensors, and other vehicles. This creates an extensive network of interconnected devices that, if not properly secured, can become potential entry points for cyberattacks.

- **Vehicle-to-Vehicle (V2V) and Vehicle-to-Infrastructure (V2I) Communication**: These communication channels are essential for enabling real-time interaction between vehicles and their surroundings, but they also present opportunities for malicious actors to intercept or manipulate data transmissions.
- **Internet of Things (IoT) Devices**: IoT sensors and devices embedded in vehicles, roads, and traffic infrastructure are designed to collect and transmit real-time data. However, these devices can be vulnerable to cyberattacks, as many are not equipped with robust security features, creating potential weak points in the system.

5.2. Vulnerabilities in Autonomous Vehicles

Autonomous vehicles (AVs) are at the forefront of AI-driven transportation innovation. These vehicles rely heavily on AI algorithms, sensors, machine learning models, and real-time data processing to navigate and make decisions. However, the complexity of autonomous systems introduces several cybersecurity risks:

Autonomous vehicles depend heavily on a range of sensors, including LIDAR (Light Detection and Ranging), radar, cameras, and GPS, to support navigation, object detection, and real-time decision-making. However, this reliance on sensor data exposes such systems to vulnerabilities such as sensor spoofing and data manipulation. Malicious actors can deliberately interfere with or falsify sensor inputs, causing the vehicle to misinterpret its surroundings. For example, GPS spoofing attacks may mislead a vehicle into navigating toward an incorrect location, while manipulated LIDAR or radar signals can prevent the detection of obstacles, pedestrians, or other vehicles, significantly increasing the risk of accidents.

In addition to sensor-level threats, autonomous vehicles are also vulnerable to adversarial attacks targeting the AI models responsible for perception and decision-making. Deep learning algorithms used for tasks such as object recognition and traffic sign interpretation can be deceived through carefully crafted adversarial inputs (Chakraborty et al., 2021). These attacks involve subtle modifications to input data—such as altering images or sensor readings—that are often imperceptible to humans but cause AI systems to make incorrect or unsafe decisions. For instance, an attacker could slightly modify a road sign so that it is misclassified by the vehicle's AI system, potentially leading to dangerous driving behavior.

Furthermore, the extensive connectivity features that enable autonomous vehicles to communicate with external systems also introduce the risk of vehicle hijacking. If onboard communication and control systems are inadequately secured, attackers may gain unauthorized remote access to the vehicle. Such access could allow malicious entities to redirect the vehicle, disable critical safety mechanisms, or deliberately cause a collision, posing serious threats to passenger safety and public trust in autonomous driving technologies.

5.3. AI Algorithms and Data Privacy Risks

AI-powered transportation systems depend on extensive data collection, encompassing traffic patterns, road conditions, vehicle sensor outputs, and personal information related to vehicle owners and passengers. This data enables AI models to make informed decisions, optimize routing, and enhance overall transportation safety. However, the collection, storage, and processing of such large volumes of sensitive data introduce significant privacy and security risks. One major concern is data breaches and unauthorized access. Connected vehicles and intelligent transportation infrastructure continuously generate and transmit data across networks, creating multiple entry points for cybercriminals. In the event of a data breach, sensitive personal information, vehicle identifiers, or real-time travel patterns may be exposed, potentially leading to privacy violations, surveillance abuse, and identity theft.

Another critical risk relates to data integrity and manipulation. AI-based transportation systems require accurate and reliable data to function correctly; however, malicious actors may deliberately alter or corrupt data inputs to mislead AI models. For example, attackers could manipulate traffic flow data to disrupt traffic management operations or interfere with vehicle sensor data to compromise navigation and decision-making processes (Hussain et al., 2021). In addition, the extensive use of surveillance cameras and monitoring sensors in AI-enabled transportation systems raises concerns regarding surveillance risks. While continuous monitoring improves traffic safety and efficiency, unauthorized access to surveillance data may enable mass surveillance or misuse of personal information without user consent, undermining public trust and raising ethical concerns.

5.4. Threats to Traffic Management Systems

AI-powered traffic management systems play a vital role in controlling traffic signals, regulating traffic flow, and ensuring the safety of both vehicles and pedestrians. These systems rely on real-time data collected from connected vehicles, sensors, and communication networks to make autonomous decisions. However, their increasing connectivity also makes them attractive targets for cyberattacks, which can have severe consequences. One significant threat is traffic signal manipulation, where attackers gain unauthorized access to AI-controlled traffic signal systems and alter signal timings. Such manipulation could cause traffic congestion, gridlocks, or even serious accidents by disrupting the normal flow of vehicles at intersections.

Another major threat is system overload through Denial of Service (DoS) or Distributed Denial of Service (DDoS) attacks. By overwhelming traffic management systems with excessive requests, attackers can cause system slowdowns or complete outages, preventing real-time traffic updates and creating hazardous road conditions. Additionally, malicious actors may exploit AI-driven traffic information and route guidance systems by injecting false or misleading data. By manipulating real-time traffic information sent to navigation applications or control centers, attackers could intentionally reroute vehicles into congested, unsafe, or restricted areas, potentially leading to accidents or widespread traffic disruption.

5.5. Challenges in Securing AI-Enabled Transportation Systems

Securing AI-enabled transportation systems presents a complex set of technical, regulatory, and organizational challenges. One of the most significant challenges is the lack of standardized security protocols across the transportation sector. The rapid advancement of AI technologies has outpaced the development of universally accepted cybersecurity standards, making it difficult to ensure consistent and effective security measures across diverse platforms and devices. Furthermore, the inherent complexity of AI algorithms, particularly machine learning and deep learning models, complicates security efforts. The opaque or "black box" nature of these algorithms makes it challenging for cybersecurity professionals to fully understand system behavior, identify vulnerabilities, or predict responses under adversarial conditions.

Another major challenge involves the integration of legacy transportation systems with modern AI-driven technologies. Many existing traffic infrastructures were not designed with cybersecurity in mind and may lack compatibility with advanced security mechanisms. Integrating new AI solutions with such legacy systems often requires significant technical effort and financial investment. Additionally, the evolving threat landscape poses continuous challenges, as cybercriminals constantly develop new attack strategies and tools. Maintaining effective security in this dynamic environment requires ongoing monitoring, regular system assessments, and adaptive defense mechanisms.

5.6. Mitigation Strategies for Cybersecurity Risks

To mitigate cybersecurity risks in AI-enabled transportation systems, a combination of technical, organizational, and regulatory strategies must be implemented. Encryption and secure communication mechanisms are essential to protect data exchanged between vehicles, roadside infrastructure, and traffic management systems. Encrypting data in transit and at rest helps prevent interception, tampering, and unauthorized access, while secure communication protocols further enhance system resilience against cyber threats (Li et al., 2020). In addition, continuous validation and testing of AI models are critical to ensure robustness against adversarial attacks. Regular audits, stress testing, and the implementation of fail-safe mechanisms can help identify weaknesses and reduce system vulnerabilities.

Strong access control measures also play a crucial role in cybersecurity mitigation. The use of multi-factor authentication (MFA) and role-based access control ensures that only authorized personnel can access or modify critical transportation systems. Regular security updates and timely deployment of software patches are equally important, as they protect systems from known vulnerabilities and emerging threats. Finally, effective cybersecurity requires collaboration and regulation among governments, transportation authorities, industry stakeholders, and cybersecurity experts. Developing comprehensive regulatory frameworks, standardized security guidelines, and best practices for the secure design, deployment, and maintenance of AI technologies is essential to ensure the long-term safety and reliability of AI-enabled transportation systems.

AI-enabled transportation systems hold great promise for revolutionizing the way we travel, manage traffic, and interact with vehicles. However, the increasing complexity and interconnectivity of these systems introduce significant cybersecurity challenges. Securing AI-powered transportation systems is crucial to protect against cyberattacks, data breaches, and other security risks that could jeopardize public safety, privacy, and the efficiency of transportation networks.

To mitigate these risks, stakeholders must prioritize robust security measures, including encryption, continuous monitoring, and regular updates. As AI technologies continue to evolve, so too must our approach to securing them. Only by addressing these cybersecurity challenges can we fully realize the potential of AI in creating safer, smarter, and more efficient transportation systems.

6. ETHICAL AND REGULATORY CONSIDERATIONS IN AI-DRIVEN TRANSPORTATION SAFETY

AI-driven transportation systems present significant ethical and regulatory challenges that need careful consideration for their safe and responsible deployment. Ethical concerns include the decision-making processes of autonomous vehicles (AVs), which must navigate life-threatening situations like the "trolley problem." AVs may face moral dilemmas, such as choosing between protecting passengers or pedestrians, prompting questions about whether AI should follow utilitarian principles (maximizing overall good) or deontological principles (following specific moral rules). Additionally, biases in AI algorithms can perpetuate societal inequalities if training data is not representative or fair, leading to biased decision-making. Privacy is another critical concern, as AI systems collect vast amounts of personal and environmental data. Ensuring transparency and robust data protection is vital to maintaining user trust and safeguarding privacy rights.

Figure 4. Ethical and Regulatory Considerations in AI-Driven Transportation Safety

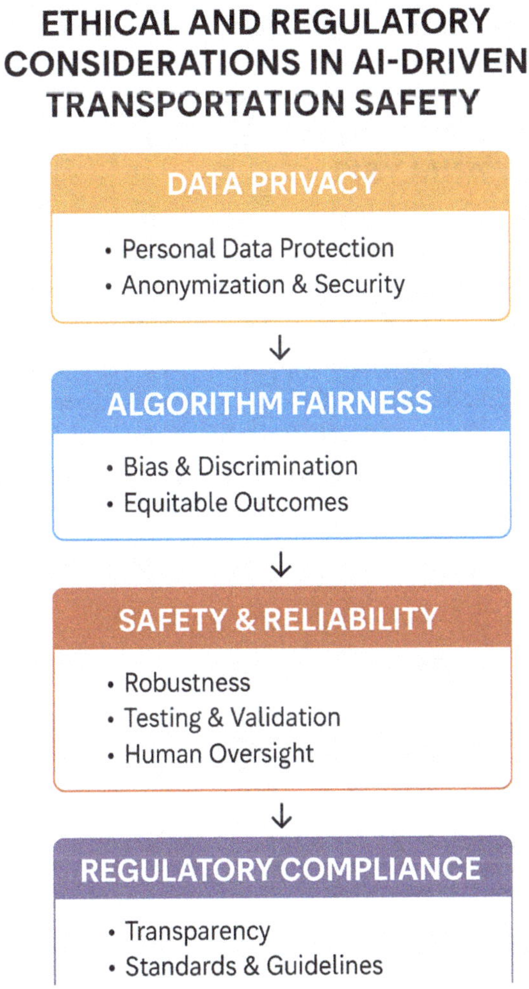

Regulatory considerations focus on creating frameworks to ensure AI systems are safe, effective, and equitable. Governments must establish comprehensive safety standards, including rigorous testing protocols for AI systems in real-world scenarios, ensuring that these technologies are reliable and secure. Liability and accountability also become complex with autonomous vehicles, as traditional methods of assigning responsibility for accidents may no longer apply. Clear legal frameworks are needed to determine accountability for accidents involving AI systems. Ethical guidelines for AI development should ensure that these systems are aligned with societal values, emphasizing fairness, transparency, and privacy. Furthermore, AI technologies often operate across borders, necessitating international cooperation and the establishment of global standards to ensure consistent safety and ethical practices worldwide. By addressing these ethical and regulatory challenges, AI-driven transportation systems can be developed responsibly, benefiting society while maintaining public trust.

7. AI FOR EMERGENCY RESPONSE AND POST-ACCIDENT ANALYSIS

The integration of Artificial Intelligence (AI) into emergency response and post-accident analysis is transforming how we approach crisis management in transportation and other sectors. AI technologies have the potential to drastically improve the speed, efficiency, and effectiveness of emergency responses, reduce the impact of accidents, and optimize post-incident investigations. These technologies range from predictive systems that can forecast potential emergencies, to real-time tools that support decision-making during crises, and advanced analytical models that assist in post-accident analysis.

Figure 5. AI for emergency response and post-accident analysis

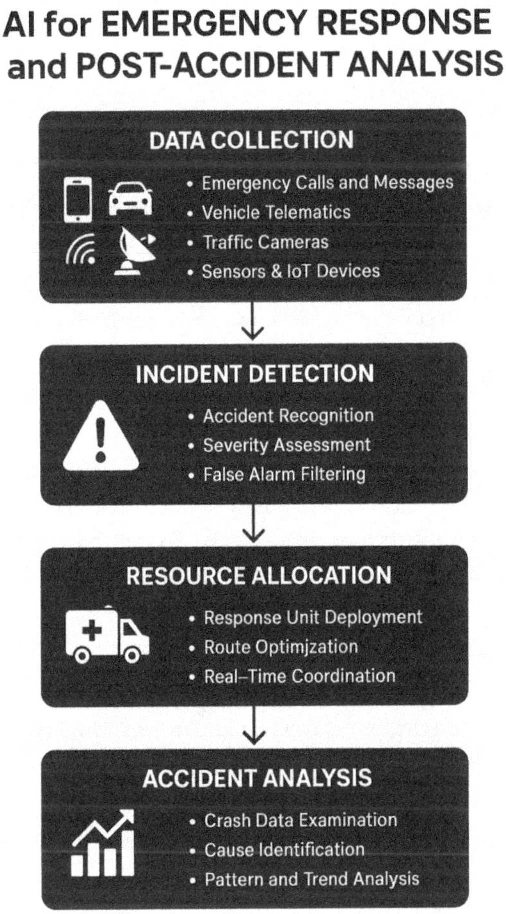

In the context of transportation, where accidents can have far-reaching consequences on both human lives and infrastructure, AI's role in emergency response and post-accident analysis is becoming increasingly crucial. This essay explores how AI is revolutionizing emergency response and the investigative process following accidents, with a focus on its applications, challenges, and potential benefits.

7.1. AI in Emergency Response: Revolutionizing Crisis Management

a. Predictive AI Systems for Risk Management

One of the most powerful applications of AI in emergency response is the use of predictive analytics. AI can analyze vast amounts of data from various sources, including traffic data, weather conditions, and historical incident reports, to identify patterns and predict potential accidents or emergencies. By using machine learning algorithms to analyze this data, AI systems can generate early warnings for potential accidents, allowing emergency response teams to prepare in advance.

For example, AI can predict traffic collisions by analyzing road conditions, speed data, and driver behavior. This predictive capability enables real-time alerts to be sent to relevant authorities, including police, ambulances, and fire departments, even before an accident occurs. Similarly, predictive models can help in disaster management, such as forecasting natural disasters like floods, hurricanes, or earthquakes, and deploying emergency services proactively. By anticipating crises, AI can help reduce response times and enhance the effectiveness of emergency preparedness.

b. Real-Time Decision Support Systems

AI can also enhance emergency response during accidents by providing real-time decision support to first responders. For example, in a car accident, autonomous vehicles (AVs) equipped with AI sensors (Zhang et al., 2023) can relay real-time data about the crash, including the location, number of vehicles involved, and potential hazards, to emergency teams. This data allows first responders to quickly assess the severity of the incident and deploy resources more effectively.

AI-powered systems can further assist in real-time decision-making during emergencies. For instance, AI can help triage victims in mass casualty incidents by assessing the severity of injuries based on available data and recommending treatment priorities. Similarly, AI can assist emergency teams in optimizing traffic routes to avoid congestion or hazardous areas, improving response times.

c. Drones and Robotics in Emergency Response

Drones and robotics are becoming integral tools in AI-powered emergency response. Drones, which are increasingly equipped with AI technologies, can be deployed to survey accident scenes, providing real-time aerial footage (Makrigiorgis et al., 2020) for first responders. These drones can help assess damage, identify hazards such as fire or structural instability, and locate victims. In cases of natural disasters, drones can be used to monitor and map disaster areas, enabling authorities to assess damage quickly and determine where to deploy aid.

Robotic systems, including autonomous vehicles and rescue robots, can also play a significant role in post-accident scenarios. In dangerous environments, such as hazardous accident scenes, robots can be sent to perform dangerous tasks, such as delivering medical supplies or assessing structural integrity, reducing the risk to human life.

7.2. Post-Accident Analysis: Enhancing Investigation and Prevention

Once an accident has occurred, AI technologies can significantly improve the investigation and analysis of the incident, helping authorities identify causes, improve safety measures, and prevent similar accidents in the future.

a. AI for Accident Data Collection and Analysis

In the aftermath of an accident, AI can assist investigators by processing vast amounts of data, including traffic camera footage, vehicle telemetry, and sensor data from the accident scene. AI algorithms can automatically identify key information from these data sources, such as vehicle speeds, vehicle trajectories, and environmental factors, providing a detailed picture of what happened.

Machine learning models can also analyze accident reports and historical accident data to identify patterns and trends. For example, AI can identify recurring issues with road infrastructure, driver behavior, or weather conditions that contribute to accidents. By providing insights into these factors, AI helps authorities understand the underlying causes of accidents and develop preventive measures.

b. AI in Forensic Analysis and Accident Reconstruction

AI is transforming the field of accident reconstruction(Guzek & Lozia, 2021). Traditionally, accident reconstructionists rely on physical evidence, witness statements, and expert analysis to determine the sequence of events leading up to an accident. AI-enhanced tools can speed up this process by automating the reconstruction using data from various sources. For example, AI can analyze data from vehicle black boxes (event data recorders), which provide information about vehicle speed, braking, and acceleration just before the crash. This data can be used in conjunction with other data sources, such as traffic camera footage, to create a detailed 3D model of the accident scene.

AI-driven accident reconstruction tools can also simulate potential crash scenarios, allowing investigators to test different hypotheses and determine the most likely cause of the incident. These tools can identify whether certain safety features, such as airbags or seatbelts, functioned as expected and assess the role of vehicle design in the accident.

c. Improving Infrastructure and Safety Measures

AI's post-accident analysis capabilities are also crucial for improving road safety and infrastructure. After accidents, AI can analyze data from the incident to determine whether the road infrastructure played a role in the crash. For example, AI can examine whether poor road design, inadequate signage, or malfunctioning traffic signals contributed to the accident. Additionally, AI systems can use historical accident data to identify "hot spots" where accidents are more likely to occur, allowing transportation departments to implement targeted safety improvements.

Machine learning models can analyze traffic flow patterns and accident rates in real time, offering suggestions for infrastructure upgrades, such as improved traffic signals, better lighting, or the addition of barriers. These AI-driven insights can guide decision-making to create safer roads and reduce accident frequency.

d. AI for Legal and Insurance Purposes

AI is also reshaping how insurance companies handle accident claims and settlements. After an accident, AI-powered tools can quickly process accident data, verify claims, and assess damage. For example, AI can analyze vehicle damage from accident photos and use this data to estimate repair costs, improving the efficiency of claims processing.

Additionally, AI can help identify fraudulent claims by cross-referencing accident data with historical patterns and using anomaly detection algorithms. This reduces the risk of fraudulent claims and ensures that insurance companies handle legitimate claims more effectively.

7.3. Challenges and Limitations

While AI offers significant potential in emergency response and post-accident analysis, there are several challenges and limitations that need to be addressed:

a. Data Privacy and Security

The collection and use of data in emergency response and post-accident analysis raise concerns about privacy and data security. Sensitive information, such as accident victims' medical records or personal travel data, must be protected to prevent misuse. Additionally, AI systems that rely on vast amounts of data are vulnerable to cyberattacks, which could compromise the integrity of emergency response systems(Saini, K., et al., 2022) and lead to disastrous consequences.

To mitigate these risks, strong data protection measures and clear data privacy regulations are essential to ensure that AI systems in emergency response and post-accident analysis are used responsibly.

b. Reliability and Accuracy

AI systems rely on large datasets to make decisions and predictions. If the data fed into these systems is incomplete or inaccurate, the outcomes may be flawed. For instance, if accident data from a particular region is not representative or if a predictive model is based on biased historical data, AI systems may provide inaccurate predictions or responses.

It is essential to ensure that AI systems are trained on diverse, high-quality data and continuously monitored for accuracy and reliability.

c. Ethical and Legal Concerns

The use of AI in emergency response and post-accident analysis may raise ethical and legal concerns. For example, in post-accident investigations, AI-driven decision-making may potentially influence legal outcomes, such as determining fault in a collision. Transparent AI systems and clear ethical guidelines must be established to ensure that these technologies are used fairly and responsibly.

AI is transforming emergency response and post-accident analysis by enhancing decision-making, improving response times, and providing deeper insights into accident causes. From predictive risk management and real-time decision support to accident reconstruction and infrastructure improvements, AI is revolutionizing how we approach crisis management and safety. However, for these technologies to reach their full potential, issues such as data privacy(Di Minin, E., et al., 2021), system reliability, and ethical considerations (Hua, S.,et al., 2024) must be addressed. By integrating AI in a responsible and transparent manner, we can ensure that these advancements benefit society by improving safety, reducing response times, and preventing future accidents.

REFERENCES

Afolayan, A., Easa, S. M., Abiola, O. S., Alayaki, F. M., & Folorunso, O. (2022). GIS-based spatial analysis of accident hotspots: A Nigerian case study. *Infrastructures, 7*(8), 103. DOI: 10.3390/infrastructures7080103

Azadani, M. N., & Boukerche, A. (2021). Driving behavior analysis guidelines for intelligent transportation systems. *IEEE Transactions on Intelligent Transportation Systems, 23*(7), 6027–6045. DOI: 10.1109/TITS.2021.3076140

Bıyık, C., Abareshi, A., Paz, A., Ruiz, R. A., Battarra, R., Rogers, C. D., & Lizarraga, C. (2021). Smart mobility adoption: A review of the literature. *Journal of Open Innovation, 7*(2), 146. DOI: 10.3390/joitmc7020146

Blokus, A., & Dziula, P. (2021). Relations of imperfect repairs to critical infrastructure maintenance costs. *Sustainability (Basel), 13*(9), 4917. DOI: 10.3390/su13094917

Chakraborty, A., Alam, M., Dey, V., Chattopadhyay, A., & Mukhopadhyay, D. (2021). A survey on adversarial attacks and defences. *CAAI Transactions on Intelligence Technology, 6*(1), 25–45. DOI: 10.1049/cit2.12028

Di Minin, E., Fink, C., Hausmann, A., Kremer, J., & Kulkarni, R. (2021). How to address data privacy concerns when using social media data in conservation science. *Conservation Biology : The Journal of the Society for Conservation Biology, 35*(2), 437–446. DOI: 10.1111/cobi.13708

Florackis, C., Louca, C., Michaely, R., & Weber, M. (2023). Cybersecurity risk. *Review of Financial Studies, 36*(1), 351–407. DOI: 10.1093/rfs/hhac024

Guzek, M., & Lozia, Z. (2021). Computing methods in the analysis of road accident reconstruction uncertainty. *Archives of Computational Methods in Engineering, 28*(4), 2459–2476. DOI: 10.1007/s11831-020-09462-w

Haghighat, A. K., Ravichandra-Mouli, V., Chakraborty, P., Esfandiari, Y., Arabi, S., & Sharma, A. (2020). Applications of deep learning in intelligent transportation systems. *Journal of Big Data Analytics in Transportation, 2*(2), 115–145. DOI: 10.1007/s42421-020-00020-1

Hua, S., Jin, S., & Jiang, S. (2024). The limitations and ethical considerations of ChatGPT. *Data Intelligence, 6*(1), 201–239. DOI: 10.1162/dint_a_00243

Hussain, B., Afzal, M. K., Ahmad, S., & Mostafa, A. M. (2021). Intelligent traffic flow prediction using optimized GRU model. *IEEE Access : Practical Innovations, Open Solutions, 9*, 100736–100746. DOI: 10.1109/ACCESS.2021.3097141

Lee, C. S., Cheang, P. Y. S., & Moslehpour, M. (2022). Predictive analytics in business analytics: Decision tree. *Advances in Decision Sciences*, *26*(1), 1–29. DOI: 10.47654/v26y2022i1p1-29

Li, H., Hua, Z., Bao, H., Zhu, L., Chen, M., & Bao, B. (2020). Two-dimensional memristive hyperchaotic maps and application in secure communication. *IEEE Transactions on Industrial Electronics*, *68*(10), 9931–9940. DOI: 10.1109/TIE.2020.3022539

Ma, M., Zhao, G., He, B., Li, Q., Dong, H., Wang, S., & Wang, Z. (2021). XGBoost-based method for flash flood risk assessment. *Journal of Hydrology (Amsterdam)*, *598*, 126382. DOI: 10.1016/j.jhydrol.2021.126382

Mahesh, B. (2020). Machine learning algorithms: A review. [IJSR]. *International Journal of Science and Research*, *9*(1), 381–386.

Makrigiorgis, R., Kolios, P., Timotheou, S., Theocharides, T., & Panayiotou, C. G. (2020, May). Extracting the fundamental diagram from aerial footage. In *2020 IEEE 91st vehicular technology conference (VTC2020-Spring)* (pp. 1-5). IEEE. DOI: 10.1109/VTC2020-Spring48590.2020.9128534

Oguntibeju, O., Adonis, M., & Alade, J. (2024). Systematic review of real-time analytics and artificial intelligence frameworks for financial fraud detection. *International Journal of Advanced Research in Computer and Communication Engineering*, *13*(9). Advance online publication. DOI: 10.17148/IJARCCE.2024.13920

Parekh, D., Poddar, N., Rajpurkar, A., Chahal, M., Kumar, N., Joshi, G. P., & Cho, W. (2022). A review on autonomous vehicles: Progress, methods, and challenges. *Electronics (Basel)*, *11*(14), 2162. DOI: 10.3390/electronics11142162

Rabbani, M. B. A., Musarat, M. A., Alaloul, W. S., Rabbani, M. S., Maqsoom, A., Ayub, S., & Altaf, M. (2021). A comparison between SARIMA and exponential smoothing models for forecasting road accidents. *Arabian Journal for Science and Engineering*, *46*(11), 11113–11138. DOI: 10.1007/s13369-021-05650-3

Ranyal, E., Sadhu, A., & Jain, K. (2022). Road condition monitoring using smart sensing and artificial intelligence: A review. *Sensors (Basel)*, *22*(8), 3044. DOI: 10.3390/s22083044

Ravish, R., & Swamy, S. R. (2021). Intelligent traffic management: A review of challenges, solutions, and future perspectives. *Transport and Telecommunication*, *22*(2), 163–182. DOI: 10.2478/ttj-2021-0013

Saini, K., Kalra, S., & Sood, S. K. (2022). Disaster emergency response framework for smart buildings. *Future Generation Computer Systems, 131*, 106–120. DOI: 10.1016/j.future.2022.01.015

Salimi, M., & Al-Ghamdi, S. G. (2020). Climate change impacts on critical urban infrastructure and urban resiliency strategies for the Middle East. *Sustainable Cities and Society, 54*, 101948. DOI: 10.1016/j.scs.2019.101948

Singh, S., Ramkumar, K. R., & Kukkar, A. (2021, October). Machine learning techniques and implementation of different ML algorithms. In *2021 2nd Global Conference for Advancement in Technology (GCAT)* (pp. 1–6). IEEE. DOI: 10.1109/GCAT52182.2021.9586806

Szeliski, R. (2022). *Computer vision: Algorithms and applications*. Springer Nature. DOI: 10.1007/978-3-030-34372-9

Vu, A. T., & Nguyen, T. T. H. (2021, November). Formulating weather-responsive traffic management strategies for Ho Chi Minh City. In *AIP Conference Proceedings* (Vol. 2428, No. 1). AIP Publishing. DOI: 10.1063/5.0071041

Zhai, X., Chu, X., Chai, C. S., Jong, M. S. Y., Istenic, A., Spector, M., & Li, Y. (2021). A review of artificial intelligence (AI) in education from 2010 to 2020. *Complexity, 2021*(1), 8812542. DOI: 10.1155/2021/8812542

Zhang, Z., Wang, L., & Lee, C. (2023). Recent advances in artificial intelligence sensors. *Advanced Sensor Research, 2*(8), 2200072. DOI: 10.1002/adsr.202200072

Zhou, Z. H. (2021). *Machine learning*. Springer Nature. DOI: 10.1007/978-981-15-1967-3

Ziakopoulos, A., & Yannis, G. (2020). A review of spatial approaches in road safety. *Accident; Analysis and Prevention, 135*, 105323. DOI: 10.1016/j.aap.2019.105323

Chapter 3

AI-Powered Traffic Management and Optimization

Kavita Kanwar
http://orcid.org/0009-0000-6661-891X
Poornima University, Jaipur, India

Ritam Dutta
http://orcid.org/0000-0003-3089-778X
Poornima Institute of Engineering and Technology, Jaipur, India

J. Rajalakshmi
http://orcid.org/0000-0002-4682-8546
Sethu Institute of Technology, India

Vivek Chamoli
http://orcid.org/0000-0002-4549-3705
Graphic Era Hill University, Dehradun, India

P. Shubha
http://orcid.org/0000-0003-1128-3838
Dr. Ambedkar Institute of Technology, Bengaluru, India

Nikhil Kumar Goyal
http://orcid.org/0009-0007-4532-8033
Noida International University, Greater Noida, India

ABSTRACT

This chapter examines how artificial intelligence (AI) has been transformative in traffic management and optimization with the view that it can help to solve the consistent issues of congestion, inefficiency, and road safety in contemporary transportation. The conventional traffic control systems are usually not capable of dynamically adjusting to evolving circumstances which lead to time wastage, pollution and loss of profits. With the incorporation of the AI methods like machine learning, computer vision, and reinforcement learning, traffic systems would be able to reach the real-time flexibility, prediction precision and enhanced efficiency. The chapter discusses AI-based applications, such as smart traffic lights, route optimization, incident

DOI: 10.4018/979-8-3373-3231-4.ch003

detection, and optimisation of public transport schedules. It further focuses on how supporting technologies such as IoT, 5G, and autonomous vehicles can be used to establish intelligent transportation ecosystems.

1. INTRODUCTION

1.1 Urbanization and the Growing Traffic Crisis

In the last century, cities have turned out to be the centre of economic and industrial growth and human habitation. As over half of the global population moves to cities, the problem of mobility has now centre-staged the debate on sustainable urban development. The roads constructed many decades ago have been overloaded with vehicles that are way more than what these roads were built to accommodate. This has rendered congestion as a norm to millions of commuters all over the world. In addition to the inconvenience of being caught in traffic, this phenomenon has some ripple effects that reach almost all spheres of life in the city. It costs businesses billions of dollars every year because of delays in deliveries and absenteeism among workers due to traffic congestion and it causes more stress and fatigue and less time to spend with family or leisure (Najat et al., 2024).

The other critical aspect of this crisis is the environmental dimensions of the crisis. Cars with stop and go traffic burn more fuel, they emit more greenhouse gases and air pollutants like carbon monoxide and nitrogen oxides. These emissions not only contribute to the rapid climate change, but directly impact on people health, particularly in highly populated urban centers, where the quality of air is already under-estimated. The children and the elderly people as well as persons having breathing problems are the ones who are highly susceptible to the health hazards of being subjected to the vehicular discharges over an extended period of time. Therefore, traffic jam is not only a mobility problem, but it overlaps with issues of health, ecological sustainability, and economic performance in the general population.

Conventional means of managing the traffic like expansion or creation of flyovers or carpool lanes have not helped much to address these problems. Although they offer a short-term solution, they tend to foster more ownership of vehicles, which eventually leads to the level of congestion being back to its former degree. This paradox has compelled the urban planners and policymakers to recognize that increasing infrastructure can not be a sole solution to the problem. Rather, a convergence towards systems that are smarter and more adaptive is emerging, systems which can dynamically react to real time conditions as opposed to having fixed rules or manually controlled interventions (Gen et al., 2024).

1.2 Evolution of Traffic Management Systems

Early on in the history of organized urban movement, traffic control was comparatively simple. Intersection control was done manually by traffic police, and fixed-timed traffic lights were used as a subsequent method to automate rudimentary control. These techniques were effective at low densities, but were quickly demonstrated to be ineffective in fast growing urban areas. The fixed signal cycles were not able to accommodate the changing traffic volumes during the course of the day thus inefficiencies that included long queue on certain approaches and underutilization of some approaches were realized.

Late in the 20 th century, the Intelligent Transportation Systems (ITS) emerged, involving the use of loop detectors, closed-circuit cameras, and primitive algorithms to scan the traffic environment. The systems were a step towards a better direction as they gave the traffic control centers the ability to manipulate the signals or even reroute vehicles according to observations made in real time. Their efficiency was, however, still limited by inflexible and prescriptive programming which was not really adaptable. Consider as an example that ITS was only able to predict and stop congestion when the congestion had already occurred on a highway even though it was able to detect congestion. Also, since there were no sophisticated data processing tools, massive amounts of data could be underutilised (Gunasekar et al., 2024).

The limitations of traditional and semi-intelligent systems underscore the need for a more transformative approach. Here is where the artificial intelligence comes into the scene. In contrast to earlier models, AI flourishes on complexity and uncertainty, thus it is especially suited to urban traffic conditions where innumerable variables, such as the behavior of drivers, weather conditions, etc, interact unpredictably. Through the ability to learn on a large scale, AI systems are able to predict issues they might face before they occur and make adjustments to their strategies on the fly. This is a paradigm shift in traffic control: reactive, predictive and preventive solutions.

1.3 Rise of Artificial Intelligence in Traffic Optimization

In the present-day setting, artificial intelligence is being viewed as a foundation of the next level of traffic control. Its capability to handle huge amount of data and produce actionable information in real-time enables cities to go beyond the bottlenecks of non-real-time systems. As an example, machine learning proxies may use years of traffic data with real-time sensor data to forecast congestion hotspots hours or even days in advance. In a similar way, adaptive traffic signal control can be realized through reinforcement learning algorithms, i.e., the timing of the signal is dynamically adjusted according to the current situation in order to reduce delays.

Such abilities not only enhance traffic flow, but also, promote safety, reduce emissions, and boost the satisfaction of the commuters (Bezek & Güre, 2024).

Figure 1. AI in Traffic Management

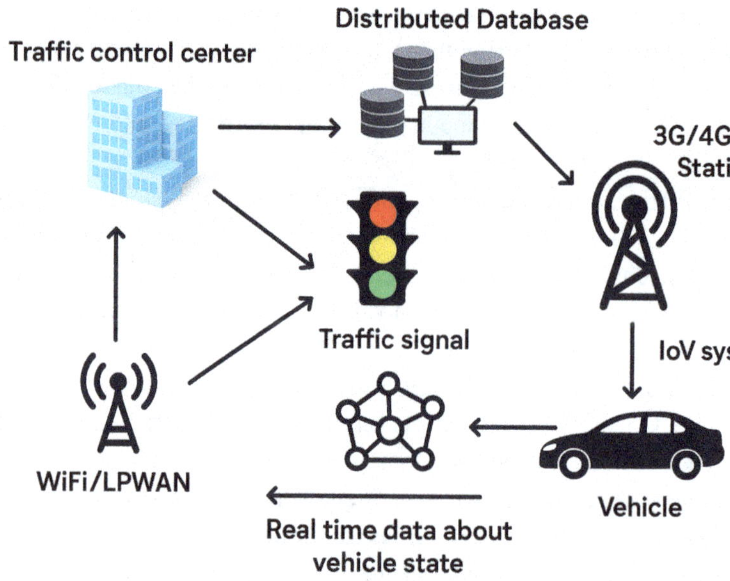

The computer vision application is one of the most effective applications of AI in that computers with AI algorithms can notice and analyze the events on the road more accurately than ever. Such systems are able to detect accidents, unlawful lane changes or jaywalking pedestrians in a matter of seconds so that police can send emergency agencies or modify the traffic controls within minutes. This degree of responsiveness is just impossible with human surveillance only. More so, by combining real-time traffic data with predictive modeling, AI-based navigation applications guide the driver on the path of lowest resistance and redistribute traffic to distribute congestion and avoid bottlenecks among networks.

Notably, AI is not used in a vacuum. Its ability to combine with the latest technologies like Internet of things (IoT), 5G communication, and cloud computing further magnifies its potential. Connected cars, such as those, may exchange location, velocity, and destination information with centralized Artificial Intelligence systems, and by such means, a holistic perspective of traffic conditions can be formed, extending beyond the scope ofroadside sensors. This integration opens the door to integrated, city-wide traffic control approaches, which are dynamically adaptive, predictive, and very efficient (Lungu, 2024).

1.4 Objectives and Scope of the Chapter

This chapter is aimed at offering an in-depth discussion of the way AI is changing traffic management and optimization. It starts off by analyzing the inherent weaknesses of the traditional systems and why the traditional ways have not managed to keep up with the pace of modern urbanization. Then the discussion shifts to the peculiarities of the strengths of AI in dealing with real-time complexity as it can be used in various fields and includes adaptive signal control, prediction of congestion, incident identification, and optimization of public transportation.

The chapter also highlights the other implications of AI-powered traffic systems on society, the economy, and the environment beyond technical use. Such issues as data privacy, algorithmic fairness, and equitable access are brought up to highlight the ethical aspects of the implementation of AI in the public. Lastly, the chapter gives a glimpse of the future specifying possible directions of AI implementation with autonomous vehicles, smart cities, and the travel policymaking systems of the world. Embracing both the technical innovation and social responsibility, the chapter aims to demonstrate how AI will help in creating sustainable, safe and efficient transportation systems in the future.

2. FUNDAMENTALS OF TRAFFIC MANAGEMENT SYSTEMS

2.1 Traditional Traffic Control Mechanisms

Historically, traffic control has been based on comparatively simple systems to control the movement of vehicles within an intersection and arterial roadways. The first was the manual control whereby traffic police controlled vehicles using their hands. Although this worked in small towns or special events, it could not be sustained in urban networks that were growing fast. To eliminate these weaknesses, there was the introduction of the traffic lights which originally had fixed periods of red, yellow, and green lights. Though this invention was a very important step in the automation of traffic control, its non-dynamic character could not always correspond to reality in the variability of traffic in motion (Kumar et al., 2024).

Fixed-time signal control is also rigid. A traffic signal programmed to give the maximum amount of thirty seconds green time on one approach can produce long lines at peak hours and waste many good minutes of green time at off-peak hours. These inefficiencies increased with the increase in traffic volumes. As a case in point, in megacities like Mumbai or New York, at times of rush-hour, intersections remain operated by a predetermined schedule of timings. The outcome is traffic jam, aggravated commuters and fuel wastage.

An effort to enhance traditional control was made by semi-actuated signals, which adjusted green time depending upon sensor inputs like inductive loop detectors in the road. These sensors would be able to detect the presence of a vehicle and green time would be extended. This was more flexible, but confined to localized intersections and unable to coordinate traffic over wider systems. In addition, malfunctions of hardware, insufficient calibration, and incompatibility with adjacent crossroads commonly diminished efficiency (Goyal et al., 2025a).

2.2 Emergence of Intelligent Transportation Systems (ITS)

The shortcomings of these primitive traffic controls led to the invention of Intelligent Transportation Systems (ITS), a collection of technologies first launched in the late 20th century. ITS was focused on improving mobility, safety and efficiency by combining sensing, communication and computing devices. Examples are closed-circuit television (CCTV) to monitor traffic, automatic number plate reader (ANPR) to enforce law, and variable message signs to tell drivers of delays or diversions.

Real-time monitoring was one of the characteristic aspects of ITS. Loop detectors, infrared sensors then GPS-enabled devices started to send live traffic data to centrally controlled centers. This gave the opportunity to respond dynamically, including the diversion of traffic in the case of an accident or a response to a special occasion by setting the signal timing. Singapore and Tokyo cities were one of the first, showing the power of a well-coordinated ITS strategies to dramatically decrease congestion (Datey & Prabhu, 2023).

Irrespective of these developments, ITS was limited in significant ways. Most systems were extremely dependent on rule based programming which failed to dynamically adjust to unforeseen events like sudden downpours, sporting events or mass protests. Moreover, the extreme magnitude of the obtained data strained even traditional information processing systems and most of the data remained unused. Although modern traffic management has its roots in ITS, its dependence on pre-programmed logic made it clear that a more intelligent approach is required, based on learning.

2.3 Core Principles of Traffic Flow Management

To see how AI develops previous systems, it is necessary to comprehend the main principles according to which traffic flows. On the most basic level, traffic management is based on the demand-capacity relationship (the quantity of vehicles wanting to utilize the road space) and the capacity (physical and operational capacity of the road infrastructure). Congestion happens when there is excess demand over the capacity and the result is delays, stop-and-go situation and higher emissions.

Mathematic mathematical models like the Greenshields model are usually used by traffic engineers in order to create a correlation between traffic density and speed. These models are used to anticipate the way traffic will act under differing circumstances and in the making of decisions like speed limits or road expansion development. Likewise, other concepts such as level of service (LOS) have offered standardized indices of road performance, between free-flowing traffic and heavy congestion (Xing et al., 2023).

Figure 2. Principles of Traffic Control

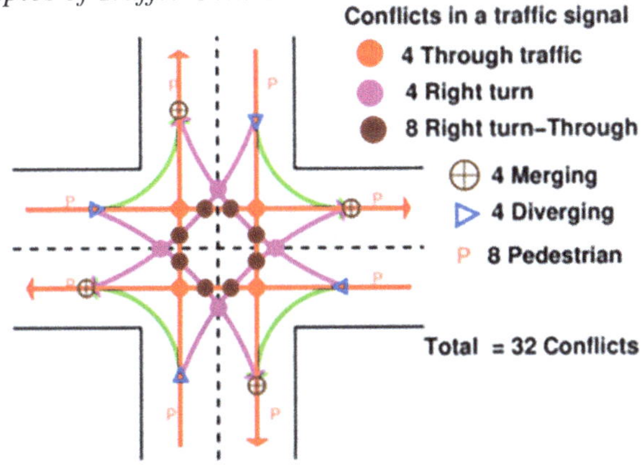

But the traffic of the real world seldom obeys nice mathematical conjectures. Human behavior creates unpredictability, and they have varied drivers, some are aggressive, some react fast and others obey the rule. Environmental factors such as weather, construction activities, and accidents further complicate flow patterns. These complications highlight the necessity of the move beyond simple models and to more adaptive systems driven by AI to solve contemporary traffic issues.

2.4 The Shift Toward Adaptive and Data-Driven Approaches

The increased understanding of traffic as a dynamic system that is both complex and adaptive opened the way to adaptive and data-driven solutions. Cities started investing in the data collection devices like GPS devices, Bluetooth sensors and mobile apps rather than basing hard infrastructure. These devices produced large amounts of data on the speed of vehicles, travel time, and selecting routes.

Data-driven traffic management is concerned with the optimization of real-time decisions with the help of this information. In an example, adaptive traffic signal control only uses live traffic data to dynamically adjust signal phases, instead of a fixed cycle. Likewise, congestion hours are predicted and given time ahead so that authorities can give preemptive warnings or reroute vehicles before jams are created (Idriss et al., 2023).

These techniques are advancements but they can only be successful with strong analytical skills. In the absence of sophisticated computational software, cities face the threat of accumulating huge volumes of data that are not fully used. Here is exactly where artificial intelligence is essential because it is not only capable of processing data at scale but also finding patterns and creating insights that remain unattainable to humans and rule-based systems (Abubeker et al., 2023).

3. ARTIFICIAL INTELLIGENCE IN TRAFFIC MANAGEMENT

3.1 Why AI Matters in Traffic Systems

Due to its capability of scaling complexity, artificial intelligence has become a innovation in the field of traffic management. Compared with the traditional approaches that have fixed rules, AI flourishes in the dynamic space where a large number of variables interact at the same time. The issue of urban traffic is exactly such an environment and it is affected not only by weather and accidents but also by human choice and events taking place in the city.

The most important advantage of AI is its ability to process big volumes of data and change strategies on the fly. As an example, machine learning systems can train on years of past history traffic and at the same time absorb real time information through sensors, cameras, and GPS systems. This bifunctional quality allows predictive precision which is many times greater than that of rule-based methods. Prior to the congestion, an AI system is able to predict it, prescribe alternative routes, and even synchronize the schedules of public transport and the movements of the private vehicles (Goyal, 2025a).

Scalability is another reason AI is of interest. Unlike human operators and fixed algorithms which fail to monitor a network within a entire city, AI systems have the ability to analyze thousands of intersections, highways, and side streets at a time. This scalability makes sure that the traffic solutions are not only effective at the intersection at local levels but also over the whole metropolitan regions.

3.2 Key AI Techniques in Traffic Optimization

Different branches of AI contribute to traffic management in distinct ways. Machine learning (ML) is fundamental as it focuses on past and current data as well as identifies trends and presents the future. As an example, a congestion level can be classified according to sensor data by using supervised learning algorithms, whereas unsupervised learning can yield concealed patterns, such as irregular traffic spikes related to particular events.

The concept of reinforcement learning (RL) is of great influence especially in adaptive traffic signal control. The AI agent in this paradigm is a continuous learner and engages with the environment. In the event where a signal adjustment decreases the total waiting time, the agent is positively reinforced whereby the decisions made are optimized over time. Practical experiments in such cities as Hangzhou, China, demonstrated that delays were reduced significantly with the use of RL-driven traffic lights (Thote et al., 2024).

Another AI method is computer vision that makes use of cameras to analyze the real-life situation. Systems are able to identify the number of vehicles, pedestrian crossing or roadblocks with great precision. To provide an example, AI-driven camera could instantly identify a traffic collision and issue rerouting procedures within seconds. Likewise, the natural language processing (NLP) can be used by the AI assistant to process feedback left by drivers in apps or social media, which offers the AI assistant another level of situational awareness (Raj, 2024).

3.2.1 Performance in the Real World Evidence

The practical analysis of AI-based traffic management technologies shows evident performance changes in several performance metrics, among which are the time spent in traffic, traffic jams, fuel usage, and emissions. An example of this is that when massive deployments of reinforcement learning-based adaptive signal control systems have been reported, they have reduced the average intersection delay by a percentage of up to 10 to 30 percent, as compared to the fixed-time signal control systems and actuated signal systems. Empirical measurements were made at Hangzhou, China, to demonstrate that the average travel time citywide declined by around 15 percent, and the diffusion of the traffic at the peak time was streamlined.

Likewise, recent research in the cities of North America like Los Angeles and Pittsburgh shows that predictive signal coordination with AI has decreased the vehicle idling time up to 25 in order to save fuel and produce significant greenhouse gas emissions. Quantitative findings also indicate that AI congestion forecasting models are more accurate, with a score of more than 85 per cent accuracy often surpassing traditional statistical prediction processes of congestion and especially non-recurring congestions like accidents and extreme weather.

In addition to efficiency, there are also safety benefits that are shown in the empirical findings. Incident detection systems involving computer vision have been demonstrated to cut the response time of an emergency by a few minutes, which is essential to ensure that there are fewer secondary accidents and better outcomes of survival. All these findings are a confirmation that AI-based traffic optimization is not just a hypothetical idea, but a provision of facts on the ground, data-driven and verifiable solutions to the problem of urbanization realities.

3.3 Integration with IoT and Smart Infrastructure

AI will have the potential to achieve its maximum potential through the Internet of Things (IoT) and smart city infrastructure. Still, IoT-based vehicles and roadside units can serve as a constant source of data, supplying AI systems with accurate information about the speed of a vehicle, its position, and its route. The ecosystem is connected in a way that provides unprecedented insight into city mobility.

As an example, connected cars can exchange their locations with centralized computer-artificial intelligence (AI) systems, allowing coordinated rerouting mechanisms that avoid congestion prior to its creation. On the same note, intelligent infrastructure like intelligent streetlights and intelligent pedestrian crossing will ensure that traffic management is not restricted to cars only but rather it also considers the cyclists, buses and pedestrians. IoT-friendly sensors and AI algorithms have already been used in Barcelona to enhance the movement of traffic and minimize idle time at crossroads and parking spaces (Ajayi & Kumkale, 2023).

Safety is also improved with the implementation of AI with IoT. Predictive algorithms are capable of detecting the high-risk crossings where accidents will tend to occur and proposing preventive measures (and actions) to avoid them, including changing signal timing or imposing enforcement patrols. Besides, AI-IoT is a significant advantage to emergency response systems because ambulances can be prioritized and placed on an AI-managed traffic route.

3.4 Challenges in AI Adoption for Traffic Systems

Although AI provides extensive potential, it is not the case with its implementation in traffic management. Data quality is one of the major obstacles. Traffic sensors in most cities are either outdated, disproportionately spread or not maintained due to which data collected is not complete or accurate. Models of AI that are trained with this data are likely to give incorrect forecasts or biased suggestions.

Computational infrastructure is another problem. To run real-time AI models on full metropolitan networks, it needs high-performance computing facilities, strong communication systems, and easy integration with older systems. Infrastructure to facilitate such advanced implementations is not always available in developing countries, and this increases the gap between the resource-constrained area and the technologically advanced one (Purwar & Chaudhry, 2023).

Ethics comes into place as well. The traffic systems based on AI should promote fairness in the distribution of resources, i.e., the road space or the priority passage. Algorithms can be biased to the detriment of some neighborhoods or socioeconomic groups without proper attention to it. Also, when cities make use of AI-driven cameras and sensors on a large scale, there are concerns regarding the privacy and surveillance of the data collected. Such problems indicate that there is a need to have thoughtful systems of governance that have the ability to balance between technological advancement and equitability and accountability (Dikshit et al., 2023).

4. CASE STUDIES OF AI-DRIVEN TRAFFIC MANAGEMENT

4.1 Hangzhou, China: Reinforcement Learning for Smart Traffic Lights

It is also one of the most prominent AI-driven traffic optimization examples in practice, which in Hangzhou. The city is characterized by a high pace of urbanization and the development of the technological sector, which led to serious congestion issues with the growing number of people and the increase in the number of privately owned vehicles. Local governments then responded by cooperating with the Alibaba research group to implement an AI-based reinforcement learning (RL) traffic controller.

This system examined camera, sensor, and connected vehicle real-time data to adjust dynamically the traffic lights at various intersections. The RL system contrasts with the traditional systems which were reliant on a pre-determined schedule, but it is capable of continually learning through traffic flow patterns. To illustrate, should a sudden surges of vehicles on a given route be noticed, the system could automatically increase the time of the green lights on that route and reduce the duration of less congested approaches (Dandotiya et al., 2026).

The results were remarkable. Research revealed that in Hangzhou, travel time reduced by as much as 15 percent and interchange waiting time was also minimized. The system also helped in environmental sustainability by lowering the amounts of fuel used, and cutting the emissions. More to the point, the success of Hangzhou proved that AI could go beyond pilot projects and be applied at the citywide level. The lessons obtained in this case still shape the efforts of smart cities across the globe (Louati, 2025).

4.2 Los Angeles, USA: Predictive Analytics for Congestion Reduction

Los Angeles, with its notorious traffic congestions, has also adopted AI applications in order to enhance mobility in the city. The city installed one of the largest adaptive traffic signal control systems in the world, called Automated Traffic Surveillance and Control (ATSAC) system. With time, this system was integrated with AI-based predictive analytics in order to improve its effectiveness.

The AI systems use historical congestion records and weather conditions and real-time sensor feedback to predict traffic movement hours ahead of time. As an illustration, in the times of large-scale events, such as concerts, or even sporting matches, the system is able to foresee traffic jams around stadiums and automatically redirect vehicles. On the same note, when an accident happens on a freeway, the AI system forecasts the ripple effect on the network and modulates signal timings to absorb the shock (Moldakhmetov et al., 2024).

Among the most remarkable advantages of the LA system, one must mention the effect on the emergency response times. Intersections are given precedence to fire trucks, ambulances, and police cars and thus, they are able to move through with minimal delays. This has not only enhanced safety of the people, but also increased belief of the people in AI-sourced systems. Although the issues like system maintenance and funding will always be apparent, the case in Los Angeles shows that AI can be used to supplement the current infrastructure to provide tangible and instant benefits.

4.3 Singapore: Integrated AI and Smart City Infrastructure

Singapore has been known to be one of the most innovative cities in the world, and implementation of AI in traffic control is one of such examples. The Land Transport Authority (LTA) of the city-state has incorporated AI as a part of an overall smart city system, which incorporates IoT devices, sensors, and cloud-based analytics.

The AI systems in Singapore handle real-time traffic by estimating the congestion hotspots and automatically sending notifications via public information boards and mobile applications. As an example, commuters receive information about delays and receive optimal route suggestions before they even start the process. In addition to individual cars, AI is used in the organization of transport routes, so that buses and trains are dispatched according to the demands (Goyal et al., 2025b).

One of the peculiarities of the approach towards Singapore is its focus on sustainability. AI will reduce the wastage of time at the intersection and fuel consumption will be low as well as carbon emission. Also, the system facilitates shared mobility programs including ride-hailing and carpooling where resources are dynamically assigned where they are highly required. Such an implementation of AI and sustainability ambitions makes Singapore an excellent example of how to shape the future urban mobility system focusing on efficiency and environmental sustainability.

4.4 Lessons Learned From Global Deployments

The case of Hangzhou, Los Angeles, and Singapore can demonstrate some of the chief lessons to the cities of the world. First, it should be scalable, and AI cannot be a pilot project but it should be incorporated into the city structures to serve as the most as possible. Second, it is important to have collaboration with stakeholders such as government agencies, technology providers, and citizens to guarantee adoption and trust. Third, we cannot ignore such ethical aspects as fairness, transparency, and privacy because the massive surveillance and the use of algorithms could raise social apprehension (Yousaf et al., 2024).

Taken together, these case studies demonstrate that AI is not a silver bullet, but the ability to learn, adapt, and optimize in real time presents the greatest opportunities to transform the mobility of the city. The future applications should leverage these achievements and deal with situation-specific issues.

4.5 Viewpoints of Emerging Economies: AI Traffic Management in Developing Economies

Whereas the use of AI-powered traffic management has been quickly introduced in developed cities, new economies pose unique challenges and opportunities. The activities in cities like South Asia, Africa and Latin America usually have heterogeneous traffic, which is characterized by mixed traffic types, informal methods of transport and weak sensing facilities. In spite of these limitations, pilot-projects show that AI-based solutions are not irrelevant sources of meaningful benefits even when localized.

In other cities across cities like Bengaluru and Nairobi, there has been an implementation of low-cost AI systems that are based on mobile GPS records and capture-based analytics that optimise signal timing and alleviate congestion at bottlenecks. Empirical evidence of those efforts shows that in some cases there have been moderate but steady declines in the number of people losing time to travels, especially during the rush hours. Notably, these projects focus on scalability and affordability and not complete automation.

These experience case studies show that no universal solution to AI traffic management exists. Rather, it thrives by adapting within a context, having institutional capacity and including planning. The lessons of the developing regions add to the worldwide assessment of smart mobility and emphasize the necessity to have more flexible, equity-based AI frameworks with the potential to accommodate a wide range of urban realities.

5. BENEFITS AND OPPORTUNITIES OF AI-POWERED TRAFFIC MANAGEMENT

5.1 Improved Traffic Flow and Reduced Congestion

Among the most obvious and direct possible advantages of AI in traffic management, one should mention its capacity to increase traffic speed and decrease congestion. Fixed schedules and reactive-only adjustments of traditional systems can be often outpaced by the variability of the real world. In comparison, AI-based systems are constantly updating the live data and changing approaches.

As an example, machine learning-driven adaptive signal control may reduce the length of idling time at the intersection of vehicles and shorten bottlenecks. Predictive analytics are used when cities anticipate congestion and interfere before it takes place. Consequently, commuters will spend less time in the traffic, productivity will be raised, and fuel consumption will be reduced. To cities that have been besieged by endless gridlocks, such advancements are not only a source of convenience but also a huge economic boost (Ashraf et al., 2024).

The same can be said about the environmental implications. Less traffic congestion translates to less traffic congestion and therefore we have a direct reduction in greenhouse gas emittance. This amounts to international sustainability and helps to enhance air quality in congested zones. Therefore, the use of AI in traffic management is not merely the technological achievement but also is the environmental need (Shrimal, 2024).

5.2 Enhanced Safety and Emergency Response

AI should be used in road safety as it helps to respond to and identify the incident more quickly. The computer vision systems are able to recognize accidents, slowed down vehicles or traffic offenses on the fly which enables the authorities to send out the emergency services faster. Algorithms of reinforcement learning can also give green lights to ambulances or fire vehicles so that essential services get their destinations as quickly as possible.

Moreover, predictive systems of AI can designate dangerous crossroads or roads that have a likelihood of accidents. Through the analysis of historical crash data, weather, and traffic density, the AI systems are able to suggest the preventive actions that can be taken; these actions may include the modification of the timing of signals, speed limit, or the intensification of the enforcement. These preventative measures cut the rates of accidents and save life (Farooqi et al., 2026).

Safety with the integration of AI and connected vehicles has another dimension. Two-way communication and sensors on vehicles can communicate with other vehicles and central traffic systems regarding road hazards or abrupt braking. This forms a networked atmosphere in which accidents can be prevented by sharing awareness. These innovations will make a significant step in the long-term aim of having zero-accident transportation networks.

5.3 Economic and Social Benefits

In addition to mobility and safety, AI-based traffic management is associated with significant economic and social advantages. Commuters will benefit directly by having less fuel money spent and having less money in vehicle maintenance as it will take less time to travel. Enterprises also benefit by having stronger logistics, with delivery vans able to follow a predictable schedule with a reduced number of delays. Such efficiency increases the reliability of supply chain and economic competitiveness of urban centers.

Socially, the traffic flows are smoother, which lowers stress among the commuters and enhances their general quality of life. Social satisfaction rises when individuals feel a noticeable positive transformations in such aspects as reduced commuting, increased safety of roads, or improved air quality. Besides, the fair allocation of traffic resources, in case provided, will eliminate inequality in cross-border movement between poor and wealthy communities, promoting the practice of inclusivity in urban planning (Almatar, 2024).

These benefits are cumulative and transformative in nature. Cities which effectively introduce AI-based systems do not only solve traffic issues; they also develop a business, tourist, and resident-attracting environment. In this respect, traffic management using AI becomes the catalyst of the larger urban development.

5.4 Environmental Sustainability and Smart City Integration

AI traffic systems have become important in ensuring that there is enhancing of environmental sustainability. They decrease fuel usage and carbon footprints by minimizing idling, and route optimization. In addition, they are compatible with environmentally friendly programs like electric cars and car-sharing schemes. As an illustration, AI can be used to prioritize charging station locations to EVs or organize carpool lanes to be as efficient as possible.

Benefits are even further enhanced by integration with larger smart city ecosystems. Traffic management is not a stand alone activity but a subset of a bigger vision that encompass energy efficiency, waste reduction and digital governance. The systems are linked to AI traffic solutions to make cities efficient and resilient as well as sustainable. To illustrate, streamlined traffic dynamics diminish pressure on mass transit and make it more attractive, and diminish the use of personal vehicles (Goyal et al., 2025c).

With the introduction of AI-driven traffic management into the smart city framework, urban planners can make sure that transportation will be a key element of sustainable development. This holistic approach transforms traffic solutions from reactive fixes into proactive enablers of a more livable future.

6. CHALLENGES AND ETHICAL CONSIDERATIONS

6.1 Data Privacy and Surveillance Concerns

The ethical dilemma between efficiency and privacy is one of the most urgent AI-related problems in traffic management. The current traffic systems are based on real-time data gathering using cameras, sensors, GPS-enabled automobiles and mobile phones. Although this information is priceless when it comes to improving routes and identifying accidents, it also provokes some serious issues of surveillance and personal privacy. As an illustration, the use of license plate recognition cameras can trace the movement of vehicles within the whole cities forming detailed mobility profiles of individuals.

Otherwise, such data may be abused by the government, businesses, or even malicious parties. Civil liberties are questionable with the possibility of the so-called function creep: as traffic surveillance systems are used to monitor other things, like political demonstrations or police profiling. Loss of anonymity in the open environment can make citizens feel threatened by the fact that otherwise positive systems are undermined. Finding the balance between efficiency based on the data and the protection of the personal freedom is one of the most topical questions that pose a threat to both policymakers and technologists (Fathima & Hariharan, 2024).

6.2 Algorithmic Bias and Fairness

The AI systems can be only as unbiased as the information they are trained on. Algorithms can be biased in traffic management in a manner that is disproportionately discriminatory to a particular group of people. As an example, predictive congestion models trained on past records can assign greater resources to richer neighborhoods with better vehicle ownership, and under-served regions are given less focus. Equally, the use of AI-based reactionary measures like automated ticketing can, when implemented unevenly, unjustly punish populations with low income levels.

These challenges underscore the need to be fair and inclusive to AI designs. An ethical system should guarantee that maximization is no longer a question of speed or economic efficiency, but also of sharing of benefits in the society equally. Cities that do not consider bias reinforcement run the risk of strengthening the existing inequalities and form a two-layered mobility system, whereby certain communities will have better access to services than others. The challenge of bias cannot be solved only with technical tools, including fairness-sensitive algorithms, but also through participatory decision-making involving a variety of community voices (Khan & Thakur, 2024).

6.3 System Reliability, Safety, and Accountability

The other significant issue in AI-based traffic management is reliability. Traffic lights that are controlled by autonomous systems, reroute cars, or emergency response priorities have to operate with near-perfect precision since any type of error can lead to accidents, economic consequences, or the lack of trust in the civilians. Indicatively, a faulty AI traffic controller, which mis-prioritizes a specific lane, may cause road accidents or traffic jams during busy time.

The question of accountability complicates this issue further. In the event that an AI traffic system commits a mistake that causes a car accident, who is liable, the city government, the provider of the technology, or the AI system itself? The existing laws fail to provide clarity and therefore stakeholders are left in uncertainty. To increase the trust of people in AI, there should be open accountability systems and strong safety guarantees that involve backup systems and human controls in critical contexts (Sharma et al., 2026).

6.4 Responsible AI Ethical Frameworks of Traffic Management

Although efficiency and optimization are the key objectives of AI-based traffic systems, there should be ethical governance frameworks to promote responsible deployment. Traffic management algorithms are making more and more an impact on access to road space, emergency priority, mobility equity, in effect forming the daily life of urban life. These systems must therefore be in line with the fundamental ethical principles such as transparency, accountability, inclusivity and proportionality.

Transparency implies that decision-making logic e.g., signal prioritization or (where suitable) rerouting strategies be exposable to policymakers and, in suitable cases, to the general public. Black-box systems can pose the threat of eroding trust, especially when traffic results seem to predispose some communities, as unfavorable. The accountability mechanisms are also essential; these cities need to have proper responsibility mechanisms that can eliminate failures or undesired harms in line with automated decisions.

There is also the presence of proportionality in ethical deployment of data. Though real-time surveillance improves accuracy of the system, data is only to be collected to the extent it is absolutely necessary and anonymised where feasible and data retention policies used which are transparent. In addition, participatory governance schemes such as the involvement of citizens, urban planners, and ethicists in the system design can be used to make sure that AI-controlled traffic solutions are designed according to the values of the society instead of being driven by pure technical goals.

Integrating these ethical frameworks makes AI traffic control a highly optimized control system into an acceptable social infrastructure that is focused on innovation without disregarding human rights and trust.

7. POLICY AND GOVERNANCE IMPLICATIONS

7.1 The Need for Regulatory Frameworks

AI-powered traffic systems blur traditional regulatory boundaries between technology, transportation, and data governance. Current traffic regulations tend to presuppose human discretion in decision-making, which does not leave any loopholes in the regulations of autonomous or AI-based systems. As an example, existing traffic regulations seldom talk about the liability of anyone in case an AI-controlled traffic light fails or that AI-controlled traffic routing algorithms lead to congestion in other places unintentionally.

Policymakers should respond to these issues by creating entire regulatory frameworks to establish accountability, safety requirements, and ethical considerations on the implementation of AI. These frameworks must not just be reactive to the current issue but also proactive to the future events, like the use of autonomous vehicles. The rules should maintain a very fine balance: on the one hand, they have to promote innovation, and on the other hand they have to protect the interest of the population (Rathore et al., 2024).

7.2 Global Standards and Interoperability

Traffic management is a globally oriented phenomenon in its own right because city centers tend to face similar issues and to share technological strategies. Nevertheless, the inability to have various AI systems interact with each other is a serious challenge. Indicatively, an AI traffic solution built on a specific city might not be compatible per se with other cities because of the differences in data standards, infrastructure, and regulations.

Across international standards of AI-based traffic management can ease the exchange of knowledge, save money, and speed up implementation. International standards like International Telecommunication Union (ITU) and International Organization for Standardization (ISO) might have a critical role in the establishment of the world standards regarding safety, transparency, and technical interoperability. In the absence of this harmonization the world will be facing the danger of developing fragmented systems that do not perform to their full capacity (Goyal et al., 2025d).

7.3 Public-Private Partnerships and Stakeholder Collaboration

Implementation and creation of AI-driven traffic systems involve a coordinated effort among various stakeholders, which are governments, commercial service providers, scientists, and civilians. Particular attention should be paid to public-private partnerships (PPP), because in the case of more complex AI systems governments simply do not have the technical experience and resources to develop them on their own. Tech companies, in turn, have the advantage of access to the real-life information and infrastructure to test their innovations.

Nevertheless, these partnerships are to be monitored rigorously in order to avoid conflicts of interests as well as to keep in mind that the common good is the main focus. The contracts ought to have explicit provisions regarding ownership of data, transparency and accountability. Moreover, inclusion of local communities in the decision making process builds confidence and makes sure that the systems are not only to the benefit of corporate interests, but also to the society at large (Hegde et al., 2024).

7.4 Governance for Ethical and Sustainable Mobility

AI-based traffic management has to go beyond efficiency indicators to adopt a more ethical and sustainable agenda. As an example, AI can be used to not only make commute time shorter but also help reduce emissions and boost the use of public transport and non-motorized ways of travel, including cycling and walking. This will involve integration of ethical principles into the systems of governance, which include fairness, accountability, and transparency.

Those cities implementing governance schemes that resonate with these values will be in a better place to attain long-term sustainability. As an illustration, policies centered on equity and fairness can be integrated with AI to monitor and mitigate carbon footprint, and policies centered on the environment can be made more comprehensive to make vulnerable groups equally enjoy the benefits of technological progress. Ethical governance and social inclusiveness are the main factors that can guarantee that AI-powered traffic management will increase the quality of life of all people living in urban areas (Zantalis et al., 2019).

8. FUTURE DIRECTIONS AND RESEARCH NEEDS

8.1 Emerging Technological Innovations

The next stage of AI-controlled traffic management is the introduction of the latest technologies that are yet to reach their maturity. The combination of AI and the Internet of Things (IoT) is one of the most promising innovations in which connected sensors, smart vehicles and roadside infrastructure constantly feed real-time data into adaptive traffic control systems. Compared to the existing systems that rely on partial databases, the future systems will process data streams of millions of nodes in real time, and allow prediction with previously unseen accuracy. This would facilitate a smooth movement of traffic thus, lowering the level of traffic jam in the overcrowded cities (Zhu et al., 2018).

The other frontier is the combination of 5G and edge computing. The latency of current centralized AI models can be an issue, especially in stakes situations like vehicle routing in a case of emergency. Edge-based AI will enable real-time split-second decisions as data processing can be done nearer to source. This will form an important component in the cities that are looking to integrate autonomous vehicles into their road networks, where cars and road systems will be required to share information in real-time to prevent accidents and streamline traffic.

The deployment of superior machine learning models like deep reinforcement learning (DRL) are also equally important. In contrast to conventional predictive models, DRL systems are trained on the strategies that are the best by simulating millions of traffic cases. These systems are able to adjust to unplanned situations over time, like sudden traffic increases by an event or accident, and provide solutions way beyond the ability of a rule based system. With the maturity of these technologies, the AI-driven traffic management will cease to be reactive and transition into proactive and self-evolving ecosystems that are able to predict and address issues before they occur (Abduljabbar et al., 2019).

8.2 Bridging Current Research Gaps

Although there is potential behind the technology, there exist a number of gaps in research that require filling to bring AI traffic systems to their full potential.

The major problem is that there are no large-scale longitudinal studies. The bulk of current studies evaluates AI traffic systems through small pilot studies, and usually these assessments concentrate on the short-term reductions in congestion or emissions. What is not understood is how these systems are working after years of providing constant service and whether or not some unknown effects like the traffic to be displaced to other unmonitored sectors will come out. These systems have to be captured in long-term research to capture their sustainability and resilience (Goyal, 2025b).

The other research gap is human-based evaluation. Recent research tends to focus on technical performance metrics, e.g., a shorter travel period or greater fuel efficiency. Nevertheless, much less focus is given on the social aspects such as satisfaction of commuters, their behavioral adjustment, or the psychological consequences of using AI systems. As an illustration, autonomy can be lost to drivers as AI systems direct them, or communities can see injustices when traffic optimization prioritizes the richer parts of a city all the time. The future studies should hence embrace the interdisciplinary methods which integrate engineering with sociology, psychology as well as the study of urban areas.

Also, standardised datasets and benchmarks are urgently required. Currently, numerous AI traffic management research articles use proprietary or city-specific data, and cross-comparability is challenging. It would be feasible to create open, standardized benchmarks so that the various algorithms can be evaluated by the researchers equally on equal terms. These efforts would propel innovation faster and the cities would implement the most efficient systems and not stumble in a disjointed market place of untested solutions (Haydari & Yılmaz, 2020).

8.3 Interdisciplinary and Participatory Approaches

Engineers and data scientists alone cannot direct the success of the AI-powered traffic systems. Rather, it should be a multidisciplinary investigation involving urban planners, ethicists, policymakers, environmental scientists, and, above all, the citizens. As an example, urban planners can match AI traffic systems with long-term development objectives, whereas ethicists can make sure that the systems do not violate privacy or fairness. Environmental scientists may give information on the effects of optimization strategies on pollution and carbon footprint.

Participatory design is especially important. The use of communities in the de sign and testing processes will make AI systems meet real world user demands as opposed to what might be assumed in laboratories. As an illustration, the residents can draw attention to safety issues with crossings that the engineers can ignore. Likewise, contextual awareness of AI systems can be added by local information on traffic hotspots during festivals or seasonal dynamics. Co-designed cities are associated with higher chances of adoption and signaled trust in the system by the community (Iyer, 2021).

Moreover, the interdisciplinary cooperation could be used to solve ethical dilemmas in the AI-governance. Through incorporating voices across disciplines, cities will be able to build common structures of transparency, accountability and inclusivity. This method will minimize the chances of coming up with systems which are technologically superior but socially inappropriate. Finally, the future of AI traffic management is not only in the more intelligent machines but in the more intelligent governance procedures that suggest combining technical skills with human virtues.

8.4 Policy-Oriented Research Priorities

As AI-driving traffic systems gain in popularity, studies should pay more attention to how to design policies that would prioritize both innovation and social good. Development of liability frameworks is one of the areas of priority. In case an AI traffic controller makes a choice resulting in an accident, it is not clear who is responsible, the municipal authority, the AI provider, or the infrastructure operator. Evidence-based research is necessary to help policymakers design regulations that are applicative and at the same time promote innovation.

The other area that is critical is data governance. As AI traffic systems rely on very large datasets, data ownership, consent, and rights to use questions should be considered. Future studies ought to consider the processes that can enable cities to utilize big data without violating the privacy of citizens; they may include federated learning, in which AI models are trained using distributed datasets, and data is not stored at a centralized location. Policy makers will be required to be guided on which models are the best in offering the most optimal trade-offs between efficiency and privacy (Kumari et al., 2026).

Lastly, policy-oriented research should look ahead and foresee the global aspect of the traffic AI systems. International logistics, global transportation networks and climate goals across countries imply that AI traffic solutions can no longer be limited to a particular city or country. International comparative studies can support the creation of best practices and guide international interoperability, data sharing and environmental standards agreements. This international view will also play a crucial role in making sure that mobility done by AI is not only locally efficient but also globally sustainable.

CONCLUSION

The incorporation of Artificial Intelligence in traffic management is something that signifies the beginning of the transformation of urban mobility. Conventional systems that are limited by strict rules and the lack of data are being substituted with dynamic, intelligent systems that are able to comprehend extensive real-time information streams. This is not just a technological change but one of transformation that redefines the way in which cities deal with congestions, maintain safety and sustainable movements. Predictive modeling up to reinforcement learning and IoT-enabled sensing, AI-powered solutions show potential to build not only efficient, but also responsive transportation ecosystems to the dynamics of human life.

Among the main lessons of this expedition is the two-fold promise and problem of AI. On the one hand, AI technologies have the potential to radically decrease the traffic jams, decrease the emission levels, and improve the experiences of commuters, predicting traffic flows and optimizing the allocation of the resources. They are able to enhance the response to emergency, facilitate easier adoption of autonomous vehicles and even promote environmental objectives by lowering fuel usage. Conversely, new problems also emerge with these systems: there is the issue of data privacy, the issue of biasing of the algorithm, the issue of unfairness, and the danger of over-reliance on automated systems. These problems illustrate that it is not the sophistication of technology that will be used to define the future of AI in managing traffic, but rather how the societies decide to govern, regulate, and embrace these systems.

The other lesson is that of the significance of human-based and multidisciplinary approaches. Traffic optimization is no less than an engineering issue- it is intensely social, economic and political. This will be influenced by the people having trust, cultural acceptance and perceptions of fairness as well as algorithmic accuracy. Similarly, the policymakers, ethicists, and community stakeholders should actively collaborate with technologists to formulate the ethical, transparent, and fair application of AI systems. Those cities that miss this balance will run the danger of implementing new high-tech gadgets that do not connect with the real life experiences of their residents.

In the future, the key element of the AI-based traffic management will be balance: between innovation and control, efficiency and equity, automation and human intervention. The development of edge computing in 5G and deep reinforcement learning and the development of participatory smart cities has the potential to transform traffic management. However, long-term sustainability will be based on strong policy frameworks, standard benchmark, and inclusive design procedures. It does not only want to transport traffic in a faster way but to create a safer, more inclusive and environmentally responsible transport system.

To conclude, AI does not only signify an optimization tool, but it also represents the driver of the reconsideration of urban mobility. When put in place in a responsible manner, it may turn the busy, polluted roads into networks that promote accessibility, resilience and sustainability. However, this change must be foresighted, and cross-disciplinary, and guided by ethics. Traffic management in the future, then, is not in AI per se, but it is in the way societies decide to make it serve wider human and environmental purposes. On this note, AI-based traffic optimization is not merely a technical novelty but a social decision, and it will determine the standards of living in urban areas over the next generations.

REFERENCES

Abduljabbar, R., Dia, H., Liyanage, S., & Bagloee, S. A. (2019). Applications of artificial intelligence in transport: An overview. *Sustainability (Basel), 11*(1), 189. DOI: 10.3390/su11010189

Abubeker, K. M., Rekhamole, G., Divya, D. T., Drishya, S. S., Mohan, M. J., & Rajimol, V. (2023). Computer vision assisted artificial intelligence enabled smart traffic control system for urban transportation management. *Journal of Transportation Engineering, 45*(3), 55–63.

Ajayi, A. O., & Kumkale, H. (2023). Optimising urban road transportation efficiency: AI-driven solutions for reducing traffic congestion in big cities. *Bournemouth University Working Paper Series.*

Almatar, K. M. (2024). Implementing AI-driven traffic signal systems for enhanced traffic management in Dammam. *International Journal of Sustainable Development and Planning, 19*(2), 134–143. DOI: 10.18280/ijsdp.190236

Ashraf, S., Shahid, T., Kim, J., Hameed, M. S., Hezam, I. M., & Jana, C. (2024). AI-powered decision making for road safety optimization under probabilistic linguistic models. *Heliyon, 10*(19), 1–12. DOI: 10.1016/j.heliyon.2024.e38594 PMID: 39687433

Bezek, O., & Güre, Ö. (2024). Predicting smart city traffic models using Adaboost regression method. *European Journal of Technique, 12*(1), 45–52. DOI: 10.36222/ejt.1436180

Dandotiya, M., Goyal, N. K., Yadav, P., Sankhla, K., & Kumari, M. (2026). GENALYTICS: A Deep Dive Into Real-Time Insights via Generative AI. In Elbaghazaoui, B., Amnai, M., & Gherabi, N. (Eds.), *Generative AI-Powered Data Architectures: From Governance to Autonomous Analytics* (pp. 105–134). IGI Global Scientific Publishing., DOI: 10.4018/979-8-3373-5616-7.ch005

Datey, V., & Prabhu, S. (2023). Smart traffic control and prediction model empowered with 5G technology, artificial intelligence, and machine learning. *International Journal of Emerging Technologies, 14*(6), 112–120. DOI: 10.1109/AIKIIE60097.2023.10389965

Dikshit, S., Atiq, A., Shahid, M., Dwivedi, V., & Thusu, A. (2023). The use of artificial intelligence to optimize the routing of vehicles and reduce traffic congestion in urban areas. *EAI Endorsed Transactions on Energy Web, 10*(1), 1–13. DOI: 10.4108/ew.4613

Farooqi, A., Das, D., Tripathi, M., Dutta, R., Goyal, N. K., & Sheikh, M. F. (2026). Navigating Ethical Frontiers in AI-Driven Geriatric Care. In Stolakis, K., & Efthymiou, I. (Eds.), *AI for Geriatric Care in an Aging Society: Ethical, Clinical, and Policy Challenges* (pp. 243–266). IGI Global Scientific Publishing., DOI: 10.4018/979-8-3373-5462-0.ch009

Fathima, M. D., & Hariharan, R., & S, E. R. (2024, December). Smart street: AI-powered traffic flow enhancement with adaptive signal control. In *International Conference on Advanced Network Technologies and Intelligent Computing* (pp. 144–158). Springer.

Gen, C., & Jia, W., & Zhang. (2024). Intelligent transportation systems: Machine learning approaches for urban mobility in smart cities. *Sustainable Cities and Society, 105*(369).

Goyal, N. K. (2025). Weaponizing Misinformation and Cyberattacks on Supply Chains. In Bukhari, S., & Zafar, I. (Eds.), *Mass Media and Impact of Fake News on Supply Chains* (pp. 177–210). IGI Global Scientific Publishing., DOI: 10.4018/979-8-3373-2175-2.ch007

Goyal, N. K. (2025). Security and Privacy in IoT, Fog, and Blockchain Networks. In Sathio, A., Rind, M., Awan, S., & Junejo, A. (Eds.), *Energy-Efficient Deep Learning Approaches in IoT, Fog, and Green Blockchain Revolution* (pp. 371–398). IGI Global Scientific Publishing., DOI: 10.4018/979-8-3373-0300-0.ch015

Goyal, N. K., Dandotiya, M., Kumari, M., Kaloria, S., & Anushya, A. (2025). AI-Powered Haptic Devices for Patient Rehabilitation. In Mamodiya, U., Kalam, A., Bossoufi, B., Awad, M., & Somwanshi, D. (Eds.), *Integrating AI With Haptic Systems for Smarter Healthcare Solutions* (pp. 143–170). IGI Global Scientific Publishing., DOI: 10.4018/979-8-3373-2307-7.ch007

Goyal, N. K., Dandotiya, M., Kumari, M., Sheikh, M. F., & Sharma, S. (2025). Emerging Job Roles in the Tourism Sector Leveraging Artificial Intelligence. In Silic, M., Luthia, M., Ahmad, F., & Chakraborty, P. (Eds.), *Empowering Sustainable Performance and Competitive Advantage in Tourism* (pp. 89–126). IGI Global Scientific Publishing., DOI: 10.4018/979-8-3373-4546-8.ch004

Goyal, N. K., Dandotiya, M., Qurashi, J. A., Kumari, M., & Sharma, S. (2025). AI-Driven Predictive Analytics for Disease Management. In Ouaissa, M., Ouaissa, M., Imad, M., Qurashi, J., & Farooq, M. (Eds.), *Utilizing AI of Medical Things for Healthcare Security and Sustainability* (pp. 231–276). IGI Global Scientific Publishing., DOI: 10.4018/979-8-3373-0690-2.ch008

Goyal, N. K., Kumari, M., Kushwah, V., Dandotiya, M., & Vijay, A. (2025). Evaluating blockchain security: Trade offs in consensus mechanisms. *IET Conference Proceedings*, 2024(38), 80–88. DOI: 10.1049/icp.2025.0775

Gunasekar, T., & Nattar, K., & Kannan. (2024). Hybrid neuro genetic causal convolution-based autoencoders for traffic prediction in smart cities. *Journal of Intelligent Transport Systems*.

Haydari, A., & Yılmaz, Y. (2020). Deep reinforcement learning for intelligent transportation systems: A survey. *IEEE Transactions on Intelligent Transportation Systems*, *23*(1), 11–32. DOI: 10.1109/TITS.2020.3008612

Hegde, S. B., Premasudha, B. G., Hooli, A. C., & Akshay, M. J. (2024, February). A review on smart traffic management with reinforcement learning. In *International Congress on Information and Communication Technology* (pp. 455–470). Springer. DOI: 10.1007/978-981-97-3305-7_37

Idriss, M., Abouchabaka, J., & Najat, R. (2023). Enhancing urban mobility: Integration of IoT road traffic data and artificial intelligence in smart city environment. *Indonesian Journal of Electrical Engineering and Computer Science*, *29*(2), 200–210.

Iyer, L. S. (2021). AI-enabled applications towards intelligent transportation. *Transportation Engineering*, *5*, 100083. DOI: 10.1016/j.treng.2021.100083

Khan, H., & Thakur, J. S. (2024). Smart traffic control: Machine learning for dynamic road traffic management in urban environments. *Multimedia Tools and Applications*, *83*(7), 12345–12367. DOI: 10.1007/s11042-024-19331-4

Kumar, A., Batra, N., Mudgal, A., & Yadav, A. L. (2024). Navigating urban mobility: A review of AI-driven traffic flow management in smart cities. *International Journal of Smart Cities*, *9*(2), 88–102. DOI: 10.1109/ICRITO61523.2024.10522206

Kumari, M., Khan, S. T., Gulati, K., Sharma, P., Goyal, N. K., & Sheikh, M. F. (2026). Innovative Procedural Content Generation (PCG) Using Generative AI: A New Paradigm in Game Design. In Sousa, J., & Tavares, R. (Eds.), *Reshaping the Video Game Landscape With AI and GenAI* (pp. 205–232). IGI Global Scientific Publishing., DOI: 10.4018/979-8-3373-4637-3.ch007

Louati, A. (2025). Machine learning framework for sustainable traffic management and safety in AlKharj city. *Sustainable Futures : An Applied Journal of Technology, Environment and Society*, *9*, 100407. DOI: 10.1016/j.sftr.2024.100407

Lungu, M. A. (2024). Smart urban mobility: The role of AI in alleviating traffic congestion. *Proceedings of the International Conference on Business Excellence*, *18*(1), 122–131. DOI: 10.2478/picbe-2024-0118

Moldakhmetov, D., Mansurova, M., Belgibayev, B., Baigarayeva, Z., Sarsembayeva, T., & Ospan, A. (2024, July). AI-powered traffic management for busy intersections. In *20th International Asian School-Seminar on Optimization Problems of Complex Systems (OPCS)* (pp. 54–59). IEEE. DOI: 10.1109/OPCS63516.2024.10720388

Najat, R., Idriss, M., Raji, F. Z., & Abouchabaka, J. (2024). Elevating smart city mobility using RAE-LSTM fusion for next-gen traffic prediction. *Indonesian Journal of Electrical Engineering and Computer Science*, *35*(1), 503–510. DOI: 10.11591/ijeecs.v35.i1.pp503-510

Purwar, S., & Chaudhry, R. (2023, October). AI-powered traffic orchestration: Reinforcement learning at traffic intersections. In *International Conference on Recent Innovations in Computing* (pp. 49–63). Springer.

Raj, M. (2024). AI-powered traffic control in IoT-based smart cities: Revolutionizing urban mobility. *Supply Chain and Operations Decision Making*, *1*(1), 13–19.

Rathore, S. A., Salam, M. H. U., Abbas, T., Irfan, M., & Ameen, K. (2024). Machine learning approach to reducing urban congestion using artificial intelligence for smart traffic management. *Dialogue Social Science Review*, *2*(5), 840–863.

Sharma, P., Kumawat, R., Jain, G., Keswani, G., Goyal, N. K., & Sheikh, M. F. (2026). Machine Learning-Based Intrusion Detection Systems for Healthcare Internet of Things (HIoT). In Shieh, C., Goel, P., Horng, M., & Pandey, H. (Eds.), *Resilient Healthcare Internet of Things Systems: Architectures, Security, and Trust* (pp. 185–214). IGI Global Scientific Publishing., DOI: 10.4018/979-8-3373-7610-3.ch007

Shrimal, H. (2024). Integration of AI-powered vehicles with smart city infrastructure to transform the future of the automotive world. *SAE Technical Paper Series, 2024-28-0028.*

Thote, R., Barhate, A., Tale, A., & Verma, P. (2024, November). Deep reinforcement learning in autonomous cars to prevent road accidents. In *2024 2nd DMIHER International Conference on Artificial Intelligence in Healthcare, Education and Industry (IDICAIEI)* (pp. 1–6). IEEE.

Xing, H., Chen, A., & Zhang, X. (2023). RL-GCN: Traffic flow prediction based on graph convolution and reinforcement learning for smart cities. *Displays*, *74*, 102–114. DOI: 10.1016/j.displa.2023.102513

Yousaf, M., Farhan, M., Saeed, Y., Iqbal, M. J., Ullah, F., & Srivastava, G. (2024). Enhancing driver attention and road safety through EEG-informed deep reinforcement learning and soft computing. *Applied Soft Computing*, *167*, 112320. DOI: 10.1016/j.asoc.2024.112320

Zantalis, F., Koulouras, G., Karabetsos, S., & Kandris, D. (2019). A review of machine learning and IoT in smart transportation. *Future Internet*, *11*(4), 94. DOI: 10.3390/fi11040094

Zhu, L., Yu, F. R., Wang, Y., Ning, B., & Tang, T. (2018). Big data analytics in intelligent transportation systems: A survey. *IEEE Transactions on Intelligent Transportation Systems*, *20*(1), 383–398. DOI: 10.1109/TITS.2018.2815678

Chapter 4
AI-Powered Traffic Management and Optimization:
Towards Safer and Smarter Urban Mobility

Mukimov Askar Shukhratovich
Alfraganus University, Tashkent, Uzbekistan

S. Aarthi
http://orcid.org/0009-0006-9064-2091
Marwadi University, Rajkot, India

Ravikumar R. N.
http://orcid.org/0009-0009-3705-1681
Marwadi University, Rajkot, India

Dilmurod Turaev
Termez University of Economics and Service, Termez, Uzbekistan

Duggirala Aravind
http://orcid.org/0009-0005-4362-7684
ASET College of Science and Technology, Chennai, India

Anorgul Ashirova
http://orcid.org/0009-0006-7271-3634
Mamun University, Khiva, Uzbekistan

ABSTRACT

Rapid urbanization and rising vehicle density have made conventional traffic management systems inadequate in addressing congestion, accidents, and environmental impacts. This chapter explores how Artificial Intelligence (AI) offers transformative solutions through predictive analytics, computer vision, and machine learning for dynamic traffic monitoring and optimization. Case studies from Los Angeles and Singapore demonstrate quantifiable improvements in travel time, delays, emissions, and public transport efficiency, emphasizing the adaptability of AI to diverse infrastructures. The discussion also highlights ethical and regulatory considerations,

DOI: 10.4018/979-8-3373-3231-4.ch004

including privacy, transparency, and fairness, while acknowledging technical and social challenges such as infrastructure readiness and public trust. Finally, the chapter proposes a scalable AI-driven framework integrating IoT, edge computing, 5G, and V2I communication to build safer, efficient, and sustainable urban transport systems.

INTRODUCTION

Traffic management in the world has been put under pressure by the presence of growing urban population and motor vehicle density. The conventional systems of traffic control built on the active use of the fixed rules and a small number of sensors based feedbacks are failing to cope with the added pressure on the safety, mobility and sustainability. The growing traffic congestions, road accidents and concentrations of pollutants have emerged as a concern of concern among government, travellers and city planners. Such issues not only sabotage road safety and air quality but also cost economy a lot of money as far as time wastage and fuel sopped. Artificial Intelligence (AI) has proved to be a gamer in this respect since it has redefined the manner in which traffic is monitored and predicted, as well as made smoother (P. Sun and Boukerche, 2021). Adaptive and dynamic traffic control can be enabled by AI since it can address the limitations of a conventional traffic system by providing data-based modelling, computer vision and machine learning. Such applications as smart traffic light control, predictive analysis, and route optimization can enhance road-based transportation with the help of AI technologies that seem to be prepared to provide real-time solutions to increase the efficiency of operations and reduce risks (Zhao and Chen, 2025). The chapter talks about the use of AI in the transport segments because it can change the security of traffic, the robustness of operations, and sustainable transportation within the urban areas.

Evolution of Traffic Management Systems

The conventional applications of traffic control have been rooted in the application of fixed signal timing, police traffic control and basic sensing technology. Though they have been effective in prior decades whereby traffic has been less congested, the systems have not been established effective in the present days of intense urbanization and complex mobility trends. The traditional infrastructure is compromised with dynamic events such as sudden congestion, accidents or other unpredictable weather changes that cause changes in traffic flows (An Intelligent, 2023). Inefficiency in route practise and travel delays as well as high production of carbon emissions will be inclined to be revealed by lack of adaptive intelligence. To address these inadequacies, the first innovations were sensor networks, loop detectors and surveillance cameras to provide real-time surveillance. However, they were as well dependent on a large extent on human intervention and rule-based systems in addition to being not scalable or responsive. Artificial Intelligence is a paradigm shift because it is capable of making the traffic systems learn, predict and adapt themselves (Tm & Cn, 2023). The vast streams of information registered by GPS trackers, Internet of Things solutions, and connected cars can be used by Artificial Intelligence not only to predict incidents but also to respond to them in advance. This innovation demonstrates the growing requirements of smarter, data-intensive systems that have the potential to ensure road safety, efficiency and sustainability in modern cities. Figure 1 demonstrates the way the view of the transportation systems changed over time, beginning with the strategy of the traditional traffic management towards real-time evaluation and consideration of the AI solutions.

Figure 1. Evolution of Traffic Management Approaches

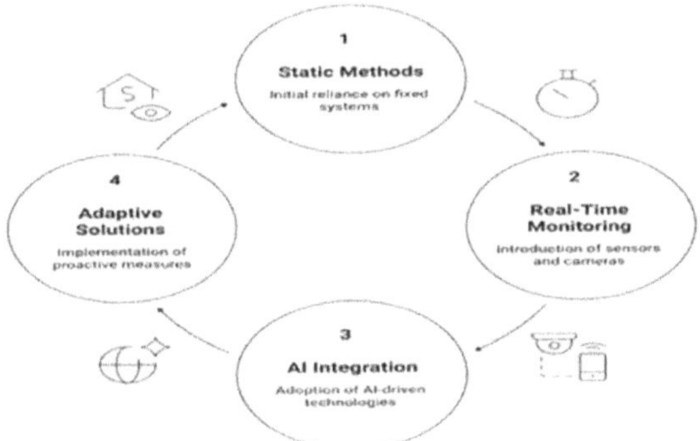

Role of Artificial Intelligence in Modern Mobility

The advanced resources of traffic optimization that contributes to safety has made AI one of the cornerstones of next-gen traffic management. With the union of machine learning algorithms, deep neural networks, and computer vision, AI systems will be capable of handling the vast volumes of real-time data introduced by sensor technologies, GPS units and connected cars. When using such systems, the traffic flows are predictable, accidents can be detected early and the timing of the traffic lights can be dynamically adjusted to minimize delay. The AI-based systems may also be used to re-direct vehicles based on the estimates of the traffic and the traffic congestion and reduce the bottlenecks and fuel consumption (Paiva et al., 2021). Besides operational additions, AI supports sustainable transport through emissions reduction through the operation of enhanced transport and promotes smart public transportation planning. Furthermore, its interrelation with other technologies that are only emerging, such as 5G, edge computing, and Vehicle-to-Infrastructure (V2I) communication ensures the significantly-faster inference and system survivability. Singapore, Los Angeles, and Beijing are the examples that demonstrate that AI may be actively used in traffic prediction, adaptive control of the signalization, and live monitoring. Everything that has been happening shows the realization that AI is not a technology of transformation but a game-changer in the path towards safe, efficient and environmentally friendly mobility systems (Moumen et al., 2023). The Figure 2 indicates how a system of AI-based traffic management continues in a loop of data gathering and forecasting, followed by readjusting parameters in real-time and enhancing congestion, fuel consumption, emissions and system resilience.

Figure 2. AI-Driven Traffic Optimization Cycle

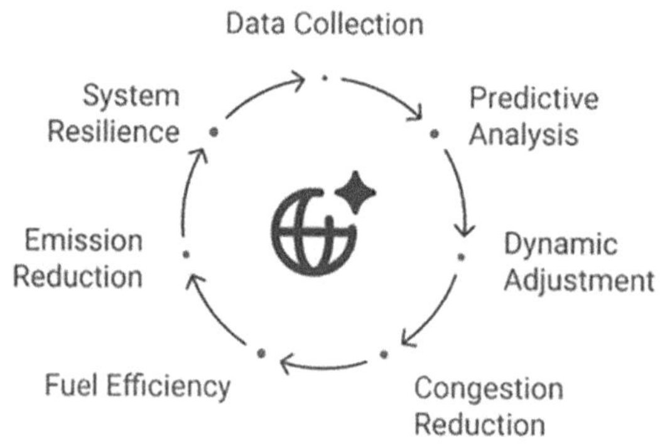

Objectives of the Chapter

This chapter aims to explore the transformative role of Artificial Intelligence in addressing critical challenges of traffic management, including congestion, accidents, and environmental concerns. By focusing on AI-driven monitoring, prediction, and optimization techniques, the chapter outlines how emerging technologies can reshape urban mobility for safer and more sustainable cities. The key objectives are:

- **To analyze the limitations of conventional traffic management systems** and highlight the pressing need for intelligent, adaptive solutions.
- **To present AI-based approaches for traffic monitoring, prediction, and control**, including machine learning, computer vision, and real-time data analytics.

- **To examine global case studies** demonstrating successful deployment of AI in urban traffic systems and their measurable benefits.
- **To discuss ethical, social, and technical challenges** such as data privacy, algorithmic bias, and infrastructure readiness in AI-driven transport.
- **To explore future directions and opportunities** including AI integration with autonomous vehicles, V2I communication, edge computing, and 5G networks.

BACKGROUND STUDY

Urbanization combined with car ownerships has put the transport infrastructure to an enormous test across the globe. A less modern approach to controlling traffic, such as long signalised patch, rigid infrastructure, and human intervention ceases to be applicable in the mobile pattern of dynamic and uncertain mobility. Traffic congestion, road accidents and air pollution are still there and besides being inconvenient to those who are affected, their consequences are disastrous both in terms of the environment and financially. Billions of dollars have also been estimated to be lost annually because of delays, wastage of fuel and health expenditures on poor air quality. The detected deficiencies lead to the fact that the ever more intelligent and adaptive systems that are able to cope with the existing requirements are needed. Another examination of the (Musa et al., 2023)past 20 years will show that as the technological advancements are being experienced, automation has entered the equation and the sensors, the surveillance cameras, and the GPS-enabled devices all have a greater view of the roadways pattern. However, such systems still heavily depend on the judgment of human beings and lack any form of predictive capacities. Artificial Intelligence (AI) is a revolutionary change that is being offered with the emergence of AI. With machine learning, computer vision and big data analytics, AI can more accurately predict traffic, maximize paths, and reduce risks like never before (M. Chen et al., 2018). The above is the background on which the narrative of AI-based solutions as safer, efficient, and sustainable transportation systems of the future can be followed.

Literature Review

The management of traffic has been identified as one of the most important areas of concern in the field of transportation engineering especially in regions that are normally congested as per the terms that require in the population exceeds provision of infrastructure. The traditional ways of traffic control rely on the pre-established rules, pre-programmed schedules and officer control (Sayed et al., 2023). These systems performed fairly well in previous decades, but have become ineffective in the face of growing traffic density and the unpredictability of travel patterns, as also due to the rising complexity of multimodal transportation. Traffic jams, delays and environmental pollution are typically caused by the lack of the opportunity to respond to unpredictable incidents such as disasters, weather crises, and massive operations (Al-Selwi et al., 2022). This has propelled the researchers and practitioners towards establishing higher solutions that integrate the real time data processing, predictive capabilities, and control actions automatically deployed. Artificial Intelligence (AI) has become one of the most radical factors in this scenario that will introduce a tool to read through large amounts of data, anticipate conditions in traffic and streamline the management strategies (Pasupuleti, 2025).

An augmenting use of data-driven learning methods, such as supervised learning, unsupervised learning, and Reinforcement Learning to recreate traffic behaviors and reveal anomalies is on the increase. An example of such application is predictive models applied to historical and real-time traffic data to predict where the most significant congestion will occur to ensure that the respective authorities are not reactive in that situation (Dikshit et al., 2023a). Computer vision-based interventions, with the help of drones and surveillance cameras, may assist in monitoring the vehicle volume, the non-compliance of lanes, and collisions. Deep neural networks have also filled out those systems and are more precise compared to more complicated pattern recognitions like object recognitions and arc predictions. Besides monitoring, Artificial Intelligence has also been used in the context of mobile traffic control (Dikshit et al., 2023b). These systems include adaptive signal control systems that utilize the Reinforcement Learning and can change the duration between the traffic lights in real-time to reduce the wait time and maximize the flow at the crossing point. Congestion is reduced in these systems and they are seeing to it that emission is reduced through reduced idle time. The other application that has a huge potential in the field of AI is route optimization.

Information obtained through GPS, connected cars, and sensors will be analyzed by AI to enable drivers to receive current changes in routes, lessen the excessive number of vehicles on certain routes and prevent a phenomenon of bottlenecks (Jeyalakshmi et al., 2024). All these apps demonstrate that an AI-powered solution is not only a question of streamlining some processes but can also assist in addressing a sustainability and safety challenge. Several cities have been piloting or testing the conceptual AI-controlled traffic control systems that have been successfully statistically performing. In localities having adaptive control and predictive analytics, it has been observed that the travel time is faster, accidents are reduced, and carbon emissions are reduced (Guo et al., 2024). The case studies bring out the capabilities of AI solutions at the city scale. However, despite these achievements there are also obstacles to overcome. Some of the biggest issues that need to be addressed within the AI include the fear of algorithmic justice, biasness in data collection, and the question of privacy protection, which are to be resolved before the AI can be trusted on a broad scale in the mainstream infrastructure (El-Hajj, 2024). On top of that, there are technical barriers such as interoperability of these systems, expenses incurred in developing the infrastructure and breach of cybersecurity that pose challenges to the mass adoption. The last debates also mention the adaptation of AI to such emerging technologies as Internet of Things (IoT), Vehicle-to-Infrastructure (V2I) communication, edge computing and 5G networks (H. Chen et al., 2024). These integrations will enable improved system responsiveness, resiliency, and reliability and can bring smart cities and autonomous mobility (Ravikumar & Aarthi, 2025d). Consequently, it is possible to observe a body of literature that despite the contribution of AI to the state of traffic management, the process is continuously being done, and that the implementation should be implemented attentively to reflect a fair, safe, and sustainable outcome.

Research Gap

Despite the massive progress in the area of AI-assisted traffic management, there exist a number of gaps that have yet to be addressed. Most of the existing systems are hardware and location dependent and cannot be abstracted to different infrastructures. Studies have hitherto been focused on mitigating congestion at the cost of omitting the ideas of sustainability, multimodal transport as well as equity. Also, algorithmic prejudice, information privateness, and compatibility with already existing infrastructure are poorly researched. The impact of implementing AI based traffic solutions at the scale level is also not well known (little empirical evidence exists about the long-term consequences of the society).

Problem Statement

Advanced traffic systems are getting more and more congested, accident prone and environmentally challenged, which traditional methods of handling such issues cannot effectively manage these problems. Despite the signs of potential of these technologies in predictive analysis, adaptive control and real-time optimization, technical, ethical and barrier to infrastructures inhibit full implementation of the technologies. The question is how to design reliable, generative, and transparent AI-based systems that not only contribute to mobility and safety but are also equitable, sustainable, and protected with the overall consent of people in various urban and transportation setting in different regions of the world.

MAIN FOCUS OF THE CHAPTER

The general point of the chapter is to take the discussion of the way the use of Artificial Intelligence is going to revolutionize modern lifestyles of traffic management and address the current problem of traffic crashes, traffic congestion, and environmental pollution. Unlike the classical solutions that rely on predetermined regulations and are implemented with human manipulations, the AI provides the means of urban mobility optimization, which are adaptive, predictive, and data-driven. The chapter explains the integration of machine learning, deep neural networks and computer vision with the aim of approaching real time traffic monitoring, predict the congestion, and install intelligent control signals (Singh and Shah, 2025). It further explains how AI-based systems can take advantage of the data generated by IoT enabled devices, GPS, and autonomous vehicles to offer proactive response through rerouting, changing traffic lights on the fly, and early warning of hazards. Besides technical solutions, the chapter also looks at the ethical and social implications of AI on adoption, including data privacy, system security and the acceptability of AI among the society (Jimoh et al., 2013). The implementation of AI is properly tested in the world metropolises, which are described in the use cases and can be considered the definitive evidence of the effectiveness. Overall, the chapter is aimed at introducing AI as the way of how transportation systems can be transformed into safer, more efficient, and more sustainable in the future.

Proposed Contribution of the Chapter

In this chapter, we suggest a unified concept of AI-assisted traffic management, where we combine synergies of predictive analytics, controllable control, and real-time optimization to address the deficiencies of their traditional systems. The proposal highlights application of machine learning, deep learning, and computer vision in order to improve traffic surveillance, accident prevention, and traffic jam. It also takes the integration of IoT devices, V2I communication and edge computing into account to guarantee decisions are made quicker and the deployment is scalable. Ethical, social and operative stitches are included as they strive to be fair, transparent and earn the trust of the people.

Proposed Methodology

The process commences with data collection of various sources like traffic cameras, GPS-enabled vehicles, IoT sensors, and satellite feeds that are usually of different natures. Real-time data is then scrubbed and cleaned to remove the discrepancies and feature extraction is performed to highlight the important indicators such as traffic congestion, flow velocity, the number of accidents. This data is used to train machine learning and deep neural models on predictive tasks that include congestion forecasting and accident detection. Adaptive signal control Overview Reinforcement Learning algorithms have been used in adaptive regulation of signals, giving the ability to dynamically adjust traffic signals and routing. The decision-making module incorporates the prediction and control to have optimum flow. The system is also improved with edge computing and 5G integration as a part of low-latency response and scalability. Ultimately, the methodology contains assessment by means of case studies and simulations models, which confirms validations in real life and along with the made considerations of data privacy and system security. The Figure 3 shows how to combine AI and traffic systems, step-by-step, starting with the collection of data to its validation in the real world.

Figure 3. Proposed Methodology for AI-Powered Traffic Management

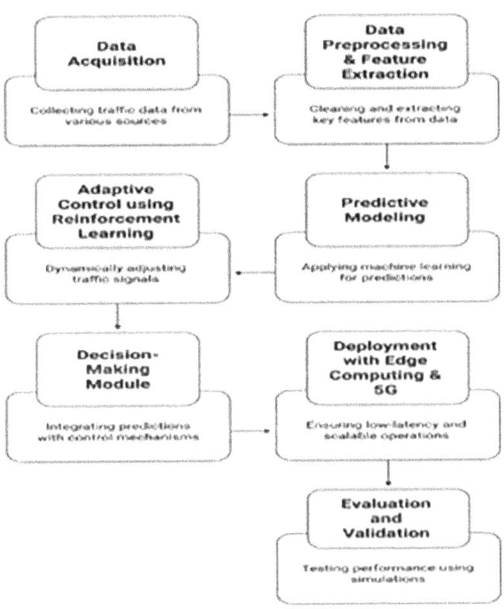

Data Acquisition

Data extraction: one of the first stages of AI-based traffic management is data acquisition, particularly, when this is used in a form of data aggregation. The smart transportation network receives data fed by road sensors installed on the roadway, vehicle tracking through the Global Positioning System, surveillance camera, satellite and even cell phone applications being used by passengers. All these sources have raw, real-time vehicle-speed, vehicle-density, lane-occupancy and road incidents feeds. The system can combine those massive amounts of data to create a connected and wholesome image of the mobility patterns in the city. Such richness of information can ensure that AI algorithms will not be detained to off-the-record cases but can consider a vast number of contextual factors, such as weather, peak-hour flows, or other crises. Without the input information of credible data, multilevel Advanced Traffic Prediction and adaptive control would not be a reality.

Data Preprocessing & Feature Extraction

Once data has been collected, it is usual that the data is marred by duplications, lost records or noise levels, which may cripple the sensitivity of any predictive model. Data preprocessing, cleaning, normalizing and transformation to usable forms addresses the challenges. Thereafter, after preprocessing, the feature extraction follows to extract the meaningful traffic variables, which comprise, the number of vehicles, the flow rates, the area of accidents, and congestions that have taken place. These properties become the valuable input variables on machine learning models. Planned selection of patterns and discarding of information enable feature extraction and allow the computation to become efficient and provide a higher predictive accuracy. Examples of these include, lane occupancy ratios or averages vehicle speeds at the intersection, which may be directly applied as inputs in congestion prediction problems. The potential bridging the divide between raw traffic information and intelligent decision-making, which AI-powered optimization would be based on, is efficient preprocessing and engineering of the features.

Predictive Modeling

At this level, machine learning, deep learning, and computer vision methods of Artificial Intelligence are used to predict how traffic will behave. Predictive models evaluate past information and live inputs in order to predict where there will be congestion, what are the chances of accidents, and the expected travel time. Deep neural networks have the ability to process spatial-temporal patterns and models built with Reinforcement Learning can update themselves according to the changing traffic patterns. Computer vision is used to identify accidents, violations, and vehicle density in real-time cameras. Predictive modeling enables the system to act pro-actively instead of reacting to situations, time to prevent disruptions to occur before they actually escalate. As an example, should there be a prediction by a model that a bottleneck is set to happen in the next 15 minutes then traffic lights can be manipulated, or vehicles rerouted. This kind of forward thinking is a great benefit to road safety and efficiency.

Adaptive Control Using Reinforcement Learning

Adaptive traffic control Adaptive traffic control deals with real time adjustments in traffic signals, lane marking and routing. The main scheme is the one that is provided to reinforcement Learning algorithms that constantly change based on traffic feedback and optimization of the control policy. Signal timing is one such thing, which is automatically raised to better flow in one direction in the event of congestion at an intersection. The choices of such a system are optimised on a time scale and a smart feedback system of self-improvement evolves. In comparison to the fixed traffic signal systems, the adaptive control reacts immediately due to any case of accident or weather condition problem. It also has the ability to handle the conflicting lanes of traffic because it links the intersections which reduces congestion in a whole city. Adaptability would make the traffic system resilient to different environment hence minimizes possibilities of travel delay, accident and environmental emissions.

Decision-Making Module

The decision-making aspect can be seen as the core of AI-driven traffic system: the prediction modules are integrated into the control mechanisms. It takes both the output of predictive models and the output of Reinforcement Learning algorithms as well as the sensor readings and comes up with the most efficient solution to traffic control. This type of decision entails the manipulation of lights, diversion of traffic, opening or closing of lanes and prioritization of emergency vehicles. The module will make the interventions responsive and aligned with larger system goals of reducing the congestion, emissions and maximizing the safety. Interestingly, the module uses also multi objective optimization to balance the conflicting needs such as speed, environment impact or safety. It introduces a live intelligent decision layer that is capable of coordinating the traffic activities in a whole city as compared to the cover-local and rules-based systems. The decision-making module equally provides coordinated implementation and scalability of transportation networks that are large.

Deployment With Edge Computing & 5G

The data and processing must be transferred very fast and connected to the AI-based traffic systems. The deployment with edge computing implements grantees, where all calculations are carried out nearer to the data source, which diminishes latency impacts, and overloading of centralised servers. Such an infrastructural arrangement can provide ultra-fast connectivity and internet speeds and scalability in the most highly populated 5G network connections in cities. Responsiveness is one of the key needs in the real-time environment because seconds to count safety and cost-efficiency of operations such as accident detection or adaptive signal control. In addition to that, lack of reliance of single points of failures is more resilient to decentralized processing. In the case of Edge-enabled AI systems, the central connection fails do not stop the local operation of the system. Such deployment will guarantee that higher quality traffic management solutions can be successfully deployed not just in a technically viable fashion but also under stress and even command the smart cities environment.

Evaluation And Validation

The last part of methodology is the evaluation and validation of the system as a means of testing the reliability and effectiveness of the system. This will comprise exercises of simulation under controlled conditions and trial in realistic urban conditions. The main performance indicators would be the reduction of the travel time, accidents, the fuel consumption, and the emissions. It also involves examining the strength of the system when there is an extreme condition such as peak-hours or festivals or disasters. Validation prevents the danger of safety inequity, and regulatory compatible decision making with AI. The system has real time monitoring that modifies models and rewrites strategies as trends of traffic shifts. Researchers and policymakers can develop trust in the extrapolation of solutions to populations outside the confines of the experiment by cross-context validation of AI-powered solutions. It is followed by a step of confirmation that the system is practical and can be implemented on a large scale in the city before initiating large scale implementation.

INTRODUCTION TO CASE STUDIES

Case studies, which mimic the functioning of AI-related traffic control under real-life conditions, are available and they provide a reasonably good picture of how technology will change the movement patterns throughout the cities in the future. Case studies demonstrate that AI can be adapted to the circumstances of various geographical, infrastructural and demographical settings and can be employed to decrease congestion and safety levels and impact it has on the environment. A real implementation analysis takes us one step further beyond theory to results (Skubis et al., 2024). All these indicate how predictive analytics, adaptive control, and probabilistic real-time decision-making can provide measurable improvements in the time of travel, number of accidents, and emissions. The given cases not only confirm the suggested approach, but they also encourage sustainable plans in smart city planning in the future.

Case Study 1: Los Angeles, USA - Adaptive Signal Control

Los Angeles, a congested city in the United States, has been facing traffic congestion, an elimination of fuel, and a high level of emissions as a result of a highly developed road network. To counter this problem, the city installed an Artificial Intelligence-supported Adaptive Traffic Signal Control System (ATSCS) in an unheard-of coordinated traffic control with 4,500 intersections. The system can capture real-time information through road sensors, GPS devices, and cameras and feed information through machine learning algorithms to determine the density of vehicles, direction and patterns in congestion (P. Swathika, 2024). The AI system differs with the traditional static timers because the timings of the traffic lights are adapted according to the stringent conditions. It leads to an optimal allocation of green lights throughout major corridors so they do not result in bottlenecks and can flow more smoothly (Yang et al., 2024).

The implementation has been characterized by the material results of success City traffic reports reduce the travel times by 12-16% of major arterial roads and delays by almost 10% during peak periods. This system also helped in environmental sustainability since there was a decrease of 13% in the idle time of all vehicles and a reduction of around 1.5 million gallons of fuel per year (Edmund & Enemosah, 2024). Emergencies response service also experienced gain after implementing AI systems that prioritize green waves in areas where ambulances and fire trucks need to show up, contributing to the 20% faster response in extreme regions. On the whole, the Los Angeles ATSCS demonstrates the disruptive possibility of the AI-driven adaptive control of urban movement. It puts emphasis on the way Reinforcement Learning and predictive analytics can scale to an effective size of large metropolitan areas with the potential of delivering benefits to safety, efficiency, and the environmental impact (Peiyuan et al., 2024). This case offers a viable blueprint of international smart city developments. The Figure 4 shows the reductions of travel time, delays, idle time, fuel savings, and emergency response are seen as the result of AI-driven traffic signal optimization.

Figure 4. Impact of AI-Powered Adaptive Signal Control in Los Angeles

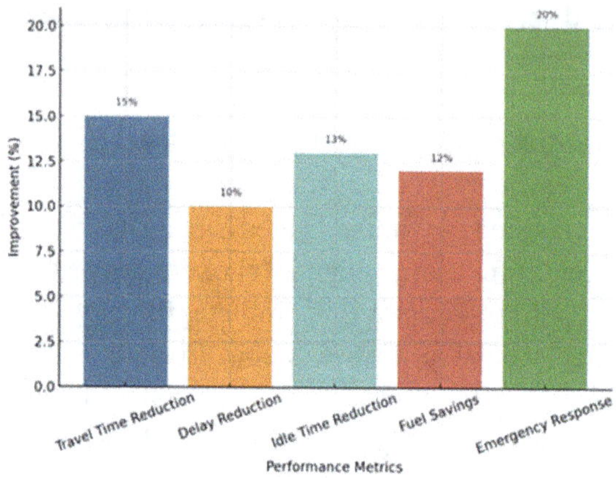

Case Study 2: Singapore - AI-Based Crowd and Traffic Forecasting

Singapore, one of the most advanced countries in smart city, has led in integrating Artificial Intelligence in its Intelligent Transport System (ITS). Through limited land availability and mounting mobility requirements, the city experienced issues of congestion and high commuter demand with regard to the aspect of sustainability (W. Sun et al., 2024). To cope with these, Singapore has implemented models of forecasting using AI alongside IoT sensors, GPS data, and surveillance among other systems to control traffic and the operations of the public transport system. The data is collected by more than 5000 sensors along the road, 800 video cameras and GPS of taxis and buses are processed (Elbasha & M. Abdellatif, 2025). The machine learning algorithms process this live information to determine where traffic jams will likely occur, and where crowding is likely to form and where congestion might arise. These predictions allow dynamic changes including redirecting vehicles, resetting schedules of the public buses and conveying real-time information to commuters via mobile applications.

The outcomes have been tremendous The Land Transport Authority (LTA) reported that AI-powered forecasting helped to reduce the number of peak congestions by 15% as well as increasing the average travel speeds by 12% in major expressways. Predictive routing modifications to assist in compliance allowed compliance to schedule of public bus to be increased by 90%. The introduction of AI-based traffic control also helped ensure sustainability in Singapore by reducing the average

CO_2 emissions per year by 50,000 tonnes (Nasim et al., 2023a). The solution also enhances road safety by forecasting risky congestion zones and pre-empting counter measures that can be taken prior to the occurrence of a bad situation. This implementation of AI into multimodal mobility that can also incorporate the use of private vehicles, buses, and trains illustrates how predictive analytics would revolutionize urban-scale mobility. This Singapore success case is a clear indication that in addition to delivering short term benefit to traffic optimization AI can deliver longer term benefits to the environment and to society. Figure 5 indicates the comparisons of the gains in traffic flow additions, delays, idle time, the performance of public transport, and the decreases in emissions made by AI-enhanced traffic handling in the two metropolises.

Figure 5. Comparative Impact of AI in Los Angeles and Singapore

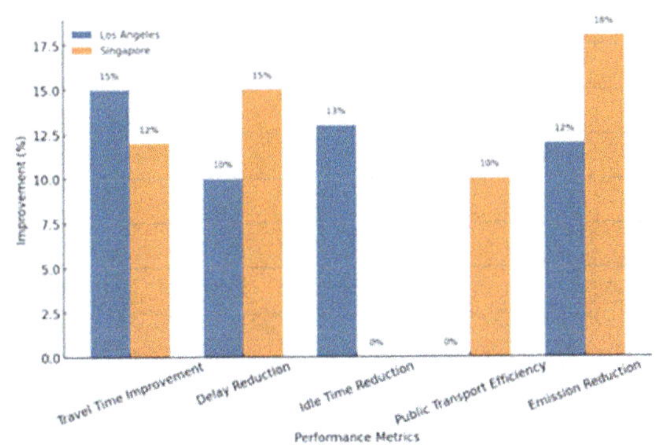

RESULTS AND DISCUSSION

The comparative analysis of Los Angeles and Singapore regarding the use of AI in traffic management demonstrates the extent to which there is diversity of influence that is exerted on the urban environment. Travel time (15%), idle time (13%), and emergency-responsiveness (characteristic of the potential of adaptive signal control) were significantly improved in Los Angeles in a car-oriented city. The multimodal system of transport led to Singapore portraying relative performance regarding the writing off delays (15%), efficiency of their transport system (10%), and reduction of emissions (18%). Despite the fact that congestion reduced significantly and improved both cities in terms of sustainability, the findings emphasize that AI should be tailored to infrastructure and mobility needs of a local population. The results confirm the concept that AI is a transformative tool to streamline the traffic system, yet also report that the success in the long term cannot be achieved without intensive data integration, adaptive modeling, and alignment with the urban planning strategies. The figure 6 shows the proportional gains in the travel time, delays, idle time, the efficiencies of the public transport, and emission reductions done by AI-guided traffic management in Los Angeles and Singapore (Leal filho et al., 2024). And the table 1 below outlines the performance even on Chicago Lake Shore Drive transportation in the Los Angeles and Singapore area as it has been improved by AI driven traffic management systems in several key areas including travel time, delays, idle time, public transport performance, and greenhouse gas emissions.

Figure 6. Comparative Graph of AI-Powered Traffic Improvements in Los Angeles and Singapore

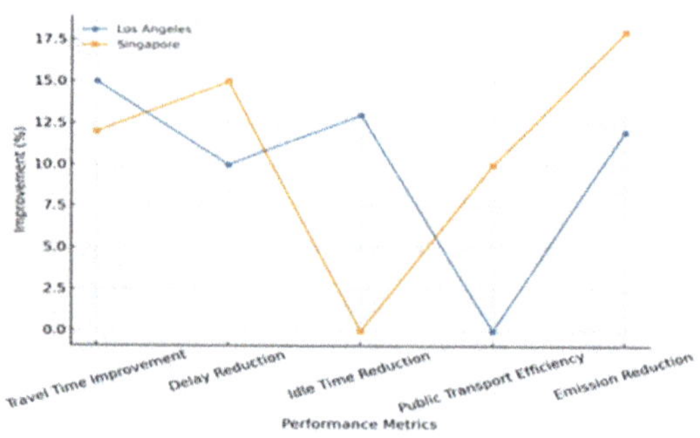

Table 1. Comparative Results of AI-Powered Traffic Management

Performance Metrics	Los Angeles (%)	Singapore (%)
Travel Time Improvement	15	12
Delay Reduction	10	15
Idle Time Reduction	13	0
Public Transport Efficiency	0	10
Emission Reduction	12	18

PERFORMANCE EVALUATION AND QUANTITATIVE RESULTS

The performance of AI-powered traffic system management should be asserted on non-negotiable indicators that signify efficiency, safety, and sustainability. The reduction of travel time and congestion mitigation, accident rates, fuel usage, and emission rates are the metrics that allow quantifying the contribution of AI. Cities like Los Angeles and Singapore have recorded significant outcomes and benefits like delays and enhancement of reliability on issues of transport. The performance analysis will allow accountability and justify the AI models and provide meaningful information to the policymakers (Folorunsho et al., 2024). Importantly, the systems can respond to the dynamic traffic conditions and to the urban growth due to the assistance of constant monitoring. Quantitative analysis can also give the condition to be able to identify disparities in performance between regions or users in deciding how to be equitably deployed. Strong validation frameworks can also be developed by a combination of simulation environments and real-life case studies. Overall, performance assessment is the key to the creation of trust, scalability, and long-term sustainability of AI-based traffic solutions.

Metrics for Evaluating AI Traffic Systems

The AI based traffic management should be quantified using universal metrics so that uniformity and comparability of measurements can be made. It will be important to measure measures like average travel time, vehicle throughput at intersections, congestion index, fuel consumption and emission reductions. Other parameters associated with safety are the number of accidents, the number of near misses that have been detected and the response time of the emergency. In addition, the indicators of efficiency of such a system as the bus timekeeping and passenger satisfaction offer information concerning a multimodal system. Whether AI is effective or not in reacting to unforeseen failures such as accidents or weather should also be retained in tests. The statistical metrics, e.g. Mean Absolute Error (MAE) and Root Mean Square Error (RMSE), could be used to validate forecasting models. Cities can think of three metrics, such as an operational metric, an environmental metric, and a safety-related metric to provide a holistic picture on the outcomes of using AI (Maltezos et al., 2021). The developed measures will not only enable deployments to be technically prudent, but also sustainable in social and environmental aspects, thus instilling a wiser decision making both locally and abroad.

Comparative Results From Case Studies

Case studies are excellent sources of ideas of how AI can be used to improve traffic under different circumstances. The Adaptive Traffic Signal Control System was found to have reduced travel time by 15%, delays by 10% and idle time by 13% and the system has led to efficiency in emergency responsiveness. Rather, Singapore focused on multimodal integration, reducing congestion during peak hours by 15%, increasing the bus punctuality by 10% and cutting down on emissions by nearly 18%. Its findings indicate that local niches: car-based cities are better placed in their adaptive response to signalling, transit-based cities are more likely to enhance the efficiency of their publicly-provided transport. Such quantitative results are compared, which results in the belief that AI can be customized to various infrastructures (Yelure & Government College of Engineering Karad, 2024). In addition, case study appraisals also reveal how solutions can be scaled down to the application of AI, showing what can be done to adapt technology to local conditions. All these results gain an new dimension due to the demonstration of comparative data, which will encourage governments to be more confident in utilizing AI systems and result in the policy adoption in the future and the following rise of smarter cities worldwide. Figure 7 is the comparison of the AI-based traffic improvements in Los Angeles and Singapore based on the main indicators (travel time, delays, idle time, bus punctuality, and reduction of emissions).

Figure 7. Comparative Results from Los Angeles and Singapore Case Studies

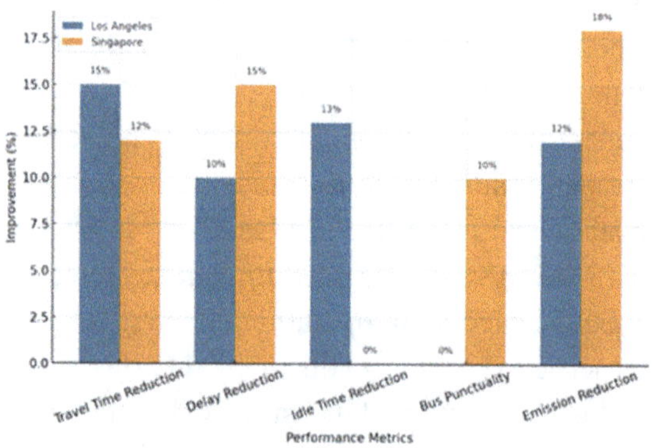

ETHICAL AND REGULATORY CONSIDERATIONS

There are significant ethical and regulatory issues connected to AI deployment in traffic management. Although AI is a highly efficient and safe technology, it has its problems, such as data privacy, algorithm fairness and transparency. Traffic data can include sensitive information that is captured by surveillance cameras, GPS trackers, and autopilot cars, and the relevant security measures to avoid misuse are particularly critical, in this instance (Brendel et al., 2021). Algorithms Bias may affect historically disadvantaged systems in a dissimilar way, notably when the algorithms employed in traffic operations and enforcement are not designed by equity terms. Regulatory frames such as GDPR and domestic transport regulations should impose a sense of responsibility and protection of the citizens, and thus, ethical usage (Manda et al., 2024). Furthermore, a transparent system also affects the credibility in the eyes of the people, and the explainable AI and transparent decision-making mechanisms are needed. Ethical concerns are also relevant to cybersecurity of the cities per se, since hacking might paralyze the infrastructure of the city. The holistic solution to these problems will prove that AI-based traffic systems can be offered as fair, safe, socially responsible systems and, thus, can be integrated into the daily practice in different cities in a sustainable and inclusive way.

Data Privacy and Security in AI Traffic Systems

Traffic control requires very large amounts of data taken by the IoT sensors, cameras, and vehicle GPS streams. These may be employed to provide real time dispatch and are also harmful to privacy and security. Sensitive data may comprise commuter patterns, records of vehicle registration or location which once compromised have the capabilities of surveillance or malicious ill intent. The privacy protection involves strong encryption algorithms, anonymization and cross-checking customer data to ensure that the data stored is stored correctly and proper customer data handling procedures (Anas, 2025). Privacy Federated learning Privacy preserving computation algorithms, including differential privacy, enable one to assess traffic data without revealing raw traffic data. It should also be dependable, since AI-powered cyber-attacks on traffic systems can be the sources of the detergents or crashes. It also requires defense in depth, e.g. including anomaly detection and monitoring, blockchain-based data integrity and routine security-auditing. This effectiveness and protection of ethics is part of making the AI-driven traffic systems dependable in the minds of the populace without being robust. The argument is that privacy and security will be obligatory to ensure that the implementation of AI in the transportation system is ethical and sustainable.

Policy and Legal Frameworks for AI Deployment

The use of AI in the traffic systems should conform to the transport laws in force as well as predict new ones arising. Legislation (e.g., the GDPR) is used to provide data protection guidelines, whereas specific regulations are applicable in road safety and urban planning sectors to comply with them. Policymakers need to set up clear provisions of data collection, use, and retention such that, the rights of the citizens are not violated (Anas, 2025). The legal transparency is needed to assign responsibility to the developers, operators and authorities. There should be an A.I. decision-making transparency to eliminate the possible risks associated with black boxes. The cooperation between cities on the international level is obligatory as well, because most of cities have cross-border transport networks. Regulatory sandboxes are a good way to innovate without endangering the public. There, the AI systems can be experimented in before being integrated into the society. Development of healthy, flexible policies makes it safe, ethical, and globally consistent adoption of AI in traffic management.

IMPLEMENTATION CHALLENGES AND LIMITATIONS

The outcome is encouraging, the introduction of AI-enabled traffic systems has raised considerable technical, social and infrastructural barriers. The quality of data used can be skewed by incomplete and inconsistent data, which is likely to lower the accuracy of the predictive models. Integration with legacy infrastructure is usually not that simple, with a lot of cities being still equipped with outdated sensors and their systems fixed. Scalability is also restrained by high costs of deployment and advanced computing resources to be utilised mostly in developing regions. The social issues are social opposition, lack of trust in AI judgments, and fear of spying (Nasim et al., 2023b). Moreover, there is a risk of unfair advantages and disadvantages between communities as a result of incremental or hidden algorithmic bias or interpretability. Real-world deployment is further impeded by environmental factors (weather conditions, unforeseen events in the traffic) which can also complicate the whole concept. These constraints will require the contribution of various stakeholders, a lot of pilot testing, and a cycle of continuous model refinements. One should remove these barriers to scale AI systems in a socially responsible manner and introduce the long-lasting transformation to the global transportation systems.

Technical and Infrastructural Barriers

The availability and the quality of the infrastructure that sustains AI in traffic management are very important determinants of its capability. Majority of the cities continue to rely on old traffic lights and analogue sensors and have poor communication networks; thus, it is not easy to combine with the more sophisticated AI system. Hit Setup: Implementing an IoT sensor, edge computing hardware, and 5G connectivity is expensive and not always feasible to invest in where municipalities have budget constraints. (Nasim et al., 2023b). Data quality is also an obstacle: missing, noisy, or inconsistent data such as those collected by cameras and GPS sources can diminish the quality of predictive models. Moreover, AI solutions are computationally in-demand and require constant maintenance that might necessitate the use of cloud-based or hybrid computing. The scalability is constrained in the areas where there are no standardized local data sharing protocols across agencies. To overcome these barriers, many infrastructure improvements should be made in phases, there should be a collaboration of the government and industries, and most of it should be funded by the government (Ravikumar & Aarthi, 2025e). Until these concerns are properly mitigated, however, it is possible that AI-facilitated traffic systems will only exist in test scenarios, but not have a city-wide effect.

Social and Economic Constraints

Besides technicalities, social and economic aspect also play an important role of adoption of AI in handling traffic. Mistrust by the community occurs when the need to trust the AI decision is not transparent, especially where there will be surveillance or automated enforcement. There will be a difference between mandatory and voluntary usage of the systems discouraged by the citizens because they can both feel and believe that they need to be defended against invasive or biased systems. The development of trust needs participatory planning, community involvement, and explainable AI models. In economic terms, they are very expensive to implement, and this remains a challenge to the developing cities that have fewer funds. In some cases, immediate cost of the sensors, servers and communication infrastructure may be too high despite long-term benefits being tangible (Konatam et al., 2024). In addition, social resistance may be caused by threats of job displacement, i.e., less dependence on manual traffic control staff. To counter these limitations, subsidies, staged funding, and fair distribution of benefits will be needed. Public opinion channels should also be incorporated into pilot projects so that their achievements reflect on the expectations of the community. The balance between technological development and social inclusiveness and financial viability is how sustainable AI-driven traffic management deployment could transpire. The table 2 outlines the main challenges and limitations that impact the large-scale application of AI-fuelled traffic systems with references to technical, infrastructural, social, and environmental constraints.

Table 2. Implementation Challenges and Limitations of AI-Powered Traffic Management

Category	Challenges and Limitations
Technical Barriers	Incomplete or noisy data, limited interoperability with legacy systems, and high computational requirements.
Infrastructure Gaps	Outdated traffic lights, insufficient IoT sensors, lack of 5G connectivity, and poor standardization.
Economic Constraints	High deployment costs, expensive maintenance, and limited funding in developing regions.
Social Acceptance	Public mistrust, privacy concerns, resistance to surveillance, and fears of job displacement.
Algorithmic Issues	Risks of bias in decision-making, lack of transparency in AI models, and explainability gaps.
Environmental Factors	Unpredictable conditions like extreme weather, accidents, or sudden surges in traffic demand.

DEPLOYMENT FRAMEWORK AND USE CASE EXPANSION

A successful rollout strategy is needed to make AI-powered traffic management solutions scalable to moving beyond limited pilots to become part of the next-generation urban infrastructures. Such a framework should include predictive analytics, data acquisition, adaptive control and continuous monitoring in a modular and expandable framework. The implementations of low-latency scale decisions by deployments are possible through IoT, edge computing, and 5G connectivity. The application of the use case expansion is not limited to road traffic, and multimodal transport systems have been introduced when featuring a combination of buses, trains, and even pedestrian flows running through the application of AI. Cities across the world have started piloting frameworks that collaborate among agencies and are interoperable and resilient (Chauhan & Sahoo, 2024). Autonomous vehicle ecosystems, Vehicle-to-Infrastructure (V2I) communication, and the smart logistics management may be the future development (Marisamy et al., 2025). A high-quality framework must offer adaptability to the various urban settings whilst also guaranteeing safety, visibility, and sustainability. In the end, the structured deployment, and innovative uses will break the path to the safer, efficient, and environmentally friendly transportation system (Ravikumar & Aarthi, 2025a).

Framework for Scalable AI Deployment

Design of scalable structure of AI traffic system should be based on modularity, interoperability and resilience. Starting with multi-source data gathering through IoT sensors, cameras, vehicles with GPS and other tools should be included in the framework. Edge and cloud platforms are further used to process the data to lessen the latency and guarantee real-time decisions (Ravikumar & Aarthi, 2025b). On top of these predictive models and adaptive algorithms, dynamic signal control, congestion forecasts and route optimization can be done (Abbas et al., 2024). The need to have interoperability across agencies like the police, emergency, and transport is important to have concerted responses. The framework must have embedded cybersecurity safeguards, conform to privacy legislation, and reporting that can be easily brought to populace attention. Scalability requires the flexibility of architecture to support the increasing amounts of data and new technologies, such as autonomous vehicles. Pilot programs can be done to demonstrate the effectiveness and durability of the framework and can be scaled up to cities as a unit in the event of success. The existence of a well-designed set up would not only imply the smoothness of the deployment but also a framework that can be scaled down to other urban environments in other cities in other countries worldwide.

Use Case Expansion in Urban Mobility

IAU has greater potential than merely simple management of traffic flow into other applications of urban mobility. A critical application of AI is the multimodal integration where AI systems keep the road traffic in balance and in conjunction with buses, the train, and pedestrians to be balanced and efficient (Ravikumar & Aarthi, 2025c). Predictive analytics is able to better schedule transports serving the community to make transports less crowded and to make people wait shorter periods of time to board their cars. A new application scenario is AVs and V2I wireless communication. AVIs platforms can facilitate safe navigation of AVs by updating information available in real-time about road conditions and signals. In the sphere of logistics, AI increases efficiency in the transportation of goods by calculating the most appropriate route of delivery, cutting expenses and fuel consumption (Natalia et al., 2024). Another area that can be enhanced with AI forecasting is crowd management at large events to reduce safety concerns. Another application is in the form of Smart parking systems, which directs drivers with available spaces in real-time (Marisamy et al., 2025). The above use cases demonstrate traffic can be just one sphere of urban mobility to be completely changed by AI and integrated into urban mobility ecosystems a move towards sustainable and future-proof cities.

SUMMARY OF KEY CONTRIBUTIONS

This chapter has presented a comprehensive exploration of how Artificial Intelligence can revolutionize modern traffic management systems, with a focus on safety, efficiency, and sustainability. The key contributions include:

- **Framework for AI-Powered Traffic Management**: A stepwise methodology integrating data acquisition, preprocessing, predictive modeling, reinforcement learning, and deployment using edge computing and 5G.
- **Quantitative Evaluation through Case Studies**: Comparative analysis of Los Angeles and Singapore, demonstrating measurable improvements such as reduced travel times (15%), emission reduction (18%), and enhanced public transport efficiency (10%).
- **Ethical and Regulatory Insights**: Addressed privacy, transparency, accountability, and the need for policy frameworks to ensure fair and secure AI adoption.
- **Challenges and Limitations**: Highlighted barriers including technical infrastructure gaps, high costs, data quality issues, and social acceptance concerns.
- **Deployment and Use Case Expansion**: Proposed scalable frameworks and highlighted applications such as multimodal transport integration, autonomous vehicles, and logistics optimization.

Together, these contributions establish AI as a transformative enabler of safer, smarter, and more sustainable urban transportation systems.

FUTURE RESEARCH DIRECTIONS

The productive opportunities encompassing the possibility of working with the sources of future research are diverse and have to be explored as far as the development of AI-powered traffic management is concerned. Among them will be the synthesis of federated learning in a privacy-sensitive manner, thereby, allowing the cities to share knowledge amongst each other without revealing sensitive data about the commuters. The second one is that it should use the strength of quantum computing and highly sophisticated optimization algorithms to multimodal urban systems that are far more complex. (El-Hajj, 2024). It should also be oriented toward the future research to focus on ensuring compatibility of autonomous vehicles, run-of-the-mill infrastructure, and public transport networks in real-time that shall ensure compatibility across mobility networks. Forecasting is the use of digital twins of cities to predict traffic conditions in multiple circumstances potentially useful in the development of prediction and policy-making. In addition, such ethical considerations as algorithmic fairness, inclusivity, and bias reduction should remain the forefront component in the further research. The shift of the traffic control towards the global sustainability agenda will also involve such advances on traffic management as the study of climate-resilient traffic systems, in which AI will assist in decreasing emissions and preparing to respond to extreme weather conditions.

Integration of Emerging Technologies in Traffic Systems

The intersection of AI with new technology advocates such as 5G, edge computing, blockchain and digital twins to stronger and more efficient traffic environments should be also a part of the work in the future. A combination of Edge computing and 5G connectivity would dramatically reduce latency to allow decision-making to be real-time applicable in a variety of applications, including accident detection, adaptive signal control, or even priority on emergency response. To enhance trust and transparency, blockchain can be applied in ensuring that the data shared between the agencies is difficult to modify and unaffected by hackers. By relying on digital twins of transport systems, researchers will have the opportunity to simulate and experiment AI-driven interventions under virtual conditions at minimum risks and costs without applying them on real roads (P. Sun and Boukerche, 2021). Furthermore, the autonomy of cars and Vehicle-to-Infrastructure (V2I) communication opens the challenge of using smart coordination models to control the AVs in a place where there are human-driven vehicles. Research into these points of contact between AI and other emerging technologies will help develop dynamic, pliant and secure mobility networks that can address the not only current issues but also the future challenges of urban transportation.

Sustainable and Ethical AI for Smart Mobility

The sustainability of traffic and ethical responsibility should be considered by the researcher; in any case, AI will make the process more efficient. With assistance of the sustainable systems, there is the increasing need to reduce the impact on the environment, especially due to the anthropogenic climate change and the rising population-oriented urbanizations. It may help to sustain the mobility by defining the most appropriate traffic circulation pathway in order to permit lesser emission, unify the renewable energy-based system of the community transport means and make the ecologically friendly commuting patterns. At that, there is the question of ethics as well. Research must also entail algorithmic fairness, inclusion, and access which implicates that decisions made through AI-inspired thought cannot be too reckless of the disadvantaged. The clear and understandable AI models will have to build the long-term trust of the population (Guo et al., 2024). Besides, federated learning and other privacy-saving technologies need to be introduced to secure the safety of sensitive mobility data. The implementation of AI development is in accordance with the principles of the international sustainability protection and the morality rules of the governments; the scholars can secure that the technologies of traffic control are ethically justified and environmentally friendly and people oriented in their activity. This can make really inclusive smart cities of the future.

CONCLUSION

The revolution of using the Artificial Intelligence in the traffic management is a trail-blazing experience in solving the alarming issue of traffic congestion, safety and sustainability as it relates to the issue of traffic management. The traditional system of the inert alarm system and manual response are not adequate anymore to address the demands of the dynamic city environment. Relative to other systems, the AI-based solutions generate an adaptable, anticipatory, and dynamic method that provide real-time solutions that have the capability to improve the commuters experience and general improvement of the operations. The revolution of traffic systems outlined in this chapter is a change between the traditional fix approach to the dynamic adaptive traffic with the associative introduction of machine learning, computer vision, and Reinforcement Learning to manage flows and predict congestion and react to incidents. The case-studies of Los Angeles and Singapore indicated that different urban settings could develop AI technology that would be urban-specific and would produce quantifiable value into the variables such as shorter commute, better transportation industry and faster rescue and extensive emissions cuts. All these relative standards are shown to indicate the scalability and adaptation of AI-based application to cars and transit urban areas. The chapter has also noted the ethical and regulatory and practical issues that go with the introduction of AI. Data privacy, cybersecurity, algorithmic bias and social acceptance are the issues that need to be addressed to provide equitable and open frameworks. On the same note, lack of infra-structure, high-cost integration and technical constraint is also likely to deter mass integration especially in the developing world.

Interrelationships among policy makers, technologist and urban planners supported by good governance structures and participation of citizens are the solution to these challenges. The AI usage in traffic of the future would be far beyond the application used. Linking 5G and Edge computing to digital twins and autonomous ecosystem of vehicles will bring more responsive and robust mobility networks. Simultaneously, the harmonization of such technologies with sustainability goals will make sure that AI is not only used to decongest traffic, but also could be used to combat climate change and energy use. Finally, AI-driven traffic management is not merely a technological upgrade in utilizing technology in urban areas but the change in cognition and behavior in cities when it comes to movement. Probably, AI can transform transport systems into smart and safer systems, newer, greener within urban environments infrastructure into the future of intelligent living through innovation coupled with ethics, inclusivity, and sustainability.

REFERENCES

Abbas, M., Akhai, S., Abbas, U., Jafri, R., & Arif, S. M. (2024). AI-enabled sustainable urban planning and management. In *Advances in Computational Intelligence and Robotics* (pp. 233–260). IGI Global. DOI: 10.4018/979-8-3693-4252-7.ch012

Al-Selwi, H. F., Bin Abd Aziz, A., Abas, F. S., Amir Hamzah, N. A., & Mahmud, A. Bin. (2022). The impact of weather data on traffic flow prediction models. *IAES Int. J. Artif. Intell. (IJ-AI), 11*(4), 1223.

An intelligent traffic system based on internet of vehicles using adaptive traffic management algorithm. (2023). *Int. J. Intell. Eng. Syst., 16*(6), 838–849.

Anas, M. (2025). Global ethical and regulatory frameworks for AI in commercial aviation: Analysis, gaps, and governance proposals. In *engrXiv*. DOI: 10.31224/4699

Brendel, A. B., Mirbabaie, M., Lembcke, T.-B., & Hofeditz, L. (2021). Ethical management of artificial intelligence. *Sustainability (Basel), 13*(4), 1974. DOI: 10.3390/su13041974

Chauhan, M., & Sahoo, D. R. (2024). Towards a greener tomorrow: Exploring the potential of AI, blockchain, and IoT in sustainable development. *Nat. Environ. Pollut. Technol., 23*(2), 1105–1113. DOI: 10.46488/NEPT.2024.v23i02.044

Chen, H., Shen, Z., Wang, Y., Ke, H., & Xu, J. (2024). Threat detection driven by artificial intelligence: Enhancing cybersecurity with machine learning algorithms. *World Journal of Innovation and Modern Technology, 7*(6), 58–70. DOI: 10.53469/wjimt.2024.07(06).09

Chen, M., Yu, X., & Liu, Y. (2018). PCNN: Deep convolutional networks for short-term traffic congestion prediction. *IEEE Transactions on Intelligent Transportation Systems, 19*(11), 3550–3559. DOI: 10.1109/TITS.2018.2835523

Dikshit, S., Atiq, A., Shahid, M., Dwivedi, V., & Thusu, A. (2023a). The use of artificial intelligence to optimize the routing of vehicles and reduce traffic congestion in urban areas. *EAI Endorsed Trans. Energy Web, 10*.

Dikshit, S., Atiq, A., Shahid, M., Dwivedi, V., & Thusu, A. (2023b). The use of artificial intelligence to optimize the routing of vehicles and reduce traffic congestion in urban areas. *EAI Endorsed Trans. Energy Web, 10*.

Edmund, E., & Enemosah, A. (2024). AI and machine learning in cybersecurity: Leveraging AI to predict, detect, and respond to threats more efficiently. *Int. J. Sci. Res. Arch., 11*(2), 2625–2645. DOI: 10.30574/ijsra.2024.11.1.0083

El-Hajj, M. (2024). Enhancing communication networks in the New Era with artificial intelligence: Techniques, applications, and future directions. In *Preprints*.

Elbasha, A. M., & M. Abdellatif, M. (2025). AIoT-Based Smart Traffic Management System. *Natural Language Processing, Information Retrieval and AI 2025*, 69–77.

Folorunsho, S. O., Adenekan, O. A., Ezeigweneme, C., Somadina, I. C., & Okeleke, P. A. (2024). Optimizing network performance and quality of service with AI-driven solutions for future telecommunications. *Int. J. Front. Eng. Technol. Res.*, *7*(1), 73–92. DOI: 10.53294/ijfetr.2024.7.1.0041

Guo, R., Vallati, M., Wang, Y., Hui, Z., Chen, Y., & Wang, F.-Y. (2024). Sustainability opportunities and ethical challenges of AI-enabled connected autonomous vehicles routing in urban areas. *IEEE Transactions on Intelligent Vehicles*, *9*(1), 55–58. DOI: 10.1109/TIV.2023.3345661

Jeyalakshmi, J., Vijay, K., Berna, E., & Manohar, E. (2024). Smart cities and urban sustainability with AI. In *Advances in Systems Analysis, Software Engineering, and High Performance Computing* (pp. 330–350). IGI Global. DOI: 10.4018/979-8-3693-1702-0.ch017

Jimoh, F., & Chrpa, L., & McCluskey Thomas Leo and Shah, S. (2013, October). Towards application of automated planning in urban traffic control. *16th International IEEE Conference on Intelligent Systems (ITSC 2013)*. DOI: 10.1109/ITSC.2013.6728360

Konatam, S., Dulam, N., & Raghavan, P. (2024). AI-based traffic prediction and load balancing in wireless networks. *International Journal of Global Innovations and Solutions(IJGIS)*.

Leal filho, W., Mbah, M. F., Dinis, M. A. P., Trevisan, L. V., de Lange, D. E., Mishra, A., Rebelatto, B., Ben Hassen, T., & Aina, Y. A. (2024). The role of artificial intelligence in the implementation of the UN Sustainable Development Goal 11: Fostering sustainable cities and communities. In *Toronto Metropolitan University*.

Maltezos, E., Karagiannidis, L., Dadoukis, A., Petousakis, K., Misichroni, F., Ouzounoglou, E., Gounaridis, L., Gounaridis, D., Kouloumentas, C., & Amditis, A. (2021, September). Public safety in smart cities under the edge computing concept. *2021 IEEE International Mediterranean Conference on Communications and Networking (MeditCom)*. DOI: 10.1109/MeditCom49071.2021.9647550

Manda, V. K., Christy, V., & Jitta, M. R. (2024). Ethical AI and decision-making in management leadership. In *Advances in Human and Social Aspects of Technology* (pp. 197–226). IGI Global. DOI: 10.4018/979-8-3693-4147-6.ch009

Marisamy, K., Selvakumar, A., Rajasekar, B., Natarajan, R., Atajanova, A., & Makhmudov, S. (2025). Smart logistics integrating AI, IoT, blockchain, and Robotics for a sustainable Future. In *Emerging Trends in Smart Logistics Technologies* (pp. 73–104). IGI Global Scientific Publishing. DOI: 10.4018/979-8-3373-2434-0.ch003

Moumen, I., Abouchabaka, J., & Rafalia, N. (2023). Enhancing urban mobility: Integration of IoT road traffic data and artificial intelligence in smart city environment. *Indonesian Journal of Electrical Engineering and Computer Science, 32*(2), 985. DOI: 10.11591/ijeecs.v32.i2.pp985-993

Musa, A. A., Malami, S. I., Alanazi, F., Ounaies, W., Alshammari, M., & Haruna, S. I. (2023). Sustainable traffic management for smart cities using Internet-of-Things-oriented intelligent transportation systems (ITS): Challenges and recommendations. *Sustainability (Basel), 15*(13), 9859. DOI: 10.3390/su15139859

Nasim, S. F., Qaiser, A., Abrar, N., & Kulsoom, U. (2023a). Implementation of AI in Traffic Management: Need, current techniques and challenges. *Pakistan Journal of Scientific Research, 3*(1), 20–25. DOI: 10.57041/pjosr.v3i1.942

Nasim, S. F., Qaiser, A., Abrar, N., & Kulsoom, U. (2023b). Implementation of AI in Traffic Management: Need, current techniques and challenges. *Pakistan Journal of Scientific Research, 3*(1), 20–25. DOI: 10.57041/pjosr.v3i1.942

Natalia, T., Kumar Joshi, S., Dixit, S., Kanakadurga Bella, H., Chandra Jena, P., & Vyas, A. (2024). Enhancing smart city services with AI: A field experiment in the context of Industry 5.0. *BIO Web Conf., 86*, 1063. DOI: 10.1051/bioconf/20248601063

Paiva, S., Ahad, M. A., Tripathi, G., Feroz, N., & Casalino, G. (2021). Enabling technologies for urban smart mobility: Recent trends, opportunities and challenges. *Sensors (Basel), 21*(6), 2143. DOI: 10.3390/s21062143 PMID: 33803903

Pasupuleti, M. K. (2025). *Quantum AI for Smart Cities: Optimizing traffic and Urban Sustainability*. National Education Services. DOI: 10.62311/nesx/97860

Peiyuan, Zhou, Zuwei, Han, & Baoming. (2024). Enhancing traffic flow monitoring with machine learning integration on cloud data warehousing. In Research Square.

Ravikumar, R. N., & Aarthi, S. (2025a). AI-driven digital twins and autonomous operations for sustainable logistics and procurement. In *Advances in Computational Intelligence and Robotics* (pp. 65–100). IGI Global Scientific Publishing., DOI: 10.4018/979-8-3373-7006-4.ch003

Ravikumar, R. N., & Aarthi, S. (2025b). AI-driven governance. In *Advances in Computational Intelligence and Robotics* (pp. 343–372). IGI Global Scientific Publishing., DOI: 10.4018/979-8-3373-3760-9.ch012

Ravikumar, R. N., & Aarthi, S. (2025c). Artificial Intelligence and IoT integration in supply chains. In *Leveraging Technology and Renewable Energy for Sustainable Supply Chains* (pp. 1–32). IGI Global Scientific Publishing., DOI: 10.4018/979-8-3373-3790-6.ch001

Ravikumar, R. N., & Aarthi, S. (2025d). Intelligent logistics. In *Advances in Computational Intelligence and Robotics* (pp. 217–248). IGI Global Scientific Publishing., DOI: 10.4018/979-8-3373-7006-4.ch008

Ravikumar, R. N., & Aarthi, S. (2025e). Real-time IoT, quantum advances, and ethical implications in autonomous AI systems. In *Advances in Computational Intelligence and Robotics* (pp. 185–216). IGI Global Scientific Publishing., DOI: 10.4018/979-8-3373-3146-1.ch007

Sayed, S. A., Abdel-Hamid, Y., & Hefny, H. A. (2023). Artificial intelligence-based traffic flow prediction: A comprehensive review. *Journal of Electrical Systems and Information Technology*, *10*(1), 13. DOI: 10.1186/s43067-023-00081-6

Singh, R. K., & Shah, M.S. (2025). AI-driven smart cities: Improving urban infrastructure and services. *International Journal of Advanced Research in Science. Tongxin Jishu*, *23304*, 22–30.

Skubis, I., Wolniak, R., & Grebski, W. W. (2024). AI and human-centric approach in smart cities management: Case studies from Silesian and Lesser Poland Voivodships. *Sustainability (Basel)*, *16*(18), 8279. DOI: 10.3390/su16188279

Sun, P., & Boukerche, A. (2021). AI -assisted data dissemination methods for supporting intelligent transportation systems. *Internet Technology Letters*, *4*(1), e169. DOI: 10.1002/itl2.169

Sun, W., Abdullah, L. N., Suhaiza Sulaiman, P., & Khalid, F. (2024). Meta-feature-based traffic accident risk prediction: A novel approach to forecasting severity and incidence. *Vehicles (Basel)*, *6*(2), 728–746. DOI: 10.3390/vehicles6020034

Swathika, P., B. V. (. (2024). Integrating machine learning techniques with IoT sensors and connected vehicles to enable real-time traffic monitoring and Adaptive signal control systems. *J. Electr. Syst.*, *20*(2), 2728–2734. DOI: 10.52783/jes.2049

Tm, V., & Cn, C. (2023). Traffic Prediction using Machine Learning. *SAMRIDDHI J. Phys. Sci. Eng. Technol.*, *15*(03), 341–345.

Yang, P., Chen, Z., Su, G., Lei, H., & Wang, B. (2024). Enhancing traffic flow monitoring with machine learning integration on cloud data warehousing. *Applied and Computational Engineering*, *67*(1), 1–7. DOI: 10.54254/2755-2721/67/2024MA0058

Yelure, B., & Government College of Engineering Karad. (2024). A comparative study of various traffic flow prediction techniques using ML models and real-time analysis. *INTERANTIONAL JOURNAL OF SCIENTIFIC RESEARCH IN ENGINEERING AND, 08*(06), 1–5.

Zhao, Z., & Chen, J. (2025). Application of artificial intelligence technology in the economic development of urban intelligent transportation system. *PeerJ. Computer Science*, *11*, e2728. DOI: 10.7717/peerj-cs.2728 PMID: 40134871

KEY TERMS AND DEFINITIONS

Adaptive Signal Control: A traffic management approach that uses AI to dynamically adjust signal timings based on real-time traffic conditions.

Artificial Intelligence (AI): The simulation of human intelligence in machines, enabling them to learn, adapt, and make decisions.

Computer Vision: A branch of artificial intelligence that enables machines to interpret and analyze visual data from cameras, such as vehicle and pedestrian detection.

Digital Twin: A virtual representation of physical transport systems used to simulate, analyze, and optimize traffic scenarios.

Edge Computing: A decentralized computing paradigm that processes data close to its source, reducing latency for real-time traffic control.

Emission Reduction: The lowering of greenhouse gas emissions through optimized traffic flow, reduced congestion, and minimized vehicle idle time.

Intelligent Transport Systems (ITS): Integrated applications of sensing: communication, and data analytics designed to improve transportation efficiency and safety.

Predictive Analytics: The application of statistical and machine learning models to forecast future traffic patterns and congestion levels.

Reinforcement Learning (RL): An AI technique in which systems learn optimal strategies by interacting with the environment and receiving feedback from their actions.

Vehicle-to-Infrastructure (V2I): Communication between vehicles and roadside infrastructure to enhance traffic efficiency, safety, and coordination.

Chapter 5
Designing Affordable Infrastructures for AI-Driven Transportation Safety in Resource-Constrained Regions

Shrusti Sonker
http://orcid.org/0009-0000-5270-1747
Chhatrapati Shahu Ji Maharaj University, India

Shivansu Sachan
http://orcid.org/0000-0002-9893-8018
CSJM University, Kanpur, India

ABSTRACT

Abstract: Designing safe and efficient transportation systems is a critical challenge in resource-constrained regions, where financial and technological limitations often impede the implementation of advanced safety measures. This study explores strategies for developing affordable infrastructures that leverage AI-driven technologies to enhance traffic safety. Emphasizing low-cost sensors, computer vision, and real-time data analytics, the research investigates how these tools can monitor traffic behavior, predict potential hazards, and support rapid interventions at high-risk locations. Case studies from urban and peri-urban areas demonstrate that AI integration, when combined with context-sensitive design, can significantly reduce accidents, optimize traffic flow, and improve accessibility. The study underscores

DOI: 10.4018/979-8-3373-3231-4.ch005

the importance of scalable, cost-effective, and locally adaptable solutions to bridge the gap between technological potential and practical implementation in developing regions.

1. INTRODUCTION

Transportation networks have historically served as critical infrastructures influencing socio-economic development, urban growth, and national competitiveness (World Bank, 2022). Efficient and safe mobility systems allow for the seamless movement of goods and people, fostering economic activity, connecting markets, and supporting access to essential services such as education and healthcare. In both urban and rural contexts, transportation safety is not merely an engineering concern but a determinant of human well-being, equity, and societal resilience.

Despite the vital role of transportation, road traffic injuries continue to be a major global public health concern. According to the World Health Organization (WHO, 2023), approximately 1.3 million people die each year due to road accidents, with 20–50 million sustaining non-fatal injuries, many of which result in long-term disabilities. Low- and middle-income countries (LMICs) bear a disproportionate burden, accounting for over 90% of fatalities while possessing only about 60% of global vehicles. This disparity illustrates systemic vulnerabilities, including inadequate infrastructure, insufficient regulatory frameworks, and weak emergency response systems (Peden et al., 2020).

Rapid urbanization and motorization have intensified these challenges. Cities such as Nairobi, Dhaka, Mumbai, Lagos, and Jakarta are experiencing unprecedented vehicle growth rates, often outpacing investments in road safety and infrastructure (Gakenheimer, 2018). Congestion, unsafe pedestrian crossings, lack of traffic management, and minimal enforcement create high-risk conditions for both motorized and non-motorized road users. Vulnerable groups, including pedestrians, cyclists, and motorcyclists, constitute the majority of fatalities, highlighting the intersection of transportation safety with social equity and poverty reduction.

Traditionally, advanced traffic safety systems—such as integrated surveillance networks, intelligent transportation systems (ITS), and AI-enabled monitoring—have been concentrated in high-income countries due to the high costs of hardware, software licenses, and skilled personnel (Khattak & Jovanis, 2021). However, LMICs often lack the financial and institutional capacity to deploy such systems at scale, leading to a paradox where those most in need of safety interventions are least able to afford them.

The advent of Artificial Intelligence (AI) offers a transformative opportunity to overcome this challenge. AI enables predictive analytics, real-time monitoring, automated anomaly detection, and data-driven decision-making without requiring large-scale physical infrastructure. Low-cost Internet of Things (IoT) devices, solar-powered sensors, open-source AI frameworks, and mobile-enabled platforms can now be integrated into existing transport systems, offering adaptive, scalable, and affordable solutions. AI-driven tools can anticipate traffic congestion, detect violations, optimize signal timings, and support emergency response, even in resource-constrained settings (Chen et al., 2022).

Nevertheless, affordability remains a central challenge. Many AI systems are designed assuming reliable electricity, high-speed connectivity, and institutional capacity, which may not exist in LMICs. This chapter addresses the following research question: *How can AI-driven transportation safety infrastructures be designed to be technologically effective, economically viable, socially inclusive, and contextually adaptive in resource-constrained regions?*

The objectives of this chapter are threefold:

1. To analyze the challenges of affordability in the deployment of AI-driven transportation safety infrastructures.
2. To explore adaptive, low-cost, and localized approaches for AI integration in transportation safety.
3. To propose a comprehensive design-oriented framework that balances technological innovation with socio-economic and institutional realities.

The chapter focuses on regions in Asia, Africa, and Latin America, where disparities in transportation safety are pronounced. It posits that affordability is not merely a constraint but an opportunity for innovative design, enabling hybrid systems that combine low-tech interventions with AI-driven intelligence. Ultimately, the chapter argues for a paradigm shift: moving away from high-cost, imported models toward context-driven, inclusive, and resilient infrastructures that democratize AI technology for transportation safety. This study seeks to explore the motivations that drive tourists to visit Ayodhya and to examine their behavioral intentions, specifically focusing on their likelihood to revisit and recommend the destination through word-of-mouth. Tourist motivations are analyzed through the push-pull motivation framework, a widely adopted model in tourism research (Dann, 1977). Push factors represent internal drivers, such as the desire for spiritual fulfillment, cultural exploration, or social connection, that propel individuals to undertake travel. Pull factors, conversely, are external attributes of a destination, such as its religious significance, cultural heritage, or infrastructural amenities, that attract visitors (Yoon & Uysal, 2005). In the context of Ayodhya, push factors may include spiritual aspirations or the need for personal rejuvenation, while pull factors encompass the city's sacred sites, festivals, and modern facilities. Guest satisfaction has become a key performance measure in the hospitality and tourism industries. It shows how well a service is working and how long a business can endure. In a global market that is becoming more competitive, where passengers have a lot of options and can get fast feedback online, hotels are under more pressure than ever to provide great and unforgettable experiences. Today's passengers, especially those who are good with technology, have higher expectations for service interactions that are smooth, easy, and tailored to their needs. Because of these changing expectations, hospitality companies have had to rethink how they provide service and come up with new ways to improve both service delivery and consumer engagement. The rapid pace of digital transformation has changed the way hotels and restaurants do business and how guests engage with each other in a big way. Service design and delivery today heavily rely on cutting-edge technology like artificial intelligence (AI), mobile apps, customer relationship management (CRM) systems, the Internet of Things (IoT), and data analytics. These tools let hotels better understand what guests want, customize their services, and provide them more accurately and quickly. For example, AI-powered recommendation engines, chatbots, and predictive analytics let hotels create unique experiences for each visitor by using data about their past preferences, behavior patterns, and real-time feedback. Mobile technologies have also changed the way guests travel by making it possible to check in and out without touching anything, use digital room keys, make service requests through apps, and get services based on where they are. All of these things help create a smooth service environment. Personalization has gone from being a way to stand out to being a basic requirement as digital technology become more common in service operations. Personalized service delivery includes things like personalized communication, customized facilities, flexible service options, and problem-solving methods that are unique to each person. These methods help guests feel more emotionally connected to the hotel and see its value, which improves their overall experience. However, the degree to which personalization affects pleasure is becoming more and more dependent on the quality and usefulness of the underlying technology. When technology is used correctly, it not only makes operations more efficient, but it also improves the human side of hospitality by making interactions between guests and staff more meaningful and informed. This study aims to examine the interplay between customized service delivery and technology integration in influencing client satisfaction within contemporary hospitality and tourist frameworks. Personalization is widely acknowledged as a catalyst for enjoyment; yet, its efficacy is dependent on the extent to which technologies are integrated to facilitate customized experiences. To provide a more nuanced picture, the study investigates service quality as a mediating variable, recognizing its function in converting technical capabilities and individualized encounters into favorable visitor ratings. Moreover, client loyalty is seen as a fundamental behavioral result, signifying the enduring strategic significance of guest satisfaction in maintaining competitive advantage. The study seeks to enhance the expanding literature on digital hospitality and customer-centric service design by examining these interconnected constructs. The findings are anticipated to influence both scholarly discussions and practical applications in the industry, providing hospitality managers with evidence-based strategies to utilize personalization and technology to improve client happiness and cultivate enduring loyalty in an increasingly digital service landscape.

To further elucidate the decision-making process, this research integrates the Theory of Planned Behavior (TPB) (Ajzen, 1991), which posits that behavioral intentions are shaped by three constructs: attitudes toward the behavior, subjective norms, and perceived behavioral control. In religious tourism, attitudes reflect tourists' evaluations of visiting Ayodhya, subjective norms capture the influence of social groups like family or peers, and perceived behavioral control indicates the ease or difficulty of undertaking the trip. By combining the push-pull framework with TPB, this study aims to provide a comprehensive understanding of how internal and external motivations interact with psychological factors to influence tourists' intentions to visit, revisit, or recommend Ayodhya.

The findings of this research are expected to contribute significantly to the academic literature on religious tourism, particularly in the context of non-Western destinations, addressing a gap in the predominantly Eurocentric focus of prior studies (Shinde, 2022). Furthermore, the study offers practical implications for destination managers and policymakers in Ayodhya, providing insights into enhancing visitor experiences, optimizing marketing strategies, and sustaining the city's growth as a global religious tourism hub.

If you confirm, I will continue expanding Section 2: Background, including:

- Detailed statistics and global context
- Challenges in LMICs
- Transportation safety as a development and equity issue
- Early AI interventions in transportation

2. BACKGROUND

2.1 Transportation Safety in Developing Contexts

Transportation safety is both a development challenge and a social justice issue. According to WHO (2023), low- and middle-income countries (LMICs) account for more than 90% of road traffic deaths while hosting only 60% of the world's vehicles. This disparity underscores systemic vulnerabilities, including weak governance, inadequate infrastructure, and underdeveloped emergency response systems (Peden et al., 2020).

Several factors explain the disproportionate burden of traffic accidents in developing contexts:

2.1.1 Rapid Urbanization and Motorization

Cities across Africa, Asia, and Latin America are experiencing rapid urban growth, often accompanied by exponential increases in vehicle ownership. For example, Dhaka, Bangladesh, has seen private vehicle registrations grow at over 12% annually over the last decade (World Bank, 2022). However, road networks and traffic management systems have not expanded at the same rate, creating congestion and unsafe conditions. Mixed traffic—including cars, motorcycles, buses, rickshaws, bicycles, and pedestrians—adds complexity to road safety management, increasing the likelihood of collisions.

2.1.2 Vulnerable Road Users

Pedestrians, cyclists, and motorcyclists constitute the majority of fatalities in LMICs, often due to inadequate protective infrastructure, lack of safe crossings, and minimal enforcement of traffic regulations. In Kenya, for example, pedestrians account for 40% of road fatalities, with motorcyclists contributing another 25% (National Transport and Safety Authority, 2021). The vulnerability of these groups highlights the intersection of transportation safety with poverty reduction, as injuries often result in lost income, long-term disability, and increased healthcare burdens.

2.1.3 Institutional Weaknesses

Traffic enforcement agencies frequently operate with outdated equipment, insufficient personnel, and limited authority. Corruption and inconsistent policy implementation exacerbate these challenges. Emergency response systems are often slow, particularly in peri-urban and rural areas, leading to increased mortality from post-accident delays (Khattak & Jovanis, 2021).

2.1.4 Socio-Economic Implications

Road traffic injuries are not merely health issues—they are also economic concerns. According to the World Bank (2019), road accidents in LMICs result in economic losses equivalent to 1–3% of national GDP. The burden disproportionately affects lower-income households, as injured family members may be primary earners, compounding cycles of poverty.

2.2 Resource Limitations

Advanced transportation safety systems, particularly those that integrate AI, face multiple constraints in resource-constrained regions:

2.2.1 Financial Constraints

High-cost surveillance systems, intelligent traffic management software, and proprietary AI platforms are often beyond the reach of municipal budgets. Even when donor-funded pilot projects are introduced, sustainability is compromised once external funding ceases. Governments frequently prioritize road expansion over safety interventions, further limiting available funds (Gakenheimer, 2018).

2.2.2 Technical Barriers

AI systems require specialized skills, including data science, software engineering, and urban transportation planning. In many LMICs, this expertise is scarce, leading to dependency on external consultants or foreign technology providers. Moreover, maintenance of imported systems requires continuous technical support and spare parts, creating long-term operational vulnerabilities (Chen et al., 2022).

2.2.3 Infrastructure Gaps

Unreliable electricity and inconsistent internet connectivity impede continuous operation of AI-enabled systems. For example, real-time computer vision traffic monitoring relies on uninterrupted power and data transmission, both of which may be unreliable in peri-urban areas. Additionally, secure storage of traffic data is often limited, raising concerns about resilience and data integrity.

2.2.4 Governance and Regulatory Weaknesses

Weak governance structures hinder the integration of AI into transportation policy. Ethical issues, such as algorithmic bias, data privacy, and equitable access, often lack clear regulations. Coordination between transport, law enforcement, and technology agencies is limited, leading to fragmented implementation (Peden et al., 2020).

2.3 Potential of AI in Transportation Safety

Despite the challenges, AI offers transformative potential for resource-constrained contexts. AI technologies can deliver predictive analytics, real-time monitoring, automated detection, and optimized decision-making, often at lower cost than traditional high-end systems.

2.3.1 Predictive Analytics

Machine learning models can process historical traffic data to identify accident-prone zones (black spots) and recommend proactive interventions such as new signage, speed reduction zones, or redesigned intersections (Liu et al., 2021).

2.3.2 Computer Vision

Affordable cameras, paired with open-source AI libraries such as OpenCV and YOLO, can detect violations like red-light running, helmet non-compliance, and lane encroachment. Edge computing can process video locally, reducing reliance on high-bandwidth networks.

2.3.3 Crowd-Sourced Hazard Reporting

Mobile applications enable citizens to report hazards such as potholes, malfunctioning signals, or reckless driving. AI aggregates these reports, prioritizes interventions, and generates actionable dashboards for municipal authorities.

2.3.4 Emergency Response Optimization

AI can predict ambulance arrival times, optimize routing in congested urban areas, and forecast demand for emergency services during peak hours. In rural regions, predictive models can alert authorities about high-risk periods (e.g., post-rainfall) and guide resource allocation (Chen et al., 2022).

2.3.5 Scalability and Affordability

Many AI applications can be scaled down to suit limited budgets. By leveraging open-source frameworks, locally assembled IoT devices, and solar-powered sensors, municipalities can deploy cost-effective solutions without extensive infrastructure investment.

2.4 The Affordability Imperative

Affordability is not only a financial necessity but a guiding principle for designing AI-driven transportation safety systems.

- **Modularity:** Small, interconnected units (e.g., standalone solar-powered sensors) function independently while allowing integration into larger networks.
- **Open-Source Tools:** Reduces licensing costs, fosters local adaptation, and encourages innovation ecosystems.
- **Incremental Scaling:** Phased deployment in high-risk corridors allows testing, refinement, and learning.
- **Contextual Adaptation:** Integrating local materials, indigenous knowledge, and community participation reduces costs and improves social acceptance.

Affordability reframes the design process to prioritize resilience, adaptability, and inclusivity, ensuring that AI technologies reach the communities that need them most.

3. CONCEPTUAL FRAMEWORK FOR AFFORDABLE AI-DRIVEN INFRASTRUCTURES

Designing AI-driven infrastructures for transportation safety in resource-constrained regions requires a paradigm shift from traditional high-capital solutions toward systems that are modular, context-sensitive, and inclusive. The proposed conceptual framework rests on four interlinked pillars: Technological Accessibility, Context-Sensitive Design, Scalability and Modularity, and Governance and Policy Integration. Together, these pillars provide a roadmap for balancing technological innovation with local realities.

3.1 Technological Accessibility

Technological accessibility is the foundation of affordable AI deployment. AI systems provide predictive safety insights, real-time monitoring, and automated control, but proprietary platforms, high-speed connectivity, and cloud infrastructure often exceed the budgets of LMICs. Ensuring accessibility involves reconfiguring hardware and software to maximize local feasibility.

3.1.1 Low-Cost IoT Devices

Affordable IoT sensors capable of measuring vehicle speed, traffic volume, and environmental conditions are now widely available. Devices such as Arduino-based traffic counters or Raspberry Pi microcomputers can be locally assembled, creating cost savings and employment opportunities (Chen et al., 2022).

3.1.2 Solar-Powered Sensors

Unreliable electricity grids in LMICs pose a major challenge. Solar-powered sensors and street devices enable continuous operation in urban and rural contexts. For example, solar-powered traffic counters installed along Kenyan highways function without connection to the electrical grid, ensuring uninterrupted data collection.

3.1.3 Open-Source AI Models

Open-source frameworks like YOLO (You Only Look Once), TensorFlow Lite, and PyTorch Mobile democratize access to advanced analytics. These models can be fine-tuned with local traffic datasets, reducing dependence on costly proprietary systems. Municipalities can adapt codebases to regional conditions, enhancing affordability and flexibility.

3.1.4 Edge Computing

Processing data locally at the device level minimizes bandwidth costs, reduces latency, and increases resilience in low-connectivity areas. Edge computing allows CCTV cameras or IoT nodes to analyze video or sensor data in real-time, transmitting only essential alerts to central servers.

3.2 Context-Sensitive Design

Technology must align with social, cultural, and environmental contexts. A technically advanced system that is socially misaligned may face resistance or underutilization. Context-sensitive design ensures AI infrastructures are embedded, adaptive, and locally relevant.

3.2.1 Use of Local Materials

Instead of importing expensive mounts and hardware, locally available materials such as bamboo poles, recycled plastics, or reclaimed metals can be used for installing IoT devices or traffic signs. This approach reduces costs while promoting environmental sustainability.

3.2.2 Cultural Compatibility

In regions dominated by informal transport modes—tuk-tuks, boda-bodas, or motorcycle delivery services—AI systems must prioritize vulnerable users rather than focusing solely on private cars. For instance, pedestrian crossings can be augmented with AI-enabled alert systems that detect non-motorized traffic.

3.2.3 Community-Based Approaches

Engaging local communities during planning, deployment, and monitoring stages increases acceptance. Mobile reporting apps can allow citizens to submit hazards, which AI systems aggregate to create dashboards for municipal decision-making. Community participation also promotes shared responsibility and reduces dependence on central authorities.

3.2.4 Environmental Adaptation

Sensors and devices must withstand local climatic conditions such as monsoon rains, high humidity, dust, or extreme temperatures. Water-resistant enclosures, dust-proof cameras, and solar-powered energy systems enhance durability and long-term functionality.

3.3 Scalability and Modularity

Scalability ensures that safety systems remain sustainable as they expand, while modularity allows phased growth and flexibility. These principles prevent pilot projects from becoming financially or operationally unsustainable.

3.3.1 Phased Implementation

AI-driven systems can be introduced in high-risk corridors or intersections with high accident rates. Lessons learned during initial deployment guide citywide or regional expansion, minimizing financial risk and implementation errors.

3.3.2 Plug-and-Play Components

Modular designs allow components to be added or upgraded without disrupting the larger network. For example, a basic traffic monitoring sensor can later incorporate AI video analytics, weather detection, or integration with centralized dashboards.

3.3.3 Hybrid Integration

AI systems should coexist with low-tech interventions like speed bumps, painted pedestrian crossings, rumble strips, and reflective signage. This approach allows for incremental technology adoption while preserving familiar safety mechanisms.

3.3.4 Scalable Financing Models

Incremental budgets, micro-financing schemes, and public–private partnerships (PPPs) enable sustainable expansion. Phased financing reduces upfront costs while ensuring long-term viability.

Scalability and modularity transform affordability from short-term cost reduction into a pathway for systemic growth, resilience, and long-term effectiveness.

3.4 Governance and Policy Integration

Institutional support is essential for sustainable AI adoption. Governance and policy integration ensure that infrastructures are not isolated experiments but integral to national and municipal transportation strategies.

3.4.1 Public–Private Partnerships (PPPs)

Collaboration between governments, private companies, and community organizations distributes financial risk, enhances innovation, and ensures maintenance. For instance, telecom operators can provide connectivity in exchange for access to anonymized traffic data.

3.4.2 Incentives for Innovation Ecosystems

Universities, research institutes, and start-ups can contextualize AI tools. Governments can incentivize localized innovation through funding for AI labs, hackathons, pilot programs, and technology transfer initiatives.

3.4.3 Regulatory Frameworks

Policies addressing ethical AI use, data privacy, and interoperability standards prevent misuse, build public trust, and ensure equitable access. Clear governance structures define responsibilities for monitoring, evaluation, and enforcement.

3.4.4 Capacity Building

Training municipal staff, engineers, and community volunteers in AI literacy ensures sustainable operation and maintenance. This reduces dependency on external consultants and strengthens local expertise.

Governance and policy integration transform technological interventions into systemic, socially accountable, and sustainable infrastructures.

3.5 Synthesis of the Framework

The four pillars—Technological Accessibility, Context-Sensitive Design, Scalability and Modularity, and Governance and Policy Integration—together provide a holistic blueprint for designing affordable AI-driven transportation safety systems. The framework prioritizes affordability as a principle of design, not merely a constraint, and emphasizes inclusivity, adaptability, and resilience.

By following this framework, municipalities and policymakers can implement AI solutions that are technically effective, economically viable, socially inclusive, and contextually adaptive, even in regions with limited resources.

4. METHODOLOGY

The development of this chapter is grounded in a mixed-method research approach, combining qualitative and quantitative methods to explore affordable AI-driven transportation safety infrastructures in resource-constrained regions. The methodology is structured into four interrelated components: literature review, comparative case studies, framework development, and stakeholder validation. Each component contributes to building a holistic and empirically grounded conceptual framework.

4.1 Literature Review

The literature review provided a theoretical foundation for identifying affordability as a critical principle in AI-driven transportation safety. A systematic review was conducted across multiple domains: transportation engineering, urban studies, computer science, and public policy.

4.1.1 Scope of Review

Key sources included:

- Peer-reviewed journals (e.g., *Transportation Research Part C: Emerging Technologies*, *Journal of Urban Technology*, *IEEE Transactions on Intelligent Transportation Systems*).
- Technical reports from global organizations (World Health Organization, World Bank, International Transport Forum).
- Policy briefs, conference proceedings, and white papers on AI, IoT, and smart transportation infrastructures.

4.1.2 Key Themes Identified

The review highlighted several recurring themes:

1. Barriers to AI Adoption: Financial limitations, technical skill gaps, and institutional weaknesses in LMICs hinder large-scale deployment.
2. Global Precedents in Low-Cost AI Applications: Case studies from India, Kenya, Brazil, and other regions demonstrate adaptive strategies using low-cost hardware, open-source software, and decentralized processing.
3. Best Practices in Modular and Incremental Deployment: Systems designed with plug-and-play components and phased scaling reduce risk and improve sustainability.
4. Ethical and Social Considerations: Equity, privacy, algorithmic bias, and community participation are crucial for socially acceptable AI deployment.

The literature review established the need for a framework that balances affordability, technological feasibility, and contextual adaptation.

4.2 Comparative Case Studies

To contextualize global precedents and operationalize insights from the literature, three pilot interventions were selected for comparative analysis:

4.2.1 Case 1: India – Low-Cost Computer Vision for Traffic Monitoring

- **Context:** Mid-sized Indian cities (Indore, Nagpur, Lucknow) with high-density, mixed traffic.
- **Intervention:** Low-cost CCTV cameras paired with open-source AI (YOLO, OpenCV) and edge computing (Raspberry Pi, NVIDIA Jetson Nano) for real-time traffic monitoring.
- **Outcomes:** Identification of accident-prone intersections, adaptive traffic signal optimization, and alerts for traffic police.
- **Challenges:** Power outages, intermittent hardware failures.

4.2.2 Case 2: Kenya – Solar-Powered IoT Road Sensors

- **Context:** Highways connecting rural towns and agricultural regions; limited urban congestion in major cities.
- **Intervention:** Locally manufactured solar-powered IoT sensors measuring traffic volume, speed, and environmental conditions; AI-driven predictive models for accident risk forecasting.
- **Outcomes:** Improved emergency response, low-cost operations, community involvement in maintenance.
- **Challenges:** Initial training requirements for local technicians.

4.2.3 Case 3: Brazil – Community Data Platforms

- **Context:** Cities such as São Paulo and Recife; informal settlements, high motorcycle use, limited municipal budgets.
- **Intervention:** Mobile applications for citizen hazard reporting; AI-based aggregation using natural language processing (NLP) and clustering algorithms.
- **Outcomes:** Low infrastructure cost, citizen empowerment, real-time responsiveness.
- **Challenges:** Report verification, digital inclusion, equitable participation.

4.3 Framework Development

The conceptual framework presented in Section 3 emerged from synthesizing insights from the literature review and case studies. The development process included:

1. **Identification of Core Principles:** Affordability, inclusivity, modularity, and context sensitivity were extracted as recurring success factors.

2. **Mapping Technology to Local Needs:** Low-cost IoT devices, solar power, edge computing, and open-source AI models were aligned with infrastructural and financial realities.
3. **Integration into Four Pillars:** Principles were organized under Technological Accessibility, Context-Sensitive Design, Scalability and Modularity, and Governance and Policy Integration.
4. **Iterative Refinement:** Preliminary versions of the framework were refined through expert feedback and validation exercises.

4.4 Stakeholder Validation

To ensure that the framework was grounded in practical realities, qualitative validation was conducted through semi-structured interviews and focus group discussions with multiple stakeholders:

4.4.1 Participants

- **Urban Planners:** Provided insights on integrating AI into broader urban mobility strategies.
- **Government Transport Agencies:** Shared perspectives on policy alignment, budget limitations, inter-agency coordination, and regulatory compliance.
- **Non-Governmental Organizations (NGOs):** Emphasized community engagement, equity, and inclusivity in AI deployment.
- **Local Technicians and End-Users:** Provided operational insights on maintainability, usability, and local adaptation.

4.4.2 Validation Process

Step 1: Presentation of preliminary framework to stakeholders.
Step 2: Collection of qualitative feedback on feasibility, contextual relevance, and scalability.
Step 3: Thematic analysis of feedback to refine framework components.
Step 4: Incorporation of recommendations into the final four-pillar model.

4.5 Summary

The mixed-method approach—combining literature review, comparative case studies, framework synthesis, and stakeholder validation—provides a robust basis for designing AI-driven transportation safety infrastructures in resource-constrained regions. By triangulating theoretical insights, empirical data, and practical feedback, the methodology ensures that the resulting framework is both evidence-based and context-sensitive, capable of guiding policymakers, urban planners, and communities in deploying affordable, sustainable, and adaptive AI interventions.

5. CASE DISCUSSIONS

This section examines three real-world pilot projects—India, Kenya, and Brazil—illustrating how AI-driven infrastructures can be deployed affordably in resource-constrained contexts. Each case highlights distinct approaches, technological adaptations, and social strategies, providing empirical support for the conceptual framework proposed in Section 3.

5.1 Case 1: India – Low-Cost Computer Vision for Traffic Monitoring

5.1.1 Context and Challenges

India's urban transport system presents a complex ecosystem characterized by:

- High vehicle density, including cars, motorcycles, buses, and auto-rickshaws.
- Mixed traffic with pedestrians and informal transport modes.
- Rapid urbanization with outdated road infrastructure and inadequate enforcement.

In mid-sized cities such as Indore, Nagpur, and Lucknow, traffic accidents frequently occur at intersections due to poor signal timing, lack of pedestrian infrastructure, and inconsistent policing. Traditional AI-driven traffic systems are often financially prohibitive due to high hardware and software costs.

5.1.2 Intervention Design

Hardware and Software

- **CCTV Cameras:** Inexpensive cameras sourced from local markets, capable of capturing 720p–1080p video.
- **Edge Computing Devices:** Raspberry Pi 4 and NVIDIA Jetson Nano units process video locally to detect violations in real time.
- **AI Models:**
  - YOLOv5 for object detection (vehicles, pedestrians, two-wheelers).
  - OpenCV for image preprocessing and vehicle tracking.

- **Data Processing:** Frames processed on edge devices reduce bandwidth dependency. Aggregated alerts sent to traffic police dashboards.

System Architecture

1. Cameras capture live video streams.
2. Edge devices analyze frames using AI models.
3. Detected violations generate real-time alerts.
4. Data aggregated to identify accident-prone zones and optimize traffic signals.

5.1.3 Outcomes

1. **Accident-Prone Junction Identification:** Data-driven identification of high-risk intersections informed interventions such as signal timing adjustments and pedestrian crosswalk installation.
2. **Adaptive Traffic Signals:** AI-enabled systems adjusted green light duration based on real-time congestion.
3. **Cost-Effectiveness:** Affordable hardware and open-source software enabled scalable deployment within municipal budgets.

5.1.4 Challenges

- **Power Supply Interruptions:** Frequent outages affected continuous operation.
- **Hardware Failures:** Low-cost cameras occasionally malfunctioned.
- **Skill Gaps:** Limited local expertise for maintaining AI systems required capacity-building initiatives.

5.2 Case 2: Kenya – Solar-Powered IoT Road Sensors

5.2.1 Context and Challenges

Kenya's road network is characterized by:

- Highways linking rural towns and agricultural regions.
- Limited night-time visibility and challenging weather conditions.
- Sparse enforcement and slow emergency response in rural corridors.

The Nairobi–Nakuru corridor was selected for piloting affordable AI-enabled interventions to enhance safety and reduce fatalities.

5.2.2 Intervention Design

Hardware and Software

- **Solar-Powered IoT Sensors:** Locally manufactured sensors capture vehicle counts, speed, temperature, and rainfall.
- **Connectivity:** GSM modules transmit data to regional traffic monitoring centers.
- **AI Models:**
  - o Predictive models analyze historical accident data combined with real-time sensor input.
  - o Statistical and machine learning models forecast high-risk periods (e.g., post-rain speeding spikes).

System Architecture

1. Sensors collect traffic and environmental data.
2. AI models run on local gateways or cloud platforms.
3. Predictive alerts sent to police units or roadside digital boards.
4. Data used for emergency response optimization and risk mapping.

5.2.3 Outcomes

1. **Improved Emergency Response:** AI-enabled alerts allowed responders to anticipate high-risk zones, reducing ambulance dispatch times.
2. **Cost Reduction:** Solar power eliminated dependence on grid electricity. Local manufacturing ensured low unit costs and accessible spare parts.
3. **Community Engagement:** Local technicians trained for sensor maintenance ensured long-term sustainability.

5.2.4 Challenges

- **Initial Training Requirements:** Technicians needed technical skills for maintaining sensors.
- **Environmental Factors:** Sensors required weatherproofing and dust-resistant casings.
- **Data Accuracy:** Ensuring reliable transmission in rural regions with intermittent GSM coverage.

5.3 Case 3: Brazil – Community Data Platforms

5.3.1 Context and Challenges

Brazilian cities, particularly São Paulo and Recife, face transportation challenges including:

- Informal urban settlements (favelas).
- Rapid growth of motorcycle-based delivery services.
- Limited municipal budgets prioritizing public transit over advanced monitoring systems.

Traditional traffic management infrastructures are financially unviable in these regions, necessitating innovative, community-driven AI solutions.

5.3.2 Intervention Design

Mobile Applications

- Citizen-accessible apps on low-cost smartphones allowed hazard reporting.
- Reports were geotagged and sent to a central server.

AI Integration

- **Natural Language Processing (NLP):** Filtered duplicate reports and categorized hazards.
- **Clustering Algorithms:** Identified high-risk zones based on report frequency and severity.
- **Dashboards:** Municipal authorities accessed real-time hazard maps for deployment of maintenance crews and enforcement officers.

5.3.3 Outcomes

1. **Low Infrastructure Cost:** Minimal reliance on physical infrastructure; leveraged widespread smartphone penetration.
2. **Citizen Empowerment:** Residents actively participated in identifying hazards, fostering shared responsibility.
3. **Real-Time Responsiveness:** AI detected sudden spikes in hazard reporting, enabling timely interventions during events such as heavy rainfall.

5.3.4 Challenges

- **Report Verification:** Ensuring accuracy and preventing misuse of reporting platforms.
- **Digital Inclusion:** Unequal access to smartphones and mobile internet required supplementary reporting strategies.
- **Sustainability:** Continuous engagement with communities needed to maintain reporting activity.

5.4 Cross-Case Analysis

Across India, Kenya, and Brazil, several common insights emerge:

1. **Affordability through Local Adaptation:** Low-cost hardware, solar power, and community participation enable AI systems to function in resource-constrained contexts.
2. **Hybrid Approaches:** Combining low-tech interventions (speed bumps, signage) with AI analytics increases adoption and reduces risks.
3. **Community Integration:** Citizen engagement and local technician training ensure sustainability beyond initial pilot phases.
4. **Context Sensitivity:** AI systems must account for local traffic patterns, cultural norms, and environmental conditions.

6. CHALLENGES IN IMPLEMENTATION

While AI-driven transportation safety interventions offer transformative potential, practical deployment in resource-constrained regions is fraught with challenges. These challenges span financial limitations, technical skill gaps, infrastructural deficiencies, and social-ethical concerns. Understanding these barriers is critical to designing scalable, sustainable, and context-sensitive solutions.

6.1 Financial Constraints

Financial limitations represent one of the most significant obstacles to AI deployment in transportation safety. Key aspects include:

6.1.1 Limited Budgetary Allocations

- Many low- and middle-income countries (LMICs) allocate modest portions of municipal budgets to transportation infrastructure. Priority is often given to road expansion rather than safety enhancements.
- Capital-intensive systems such as high-definition CCTV networks, AI-enabled traffic management centers, or autonomous monitoring units are often beyond reach.

6.1.2 Sustainability of Donor-Funded Pilots

- While donor-funded pilot projects introduce innovative solutions, long-term sustainability is uncertain once external funding ends.
- Pilots risk discontinuation without financial plans for maintenance, hardware replacement, or software updates.

6.1.3 Mitigation Strategies

1. **Phased Implementation:** Deploy AI systems incrementally in high-risk corridors to reduce upfront costs.
2. **Open-Source Platforms:** Using open-source AI reduces software licensing expenses and fosters local innovation.
3. **Public–Private Partnerships (PPPs):** Shared investment between governments and private sector reduces financial burden and incentivizes innovation.
4. **Community Engagement:** Crowdsourced funding or local participation can supplement municipal budgets.

6.2 Technical Skill Gaps

AI-driven infrastructures require specialized expertise that is often scarce in resource-limited regions.

6.2.1 Human Capital Limitations

- Shortage of data scientists, AI engineers, transportation planners, and IT support personnel.
- Over-reliance on external consultants introduces operational vulnerabilities and increases costs.

6.2.2 Maintenance and Operations Challenges

- Imported hardware requires specialized maintenance.
- Software updates, model retraining, and system calibration demand technical knowledge that is frequently unavailable locally.

6.2.3 Mitigation Strategies

1. **Capacity Building Programs:** Train local engineers, municipal staff, and community volunteers in AI literacy and basic maintenance skills.
2. **Local Innovation Hubs:** Foster collaboration between universities, start-ups, and municipal agencies to develop context-specific expertise.
3. **Simplified System Designs:** Use modular, plug-and-play hardware and AI models that are easier to maintain with minimal technical intervention.

6.3 Infrastructure Gaps

The functionality of AI systems is closely tied to the availability of reliable electricity, internet connectivity, and data storage. Resource-constrained contexts often face significant limitations in these domains.

6.3.1 Power Supply Issues

- Frequent outages disrupt continuous operation of traffic cameras, IoT sensors, and edge devices.
- Backup power solutions (batteries, solar panels) increase upfront costs but are essential for system resilience.

6.3.2 Connectivity Challenges

- Low broadband penetration and intermittent internet access hinder centralized AI processing and cloud-based analytics.
- Remote areas may rely on GSM modules or offline edge computing, which introduces limitations in real-time analytics.

6.3.3 Data Storage and Security Concerns

- Limited availability of secure data storage infrastructure raises risks for data loss or breaches.

- Inadequate policies on data privacy and access control may undermine public trust in AI systems.

6.3.4 Mitigation Strategies

1. **Edge Computing:** Local data processing reduces reliance on continuous internet connectivity.
2. **Renewable Energy Integration:** Solar-powered sensors and backup batteries mitigate electricity interruptions.
3. **Decentralized Storage Solutions:** Use local storage nodes with redundancy to ensure data integrity and accessibility.

6.4 Ethical and Social Concerns

The deployment of AI systems in public spaces raises significant ethical and social challenges.

6.4.1 Algorithmic Bias

- AI models trained on limited or non-representative data may underreport risks for vulnerable road users such as pedestrians, cyclists, and motorcyclists.
- Bias can exacerbate social inequalities if certain communities receive less attention in traffic safety interventions.

6.4.2 Surveillance and Privacy Concerns

- Continuous monitoring may be perceived as intrusive, particularly in residential areas.
- Lack of transparent policies on data collection and usage erodes public trust.

6.4.3 Equitable Access and Inclusion

- Mobile-based or IoT-enabled systems may exclude communities lacking smartphones or digital literacy.
- Citizen engagement platforms require deliberate strategies to ensure representation across socio-economic groups.

6.4.4 Mitigation Strategies

1. **Ethical AI Guidelines:** Establish policies for fair, transparent, and accountable use of AI in public infrastructure.
2. **Inclusive Design:** Ensure that AI systems and reporting platforms are accessible to marginalized communities.
3. **Community Participation:** Engage residents in the design, deployment, and monitoring phases to build trust and legitimacy.
4. **Regular Model Audits:** Continuously evaluate AI algorithms for bias, accuracy, and fairness.

6.5 Summary

The challenges outlined—financial, technical, infrastructural, and ethical—underscore the complexity of implementing AI-driven transportation safety systems in resource-constrained regions. While these barriers are significant, they are not insurmountable. Strategies such as modular design, local capacity building, edge computing, renewable energy integration, community participation, and ethical governance provide pathways to overcome these challenges.

By addressing these constraints proactively, policymakers, municipal authorities, and communities can design AI-driven infrastructures that are affordable, resilient, inclusive, and contextually adaptive, laying the groundwork for safer and more sustainable transportation systems.

7. RECOMMENDATIONS FOR SUSTAINABLE DEPLOYMENT

The successful deployment of AI-driven transportation safety infrastructures in resource-constrained regions requires strategies that are technologically viable, socially inclusive, economically sustainable, and institutionally supported. Drawing on lessons from India, Kenya, and Brazil, as well as the conceptual framework outlined in Section 3, this section presents practical recommendations for achieving sustainable AI adoption.

7.1 Promote Open-Source Ecosystems

Open-source AI frameworks and software platforms reduce dependency on costly proprietary solutions while fostering local innovation.

7.1.1 Benefits

- **Cost Reduction:** Eliminates software licensing fees, lowering the financial barrier to entry.
- **Customizability:** Local developers can adapt models to regional traffic patterns, road conditions, and user behaviors.
- **Community Collaboration:** Global developer networks support troubleshooting, model improvement, and knowledge sharing.

7.1.2 Implementation Strategies

1. **Adopt Open-Source Models:** Use frameworks such as TensorFlow Lite, PyTorch Mobile, and YOLO for traffic monitoring and predictive analytics.
2. **Contribute to Local Knowledge Hubs:** Encourage municipal and academic institutions to document AI adaptations and share best practices.
3. **Integrate with Edge Computing:** Open-source software can be deployed on low-cost edge devices to enable decentralized processing.

Example: In India, YOLO-based object detection was deployed on Raspberry Pi devices, combining open-source software with affordable hardware to monitor traffic intersections effectively.

7.2 Establish Local Innovation Hubs

Local innovation hubs serve as incubators for context-specific AI solutions, linking universities, start-ups, and municipal authorities.

7.2.1 Benefits

- Encourages experimentation with low-cost AI applications tailored to local conditions.
- Creates local employment opportunities in AI development, data management, and system maintenance.
- Strengthens knowledge transfer between research institutions and government agencies.

7.2.2 Implementation Strategies

1. **University–Municipal Partnerships:** Establish collaborative labs for prototyping AI-driven traffic solutions.

2. **Hackathons and Innovation Challenges:** Invite students, start-ups, and local entrepreneurs to design affordable safety interventions.
3. **Technology Demonstration Zones:** Use pilot corridors to test and refine AI solutions before citywide or regional deployment.

Example: Kenya's solar-powered IoT sensor pilots involved local technicians in both assembly and deployment, ensuring technical knowledge remained within the community.

7.3 Capacity Building

Human capital development is crucial for sustainable AI deployment. Training programs should target engineers, municipal staff, and community volunteers.

7.3.1 Key Components

- **Technical Training:** System installation, AI model maintenance, sensor calibration, and edge computing operations.
- **Policy and Management Training:** Traffic management planning, data governance, and regulatory compliance.
- **Community Training:** Engaging residents in hazard reporting and participatory monitoring platforms.

7.3.2 Implementation Strategies

1. **Modular Training Programs:** Offer tiered courses for technical staff, policy makers, and community participants.
2. **On-the-Job Learning:** Embed training within pilot projects for hands-on experience.
3. **Certification Programs:** Establish local accreditation to recognize expertise in AI-enabled transportation safety.

Example: In Brazil, citizen-focused hazard reporting platforms were paired with training workshops to teach users how to submit accurate reports, improving both participation and data reliability.

7.4 Phased Policy Implementation

AI-driven safety infrastructures should be deployed gradually, prioritizing high-risk areas and iteratively scaling based on performance.

7.4.1 Benefits

- Minimizes financial risks by spreading costs over multiple phases.
- Allows iterative learning to refine AI models, operational protocols, and community engagement strategies.
- Enhances institutional capacity, ensuring municipal agencies adapt to new technologies effectively.

7.4.2 Implementation Strategies

1. **Pilot High-Risk Corridors:** Start with intersections, highways, or neighborhoods with high accident rates.
2. **Data-Driven Expansion:** Use performance metrics to guide scaling decisions, including hardware deployment and staffing.
3. **Feedback Loops:** Regularly evaluate AI system performance and incorporate community feedback into design improvements.

Example: India's low-cost computer vision pilot began with a few high-incident intersections before expanding to multiple mid-sized cities.

7.5 Hybrid Models

Hybrid approaches combine traditional low-tech safety interventions with AI-driven enhancements, balancing cost, effectiveness, and scalability.

7.5.1 Components of Hybrid Models

- **Low-Tech Measures:** Painted crosswalks, rumble strips, speed breakers, signage, and reflective road markings.
- **AI Enhancements:** Predictive analytics, adaptive traffic signals, automated violation detection, and hazard prioritization.
- **Community Integration:** Citizen reporting and participatory monitoring feed data into AI models for real-time responsiveness.

7.5.2 Implementation Strategies

1. **Assess Existing Infrastructure:** Identify which low-tech interventions are already in place and can complement AI systems.
2. **Integrate Gradually:** Introduce AI-driven features without replacing effective low-tech measures.

3. **Measure Impact:** Monitor safety outcomes and cost-effectiveness to optimize the hybrid system.

Example: Kenya's solar-powered IoT sensors augmented traditional roadside signage and enforcement practices, creating a combined technological and low-tech solution.

7.6 Governance and Policy Integration

Strong governance frameworks are essential for ensuring accountability, ethical AI use, and sustainability.

7.6.1 Policy Recommendations

- **Public–Private Partnerships (PPPs):** Encourage collaboration between municipal authorities, private technology providers, and community organizations.
- **Ethical Guidelines:** Establish frameworks for data privacy, algorithmic transparency, and equitable treatment of vulnerable road users.
- **Regulatory Compliance:** Align AI deployment with national and regional transportation safety policies.
- **Institutional Support:** Develop dedicated municipal units responsible for AI system oversight, maintenance, and policy integration.

7.6.2 Implementation Strategies

1. **Incentivize Innovation Ecosystems:** Offer grants, tax incentives, or recognition programs for AI labs, start-ups, and research institutes focusing on affordable traffic safety.
2. **Data Governance Frameworks:** Ensure collected data is securely stored, anonymized, and used in line with privacy laws.
3. **Capacity for Policy Enforcement:** Train municipal staff to monitor AI system compliance and respond to ethical or technical issues.

Example: Brazil's municipal authorities created dashboards that integrated citizen-reported data with official enforcement policies, allowing for evidence-based allocation of resources.

7.7 Summary

Sustainable deployment of AI-driven transportation safety systems requires multi-dimensional strategies that combine technology, community participation, phased planning, and robust governance. Key recommendations include:

1. **Open-Source Ecosystems** – reduce costs and encourage local innovation.
2. **Local Innovation Hubs** – foster experimentation and technical skill development.
3. **Capacity Building** – train staff and communities to sustain AI systems.
4. **Phased Policy Implementation** – deploy incrementally to minimize risks and optimize learning.
5. **Hybrid Models** – combine low-tech safety interventions with AI enhancements.
6. **Governance and Policy Integration** – ensure ethical, equitable, and sustainable AI deployment.

By following these recommendations, resource-constrained regions can implement AI-driven infrastructures that are affordable, adaptive, inclusive, and future-ready, ultimately reducing road fatalities, enhancing mobility, and improving quality of life.

8. CONCLUSION

Transportation safety in resource-constrained regions presents a multifaceted challenge, encompassing technical, financial, institutional, and social dimensions. Traditional approaches, which rely on high-cost, capital-intensive systems, are often inaccessible to low- and middle-income countries, leaving vulnerable populations—pedestrians, cyclists, and motorcyclists—at disproportionate risk.

This chapter has argued that affordability need not be a limitation but can serve as a design principle, guiding the development of AI-driven infrastructures that are both effective and inclusive. By integrating technological accessibility, context-sensitive design, scalability and modularity, and governance and policy alignment, resource-limited regions can implement solutions that balance innovation with practical feasibility.

8.1 Key Insights

8.1.1 Affordability as a Catalyst for Innovation

- Affordability compels the design of modular, context-aware systems rather than wholesale adoption of high-end technologies.

- Low-cost hardware, open-source AI software, decentralized edge processing, and locally sourced materials collectively reduce costs while maintaining functionality.
- Case studies in India, Kenya, and Brazil illustrate how affordability can drive creativity, community engagement, and sustainable technology adoption.

8.1.2 Context-Sensitive Approaches are Essential

- AI systems must be aligned with social, cultural, and environmental realities.
- Community participation, local manufacturing, and climate-adapted designs ensure that technology is usable, accepted, and resilient.
- Context-sensitive strategies enable interventions that prioritize vulnerable road users, integrate local knowledge, and reflect region-specific mobility patterns.

8.1.3 Phased and Modular Implementation Enhances Sustainability

- Incremental deployment allows municipalities to test, refine, and scale interventions without overwhelming limited budgets or institutional capacities.
- Modular AI systems, plug-and-play hardware, and hybrid approaches combining low-tech and AI-enabled measures ensure gradual adoption and long-term sustainability.

8.1.4 Governance and Policy Integration Strengthen Impact

- Successful deployment requires coordinated policy frameworks, ethical AI guidelines, and institutional capacity building.
- Public–private partnerships, community engagement, and incentive structures facilitate innovation while ensuring accountability.
- Governance structures provide the backbone for sustainable AI-driven transportation safety infrastructures.

8.2 Implications for Global Practice

While the chapter focuses on Asia, Africa, and Latin America, the lessons are globally applicable, particularly in regions where financial, technical, and institutional constraints are significant. Key implications include:

1. **Scalability Beyond Pilots:** Lessons from small-scale pilots can inform national strategies for urban and rural traffic safety.

2. **Democratization of Technology:** Open-source platforms and locally assembled devices reduce dependence on foreign vendors and democratize AI innovation.
3. **Community-Centered Design:** Citizen engagement platforms, hazard reporting apps, and participatory monitoring systems promote shared ownership of transportation safety initiatives.
4. **Resilience and Equity:** Affordable, adaptive systems prioritize both technical performance and social inclusivity, reducing disparities in access to safe mobility.

8.3 Future Directions

The future of AI-driven transportation safety in resource-constrained contexts lies in continuous innovation, cross-sector collaboration, and policy evolution:

- **Integration of Emerging Technologies:** Machine learning, computer vision, IoT, and predictive analytics will evolve to provide more precise, context-aware interventions.
- **Global Knowledge Sharing:** Open-source communities and international collaborations can accelerate the dissemination of affordable AI solutions.
- **Adaptive Policy Frameworks:** Governments and municipalities will need to refine regulations to ensure ethical, equitable, and sustainable AI deployment.
- **Holistic Safety Ecosystems:** AI-driven systems should complement low-tech safety measures, emergency response frameworks, and urban planning strategies to create integrated safety ecosystems.

8.4 Final Reflection

This chapter demonstrates that AI-driven transportation safety is not the exclusive domain of high-income countries. By embracing affordability as a core principle, regions with limited resources can develop infrastructures that are technologically intelligent, socially inclusive, and economically sustainable.

The four-pillar framework—Technological Accessibility, Context-Sensitive Design, Scalability and Modularity, and Governance and Policy Integration—serves as a roadmap for policymakers, urban planners, researchers, and communities seeking to implement AI-driven solutions that save lives, enhance mobility, and foster sustainable development.

Ultimately, the chapter advocates for a paradigm shift: moving away from imported, high-cost systems toward context-driven, community-embedded, and adaptable AI infrastructures. In doing so, resource-constrained regions can create equitable, resilient, and future-ready transportation networks, ensuring that the benefits of AI innovation are shared by all segments of society, particularly those most at risk.

REFERENCES

Ajzen, I. (1991). The theory of planned behavior. *Organizational Behavior and Human Decision Processes*, *50*(2), 179–211. DOI: 10.1016/0749-5978(91)90020-T

Chen, C., Wang, Y., Li, L., Hu, J., & Zhang, Z. (2022). Artificial intelligence-enabled smart transportation systems: A review. *IEEE Transactions on Intelligent Transportation Systems*, *23*(5), 1–16. DOI: 10.1109/TITS.2021.3089836

Dann, G. M. S. (1977). Anomie, ego-enhancement and tourism. *Annals of Tourism Research*, *4*(4), 184–194. DOI: 10.1016/0160-7383(77)90037-8

Gakenheimer, R. (2018). Land use and transport in rapidly motorizing cities. *Journal of Urban Technology*, *25*(1), 3–22. DOI: 10.1080/10630732.2017.1401139

Khattak, A. J., & Jovanis, P. P. (2021). Safety impacts of intelligent transportation systems. *Transportation Research Part C, Emerging Technologies*, *127*, 103114. DOI: 10.1016/j.trc.2021.103114

Liu, X., Wang, Y., & Zhang, L. (2021). Machine learning applications in road safety analysis: A systematic review. *Accident; Analysis and Prevention*, *151*, 105949. DOI: 10.1016/j.aap.2020.105949

National Transport and Safety Authority. (2021). *Road safety status report*. Government of Kenya.

Peden, M., Scurfield, R., Sleet, D., Mohan, D., Hyder, A. A., Jarawan, E., & Mathers, C. (2020). *World report on road traffic injury prevention*. World Health Organization.

Shinde, K. A. (2022). Religious tourism in Asia: Traditions, transformations and trajectories. *Tourism Geographies*, *24*(4–5), 645–661. DOI: 10.1080/14616688.2021.1967687

World Bank. (2019). *Guide for road safety opportunities and challenges: Low- and middle-income countries*. World Bank Group.

World Bank. (2022). *Transport for development: Global mobility and safety report*. World Bank Group.

World Health Organization. (2023). *Global status report on road safety 2023*. World Health Organization.

Yoon, Y., & Uysal, M. (2005). An examination of the effects of motivation and satisfaction on destination loyalty: A structural model. *Tourism Management*, *26*(1), 45–56. DOI: 10.1016/j.tourman.2003.08.016

Chapter 6
Autonomous and AI-Enabled Vehicles for Green Transportation Systems

Sitesh Kumar Singh
http://orcid.org/0000-0002-7108-0808
College of Engineering, National University of Science and Technology, Muscat, Oman

Jaishri Gothania
http://orcid.org/0000-0002-2656-642X
Lingaya's Vidyapeeth, Faridabad, India

Shyam Agasthya H. R.
http://orcid.org/0009-0001-9382-9971
College of Engineering, National University of Science and Technology, Muscat, Oman

ABSTRACT

This chapter reviews how AI-enabled and AI-autonomous vehicles will have a transformational impact on meeting sustainable transportation and low-carbon transportation demands. It looks at the examples in vehicle navigation, traffic prediction, energy-efficient propulsion, and fleet optimization, the examples coming forward that propose to cut down emissions, traffic efficiency, and urban mobility. The combination of AVs with smart city technology and communication with elements of V2V and V2I open communication are described, together with the issues of safety, compliance, data protection, and infrastructure preparedness. The examples of the autonomous electric bus, the autonomous taxi, and the AI-driven logistics show the

DOI: 10.4018/979-8-3373-3231-4.ch006

feasible results of sustainability. Future research directions are outlined at the end of the chapter and include a focus on scalable, ethical, and integrated solutions towards AI-driven transport systems for sustainable mobility in urban areas.

1. INTRODUCTION

The world transportation industry is currently experiencing a paradigm shift due to the twin imperatives of making transportation sustainable and technological enhancement. Since transport only accounts for a small portion of many countries, emissions accounting contributes nearly a quarter of the global carbon dioxide (CO_2) emissions and thus the necessity of the development of solutions which guarantee mobility efficiency and minimize greenhouse gas (GHG) emissions (International Energy Agency [IEA], 2023). Over the onset of the past decade, AI and AVs have been discussed as transformative technologies in this respect. They hold promises of reducing emissions, better energy use and for intelligent, sustainable mobility.

The availability of more intuitive machine learning models, computer-vision, and intelligent transport systems in the last ten years has provided a supportive background to autonomous and AI-powered transport solutions (Kellerman, 2022). Such systems are not limited to mere automation and instead use data to inform their decision to improve navigation, energy management and logistics operations. With cities around the globe increasingly under environmental pressures, AI-powered transport technologies can serve as a booster in the attempt to reach carbon neutrality targets and make cities traffic well again.

This chapter examines interconnections among autonomous mobility, AI, and sustainability. It identifies the contribution of AI-driven automobiles in emissions reduction, shows the various issues faced globally in reaching sustainable transportation, and shows objectives of the chapter. The discussion also appeals to practical understanding and examples to show how the AI-enabled cars can help build the low-carbon urban transport sustainability.

Figure 1. AI-Enabled and Autonomous Vehicles in Sustainable Urban Mobility

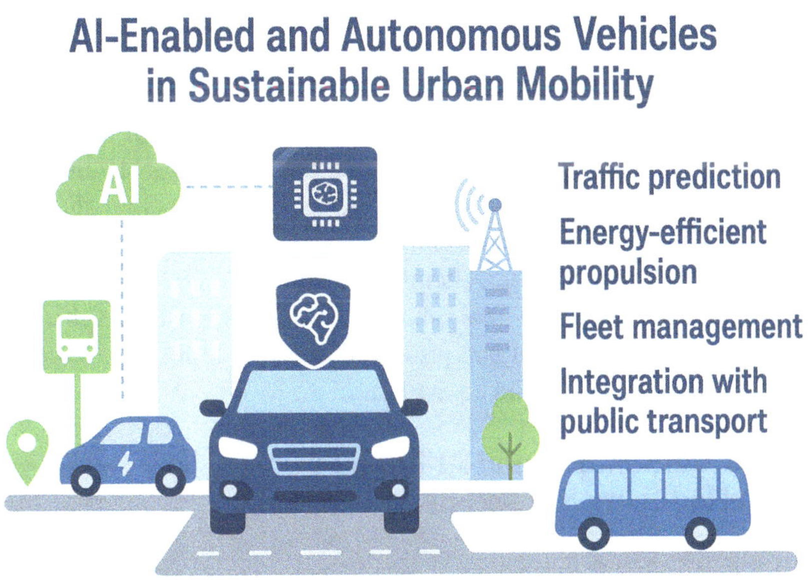

The concept of AI-enabled and autonomous vehicles/sustainable urban mobility is displayed in Figure 1. It illustrates vehicle-vehicle, vehicle-smart city infra and vehicle-AI technology interactions and identifies smart components cooperated in it as traffic prediction, energy-efficient propulsion, fleet management and public transport integration. This graphical summary illustrates the role of AI-based transport networks in decreasing emissions, optimizing traffic flows, and boosting energy consumption, which makes it possible to picture the discussions represented in the following sections of the chapter.

1.1 Global Sustainability Challenges in Transport

Transport is one of the biggest risk factors to GHG emissions with an estimated overall energy-related emission of CO_2 of 37% in 2022 (IEA, 2023). Road transport, specifically, takes the largest share, and both passenger cars and freight vehicles have a significant share in the use of energy and air pollution (The World Bank, 2025). In addition to climate effects, transportation also increases other challenges such as worsening air quality, urban congestion, land-use inefficiencies, and social disparities in mobility access.

Urbanization interest puts more strain on the sustainability objectives. As close to 70 percent of the global population is expected to live in cities by the year 2050, the sophistication of the transport services provided to them is also set to grow enormously, counterbalancing the benefits of decarbonization policies in case nothing changes (United Nations, 2022). The current mobility systems are mostly based on fossil fuels, and it is not well adapted at fulfilling this need, not to mention the fact that it will exacerbate the degradation of the environment.

Furthermore, the decentralization characteristic of the transport infrastructure e.g., transport infrastructure in developing economies poses a problem because it restricts the scalability of sustainable transport processes. Even in cities, there are no comprehensive public transport mechanisms, high-performance traffic management systems, or sustainable car EV use systems (Creutzig et al., 2021). These deficits underline the need to implement new solutions, such as AI-assisted vehicles and intelligent transport systems, that can provide decreasing levels of emissions, as well as resolve capacity, safety, and equity issues.

1.2 Role of Autonomous and AI-Enabled Vehicles in Reducing Emissions

AI-enabled and autonomous vehicles are promising contributions in mitigating most, as it increases efficiency and lets vehicles be less carbon-intensive. Even at the operational level, AI-powered navigation and traffic projections tools can save fuel by reducing idling as well as re-routing more efficient ways to bypass congestion (Bengler et al., 2014). Research indicates that an AI-based route planner can save 10 - 15 percent in fuel usage which translates into quantifiable carbon reduction (Kellerman, 2018).

Combining these with EV and hybrid propulsion systems further increases these values. In conjunction with the use of renewable energy sources, autonomous EV fleets can become a solution to near-zero emission transport in cities (Liao et al., 2021). Predictive maintenance can also be aided with AI algorithms, which, when vehicles are used, will ensure that they live as long as possible without replacing their battery.

In logistical processes, AI-enabled freight management says goodbye to previous load optimization and last-mile delivery minimizing the amount of vehicle kilometers traveled (VKT) and resulting emissions (Shuaibu et al., 2025). The case studies of the two major companies DHL and UPS show how machine learning has already helped minimize empty truck runs and increase delivery efficiency and subsequently driving down greenhouse gas emissions in the supply chain environment.

AVs also help to achieve safety and urban livability, in addition to emissions reduction. They minimize accelerating and braking traffic and allow the car to drive through more relaxed patterns work, decreasing particulate emission and noise pollution. The potential use of AVs in multimodal transport integration, namely, a connection with the public transportation and active mobility, also contributes to making cities more sustainable.

According to Singh, Alasali, and Belbase (2025), deployment of AI within the transportation planning process is essential to move towards low-emission mobility systems. The results confirm the importance of artificial neural networks in the optimization of traffic in the city, which is a major step to mitigating congestion-related emissions.

1.3 Chapter Objectives

This chapter will:

- Research the possibilities of autonomous and AI-based transportation that would encourage low-carbon and sustainable transportation.
- Investigate artificial intelligence related applications in vehicle navigation, traffic optimization and energy efficiency, and fleet management.
- Respond to fundamental questions of safety, regulation, infrastructure, and data ownership as part of AVs adoption.
- The current case studies showed that AI-enabled vehicles can be successfully integrated into urban mobility and logistics.
- Determine future research possibilities and roadmap on how AVs can be integrated within the sustainable transport network.

By integrating theory with practical and case-based research combined with critique, the chapter will manage to create an all-around understanding of how cars enabled with AI can be used to help sustain the world.

2. BACKGROUND AND RELATED WORK

The emergence of driverless vehicles (AVs) and transport systems enhanced through analytics intelligence (AI) can be listed as one of the most important new technological advances in transportation. These technologies have transformed themselves into practical solutions, which have already reproduced passenger and freight transport. This section gives a history of AV and AI in the applications in transportation, descriptions of similarities and differences between traditional and AI-based systems, and insight into the academic research on sustainability and low-carbon mobility research.

2.1 Evolution of Autonomous Vehicles and AI in Transportation

Autonomous driving technologies have a long history, as similar technologies were being developed in robotics and computer vision in the middle of the XX century. Subsequent missions, including the General Motors models of the Firebird concept cars of the 1950s and the Navlab project of Carnegie Mellon University in the 1980s, laid down the blocks of machine-perception-based movement (Anderson et al., 2016). Artificial intelligence has advanced because of machine learning and deep neural networks, which allow more advanced applications in sensing, perception, and decision-making in autonomous navigation (Badue et al., 2020).

During the 2000s, the Defense Advanced Research Projects Agency (DARPA) challenges became the driver, speeding up the work on AV prototypes with complex navigation in unstructured settings. Ever since, corporations like Waymo, Tesla, and Baidu have taken a dominant position in the market of commercially used autonomous driving systems based on LiDAR, radar, and image-based sensors coupled with an AI-based decision-making architecture (Fagnant & Kockelman, 2015).

AI is used to advance more than just autonomy. The next generation of analytical techniques, such as predictive analytics, swarm intelligence, and reinforcement learning, are currently being employed in getting the traffic systems into optimized form, improving safety, and cutting down on the emissions wrapped up in the context of cars or vehicles. This interconnection marks a greater paradigm shift, that is, human-controlled mobility to self-learning transportation systems to facilitate sustainability goals on the planetary level.

2.2 Traditional vs. AI-Enabled Transport Systems

Conventional transport systems are hugely dependent on human resources, fixed infrastructure and management as a reactive process. These systems are efficient as they offer fundamental mobility services but have drawbacks of inefficiency due to congestion, driver error, and lack of integration with environmental goals (Qin et al., 2022). Reliance on fossil fuels in traditional road transport is an added liability as it forms a significant part of emissions of GHGs and air pollution in cities.

Given the difference, real-time information on vehicles, infrastructures, and users is exploited by AI-enabled transport systems to streamline decision-making. New features like adaptive cruise control, automated lane keeping and predictive traffic routing are just a few examples of how machine intelligence can outperform human abilities in efficiency and safety. Contrary to the conventional systems, AI-based systems are foresighted being able to predict traffic congestion, idling time, and adhering to EV-driven propulsion (Pimenta et al., 2023).

Also, the merging of AI and electric and hybrid vehicle functions contribute to synergies enabling sustainability. By optimizing energy usage and allowing vehicle-to-grid connections, AI improves the eco-friendliness of electrification as well as alleviating the problems related to choosing the optimal time to charge an electric vehicle and coordinating charging stations on the demand-side of the spectrum.

2.3 Review of Literature on AVs for Sustainability and Low-Carbon Mobility

A review of the existing literature shows that there is increasing interest in the capability of AVs and AI-based transport systems to achieve low-emissions mobility. Studies indicate that AVs have a potential to curb the use of vehicle kilometers traveled (VKT) and fuel consumption by creating more eco-friendly movements on the road, vehicle platooning, and shared mobility model (Yigitcanlar et al., 2019). Such efficiencies can be converted to substantial decrease in emissions and enhancement of urban air quality.

Literature shows systemic consequences of AVs on urban planning as well. As another example, shared autonomous fleets with AI may make it possible to eliminate common vehicle ownership by occupying city spaces with verdant lands and lessening congestion (Golbabaei et al., 2020). Autonomous logistics system also enhances effectiveness in the freight, having the ability to reduce empty runs and maximize the efficiency of schedules, fostering sustainability in the supply chain (Coppola & Silvestri, 2019).

Scholars warn, however, of unintended consequences. The apparent increase in convenience of AVs is likely to cause induced demand, whereby less costly travel would lead to an increase in travel (Pimenta et al., 2023). Policy frameworks and governance structure are thus vital in supporting the implementation of AVs that are in line with the goals of sustainability.

Singh and Jain (2025) also note that the integration of AI in the domain of sustainable mobility is largely attributed to transport demand forecasting. Their activities demonstrate how the AI-powered forecasting could increase the effectiveness of designing road infrastructure in the long-term perspective to ensure decreased emissions and resources to use.

There is a cumulative emphasis in the literature that AI-enabled transport can be immensely beneficial on sustainability but only when this technology is adopted in a coordinated manner and when it is cross-indexed with other approaches to climate change management.

3. AI APPLICATIONS IN GREEN TRANSPORTATION SYSTEMS

Artificial intelligence (AI) has emerged as an all-important facilitator of green transportation systems. By utilizing data in real-time, machine intelligence, and automatized decision-making, AI technologies help in cutting emissions, improving energy efficiency, and rationalizing mobility processes. The following parts discuss the uses of AI in some of its most important applications, such as vehicle navigation and control, energy efficiency due to combining with an electric and hybrid propulsion, predicting and managing traffic, and optimizing fleets and logistics. Figure 2 presents the key AI deployed in sustainable transportation such as vehicle navigation and control applications, electric and hybrid propulsion energy-efficient applications, traffic prediction and management applications, and fleet and logistics optimization applications. The schematic displays the interaction between AI technologies and the interaction with vehicles, infrastructure, and data systems to deliver reduced congestion, emissions, and resources used more optimally. These applications as shown in Figure 2 form the backbone of mobility system platforms enabled by AI that have a direct result in environmental sustainability.

Figure 2. AI Applications in Green Transportation Systems

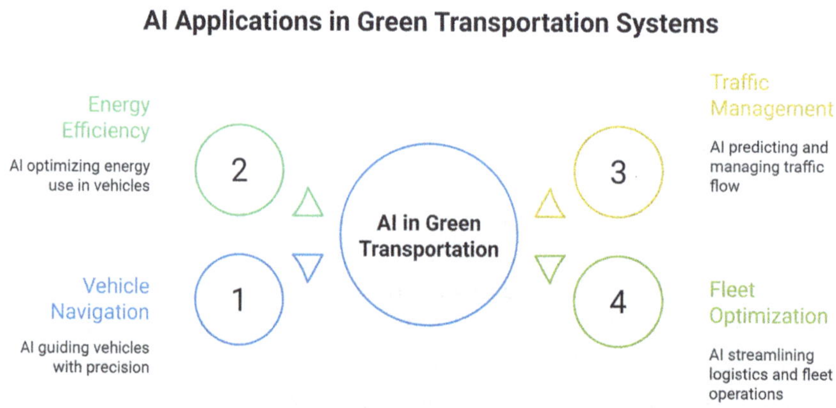

3.1 Vehicle Navigation and Control: AI for Route Optimization, Adaptive Cruise Control, and Traffic-Aware Navigation

Transport and logistics are also the spheres in which AI shows impressive ecological advantages. The AI instigated route optimization algorithms are used to minimize the consumption of unnecessary fuel and find the most efficient route possible so that the vehicles respond to the dynamics of traffic. In contrast to the traditional GPS, which uses data that is not constantly updated, AI-enhanced platforms have direct feeds of sensors, traffic cameras, and mobile devices to make sure that there is constant recalculation of routes (Dikshit et al., 2023).

Adaptive cruise control-(ACC) uses machine learning and sensor fusion to control the vehicle speed to ensure safe distances and this helps to reduce aggressive driving, which increases fuel consumption. Exported to fleet, ACC will minimize stop-and-go traffic and minimize fuel waste and greenhouse gas emissions (Jeng & Chiang, 2023). On the same note, the traffic-aware navigation also incorporates reinforcement learning into it to predict congestion patterns that can be used to avoid delays in times of travel as well as carbon footprints.

Cases with such companies as Tesla and Waymo demonstrate that AI-based navigation systems may do more than help a vehicle be even safer- they may actively contribute to the goals of sustainability by minimizing the use of vehicles at idle and allowing drivers to ride more smoothly. These systems are good fit for the giving-way trend of the global transition to the energy-efficient model of mobility (Driessen et al., 2024).

3.2 Energy Efficiency: Integration with Electric and Hybrid Propulsion Systems

The decarbonization of transport has the potential to be the most significant way forward and AI is a critical part in ensuring that the positive outcome of electric and hybrid drive is fully realized. Fuel efficiency in electric vehicles is also improved through energy management systems that are enabled by AI, which optimizes battery power, charges, braking, and regenerative braking (Rauf et al., 2024).

AI-powered battery management systems (BMS) determine energy requirements and dynamically distribute power among the electric engines and internal combustion engine in a hybrid car to ensure minimal fuel use. In addition to that, predictive algorithms also enable smart charging, where charging is synchronized with the state of the grid, the renewable energy generation, and demand-response. Such a strategy not only minimizes energy expenditure, but it also enables the renewable energy sources integration to minimize the carbon intensity of the transport electrification. Electric and hybrid vehicle technologies are examples of ICT-enabled vehicles with some of the highest levels of energy conservation, and a considerable reduction of greenhouse gases compared to the use of conventional vehicles. As Table 1 demonstrates, electric cars and buses that are run with the use of AI consume significantly less energy per one kilometer and emanate less CO_2 emissions than regular cars on gasoline and diesel. The ability to take advantage of technologies that use AI, such as adaptive cruise control, predictive traffic management, and traffic-aware navigation, enables AI-enabled cars to optimize the route, limit idle time, and maximize overall fuel efficiency. These improvements demonstrate the pivotal importance of AI in ensuring sustainable urban transport and reaching low-carbon mobility goals.

Table 1. Comparative Energy and Emission Benefits of Conventional vs. AI-Enabled Vehicles

Vehicle Type	Energy Consumption (kWh/km)	CO_2 Emissions (g/km)	Fuel Efficiency Improvement	Key AI Feature	References
Conventional Gasoline Car	0.85	195	–	–	Post et al., 2024
Conventional Diesel Bus	2.1	350	–	–	Nordelöf et al., 2019
AI-Enabled Electric Car	0.35	0	59%	Adaptive Cruise Control, Route Optimization	Yu et al., 2025
AI-Enabled Electric Bus	1.2	0	43%	Predictive Traffic Management	Amrani et al., 2025
AI-Enabled Hybrid Taxi	0.55	80	35%	Traffic-Aware Navigation	-

Moreover, AI plays a critical role in supporting vehicle-to-grid (V2G) technologies in which electric vehicles are utilized as distributed energy storage systems. By making energy flow two-way, AI systems facilitate load balancing and robustness and lead to less polluting energy. With this, there is a synergistic possibility of the green and digital transitions in transportation, as manifested by the combination of AI and electrification.

3.3 Traffic Prediction and Management: Machine Learning for Congestion Reduction and Improved Traffic Flow

Congestion is one of the worst emissions contributors to towns as backed up cars consume high amounts of fuel through idling and start-stop. Traffic prediction and management systems are based on AIs that are used to manage the traffic, predict potential traffic jams and avoid them before they occur (Lv et al., 2014).

Models, like LSTM networks and GNN are models of learning that allow forecasts with high precision traffic flows in changing conditions (Polson & Sokolov, 2017). These predictive findings are valuable when it comes to pro-active measures that include dynamic traffic lights, rerouting, and even congestion prices. Such cities as Singapore and Amsterdam have already implemented AI-based traffic management systems that can minimize traffic congestion and emissions and increase mobility efficiency.

With the capability of transitioning traffic management, AI can be used to transform traffic management into one that is predictive as opposed to reactive and therefore eliminate energy and time wastage. It helps achieve Sustainable Development Goal 11 (sustainable cities and communities) as it provides better city air quality and equity on transportation.

3.4 Fleet and Logistics Optimization: AI for Green Freight Routing, Load Optimization, and Last-Mile Delivery

The logistics industry is a very important part of the transportation process, yet it contributes to emissions as well. The AI directly in fleet and logistics optimization aims at reducing the environmental impact and, simultaneously, streamlining operations.

Freight routing is enhanced by AI since fuel usage, emissions, weather conditions, and delivery limits are factors that are considered (Lin et al., 2013). Machine Learning- enabled cargo optimization systems optimize the use of cargo and the capacity of vehicles by minimizing empty moves and maximizing vehicle capacity. In the urban scenario, demand forecasting and routing optimization on AI-enabled last-mile delivery platforms reduce the number of kilometers traveled by vehicles as well as carbon emissions (Amrani et al., 2025).

Freight transport is another area influenced by AI: the deployment of autonomous delivery robots and drones. Operating smaller, electric-powered driverless logistics vehicles, companies will eliminate the overuse of heavy vehicles in the highly populated cities. Such innovations agree with international net-zero logistics and eco-friendly logistical commitments.

A good case in point is DHL, utilizing AI-based fleet-management solutions to track fuel and carbon efficiency in their operations worldwide. These types of initiatives demonstrate that AI does not only bring down costs but also has a tangible sustainability impact.

4. INTEGRATION WITH SMART CITIES AND ITS

Artificial intelligence and autonomous vehicles are not technologies in a vacuum, as their potential is optimized when propelled through smart city and intelligent transport systems (ITS). Smart cities apply digital infrastructure, data driven governance and smart networks to finetune operations of the city and deliver desirable sustainability outcomes. In this respect, self-driving and AI-powered cars act as autonomous elements of data networks and fuel-efficient transportation vehicles that both generate and consume interconnected infrastructure. In this part, the importance of vehicle-to-infrastructure (V2I), vehicle-to-vehicle (V2V) communication, mutual coordination with public transport and active mobility and the relevance of data-based decision making in urban mobility is discussed. Figure 3 represents positioning autonomous vehicles that are AI-powered within smart cities and Intelligent Transport System (ITS). It demonstrates vehicle-to-infrastructure (V2I), vehicle-to-vehicle (V2V) communication, integration with the public transport and data-driven decision-making procedures. Through this encompassment, urban mobility within cities can be considered free-flowing, multi-modal transportation systems, and sustainable city growth.

Figure 3. Integration of AI-Enabled Vehicles with Smart Cities and Intelligent Transport Systems

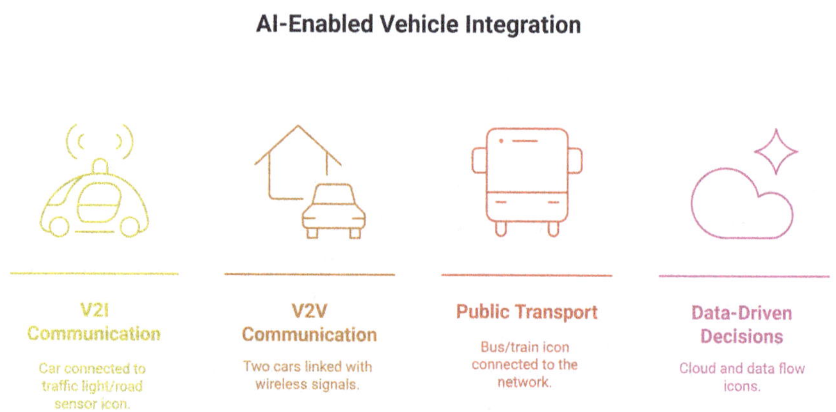

4.1 Vehicle-to-Infrastructure (V2I) and Vehicle-to-Vehicle (V2V) Communication

V2I and V2V communication feature an important establishing block of intelligent transport systems. The technologies provide vehicles with real-time data sharing with traffic lights, road sensors, and other cars, which are used to make predictive decisions based on a complex algorithm that improves congestion and safety. AI will reinforce these communication mechanisms with the capabilities to process such large amounts of data and turn them into usable information within a fraction of a second (Yusuf et al., 2023).

AI assists the green wave systems where the AI coordinates the traffic signals to decrease vehicle idling and the consumption of fuel. VDV communication also enables cooperative adaptive cruise control (CACC) where vehicles can travel very closely together and reduce drag and improve fuel effectiveness (Talluri et al., 2025). Combined with the results of the number fleets, such systems will lead to overall energy savings and emission reduction, increasing the sustainability benefits of autonomous vehicles.

4.2 Coordination With Public Transport and Active Mobility Networks

AVs and AI-enabled transport cannot have sustainable policies without integrating with other, more sustainable, means of transportation such as active mobility (cycling, walking) and public transport. Autonomous shuttles and feeder services can expand mass transit systems by connecting areas that are currently underserved to the metro and bus systems (Carrese et al., 2023). Ride-sharing platforms AI tools also make it easier to cut down on the dependability on their personal cars by linking users with effective shared journeys.

In addition, the introduction of AVs with active mobility guarantees that the focus of urban mobility plans is set on low-carbon forms of mobility. An example would be to elicit some European cities piloting AI-powered multimodal apps that will make recommendations to users about the greenest trip choice combining a mix of AVs with cycling, walking, and transit. This kind of coordination empowers efficiency in the system, bolsters health and equity goals, and climate goals.

4.3 Data-Driven Decision-Making for Urban Mobility and Sustainability

Smart cities rely on data, and AI in processing, analyzing and interpreting this data is essential to smart decisions. The AIs can enable city planners to simulate human movement and to forecast energy demands and scaling up and down the policies to meet the demands of sustainable urban development (Batty et al., 2012).

Dissemination of real-time data collected by AVs, traffic cameras, mobile devices, IoT sensors, enables the urban authorities to optimize traffic streams, apply low-emission zones, and assess the environmental effects of mobility interventions. As an example, predictive analytics allow cities to test the possible reduction of emissions due to congestion pricing before carrying it out at all.

In addition, data-driven governance accomplishes accountability and transparency in the urban mobility decision. Traffic Management Dashboards powered by AI to visualize how traffic flows, what are the current emission levels, and transport equity meters enable stakeholders that include policymakers and citizens to engage in governance towards sustainability (Jiang et al., 2021).

5. CHALLENGES AND LIMITATIONS

Although the autonomous vehicles powered with AI (AVs) can be a bright prospect in terms of sustainability, numerous obstacles must be overcome before it can become a mainstream product. Such challenges are spread into the technical, regulatory, ethical, and infrastructural domain, and social domain. These limitations are important to address when finding the solution to the effective AVs contribution to the sustainable and low-carbon transportation system.

5.1 Safety, Liability, and Regulatory Issues

Safety is of primary consideration in the introduction of AVs. Even though AI systems will help eliminate human error as the most common cause of car accidents, the performance of the algorithms on unpredictable circumstances is questionable (Myllyaho et al., 2021). Issues of determining liability in the case of an accident create legal and ethical tensions as well. Are the manufacturers to blame, the software developers or the operator? Such topics are confronting policymakers in every corner of the globe, and the need exists to harmonize international regulatory systems to address safety at the same time as promoting innovation.

5.2 Data Privacy, Cybersecurity, and Ethical Considerations

The data that AVs captures and processes massively and is sensitive: location history, behavioral patterns, biometrics data. Data protection and cybersecurity must be robust to prevent the occurrence of any breach that may jeopardize the privacy of the users or even lead to the compromise of security in the general societal settings

(Iftikhar et al., 2023). Besides the security concerns, there is an unresolved question of how AI should prioritize the safety of various stakeholders in situations involving unavoidable collision. Such issues demand explicit ethics codes and data stewardship models that support innovation and do not limit rights of users.

5.3 Infrastructure and Technology Adoption Barriers

The corresponding infrastructure to upgrade this transformation to AI-enabled transport systems is that of road-side sensors, 5G networks, and smart traffic oversight frameworks (Tengilimoglu et al., 2024). Large cities in many developing economies, however, cannot afford such technologies, and acceptance has been variable. Moreover, in the case of AVs, the integration of both motor vehicles and AVs will be transitional, and its regulation through adaptive policies is necessary.

5.4 Scalability and Interoperability Challenges

The factors include scaling the AVs of pilot projects to full-scale implementation are associated with the problem of interoperability across various makers, standards, and jurisdictions. The restriction of proprietary systems and variances in standards prevents a smooth transition of AVs to the existing transport infrastructure. They will not necessarily be able to use AI trained in another geographical or cultural location and as such, concerns about adaptability and inclusivity come into play. Global standards must be established to achieve interoperability in terms of standards, protocols and testing procedures (Arifuzzaman & Amin, 2025).

6. CASE STUDIES AND PRACTICAL APPLICATIONS

Essentially, the combination of AI and autonomous vehicles (AVs) can bring about positive outcomes theoretically; in reality, however, a real-world implementation can inform towards determining how effective they are, what the challenges are, and what the effects of their sustainability new the environment will be. This section identifies case studies on AV implementation in urban transport systems and green logistics, and on pilot project results and lessons.

6.1 Autonomous Electric Buses and Taxis in Leading Smart Cities

Singapore, Helsinki and Amsterdam are some of the cities on the front line in the deployment of autonomous electric vehicles (AEVs) to develop more sustainable urban mobility. These early efforts reveal how smart infrastructure, electrification, and artificial intelligence (AI) can complement one another to mitigate environmental effects but also combine to make transportation safer, faster, and more convenient.

In Singapore, Land Transport Authority (LTA) has undertaken pilot programs involving autonomous electric shuttles in a controlled city environment. These shuttles also utilize route planning and predictive traffic management, which is based on AI to make their travel as efficient as possible. The pilot projects have delivered a reduction of 15-20% of energy used by a diesel-powered bus (LTA, Autonomous Vehicles, n.d.), likely through the adjustment of acceleration and deceleration patterns and synchronization with the system of traffic communications. On top of this, the shuttles will dynamically manage congestion using traffic analytics in real-time extending to more efficient passenger travel and a decrease in idle time, thereby minimizing emissions even further. The initiative carried out by Singapore shows that a specific implementation of AEVs on high-density urban corridors is possible to achieve both environmental and performance gains and make a more appealing public transport available to commuters.

Another strong case can be seen in Helsinki with the Sensible 4 autonomous taxi. These electric taxis are linked to AI-enabled ride-hailing systems that adapt every route to changing traffic, passenger needs, and current conditions in the air like weather and air quality. With an overall reduction in the time vehicles spend waiting and wasting fuel, a significant efficiency and convenience margin is evident in the service. More importantly, Helsinki combines the autonomous taxis with the existing transport hubs, and that is, the AEVs will be complementary to the current transport systems, improving the multimodal mobility environment. Such implementation will enhance accessibility as well as cause people to shift the mode of transport, which will directly help the city reach its overall sustainability objectives (Sauliala, 2020).

Amsterdam, however, focuses on fleet level cooperation in the form of an interrelated AI and IoT infrastructural capability. Self-driven cars are always connected to roadside sensors and central traffic control systems, and as a result, traffic flow and route selection can be optimized in real-time (Kidmose, 2024). This is an example of the city-wide implementation of AI that increases functioning efficiency and safety levels. The system lets the multiple vehicles that coordinate themselves at the same time avoid stop-and-go traffic, conserve energy, and reduce urban air pollution. Although the experience of Amsterdam shows that installed autonomous electric fleets and smart infrastructure scale may provide a visible positive result in the form of environmental benefits, to support a seamless system of mobility in an urban environment.

The case studies as a group demonstrate that in combination with intelligent infrastructure and management systems driven by AI, autonomous electric vehicles offer significant potential to increase urban transport sustainability. The advantages of this would be energy and greenhouse gas reduction, better air quality and fees reliable and convenient mobility services. Besides the environmental benefits, what these deployments show is a more macro-level shift in the trend: cities can use emerging technologies in order not only to become more efficient but also to implement resilient, adaptable transportation ecosystems. Beyond conducting research into marginally different hybrid electric mobility solutions, self-driving electric-powered vehicles combined with AI and IoT capabilities are also a scalable means of making cities greener, smarter, and ultimately more sustainable.

6.2 AI-Enabled Fleet Management for Green Logistics

The use of artificial intelligence (AI) in the logistical industry has become a major new technology that empowers companies to optimize their fleet bands movements, minimize their environmental footprints on the environment, and increase the efficiency of their deliveries. With the use of machine learning, predictive analytics, and real-time data processing, AI-based fleet management software can considerably enhance the operation efficiency and sustainability results.

One of the examples of an AI platform applied in logistics management is the DHL Resilience360 platform. The system uses machine learning to optimize the freight routes in real time, forecast possible delays and check vehicles performance within its transportation network. By studying traffic patterns, weather conditions and prioritizing shipments in real-time, the platform can be used to make proactive steps that minimize waiting and resource underutilization. In pilot studies of Resilience360, there is a potential reduction in fuel usage of around 10%, with a corresponding reduction of around 50% of greenhouse gaseous emissions (DHL, 2015). In addition to environmental impacts, the visibility into fleet operations that are enabled by AI can also optimize supply chain resilience and help DHL navigate disruptions quickly and maintain on-time delivery performance.

Likewise, data laboratory technology in UPS has been used to enhance the use of load optimization, predictive delivery timelines and schedules. Using a combination of historical delivery data and vehicle capacities and responding to real-time traffic conditions, AI-based algorithms determine the best and most efficient routing decisions and optimize load across the vehicles. Newer means that one has a lesser number of empty running, maximized payload and less wastage of energy. The introduction of AI into UPS logistics planning brought real sustainability improvements since it revealed that massive logistics systems can be more efficient and have a decreased impact on the environment (Lysenko et al., 2023). Further, predictive analytics can reduce traffic and gasoline consumption within city distributions that can assist with all-city sustainability impacts.

One of the brightest examples of AI-driven green logistics is offered by Amazon, especially last-mile delivering activities. The company juxtaposes electric delivery vans, autonomous delivery robots, and routing algorithms to build a system that fulfills each of these requirements by balancing them. AI capabilities configure shipping routes and adjust real-time delivery schedules to real-world traffic conditions, vehicle battery charges and package urgency. Such comprehensive strategy allows Amazon to cut expendable costs on fuel consumption, emissions and enhance service quality overall. Notably, the scheme reflects the idea of the concept of sustainable logistics, when environmental impact is brought into balance with operational efficiency. Amazon is an example of how logistics service providers can make logistics efficient and sustainable by integrating electrification, automation, and AI analytics to handle the expectations of emerging e-commerce markets on scale (Luo, 2025).

Taking together, these case studies indicate the potential of AI in creating groundbreaking sustainable logistic change. Fleet management systems that rely on AI enable more intelligent route planning, more dynamic vehicle utilization, and predictive maintenance, which subsequently lowers cost of operations and environmental pollutants. Moreover, with the advantages of real-time data and predictive modeling, logistics companies may provide proactive solutions to reduce disruptions, optimal energy, and to support emissions reduction goals. The emergence of AI, electrification, and autonomous systems in logistics has become a paradigm shift as it makes the case that digital intelligence can be used to bring both efficiency and/or sustainability in challenging supply chains. With the volume of freight movement remaining on an upsurge despite the soaring cost of energy, green logistics that is driven with AI has the potential to achieve operational perfection and serve global sustainability and societal objectives.

6.3 Results and Lessons Learned From Pilot Projects

The results of pilot projects in various cities and logistics chains enable the evaluation of the potential of large-scale implementation of vehicles with AI technology and the directions of the formation of sustainable urban transportation. These solutions, including AEVs, AI-powered fleet control, and logistics systems as they are, demonstrate certain positive results along with challenges that are to be overcome on the way to successful scale-up.

- **Energy Efficiency Improvements:** the autonomous electric buses and taxis have consistently exhibited substantial improvements in energy consumption over the conventional ones across pilot programs. The sources of such efficiency gains include AI-enabled route optimization, predictive traffic lights and synchronized patterns of vehicle acceleration and deceleration. To give an example, in cities like Singapore and Helsinki, there is integration of real-time traffic data and predictive algorithms that have resulted in 15-20% reduction in energy consumption in autonomous shuttles, compared to traditional transit cars. Similarly, logistic AI-enhanced route planning and load optimization save fuel, and that translates to cost savings as well as environmental gains. These results validate that electrification and intelligent navigation systems have the ability of vastly improve the efficiency of energy consumption in passenger and freight transport.

- **Emissions Reduction:** AI-and-electrified fleets contribute to major decreases in greenhouse gases and air pollutant emissions, including CO_2 and NOx. Urban areas, where vehicular traffic and air pollution are most severe problems, avail of such measures. As another example, route optimization of logistics networks facilitated by AI can reduce emissions dramatically due to lower idling and the absence of unnecessary dents caused by route optimization. Taken together, these actions will help cities reach their climate and air quality goals and prove that sustainable transport solutions can yield practical environmental results (Miller et al., 2024).
- **Ongoing operational difficulties:** Traffic conditions mixing traffic, uncertainty in human driver behavior and shortcomings in existing infrastructure may impact the performance of vehicles and their safety. The controlled zones or dedicated corridors in early deployments are typical ways of overcoming the risks of having autonomous systems operating reliably, whilst learning information that can later be used to add functionality and capabilities. The possible operational barriers are fixed by solving human-machine interaction protocols, formidable sensor network, and adaptive traffic management that is vital in more challenging, city-wide settings.
- **Data and cybersecurity:** In real-time, data sharing among the AI algorithms are the only way they can optimize vehicle routing, predictive maintenance, and fleet management. Being more connected, however, poses a threat of cybersecurity that may jeopardize safety or privacy. As pilot projects reveal, it is instrumental to align in-depth encryption, secure data protocols, and constant surveillance to protect AI-based transport systems. Good governance and a good regulatory framework also need to be in place so that data driven mobility solutions are safe and trustworthy.
- **Scalability and Standardization:** Another major lesson learned in these pilots is that of interoperability among vehicle manufacturers, traffic management systems and municipal agencies. This is essential to ensure the standardization of the communication protocols, vehicle-to-infrastructure (V2I) interfaces as well as data provisions so that the features can be scaled to the city-wide networks. The lack of coordinated standards will leave less fully

the rewards of AI-enabled mobility, such as energy savings, carbon emissions reduction, and reliable services, to be achieved on a limited scale only.

International pilot schemes have demonstrated how the real-world applications and utility of autonomous vehicles in less centralized settings. Table 2 represents a comparative outlook of some of the case studies, mentioning the type of AVs that were deployed, the AI applications that were relied upon, its fleet size, and energy and emissions savings made. Particularly, the energy savings and efficiency of traffic were proven by such experiments as electric shuttle buses in Singapore and autonomous taxis in Helsinki. Equally, AV fleets in Dubai and Amsterdam have been optimizing fleet operations and lowering environmental effects through prediction traffic and energy management. These examples show that related to smart city infrastructure, AI enabled vehicles can play an important part in sustainable urban mobility.

Table 2. Summary of Global Case Studies of Autonomous Vehicle Deployment

City	AV Type	AI Application	Fleet Size	Energy Savings	Emission Reduction	Key Outcome	References
Singapore	Electric Shuttle Bus	Route Optimization, Traffic Prediction	20	18%	15%	Reduced energy use and improved traffic flow	LTA, Autonomous Vehicles, n.d.
Helsinki	Electric Autonomous Taxi	Ride-Hailing Optimization, Demand Forecasting	15	12%	10%	Reduced idle time and waiting periods	Sauliala, 2020
Amsterdam	Electric AV Fleet	V2I/V2V Coordination, IoT Integration	25	20%	18%	Enhanced fleet efficiency and air quality	Kidmose, 2024

As a conclusion, these case studies undeniably prove that autonomous electric vehicles and AI-driven transport systems can contribute heavily to the sustainability goals. By solving the operational, cybersecurity and standardization challenges, cities and logistics operators will be able to leverage established urban transport and freight networks by integrating these new technologies, generating efficiencies, cutting emissions and creating resilient and adaptive urban mobility ecosystems. The learnings of pilot projects hence give a guide on how AI-driven sustainable transportation should be adopted responsibly at large scales.

7. CONCLUSION AND FUTURE DIRECTIONS

7.1 Conclusion

AI and autonomous vehicles have massive potential to transform the way cities move as well as sustainable transportation. It has discussed how they can minimize emissions of greenhouse gases, optimize traffic conditions, and improve energy efficiency along passenger and freight networks. By combining with smart city infrastructures in the form of intelligent traffic, V2V and V2I communications and multimodal transport networks, autonomous vehicles have the potential to not only act as mobility creators but data-filled entities that can enable cities to be more successful in meeting their sustainability goals.

The use of AI in navigation, traffic forecasting, fleet and energy optimization has real potential in operational efficiency and environmental improvement. Case examples of automated electric buses, taxis, and a logistics company assisted with AI indicate that it is possible to attain measured improvements in reduction of emissions, energy consumption, and quality of service. These, however, are dependent on how successful physical planning in cities, control of the data, data management in the infrastructure and vehicle technology is coordinated.

Notwithstanding all the achievements, there are still challenges. These concerns are safety, ethical considerations, data privacy, cybersecurity, infrastructure and scale constraints which define the future of adoption. Pilot projects show that an early introduction in small-scale environments, as well as controlled interaction with other transport networks, can contain risks and demonstrate insight into a broader implementation. In the chapter, the importance of technological solutions is not in their exclusivity but rather their presence in regulatory frameworks, civic engagement, and interdisciplinary collaboration that will ensure the best outcome of AI-enabled transport system.

7.2 Future Directions

In terms of the future, the future of sustainable transportation is notably tied together with the development of artificial intelligence and automatic technologies. Future research and development need to be conducted on the areas of enhancing algorithm robustness and flexibility to various traffic, environmental and cultural conditions. To solve this integration challenge, interoperability between autonomous systems, manufacturers and urban infrastructures would need to be enhanced to operate at city-wide scale.

The additional studies of multimodal transport coordination, such as last-mile solutions of automation, coordination of the public transport system, and contribution to active mobility, will increase the sustainability and accessibility impact of these technologies. A combination of renewable energy sources, Dynamic charging networks supported by AI-driven energy management systems will further contribute to the transportation carbon footprint reduction.

Ethical, social and policy aspects will remain issues that will help shape adoption strategies. It is vital to establish effective frameworks of liability, cybersecurity and data governance, so that any new technology application resonates with the perception and trust of the society. Besides, urban planning will have to be modified in favor of autonomous and shared mobility solutions, and a balance between efficiency and inclusiveness and resilience will need to be achieved.

Finally, automatic and self-driving vehicles are also about to become prime in terms of reshaping sustainable mobility in urban areas. Through solutions to technological, operational and regulatory problems, as well as enhanced cooperation between governments, industry and end users, such systems have the potential to make transport systems safer, greener, and more efficient. The future of mobility arrived when AI, connectivity and smart infrastructure came together, creating a future where mobility is not only smart but sustainable, resilient, and responsive to the increasingly complex needs of global cities.

REFERENCES

Amrani, A. M. E., Fri, M., Benmoussa, O., & Rouky, N. (2025). A Deep Reinforcement Learning Framework for Last-Mile Delivery with Public Transport and Traffic-Aware Integration: A Case Study in Casablanca. *Infrastructures, 10*(5), 112. DOI: 10.3390/infrastructures10050112

Anderson, J. M., Kalra, N., Stanley, K. D., Sorensen, P., Samaras, C., & Oluwatola, O. A. (2016). *Autonomous vehicle technology: A guide for policymakers*. RAND Corporation., DOI: 10.7249/RR443-2

Arifuzzaman, M., & Amin, S. (2025). Automated vehicles: Are cities ready to adopt AVs as the sustainable transport solution? *Sustainability (Basel), 17*(5), 2236. DOI: 10.3390/su17052236

Badue, C., Guidolini, R., Carneiro, R. V., Azevedo, P., Cardoso, V. B., Forechi, A., Jesus, L., Berriel, R., Paixão, T. M., Mutz, F., De Paula Veronese, L., Oliveira-Santos, T., & De Souza, A. F. (2020). Self-driving cars: A survey. *Expert Systems with Applications, 165*, 113816. DOI: 10.1016/j.eswa.2020.113816

Batty, M., Axhausen, K. W., Giannotti, F., Pozdnoukhov, A., Bazzani, A., Wachowicz, M., Ouzounis, G., & Portugali, Y. (2012). Smart cities of the future. *The European Physical Journal. Special Topics, 214*(1), 481–518. DOI: 10.1140/epjst/e2012-01703-3

Bengler, K., Dietmayer, K., Farber, B., Maurer, M., Stiller, C., & Winner, H. (2014). Three Decades of Driver Assistance Systems: Review and Future Perspectives. *IEEE Intelligent Transportation Systems Magazine, 6*(4), 6–22. DOI: 10.1109/MITS.2014.2336271

Carrese, F., Sportiello, S., Zhaksylykov, T., Colombaroni, C., Carrese, S., Papaveri, M., & Patella, S. M. (2023). The Integration of Shared Autonomous Vehicles in Public Transportation Services: A Systematic review. *Sustainability (Basel), 15*(17), 13023. DOI: 10.3390/su151713023

Coppola, P., & Silvestri, F. (2019). Autonomous vehicles and future mobility solutions. In Elsevier eBooks (pp. 1–15). https://doi.org/DOI: 10.1016/B978-0-12-817696-2.00001-9

Creutzig, F., Niamir, L., Bai, X., Callaghan, M., Cullen, J., Díaz-José, J., Figueroa, M., Grubler, A., Lamb, W. F., Leip, A., Masanet, E., Mata, É., Mattauch, L., Minx, J. C., Mirasgedis, S., Mulugetta, Y., Nugroho, S. B., Pathak, M., Perkins, P., & Ürge-Vorsatz, D. (2021). Demand-side solutions to climate change mitigation consistent with high levels of well-being. *Nature Climate Change*, *12*(1), 36–46. DOI: 10.1038/s41558-021-01219-y

DHL. (2015). DHL RESILIENCE 360. In Corporate Solutions & Innovation. https://tapcrowdstatic-ie.s3-eu-west-1.amazonaws.com/5544/confbagfiles/event/10890/CSI_DHL_Resilience360_1474895712_1507195876.pdf

Dikshit, S., Atiq, A., Shahid, M., Dwivedi, V., & Thusu, A. (2023). The use of artificial intelligence to optimize the routing of vehicles and reduce traffic congestion in urban areas. *EAI Endorsed Transactions on Energy Web*, *10*. Advance online publication. DOI: 10.4108/ew.4613

Driessen, T., Siebinga, O., De Boer, T., Dodou, D., De Waard, D., & De Winter, J. (2024). How AI from Automated Driving Systems Can Contribute to the Assessment of Human Driving Behavior. *Robotics (Basel, Switzerland)*, *13*(12), 169. DOI: 10.3390/robotics13120169

Fagnant, D. J., & Kockelman, K. (2015). Preparing a nation for autonomous vehicles: Opportunities, barriers and policy recommendations. *Transportation Research Part A, Policy and Practice*, *77*, 167–181. DOI: 10.1016/j.tra.2015.04.003

Golbabaei, F., Yigitcanlar, T., & Bunker, J. (2020). The role of shared autonomous vehicle systems in delivering smart urban mobility: A systematic review of the literature. *International Journal of Sustainable Transportation*, *15*(10), 731–748. DOI: 10.1080/15568318.2020.1798571

Iftikhar, A., Qureshi, K. N., Shiraz, M., & Albahli, S. (2023). Security, trust and privacy risks, responses, and solutions for high-speed smart cities networks: A systematic literature review. *Journal of King Saud University. Computer and Information Sciences*, *35*(9), 101788. DOI: 10.1016/j.jksuci.2023.101788

International Energy Agency. (2023). *CO_2 emissions from fuel combustion: Highlights 2023*. IEA Publications. https://www.iea.org/reports/co2-emissions-in-2023/executive-summary

Jeng, S., & Chiang, C. (2023). End-to-End Autonomous Navigation Based on Deep Reinforcement Learning with a Survival Penalty Function. *Sensors (Basel)*, *23*(20), 8651. DOI: 10.3390/s23208651

Kellerman, A. (2018). Automated and autonomous spatial mobilities. In Edward Elgar Publishing eBooks. https://doi.org/DOI: 10.4337/9781786438492

Kidmose, B. (2024). A review of Smart Vehicles in Smart Cities: Dangers, impacts, and the threat landscape. *Vehicular Communications*, *51*, 100871. DOI: 10.1016/j.vehcom.2024.100871

Liao, Z., Taiebat, M., & Xu, M. (2021). Shared autonomous electric vehicle fleets with vehicle-to-grid capability: Economic viability and environmental co-benefits. *Applied Energy*, *302*, 117500. DOI: 10.1016/j.apenergy.2021.117500

Lin, C., Choy, K., Ho, G., Chung, S., & Lam, H. (2013). Survey of Green Vehicle Routing Problem: Past and future trends. *Expert Systems with Applications*, *41*(4), 1118–1138. DOI: 10.1016/j.eswa.2013.07.107

LTA | Autonomous Vehicles. (n.d.). https://www.lta.gov.sg/content/ltagov/en/industry_innovations/technologies/autonomous_vehicles.html

Luo, T. (2025). The application of artificial intelligence in supply chain management: A case study of Amazon. *Advances in Economics Management and Political Sciences*, *195*(1), 193–199. DOI: 10.54254/2754-1169/2025.BL24587

Lv, Y., Duan, Y., Kang, W., Li, Z., & Wang, F. (2014). Traffic flow prediction with Big Data: A Deep learning approach. *IEEE Transactions on Intelligent Transportation Systems*, •••, 1–9. DOI: 10.1109/TITS.2014.2345663

Lysenko, S., Makovoz, O., & Perederii, T. (2023). The impact of artificial intelligence in logistics management on sustainability development of e-business. In *V International Conference on European Dimensions of Sustainable Development*, Kyiv 2023 (pp. 99–110). https://doi.org/DOI: 10.24263/EDSD-2023-5-12

Miller, T., Durlik, I., Kostecka, E., Łobodzińska, A., & Matuszak, M. (2024). The emerging role of artificial intelligence in enhancing energy efficiency and reducing GHG emissions in transport systems. *Energies*, *17*(24), 6271. DOI: 10.3390/en17246271

Myllyaho, L., Raatikainen, M., Männistö, T., Mikkonen, T., & Nurminen, J. K. (2021). Systematic literature review of validation methods for AI systems. *Journal of Systems and Software*, *181*, 111050. DOI: 10.1016/j.jss.2021.111050

Nordelöf, A., Romare, M., & Tivander, J. (2019). Life cycle assessment of city buses powered by electricity, hydrogenated vegetable oil or diesel. *Transportation Research Part D, Transport and Environment*, *75*, 211–222. DOI: 10.1016/j.trd.2019.08.019

Pimenta, A. R., Kamruzzaman, M., & Currie, G. (2023). Long-term effects of autonomous vehicles on the built environment: A systematic scoping review towards conceptual frameworks. *Transport Reviews, 43*(6), 1083–1117. DOI: 10.1080/01441647.2023.2189325

Polson, N. G., & Sokolov, V. O. (2017). Deep learning for short-term traffic flow prediction. *Transportation Research Part C, Emerging Technologies, 79*, 1–17. DOI: 10.1016/j.trc.2017.02.024

Post, J. M., Ünal, A. B., Veldstra, J. L., De Waard, D., & Steg, L. (2024). Acceptability of connected automated vehicles: Attributes, perceived behavioural control, and perceived adoption norm. *Transportation Research Part F: Traffic Psychology and Behaviour, 102*, 411–423. DOI: 10.1016/j.trf.2024.03.012

Qin, X., Ke, J., Wang, X., Tang, Y., & Yang, H. (2022). Demand management for smart transportation: A review. *Multimodal Transportation, 1*(4), 100038. DOI: 10.1016/j.multra.2022.100038

Rauf, M., Kumar, L., Zulkifli, S. A., & Jamil, A. (2024). Aspects of artificial intelligence in future electric vehicle technology for sustainable environmental impact. *Environmental Challenges, 14*, 100854. DOI: 10.1016/j.envc.2024.100854

Sauliala, T. (2020, April 8). Sensible 4 adapts its autonomous driving pilot in Helsinki during COVID-19 - Sensible 4. Software for Automated Driving in Any Weather. https://sensible4.fi/2020/04/08/sensible-4-adapts-its-autonomous-driving-pilot-in-helsinki-during-covid-19/

Shuaibu, A. S., Mahmoud, A. S., & Sheltami, T. R. (2025). A review of Last-Mile Delivery Optimization: Strategies, technologies, drone integration, and future trends. *Drones (Basel), 9*(3), 158. DOI: 10.3390/drones9030158

Singh, S. K., Alasali, B. E., & Belbase, Y. (2025). Urban Transportation Planning: Artificial Neural Network Applications. In Syam Sundar, L., Balakrishnan, D., & Sousa, A. C. M. (Eds.), *Expert Artificial Neural Network Applications for Science and Engineering* (pp. 469–504). IGI Global Scientific Publishing., DOI: 10.4018/979-8-3693-7250-0.ch017

Singh, S. K., & Jain, R. (2025). Future of mobility: potential challenges and ethical considerations. In AIoT for Smart Transportation: Transforming the future of mobility (pp. 235–253). https://doi.org/DOI: 10.1049/PBTR047E_ch12

Talluri, K. K., Stang, C., & Weidl, G. (2025). Green wave as an integral part for the optimization of traffic efficiency and safety: a survey. *2022 IEEE Intelligent Vehicles Symposium (IV)*, 2422–2429. https://doi.org/DOI: 10.1109/IV64158.2025.11097706

Tengilimoglu, O., Carsten, O., & Wadud, Z. (2024). Are current roads ready for highly automated driving? A conceptual model for road readiness for AVs applied to the UK city of Leeds. *Transportation Research Part A, Policy and Practice, 186*, 104148. DOI: 10.1016/j.tra.2024.104148

The World Bank. (2025). Enhancing transport decarbonization in the European Union. In The World Bank. https://documents1.worldbank.org/curated/en/099012125085536598/pdf/P5016401b3ed380a618fac1940454097e66.pdf

United Nations. (2022). *World urbanization prospects: The 2022 revision*. United Nations Department of Economic and Social Affairs. https://population.un.org/wup/

Yigitcanlar, T., Wilson, M., & Kamruzzaman, M. (2019). Disruptive impacts of automated driving systems on the built environment and land use: An Urban Planner's perspective. *Journal of Open Innovation*, *5*(2), 24. DOI: 10.3390/joitmc5020024

Yu, T., Tang, Y., Chen, R., & Zhao, S. (2025). Optimization of adaptive cruise control strategies based on the Responsibility-Sensitive Safety Model. *Vehicles (Basel)*, *7*(2), 28. DOI: 10.3390/vehicles7020028

Yusuf, S. A., Khan, A., & Souissi, R. (2023). Vehicle-to-everything (V2X) in the autonomous vehicles domain – A technical review of communication, sensor, and AI technologies for road user safety. *Transportation Research Interdisciplinary Perspectives*, *23*, 100980. DOI: 10.1016/j.trip.2023.100980

Chapter 7

Revolutionizing Vehicle Safety Through AI-Driven V2X and Digital Twin Technologies

Ismatulla Khayrullayev
Termez University of Economics and Service, Uzbekistan

Shruthishree S. H.
Global Academy of Technology, India

Abdullayev Abdulla Fayzulla Ugli
Tashkent State University of Economics, Uzbekistan

Nethravathi H. M.
BGS Institute of Technology, Adichunchanagiri University, India

Anorgul Ashirova
http://orcid.org/0009-0006-7271-3634
Mamun University, Uzbekistan

G. Swetha
http://orcid.org/0000-0001-8620-7149
RR Institute of Technology, India

N. Raghu
http://orcid.org/0000-0002-2091-8922
Jain (Deemed to be) University, India

Kamal Narayanan
University of Nevada, USA

Arjun Kumar G. B.
http://orcid.org/0000-0002-8120-216X
Dayananda Sagar College of Engineering, India

Qosimbetov Batirbay
Karakalpak State University, Uzbekistan

ABSTRACT

Connected car technologies are redefining vehicle safety by enabling cars to communicate, sense their surroundings, and respond to real-time conditions. Today's vehicles go beyond transportation they act as intelligent systems working to prevent

DOI: 10.4018/979-8-3373-3231-4.ch007

accidents before they happen. Features like Vehicle-to-Vehicle and Vehicle-to-Infrastructure communication, advanced driver assistance systems and automated alerts help drivers stay aware of potential dangers and make quicker, safer decisions. The support of fast networks like 5G, cloud computing, and edge data processing, these innovations are becoming more effective and accessible. Automakers such as Tesla, Volvo, and Ford are already integrating these technologies into mainstream vehicles, making safety smarter and more proactive. Concerns about cybersecurity, data privacy, and uneven infrastructure still pose hurdles. Legal and regulatory frameworks are evolving but often struggle to keep pace with the technology.

1. INTRODUCTION

1.1 Background of Connected Vehicles

The journey of automobiles has been one of constant innovation. What started as simple mechanical machines has evolved into sophisticated systems powered by software, sensors, and data. A major leap in this evolution is the rise of connected vehicles cars that can "talk" not just to each other but also to traffic systems, road infrastructure, and even pedestrian devices (Ouaissa et al., 2022). These vehicles are reshaping how transportation works, aiming to make driving safer, smoother, and smarter. Connected vehicles are part of a broader shift toward intelligent transportation systems (ITS). These systems use real-time data to improve everything from traffic flow to accident prevention. With the help of Vehicle-to-Vehicle (V2V), Vehicle-to-Infrastructure (V2I), and Vehicle-to-Everything (V2X) technologies, connected cars share information about speed, location, road hazards, and more. This kind of communication means drivers can act on important updates almost instantly.

Imagine a scenario where a car approaching an intersection receives a signal that another vehicle is about to run a red light. That warning, delivered in real time, could be the difference between a near miss and a serious accident as in figure 1. The power of connected cars lies in their ability to gather, process, and act on data before danger becomes visible. The technologies making this possible include 5G networks, advanced sensors, cloud computing and AI. Together, they create a digital ecosystem where cars aren't just tools for travel they become proactive safety partners (Taleb et al., 2017).

Figure 1. Interconnection missing between the vehicles.

1.2 Importance of Vehicle Safety in Modern Transportation

Every year, road accidents lead to the loss of millions of lives and leave countless others injured. Despite progress in vehicle design and traffic regulations, the challenge of road safety remains significant especially as urban areas become more congested and fast-paced. Traditional safety measures, like airbags and crumple zones, help after an accident occurs (Liu et al., 2023). But today's needs demand more than that they require tools that can prevent accidents from happening in the first place.

That's where active safety systems come in. These include technologies that assist drivers in real time, such as lane-keeping assistance or automatic braking. When combined with connectivity, these tools become even more powerful (Alam, 2024). The idea is to shift from reactive safety to predictive safety spotting potential issues before they turn into real threats. Safer vehicles don't just benefit the individual driver. They reduce the burden on emergency services, improve traffic flow, and even lower environmental impact by reducing stop-and-go driving caused by accidents. As roads grow more complex, safety through connected technology becomes less of a luxury and more of a necessity for individuals and communities alike (Sarsam, 2023).

1.3 Role of Connectivity in Enhancing Safety

Connectivity changes the way vehicles see the world. Instead of relying solely on onboard cameras and sensors, connected cars gain access to a much wider pool of information from nearby vehicles, road signs, traffic lights, and even weather updates. This broader awareness helps cars respond faster and more accurately to potential hazards (Rupanagudi, 2020). For instance, through V2V communication, two vehicles can exchange data about their position and speed. If one vehicle brakes suddenly, the other can receive a warning and react immediately even if the driver hasn't noticed yet. Similarly, V2I systems can alert a driver about a sharp curve ahead, icy conditions, or a traffic jam, allowing for smoother and safer navigation (Ke et al., 2020).

Cloud platforms and AI add another layer of intelligence. They can analyze patterns from thousands of vehicles to predict trouble spots before problems arise. This ability to forecast and prevent issues marks a big step forward from traditional safety methods. What's truly transformative is how connectivity turns safety from something passive into something active and intelligent. The car becomes more than just a vehicle it becomes a co-pilot, helping to make every journey safer.

1.4 Objectives and Structure of the Chapter

This chapter delves into the evolving world of connected vehicle technologies and their powerful role in making our roads safer. As vehicles become smarter and more connected, they are no longer just modes of transport—they are active participants in a broader network. Through systems like Vehicle-to-Vehicle (V2V) and Vehicle-to-Infrastructure (V2I) communication, cars can now share real-time data about traffic conditions, potential collisions, and other critical events. These advancements are already making a difference by helping drivers avoid accidents and improving overall traffic flow. Beyond current applications, the chapter also looks ahead to the future, where connected technologies could work hand-in-hand with automation and artificial intelligence to create even safer, more efficient transportation systems. The aim is to showcase both the real-world impact and the exciting potential of connected cars in transforming road safety.

The chapter is organized into several sections. It begins by introducing the core technologies behind connected cars, followed by a closer look at safety-specific features like advanced driver-assistance systems and emergency alerts. Infrastructure and communication frameworks are covered next, along with real-world case studies (Zhang et al., 2023). There's also a discussion on regulations, implementation challenges, and emerging trends like AI and digital twins. The chapter wraps up with a summary of key insights and what lies ahead for connected vehicle safety.

1.5 Literature Survey

Ouaissa et al. (2022) highlights how Vehicle-to-Everything (V2X) technologies such as Vehicle-to-Vehicle (V2V) and Vehicle-to-Infrastructure (V2I) are transforming road safety when combined with advanced driver assistance systems (ADAS). Unlike onboard sensors that struggle with blind spots or hidden obstacles, V2X allows vehicles to "see" beyond their immediate surroundings by sharing real-time data with other vehicles and traffic infrastructure. This makes warnings like sudden braking alerts or intersection collision prevention faster and more reliable. The study emphasizes the importance of 5G connectivity for meeting the low-latency requirements of these systems. However, challenges remain, particularly in ensuring interoperability across car manufacturers, handling dense traffic communication, and overcoming interference in real-world conditions. The authors conclude that while V2X is not a replacement for ADAS, it significantly strengthens it, provided standardization and robust testing continue to evolve.

Alam et al. (2024) explore how connected and autonomous vehicles (CAVs) can be made more secure by combining blockchain with federated reinforcement learning (FRL). Since these vehicles rely on exchanging sensitive data for navigation and decision-making, privacy and security are constant concerns. Traditional centralized data sharing is risky, as it exposes personal and operational information to leaks or cyberattacks. The proposed solution takes a decentralized approach by allowing vehicles to train models locally, while sharing only encrypted updates through a blockchain network. This design builds trust and protects against tampering, ensuring that sensitive data never leaves the vehicle. It reduces risks like data theft or malicious interference with the models. The authors also connect these security measures to real-world safety improvements, such as fewer false alarms in hazard detection systems. At the same time, they acknowledge important challenges like the added latency, higher energy consumption, and the difficulty of achieving widespread industry collaboration. Ultimately, the study highlights that the future of connected vehicle safety will depend on finding the right balance between real-time performance and robust security.

Sarsam's (2023) study takes a closer look at the growing gap between existing laws and the cybersecurity needs of connected and autonomous vehicles. It highlights how current regulations such as ISO/SAE 21434 and UNECE WP.29 struggle to keep pace with fast-evolving threats like spoofed V2X messages, hacked over-the-air updates, and weaknesses in the supply chain. One of the biggest challenges they point out is the mismatch between traditional safety standards, such as ISO 26262, and the ongoing cybersecurity monitoring that vehicles require long after they leave the factory. The researchers also raise important questions about accountability for instance, who is liable if a car makes a decision based on compromised external data with their argument is that regulations must evolve toward continuous risk assessment, shared responsibility among manufacturers and operators, and certification frameworks that link cybersecurity directly to safety outcomes. Ultimately, the work underscores that cybersecurity is not just a technical necessity, but a core element of ensuring road safety in the connected era.

Ke et al. (2020) this research explores how predictive safety can be achieved through machine learning models trained on vast datasets that capture driving behavior, environmental conditions, and traffic incidents. By using ensemble methods like random forests and gradient boosting, the system is able to forecast high-risk situations such as sudden braking ahead, unexpected lane changes, or pedestrians crossing the road with impressive accuracy. What makes the approach even stronger is its adaptability: the model continuously learns from new data, refining its predictions and cutting down on false alarms. Real-time processing is a key focus, ensuring that drivers receive timely warnings without being overloaded with unnecessary alerts. The study also looks at how these predictive models can be translated into practical safety tools, from preemptive braking to subtle steering corrections. At the same time, it acknowledges the challenges, including striking the right balance between sensitivity and specificity, managing unbalanced datasets, and making the system compatible across different vehicle platforms. Ultimately, the work shows how predictive analytics can shift vehicle safety from being purely reactive to truly proactive, and argues for integrated frameworks that bring together real-time sensing with historical risk profiling.

Zhang et al. (2023) a novel approach on collision avoidance system that adapts its control strategy based on the driver's behavior. By combining fuzzy logic with predictive control algorithms, the system can distinguish between aggressive and conservative driving styles, adjusting braking and lane-change interventions to match. Simulation results across varying speeds demonstrate that this adaptive approach not only increases safety margins but also delivers smoother responses, particularly during sudden maneuvers. By accounting for human variability, the system minimizes unnecessary interventions in calm driving situations while still taking decisive action when risks are high. The findings highlight improvements in lateral stability and faster response times compared to traditional, fixed collision avoidance systems. However, challenges remain—such as the difficulty of accurately profiling driver behavior and the ethical questions that arise when tailoring safety features to individual styles. Even so, this adaptive framework points to a promising direction for future safety technologies: systems that not only protect but also respect the unique ways in which people drive.

Islam & Karimoddini (2022) This literature review examines recent progress in improving pedestrian safety through sensor fusion and infrastructure-assisted detection systems. By combining data from lidar, cameras, and radar, vehicles can achieve more reliable detection, even in challenging non-line-of-sight situations such as around corners or behind parked cars. When integrated with roadside units and smart city infrastructure, these systems further enhance accuracy and reduce response times, allowing vehicles to anticipate risks and react proactively. The review explores different use cases across urban and suburban environments, with particular attention to the role of edge computing at intersections where pedestrian activity is high. At the same time, it highlights ongoing challenges, including sensor occlusion, biases in datasets that may underrepresent certain populations, and the heavy processing demands of real-time analysis. Taken together, the findings underscore the potential of infrastructure-assisted approaches to pedestrian safety, while also stressing the importance of standardized sensor placement and cooperative strategies that enable vehicles to share the detection workload across a connected network.

Li & Zhang's (2010) research focuses on developing autonomous vehicle control systems that remain safe even when sensors or subsystems fail. Built on fault-tolerant architecture, redundancy, and anomaly detection, the system ensures that critical functions such as emergency braking and steering support continue to operate under partial failure. By combining machine learning classifiers with logic-based protocols, it can quickly detect inconsistencies, isolate faulty components and trigger fallback mechanisms that keep the vehicle operating safely. Tests across different failure scenarios, from lost GPS signals to mis-calibrated cameras, demonstrate the system's resilience in real-world conditions. The study emphasizes that autonomous driving cannot depend on perfect environments safety must be maintained despite inevitable imperfections. The findings highlight the importance of embedding resilience directly into system design, creating vehicles that are inherently fault-aware, adaptive, and capable of responding effectively when things go wrong.

Martinez et al. (2011) This simulation study explores a coordinated collision avoidance strategy that leverages V2X communication to improve multi-vehicle safety. These sharing information such as intent, braking decisions, and trajectory data, vehicles can work together to execute cooperative maneuvers like synchronized braking or lateral adjustments that prevent chain-reaction crashes. Tested in realistic traffic conditions, including blind corners and tightly spaced vehicles, the system showed a marked reduction in collision probability. The algorithm is designed to account for real-world communication challenges, such as delays and message loss, by optimizing timing and adding redundancy to preserve safety margins. While the work is largely simulation-based, the researchers outline practical pathways for deployment through hardware-in-the-loop testing, demonstrating how the approach could improve convoy dynamics and distributed decision-making on the road. Ultimately, the study underscores the power of collaboration: by sharing the safety responsibility across a network of vehicles, rather than leaving each one to act in isolation, connected systems can create a stronger, more resilient road safety framework.

2. OVERVIEW OF CONNECTED CAR TECHNOLOGIES

2.1 Definition and Evolution of Connected Cars

Connected cars represent an exciting fusion of automotive engineering and digital technology. At their heart, these vehicles are equipped with internet connectivity and communication networks that allow them to interact not only with other vehicles but also with infrastructure, cloud services, and even pedestrians' smartphones (Alsudani, 2023). This constant flow of information is transforming traditional cars into intelligent mobility systems, enabling them to perceive, respond, and adapt to their surroundings in ways that were once impossible, as illustrated in Figure 2.

The idea of car connectivity has been around for quite some time. Early implementations focused on basic features like navigation systems and onboard diagnostics. Over the years, these tools have grown more sophisticated, evolving into systems that provide real-time traffic updates, emergency assistance, and even remote vehicle monitoring. What once began as simple conveniences has gradually transformed vehicles into connected platforms capable of interacting with the world around. Recent advancement in wireless technologies and the rise of cloud computing, today's connected vehicles are capable of far more than simple GPS navigation, they can share and receive information critical to safety, efficiency, and convenience. The journey from analog to connected vehicles reflects broader changes in how society uses technology. As cars become part of a larger digital ecosystem, they open up opportunities not only for safer roads but also for smarter cities and cleaner mobility. The connected car is no longer a futuristic idea, it's a fast-moving reality that is reshaping the transportation landscape (Gupta et al., 2023).

Figure 2. Technology enabled vehicle.

2.2 Key Communication Types: V2V, V2I, V2P, V2N

The strength of connected vehicle systems lies in their ability to communicate. This communication happens through several primary channels each offering a different type of awareness and functionality.

- Vehicle-to-Vehicle (V2V) communication allows cars to share data like speed, direction, and braking status with each other. This peer-to-peer exchange helps vehicles anticipate potential collisions, even before drivers can react.
- Vehicle-to-Infrastructure (V2I) connects vehicles with elements of the road environment, such as traffic lights, lane markers, and toll booths. This communication allows drivers to receive timely alerts like an approaching red light or suggested detours during congestion bringing city infrastructure directly into their driving decisions.
- Vehicle-to-Pedestrian (V2P) focuses on interactions with vulnerable road users, including pedestrians and cyclists who carry mobile devices. By sending alerts to both drivers and pedestrians, it creates a shared awareness that enhances safety, particularly in crowded urban settings.

- Vehicle-to-Network (V2N) links vehicles to cloud-based services, providing access to real-time traffic data, weather updates, software improvements, and broader insights from a connected intelligence network. This allows vehicles to make smarter decisions based on both local and global information.

Together, these three communication modes create a robust safety ecosystem, equipping connected vehicles with the awareness and responsiveness needed to navigate complex, dynamic road environments effectively.

2.3 Components of a Connected Vehicle Ecosystem

A connected vehicle exists within a complex ecosystem of technologies and infrastructure, where every component contributes to its performance and safety. At the center are embedded sensors and electronic control units (ECUs) that constantly monitor the car's surroundings—tracking speed, distance, lane position, and obstacles (Alshamrani et al., 2021). Communication modules, such as DSRC or cellular networks like LTE and 5G, act as the vehicle's link to the outside world, enabling it to exchange data with other vehicles, infrastructure, and cloud systems. These modules are vital for ensuring fast, reliable communication, particularly in critical situations like collision avoidance. Onboard processors and edge computing units interpret the incoming data in real time, supporting driving decisions and triggering alerts as needed. AI algorithms further enhance these systems, allowing them to learn from driving behavior and environmental patterns to improve responsiveness over time (Raghu et al., 2023).

Outside the vehicle, infrastructure elements like smart traffic signals, roadside units (RSUs), and cloud platforms help manage and aggregate information, creating a connected environment that spans neighborhoods, cities, and regions. Within this network, vehicles no longer function in isolation; they interact intelligently with their surroundings and each other. Together, these technologies form a collaborative system where each vehicle contributes to a safer, smarter, and more efficient transportation ecosystem.

2.4 Integration with IoT and 5G Networks

The integration of connected cars with IoT and 5G networks marks a transformative step in modern transportation. IoT acts as a vast data network, linking road sensors, traffic cameras, navigation apps, and vehicles to enable seamless information sharing. This connectivity allows cars to perceive and respond to their environment with greater awareness and precision. Equipped with IoT-enabled systems, vehicles can track tire pressure, fuel levels, engine performance, and even signs of driver fatigue in real time (Vanjire & Naveen, 2024). Far from being confined to the vehicle, this data can be transmitted to the cloud, supporting predictive maintenance, fleet optimization, and other smart transportation services. By bridging vehicles, infrastructure, and cloud intelligence, IoT and 5G together create a more responsive, efficient, and safer driving ecosystem. For instance, a delivery company can track its entire vehicle fleet, identify which cars need servicing, and avoid costly breakdowns all thanks to IoT integration (Raghu et al., 2023).

5G, on the other hand, brings speed and reliability to the system. Unlike earlier cellular networks, 5G offers ultra-low latency (as low as 1 millisecond) and high bandwidth, which are crucial for safety-critical communication. This ensures that alerts like "brake now" or "pedestrian ahead" are delivered almost instantly, without delay. For high-speed vehicles or autonomous driving systems, even a fraction-of-a-second lag can be the difference between safety and disaster. Moreover, the combination of 5G and IoT supports the development of vehicle platooning, where a convoy of connected trucks or cars can travel closely together to reduce fuel use and congestion as shown in figure 3. It also enables real-time HD mapping, dynamic rerouting, and faster software updates over the air. As vehicles become more connected, they also become more responsive capable of navigating not just physical roads, but digital networks that offer foresight and real-time insight. The collaboration between IoT and 5G is paving the way for a transportation system that is intelligent, adaptive, and, above all, safer for all road users (Lee & Hess, 2022).

Figure 3. IoT and 5G communication enabled car.

3. SAFETY-CENTRIC TECHNOLOGIES IN CONNECTED CARS

Connected cars are not just about convenience or entertainment, they are fundamentally reshaping how safety is approached on roads. By embedding intelligence into vehicles and integrating them with surrounding environments (S. R. Rupanagudi, 2021), these technologies serve as vigilant co-pilots. The following sections explore the key safety innovations that help prevent accidents, protect passengers, and reduce fatalities.

3.1 Advanced Driver Assistance Systems (ADAS)

Modern vehicles have become more than just mechanical machines, they're intelligent systems designed to actively assist the driver. At the forefront of vehicle safety innovation are Advanced Driver Assistance Systems (ADAS), which provide drivers with real-time support through features like adaptive cruise control, automatic parking, and traffic sign recognition. These systems help ease the mental load on drivers by managing speed, maintaining lane position, and even monitoring for signs of fatigue (Molinaro et al., 2020). The ADAS usage particularly powerful is its proactive approach. A driver might miss a traffic sign or fail to notice a slowing vehicle ahead, but the system can step in to alert or assist before a dangerous situation arises. In congested traffic, adaptive cruise control keeps a safe distance from other vehicles, reducing the risk of collisions. Many of these features operate seamlessly in the background, providing safety without taking control away from the driver.

As vehicles continuously gather data through radar, LiDAR, cameras, and other sensors, ADAS becomes increasingly precise and adaptive. Beyond convenience, these systems play a crucial role in making roads safer, especially in complex driving scenarios where human reaction alone may not be enough. The full automation is still evolving, ADAS is a strong step toward a future where vehicles and humans work together for safer travel.

3.2 Collision Avoidance Systems

Accidents often happen in a split second just enough time for a distracted glance or misjudged distance. Collision Avoidance Systems aim to step in during those crucial moments. These systems continuously monitor the surroundings using radar, cameras, and ultrasonic sensors, ready to intervene when a crash seems likely. They're designed to give warnings first through sound, visual alerts, or seat vibrations and then apply emergency braking or steering if needed as shown in figure 4. These technologies are especially useful in urban settings, where traffic is dense and unpredictable. Imagine a pedestrian stepping into the road suddenly or a cyclist swerving unexpectedly. A collision avoidance system can detect these changes faster than most drivers can react. In highway driving, it can identify rapid deceleration in the vehicle ahead, helping prevent high-speed pileups. (Bagheri et al., 2021).

The system's intelligence lies in how it prioritizes threats. For example, it can differentiate between a stationary vehicle and a plastic bag blowing across the road. It doesn't just look ahead it considers the speed, direction, and movement patterns of nearby objects. This decision-making capability is built through algorithms trained on countless driving scenarios. As the technology matures, it's also becoming more affordable and available in mid-range vehicles not just luxury models. That means more drivers benefit, and more lives are potentially saved. Collision avoidance is no longer a futuristic idea it's becoming a standard expectation in responsible car design.

Figure 4. Decision enabled vehicle to avoid collision.

3.3 Lane Departure and Blind Spot Monitoring

Drifting out of a lane without realizing or missing a vehicle lurking in the blind spot can easily lead to a serious accident. The lane departure warning (LDW), lane-keeping assist (LKA), and blind spot monitoring (BSM) systems come into play. These features act as extra eyes on the road, helping drivers maintain awareness and control, especially during highway travel or in dense traffic (Raghu et al., 2023). Lane departure systems use cameras mounted on the windshield to monitor road markings. The vehicle begins to veer without signaling, the driver receives an alert maybe a beep, a steering wheel vibration, or a gentle push back into the lane. This small nudge can be enough to prevent a collision caused by distraction or drowsiness (Verma et al., 2024).

Blind spot monitoring uses radar sensors to detect vehicles approaching from behind or alongside, in areas a driver may not easily see through mirrors. A visual indicator often a small light on the side mirror warns the driver not to change lanes. Some systems also offer active intervention by gently applying steering correction. These technologies are particularly important during long drives or night travel, when attention can waver. They're not just add-ons for comfort, they're key safety elements. And they're becoming more precise with every software upgrade, helping reduce lane-related accidents across vehicle types.

3.4 Real-time Traffic and Hazard Alerts

Getting caught in a sudden traffic jam or heading straight into a dangerous road condition can be both frustrating and risky. Real-time traffic and hazard alert systems are designed to keep drivers informed ahead of time. These systems collect and process data from GPS, traffic control centers, and even other connected vehicles to provide live updates about congestion, accidents, roadblocks, or weather-related hazards. The alerts are delivered through the vehicle's infotainment system or dashboard display, often paired with recommended alternative routes. This not only saves time but also reduces the likelihood of last-minute lane changes or sudden braking that might lead to accidents. In more advanced systems, alerts are tied to the vehicle's sensors so if slippery conditions are detected ahead, the car can adjust braking response automatically. Real-time information enhances decision-making (Hasan et al.,2020).

3.5 Autonomous Emergency Braking (AEB)

One of the most critical last-line defenses in vehicle safety is Autonomous Emergency Braking (AEB). When a driver doesn't react in time to a looming obstacle, a stopped vehicle, a crossing pedestrian, or a cyclist AEB steps in. The system calculates closing speed and distance using radar and cameras, and if a collision seems imminent, it applies the brakes automatically to avoid or reduce impact. What makes AEB so valuable is its speed and precision. The average human reaction time may not be fast enough to prevent a crash, especially at higher speeds or in unexpected scenarios (Raghu et al., 2024). AEB operates in milliseconds, detecting the threat and acting on it sometimes before the driver even realizes the danger. This feature has already shown its potential in real-world use, with many insurers and regulators noting significant reductions in rear-end collisions in vehicles equipped with AEB. Many automakers now include AEB as a standard feature, recognizing it not just as a luxury, but as a safety essential. It's a quiet guardian, always watching the road ahead (Gürses et al., 2025).

4. DATA AND COMMUNICATION INFRASTRUCTURE

4.1 Vehicle-to-Everything (V2X) Protocols

Modern connected vehicles operate within a complex web of communication protocols that enable seamless data exchange between different actors on the road. Among the most transformative is Vehicle-to-Everything (V2X) communication, a framework that allows vehicles to communicate with other vehicles (V2V), infrastructure (V2I), pedestrians (V2P), and networks (V2N). This holistic approach to vehicular communication enhances situational awareness and supports proactive safety mechanisms as shown in figure 5.

V2X communication functions by transmitting data packets between nodes in real time, covering aspects such as speed, location, braking patterns, and traffic signal timing. By continuously analyzing this information, vehicles can anticipate potential hazards like another car running a red light or a pedestrian stepping onto the road long before visual contact is made. This proactive layer of intelligence strengthens the foundation of safety-oriented driving environments. Importantly, V2X isn't merely a technological novelty. Its deployment is a strategic response to rising urban congestion, increasing traffic fatalities, and the growing demand for intelligent transport systems. As adoption spreads, V2X has the potential to transform transportation ecosystems into adaptive, responsive networks where vehicles not only move but think, predict, and act in unison (Samak et al., 2024).

Figure 5. Holistic approach to vehicular communication

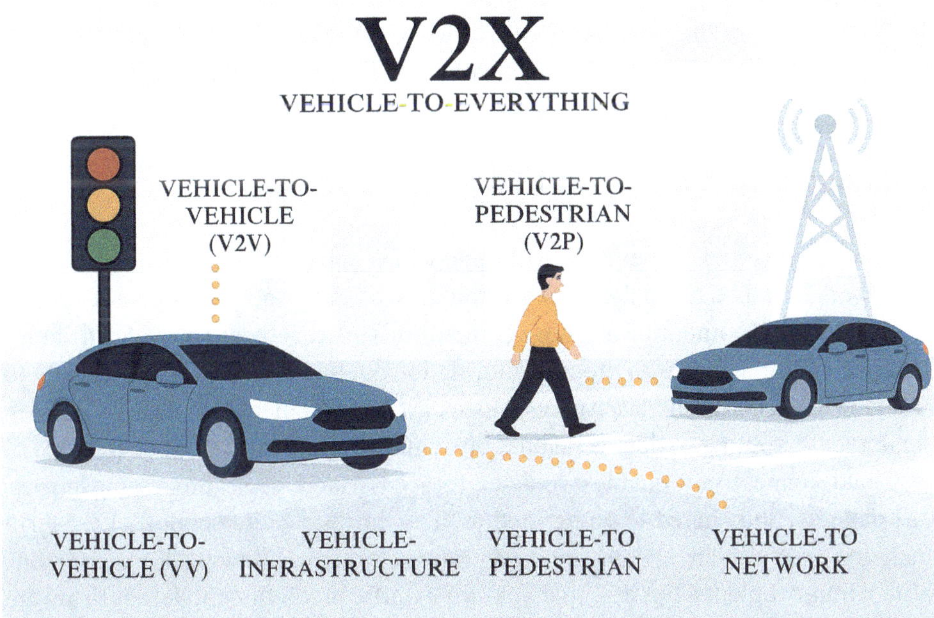

4.2 DSRC vs. C-V2X: A Comparative Analysis

Two primary technologies currently dominate the V2X communication landscape: Dedicated Short-Range Communication (DSRC) and Cellular Vehicle-to-Everything (C-V2X). Both are designed to enable fast, low-latency communication between vehicles and infrastructure, yet they differ significantly in architecture, capabilities, and scalability. DSRC operates over the 5.9 GHz band and is based on IEEE 802.11p standards. It allows vehicles to communicate directly without relying on cellular networks, which proves advantageous in areas with poor coverage. DSRC is mature, well-tested, and widely adopted in pilot programs, particularly in North America. Its key strength lies in low latency and dependable performance under fixed scenarios (Shoukat et al., 2024).

C-V2X, on the other hand, relies on cellular networks such as 4G LTE and 5G to enable both direct vehicle-to-vehicle communication and network-based connections. This design provides greater range, better scalability, and easy compatibility with existing mobile infrastructure. The arrival of 5G has further strengthened C-V2X's capabilities, offering higher data speeds and lower latency both essential for applications like autonomous driving and remote vehicle management. The DSRC remains highly reliable for immediate, short-range communication, C-V2X offers a more versatile, future-ready solution that fits seamlessly into the broader digital ecosystem. As a result, regulators and manufacturers are now at a crossroads: should they continue with the established reliability of DSRC, or invest in the long-term potential of C-V2X. The decision made today will influence the evolution of connected mobility infrastructure for decades to come.

4.3 Cloud Connectivity and Edge Computing

Connected vehicles operate in real time, which calls for a strong computational framework capable of analyzing vast amounts of data and making split-second decisions. Cloud computing and edge computing work together to provide this foundation, enabling vehicles to process information quickly and respond effectively to changing road conditions. Each plays a distinct yet complementary role in processing the immense volume of data generated by connected vehicles (Polat et al., 2024).

Cloud connectivity enables vehicles to access large-scale processing power, vast datasets, and shared learning models. This infrastructure supports long-term analytics, over-the-air updates, and centralized storage, allowing for continuous system improvements. For example, real-time traffic prediction models or firmware upgrades are typically handled via cloud platforms. However, reliance solely on the cloud presents latency and connectivity challenges, especially in fast-moving scenarios (Sadaf et al., 2023). This is where edge computing becomes essential. By processing data locally within the vehicle itself or nearby roadside units edge computing ensures immediate responses to critical safety situations. Functions like collision warnings or emergency braking cannot afford the delay of cloud transmission. The integration of cloud and edge computing creates a balanced ecosystem. While cloud systems provide strategic intelligence and historical learning, edge nodes deliver rapid, mission-critical execution. Together, they empower connected vehicles with the speed and intelligence needed to function safely and efficiently in diverse, dynamic environments (Annabi et al., 2024).

4.4 Data Security, Privacy, and Cyber Safety

As vehicles evolve into mobile data hubs, the issue of cybersecurity becomes paramount. Every sensor, application, and network connection represent a potential point of vulnerability. Ensuring the integrity, confidentiality, and availability of vehicular data is not only a technological concern but also a societal one, deeply tied to public trust and regulatory compliance (Mudhivarthi et al., 2023). Connected cars collect and transmit sensitive information ranging from location history to driver behavior. Without proper safeguards, this data could be exploited by malicious actors for identity theft, surveillance, or remote control of vehicle functions. High-profile incidents of automotive hacking have highlighted the real risks involved, prompting an urgent need for comprehensive cybersecurity strategies (Ibrahim et al., 2023).

Data encryption, secure authentication protocols, and intrusion detection systems form the core of vehicle cybersecurity frameworks. Regular software updates and patches, delivered via over-the-air (OTA) channels, are essential to address evolving threats (Shetty & Raghu, 2025). In parallel, manufacturers are increasingly adopting hardware security modules (HSMs) and secure communication standards like Transport Layer Security (TLS) to protect data in transit and at rest Privacy protection also requires clear policy frameworks. Drivers should have transparent control over how their data is used and shared. Regulations such as the General Data Protection Regulation (GDPR) in Europe and emerging automotive-specific standards globally are shaping data governance practices in the industry (Nagpal et al., 2025). Cybersecurity in connected vehicles is not a one-time fix but an ongoing commitment. With vehicles becoming increasingly autonomous and interconnected, the stakes for cyber safety have never been higher. Ensuring resilience against threats must go hand-in-hand with technological innovation to protect lives, property, and public confidence in connected mobility systems.

5. CASE STUDIES AND INDUSTRY APPLICATIONS

5.1 Volvo's Safety-First Connected Features

Volvo has long been recognized for its unwavering commitment to vehicle safety, and its approach to connected technologies reflects this legacy. Volvo's connected safety features go beyond driver convenience they are designed to actively prevent accidents and lessen the impact of collisions. These innovations rely on real-time data sharing and cooperative vehicle behavior, turning each car into a proactive participant in road safety. One standout feature is the Slippery Road Alert and Hazard Light Alert, which allow Volvo vehicles to automatically share warnings about road conditions with nearby Volvos via a cloud-based network. This early warning system gives drivers advanced situational awareness, particularly in challenging weather or unpredictable road conditions.

Through this networked approach, vehicles don't operate in isolation they contribute to a shared safety ecosystem. Volvo has also integrated real-time traffic data into its navigation systems, offering adaptive routing and predictive analytics to help drivers make smarter decisions. Pilot projects with Vehicle-to-Infrastructure (V2I) communication, such as smart traffic light coordination, further aim to reduce risks at intersections. Together, these efforts illustrate Volvo's vision of safety as a collaborative, connected outcome rather than a feature confined to a single vehicle an intelligent approach that harmonizes cars with their environment to make roads safer for everyone.

5.2 Tesla's Autopilot and Sensor Fusion

Tesla has led the way in commercially adopting advanced driver-assistance technologies, blending connectivity, sensor fusion, and machine learning into an integrated driving system. Its Autopilot platform demonstrates how real-time data and intelligent decision-making can enhance safety on the road. By combining inputs from cameras, radar, ultrasonic sensors, and GPS, the system builds a complete understanding of the vehicle's surroundings, enabling automatic lane keeping, adaptive cruise control, and even lane changes with minimal driver intervention.

A key differentiator for Tesla is its ability to continually improve these systems through over-the-air (OTA) updates. Data collected from the entire fleet helps refine algorithms and enhance performance, ensuring that every vehicle benefits from the collective learning of millions of miles driven. Tesla's ecosystem also relies heavily on cloud connectivity, allowing data from individual vehicles to be aggregated and analyzed in real time. This networked intelligence supports immediate enhancements to driver-assist functions and contributes to Tesla's long-term vision of fully autonomous vehicles. Safety remains central, with redundancies and emergency override protocols designed to maintain control in unpredictable situations. Tesla's approach shows how connectivity, centralized learning, and real-time environmental awareness can work together to push the boundaries of vehicle safety and redefine the driving experience.

5.3 Ford's SYNC and Vehicle Safety Apps

Ford's SYNC platform illustrates how connectivity can enhance safety while improving the overall driving experience. What began as a simple infotainment system has grown into a central hub for communication, navigation, and safety applications. A key element of SYNC is its integration with Ford Co-Pilot360, which combines driver-assist technologies like automatic emergency braking, lane-keeping assistance, and blind spot detection. With real-time updates on traffic and maps, drivers can make better decisions, navigate safely through congested roads, and stay prepared for changing conditions.

SYNC also connects with mobile safety apps such as Life360 and features emergency SOS functions that can automatically alert responders with crash data and GPS location, helping reduce response times. Voice control adds another layer of safety by allowing hands-free operation of critical systems, minimizing distractions. Ford's approach highlights how connected technologies can make a real difference in everyday driving. Rather than focusing solely on fully autonomous features, SYNC emphasizes practical, reliable, and proactive safety support making vehicles not just smarter, but safer, for drivers and passengers alike.

5.4 Smart City Initiatives With Connected Infrastructure

Connected car safety is not limited to individual vehicles it extends into entire urban environments. Smart city initiatives are increasingly using vehicle connectivity to ease traffic congestion, improve emergency response, and enhance overall road safety. These collaborations between automakers, city planners, and technology providers represent a new era of cooperative traffic management. Cities such as Singapore, Amsterdam, and San Diego have deployed integrated traffic systems that communicate directly with vehicles. Smart traffic signals, pedestrian alert systems, and adaptive lighting work together through Vehicle-to-Infrastructure (V2I) networks. As vehicles approach intersections, signals can be adjusted to improve flow or warn drivers of potential hazards, reducing the risk of collisions.

Connected infrastructure also helps protect vulnerable road users. AI-powered cameras and crosswalk sensors can detect pedestrians and cyclists, sending real-time alerts to nearby vehicles to enable automatic braking or lane adjustments. These systems are particularly effective in busy or low-visibility areas. The embedding intelligence into the urban environment, smart cities make roads safer and more responsive. The integration of connected vehicles and infrastructure demonstrates how these technologies can scale beyond individual cars, creating a coordinated, community-wide approach to traffic safety.

6. CHALLENGES IN IMPLEMENTING CONNECTED SAFETY SYSTEMS

6.1 Technical Challenges

The connected car technologies hold great promise, their real-world deployment faces a host of technical challenges. One of the most critical is latency, the small but crucial delay between sending and receiving data. In high-stakes situations like emergency braking or collision avoidance, even a few milliseconds can determine whether an accident is avoided or occurs. Achieving consistently ultra-reliable, low-latency communication (URLLC) across different driving environments remains a major technical hurdle. Interference is another pressing issue. In busy urban areas, the electromagnetic spectrum is crowded with numerous devices and networks. V2X communications, whether through Dedicated Short-Range Communications

(DSRC) or Cellular-V2X (C-V2X), must compete for clear channels. Signals can be disrupted by other vehicles, buildings, or weather, potentially compromising critical safety alerts when they are needed most.

Cross-platform compatibility adds further complexity. Vehicles from different manufacturers may use different software systems, communication protocols, and data formats. Without standardization, the promise of seamless interaction between vehicles is limited, leaving drivers with fragmented systems rather than a unified safety network. Engineers and regulators are working on establishing international standards and unified protocols, but legacy technologies and proprietary systems can slow adoption. Building vehicles that communicate reliably, instantly, and universally remains a work in progress. Achieving this goal requires collaboration across industries, innovative engineering, and globally coordinated efforts to transform the potential of connected safety into everyday reality.

6.2 Infrastructure Limitations and Urban-Rural Divide

Effective connected vehicle safety relies heavily on robust and intelligent infrastructure. Cities with smart traffic signals, widespread 5G coverage, and dense sensor networks can fully support advanced vehicle communication systems. However, such infrastructure is far from universal. In many rural and semi-urban areas, essential digital systems are limited or inconsistent, creating a significant urban–rural divide in the reach and effectiveness of connected technologies. Roads outside major urban centers often lack roadside units (RSUs), high-resolution maps, or reliable cellular coverage components vital for features like real-time hazard alerts or adaptive cruise control. Even in regions with more developed networks, implementing infrastructure upgrades is costly and logistically challenging, from updating traffic systems to embedding sensors and laying fiber-optic cables.

Drivers in underserved areas may not receive timely alerts about road conditions or traffic, reducing safety. Automakers may also hesitate to deploy features that rely on infrastructure available only in select locations, slowing innovation. Bridging this divide requires coordinated investment from public agencies, local governments, and private stakeholders. Public-private partnerships and targeted subsidies can accelerate deployment, while scalable technologies such as satellite communication or edge computing can provide alternatives that do not rely entirely on fixed infrastructure, helping ensure that connected vehicle safety reaches every driver.

6.3 Data Handling, Trust, and Privacy Concerns

Connected safety systems rely on the continuous exchange of information between vehicles, infrastructure, and cloud services. This data flow is essential for real-time alerts and predictive decision-making, but it also raises important questions about privacy, security, and public trust. Managing this data responsibly is one of the biggest challenges facing the connected mobility landscape. Modern vehicles collect far more than location and speed. They can gather biometric data, driving habits, and even live video feeds. Without proper protections, this information could be misused, exposing users to security risks or unauthorized surveillance. There is also ongoing debate about data ownership does it belong to the manufacturer, the cloud provider, or the driver.

Cybersecurity adds another layer of complexity. Connected vehicles are potential targets for hacking, ransomware, or system manipulation. A breach could compromise personal data and even endanger physical safety through hijacked braking systems or altered navigation. Ensuring robust cybersecurity is therefore not optional it is essential. Transparent data policies, clear consent mechanisms, and techniques like encryption and anonymization help build trust. Regulatory frameworks, such as Europe's GDPR, set important standards, but global consistency is still needed to ensure secure and reliable data handling. Ultimately, trust depends on communication. Drivers need to know what data is collected, why it's needed, and how it is protected. When transparency and security are in place, people are more likely to adopt connected technologies, creating safer, smarter, and more reliable roads for everyone.

6.4 Cost and Industry Adoption Barriers

The high cost of developing, testing, and deploying connected safety technologies can pose a significant barrier to widespread adoption, particularly for smaller automakers, logistics operators, or consumers with limited budgets. Advanced sensors, specialized communication hardware, and sophisticated software platforms require substantial investment. On top of that, maintaining real-time data processing systems, cloud infrastructure, and robust cybersecurity adds ongoing operational expenses. For vehicle manufacturers, these costs often translate into higher prices, which premium brands may pass on to tech-savvy buyers. For the broader market, however, the additional expense can seem prohibitive, especially in regions where safety regulations do not mandate such technologies. This could create a two-tiered system, where only certain consumers benefit from advanced connected safety features.

Widespread adoption also depends on collaboration across the industry. Technology providers, telecom companies, governments, and automakers need to coordinate their systems, timelines, and objectives. Yet differences in market readiness, competitive priorities, and regulatory frameworks often create friction. Uncertainty about return on investment and evolving standards can make stakeholders hesitant to commit. Financial incentives, subsidies, and clear policy guidance can help address these challenges. Pilot programs, public sector procurement, and phased rollout strategies offer pathways for sustainable growth. As the technology matures and economies of scale improve, the cost of implementation is expected to decrease, making connected safety features not just a luxury, but a fundamental component of everyday mobility.

7. FUTURE TRENDS AND INNOVATIONS

7.1 Integration with AI and Machine Learning

Artificial Intelligence (AI) and Machine Learning (ML) are revolutionizing the way vehicles understand and respond to the world around them. As connected cars become more intelligent, these technologies are no longer just optional enhancements they are becoming central to the evolution of automotive safety. Artificial intelligence (AI) and machine learning (ML) enable vehicles to "learn" from their environment. Connected cars can process massive amounts of sensor data ranging from road conditions and traffic patterns to driver behavior and make intelligent decisions in real time. For example, rather than simply detecting a pedestrian crossing, an AI system can anticipate their likely path and adjust speed or trajectory accordingly, significantly reducing the risk of collisions.

Machine learning is also essential for adaptive systems. Features like adaptive cruise control or lane-keeping assist can tailor their responses to a driver's habits or changes in the environment. Over time, these systems become increasingly accurate and reliable, drawing insights from millions of real-world scenarios encountered by vehicles worldwide. AI also strengthens cybersecurity. Threat detection algorithms and anomaly recognition systems help connected vehicles identify potential attacks or malfunctions before they escalate, keeping both passengers and vehicle systems safe. Looking ahead, AI is poised to integrate more deeply across every aspect of vehicle operation from engine performance and energy management to voice-based driver support. The connected vehicle of the future will be less reactive and more predictive, using continuous learning to anticipate challenges, enhance safety, and provide a smarter, more responsive driving experience.

Figure 6. Real time monitoring of Road environment.

7.2 Predictive Safety Algorithms and Big Data

Predictive safety is emerging as one of the most exciting advancements in the connected vehicle space. Instead of merely reacting to incidents after they occur, future cars will be able to anticipate risks before they happen, thanks to predictive algorithms and big data analytics. These systems draw on massive volumes of data from a variety of sources onboard sensors, vehicle-to-vehicle (V2V) communication, road infrastructure, weather reports, and even mobile apps. By analyzing these inputs, predictive models can identify potential hazards. For example, if a stretch of road is known for sharp curves, frequent accidents, or dense fog, the system can alert the driver in advance, even if conditions initially seem normal. Machine learning takes this a step further. By processing historical crash data, vehicle behavior logs, and real-time feedback, the system continuously improves its predictive accuracy. It's like having a personal data scientist riding alongside you, constantly evaluating risk and offering guidance tailored to your environment.

Big data also enables fleet-level insights. Aggregated information from multiple vehicles can help manufacturers and city planners redesign dangerous intersections, adjust speed limits, or deploy emergency services more efficiently. In this way, each vehicle becomes a sensor, contributing to a smarter and safer transportation network. Ultimately, predictive safety transforms driving from a reactive activity into a proactive one. Drivers are not just receiving more information they're being guided by intelligent recommendations that help them avoid danger before it arises.

7.3 Roadmap to Fully Autonomous, Safe Vehicles

The vision of fully autonomous vehicles cars that drive themselves with no human intervention has captured the imagination of engineers, investors, and everyday commuters alike. While remarkable progress has been made, achieving complete autonomy remains a complex challenge, especially when safety is the ultimate goal. Autonomous vehicles are classified into five levels. Level 1 and 2 offer driver-assistance features such as steering support or adaptive cruise control. Level 3 introduces conditional automation, where the car can manage certain tasks independently but still requires human oversight. Levels 4 and 5 represent full autonomy, with Level 5 requiring no driver at all. Reaching these higher levels demands an intricate blend of technologies: high-definition mapping, lidar and radar sensors, V2X communication, AI-driven decision-making, and redundant safety systems. One of the toughest challenges is ensuring that these vehicles can navigate unpredictable real-world environments heavy traffic, adverse weather, or erratic human drivers with consistent reliability.

Testing and validation are critical along this journey. Companies like Waymo, Cruise, and Tesla are leveraging billions of miles of driving data to train and refine their autonomous platforms. Simulations and digital twin's virtual replicas of vehicles and road networks allow engineers to test rare but crucial "edge cases" that might not appear during real-world driving. Equally important is public trust. Even the most advanced systems will struggle to gain adoption if people don't feel safe riding in a driverless car. Transparent reporting, regulatory oversight, and a focus on passenger experience are essential to bridging this gap. While the path to fully autonomous, safe vehicles is far from straightforward, each innovation brings us closer to a future where self-driving cars are not only possible but reliable and secure.

7.4 Role of Digital Twins in Vehicle Safety

Digital twins are quickly becoming a game-changer in the automotive industry. These virtual counterparts of physical vehicles, systems, or environments allow engineers to explore and analyze real-world behavior without leaving the lab. In terms of safety, digital twins offer a proactive approach to prevention, prediction, and continuous improvement. Imagine a connected car mirrored in real time in the cloud: every maneuver braking, accelerating, or turning is tracked virtually. If the digital twin detects abnormal brake stress or unusual fuel consumption, it can alert engineers or initiate maintenance before a problem ever reaches the road. Crash testing has been transformed as well. Engineers can simulate thousands of crash scenarios digitally, testing and refining features like airbags and crumple zones without costly physical prototypes. When combined with AI, digital twins gain predictive capabilities, such as forecasting tire wear or aquaplaning risk for vehicles frequently exposed to rain. This allows for targeted interventions, enhancing safety in a personalized way.

Beyond individual vehicles, digital twins are also reshaping urban planning. By modeling traffic flow, intersection layouts, and signal timings, city planners can design safer road networks and optimize traffic management before construction begins. Essentially, digital twins bridge the physical and digital worlds, giving manufacturers, regulators, and drivers unparalleled insight. As these tools evolve, they promise safer vehicles not only at the point of sale but throughout their entire operational life.

Table 1. Comparative Analysis of Research Papers on Connected Car Safety Technologies

Author & Year	Methodology	Key Findings	Limitations
Petrov et al., 2022	Simulation benchmarking LTE-based vs 5G C-V2X in urban V2I scenarios	5G C-V2X yielded significantly lower end-to-end latency and higher packet delivery ratio (PDR) than 4G LTE	Based entirely on simulations; may lack real-world variability
Coll-Perales et al., 2022	Analytical modeling of latency for edge vs cloud deployments in V2X (V2N2V)	Edge deployment of 5G reduced latency drastically compared to centralized cloud models	Based on theoretical frameworks with limited field validation
Tahir et al., 2022	Real-world tests comparing IEEE 802.11p vs 5G Test Network (5GTN)	5GTN demonstrated higher throughput and range; DSRC (802.11p) maintained more stable, lower-latency performance	5GTN showed higher jitter, and robustness may vary by deployment
Ali et al., 2024	Survey and analysis of V2X communication technologies	5G C-V2X offers comprehensive V2X support (V2V, V2I, V2P, V2N) with high throughput and low latency	Early-stage survey; technology in evolving deployment stage

The comparative analysis presents a spectrum of strategies aimed at improving vehicle safety through connected car technologies, as shown in Table 1. Each approach contributes a unique piece to the overall puzzle enhancing real-time communication between vehicles and their environment, embedding intelligent sensors for quicker and more accurate responses, or using AI-driven predictive tools to foresee potential hazards. A recurring theme across these methods is the reliance on V2X communication, which enables vehicles to share critical information and react swiftly to changing conditions. Technologies like 5G are accelerating these capabilities, allowing faster, more reliable data exchange that helps vehicles make smarter, faster decisions. Yet, alongside these technological gains, challenges such as ethical considerations, regulatory gaps, and public trust remain significant.

The analysis makes clear that achieving safety in connected cars goes beyond innovation alone. It requires a well-integrated system that balances cutting-edge technology with sound policy and a focus on human needs. By combining technical, regulatory, and ethical layers, the journey toward safer roads becomes not only smarter but more inclusive for all road users.

CONCLUSION

Connected car technologies are redefining safety in modern transportation. With advanced communication systems like Vehicle-to-Vehicle (V2V), Vehicle-to-Infrastructure (V2I), Vehicle-to-Pedestrian (V2P), and Vehicle-to-Network (V2N), vehicles can now interact with each other and their surroundings in real time. This connectivity allows cars to anticipate hazards, respond to changing road conditions, and navigate traffic more intelligently. Core components such as embedded sensors, onboard control units, and cloud-based analytics work together to support decision-making that prevents accidents and assists drivers. The integration of IoT frameworks with 5G networks ensures low latency, reliable data exchange, and seamless communication. Safety-focused technologies including Advanced Driver Assistance Systems (ADAS), autonomous emergency braking, and collision avoidance systems have proven effective in reducing the likelihood and severity of accidents.

Industry examples from Tesla, Volvo, and Ford highlight real-world deployments, while smart city initiatives demonstrate how urban infrastructure can enhance connected vehicle performance, creating safer and more responsive mobility environments. Despite these advances, challenges remain. Data privacy, cybersecurity, inconsistent infrastructure, and latency concerns continue to pose obstacles. Overcoming these issues requires coordinated efforts among automakers, regulators, technology providers, and city planners. By combining innovation with strategic collaboration, connected car technologies promise a future of safer, smarter, and more inclusive roadways.

REFERENCES

Alam, T. (2024). Data privacy and security in autonomous connected vehicles in smart city environment. *Big Data and Cognitive Computing*, *8*(9), 95. DOI: 10.3390/bdcc8090095

Ali, G. G. M. N., Sadat, M. N., Miah, M. S., Sharief, S. A., & Wang, Y. (2024). A Comprehensive Study and Analysis of the Third Generation Partnership Project's 5G New Radio for Vehicle-to-Everything Communication. *Future Internet*, *16*(1), 21. DOI: 10.3390/fi16010021

Alshamrani, S. S., Jha, N., & Prashar, D. (2021). B5G ultrareliable low latency networks for efficient secure autonomous and smart Internet of vehicles. *Mathematical Problems in Engineering*, *2021*(1), 3697733. DOI: 10.1155/2021/3697733

Alsudani, M. (2023). The wireless communication between vehicles: exploring the potential of V2V and V2X communication for improved efficiency, safety, and sustainability: wireless communication between (V2V) and (V2I, V2X). Journal of Millimeterwave Communication. *Optimization and Modelling*, *3*(1), 9–13.

Annabi, M., Zeroual, A., & Messai, N. (2024). Towards zero trust security in connected vehicles: A comprehensive survey. *Computers & Security*, *145*, 104018. DOI: 10.1016/j.cose.2024.104018

Bagheri, H., Noor-A-Rahim, M., Liu, Z., Lee, H., Pesch, D., Moessner, K., & Xiao, P. (2021). 5G NR-V2X: Toward connected and cooperative autonomous driving. *IEEE Communications Standards Magazine*, *5*(1), 48–54. DOI: 10.1109/MCOMSTD.001.2000069

Coll-Perales, B., Lucas-Estañ, M. C., Shimizu, T., Gozalvez, J., Higuchi, T., Avedisov, S., Altintas, O., & Sepulcre, M. (2022). End-to-end V2X latency modeling and analysis in 5G networks. *IEEE Transactions on Vehicular Technology*. Advance online publication.

Gupta, P., Wang, R., Ard, T., Han, J., Karbowski, D., Vahidi, A., & Jia, Y. (2023, October). An x-in-the-loop (xil) testing framework for validation of connected and autonomous vehicles. In *2023 IEEE International Automated Vehicle Validation Conference (IAVVC)* (pp. 1-6). IEEE. DOI: 10.1109/IAVVC57316.2023.10328040

Gürses, A., Reddy, G., Masrur, S., Özdemir, Ö., Güvenç, İ., Sichitiu, M. L., Şahin, A., Alkhateeb, A., Mushi, M., & Dutta, R. (2025). Digital twins and testbeds for supporting AI research with autonomous vehicle networks. *IEEE Communications Magazine*, *63*(4), 56–62. DOI: 10.1109/MCOM.001.2400222

Hasan, M., Mohan, S., Shimizu, T., & Lu, H. (2020). Securing vehicle-to-everything (V2X) communication platforms. *IEEE Transactions on Intelligent Vehicles, 5*(4), 693–713. DOI: 10.1109/TIV.2020.2987430

Ibrahim, M., Rjabtšikov, V., & Gilbert, R. (2023). Overview of digital twin platforms for EV applications. *Sensors (Basel), 23*(3), 1414. DOI: 10.3390/s23031414 PMID: 36772453

Islam, M. M., & Karimoddini, A. (2022). Pedestrian detection for autonomous cars: Inference fusion of deep neural networks. *IEEE Transactions on Intelligent Transportation Systems, 23*(12), 23358–23368. DOI: 10.1109/TITS.2022.3210186

Ke, R., Cui, Z., Chen, Y., Zhu, M., Yang, H., & Wang, Y. (2020). Edge computing for real-time near-crash detection for smart transportation applications. arXiv preprint arXiv:2008.00549

Lee, D., & Hess, D. J. (2022). Public concerns and connected and automated vehicles: Safety, privacy, and data security. *Humanities & Social Sciences Communications, 9*(1), 1–13. DOI: 10.1057/s41599-022-01110-x PMID: 36187845

Li, X., & Zhang, W. (2010). An adaptive fault-tolerant multisensor navigation strategy for automated vehicles. *IEEE Transactions on Vehicular Technology, 59*(6), 2815–2829. DOI: 10.1109/TVT.2010.2050014

Liu, S., Gao, C., Chen, Y., Peng, X., Kong, X., Wang, K., & Wang, M. (2023). Towards vehicle-to-everything autonomous driving: A survey on collaborative perception. arXiv preprint arXiv:2308.16714.

Martinez, F. J., Toh, C. K., Cano, J. C., Calafate, C. T., & Manzoni, P. (2011). A survey and comparative study of simulators for vehicular ad hoc networks (VANETs). *Wireless Communications and Mobile Computing, 11*(7), 813–828. DOI: 10.1002/wcm.859

Molinaro, A., Campolo, C., Härri, J., Esteve Rothenberg, C., & Vinel, A. (2020). 5g-v2x communications and networking for connected and autonomous vehicles. *Future Internet, 12*(7), 116. DOI: 10.3390/fi12070116

Mudhivarthi, B. R., Thakur, P., & Singh, G. (2023). Aspects of cyber security in autonomous and connected vehicles. *Applied Sciences (Basel, Switzerland), 13*(5), 3014. DOI: 10.3390/app13053014

Nagpal, M., Chakraborty, S., & Gupta, M. R. N, S. Vigneshwari and P. Chaudhary, "Integrating Blockchain Technology for Secure and Decentralized Network Communication," *2025 International Conference on Automation and Computation (AUTOCOM)*, Dehradun, India, 2025, pp. 512-517. DOI: 10.1109/AUTOCOM64127.2025.10957352

Ouaissa, M., Ouaissa, M., Houmer, M., El Hamdani, S., & Boulouard, Z. (2022). A secure vehicle to everything (v2x) communication model for intelligent transportation system. In *Computational intelligence in recent communication networks* (pp. 83–102). Springer International Publishing. DOI: 10.1007/978-3-030-77185-0_6

Petrov, T., Pocta, P., & Kovacikova, T. (2022). Benchmarking 4G and 5G-Based Cellular-V2X for Vehicle-to-Infrastructure Communication and Urban Scenarios in Cooperative Intelligent Transportation Systems. *Applied Sciences (Basel, Switzerland)*, *12*(19), 9677. DOI: 10.3390/app12199677

Polat, A. O., Erden, B. C., Kul, S., & Nasiroglu, F. (2024). Light electric vehicle performance with digital twin technology: A comparison of motor types. *Arabian Journal for Science and Engineering*, *49*(5), 7209–7222. DOI: 10.1007/s13369-023-08668-x

Raghu, N., Mahajan, R. A., Sivanantham, A., Bareja, L., Rayaguru, N. K., & Ganesh, R. "Optimizing Real-Time Process Control through Cloud-Based Automation and Big Data Analysis," 2024 15th International Conference on Computing Communication and Networking Technologies (ICCCNT), Kamand, India, 2024, pp. 1-6. DOI: 10.1109/ICCCNT61001.2024.10725875

Raghu, N., Phadke, A., Kumar, R., & Joshi, K. (2023). Smart Traffic Light and Dynamic Navigational System for Emergency Vehicles. In Ranganathan, G., Fernando, X., & Piramuthu, S. (Eds.), *Soft Computing for Security Applications. Advances in Intelligent Systems and Computing* (Vol. 1428). Springer. DOI: 10.1007/978-981-19-3590-9_30

Rupanagudi, S. R.. (2020). A Novel Air Gesture Based Wheelchair Control and Home Automation System. In Abraham, A., Cherukuri, A., Melin, P., & Gandhi, N. (Eds.), *Intelligent Systems Design and Applications. ISDA 2018 2018. Advances in Intelligent Systems and Computing* (Vol. 941). Springer., DOI: 10.1007/978-3-030-16660-1_71

Rupanagudi, S. R., ... "Optic Disk Extraction and Hard Exudate Identification in Fundus Images using Computer Vision and Machine Learning," *2021 IEEE 11th Annual Computing and Communication Workshop and Conference (CCWC)*, NV, USA, 2021, pp. 0655-0661, DOI: 10.1109/CCWC51732.2021.9376018

Sadaf, M., Iqbal, Z., Javed, A. R., Saba, I., Krichen, M., Majeed, S., & Raza, A. (2023). Connected and automated vehicles: Infrastructure, applications, security, critical challenges, and future aspects. *Technologies, 11*(5), 117. DOI: 10.3390/technologies11050117

Samak, T. V., Samak, C. V., & Krovi, V. N. (2024, July). Towards Validation of Autonomous Vehicles Across Scales using an Integrated Digital Twin Framework. In *2024 IEEE International Conference on Advanced Intelligent Mechatronics (AIM)* (pp. 1068-1075). IEEE. DOI: 10.1109/AIM55361.2024.10637205

Sarsam, S. M. (2023). Cybersecurity challenges in autonomous vehicles: Threats, vulnerabilities, and mitigation strategies. *SHIFRA, 2023*, 34–42. DOI: 10.70470/SHIFRA/2023/005

Shetty, G. S., & Raghu, N. (2025). Waterwheel Archimedes Optimization and Resilient Consensus BiLSTM for Efficient Data Aggregation to Detect and Discard Malicious Nodes and Packets in IoT-Based Healthcare Applications. *SN Computer Science, 6*(4), 338. DOI: 10.1007/s42979-025-03873-x

Shoukat, M. U., Yan, L., Yan, Y., Zhang, F., Zhai, Y., Han, P., Nawaz, S. A., Raza, M. A., Akbar, M. W., & Hussain, A. (2024). Autonomous driving test system under hybrid reality: The role of digital twin technology. *Internet of Things : Engineering Cyber Physical Human Systems, 27*, 101301. DOI: 10.1016/j.iot.2024.101301

Tahir, M. N., Katz, M., & Rashid, U. (2022). Analysis of collaborative wireless vehicular technologies under realistic conditions. *Journal of Engineering, 2022*(2), 201–209.

Taleb, T., Samdanis, K., Mada, B., Flinck, H., Dutta, S., & Sabella, D. "On Multi-Access Edge Computing: A Survey of the Emerging 5G Network Edge Cloud Architecture and Orchestration," IEEE Communications Surveys & Tutorials, vol. 19, no. 3, pp. 1657–1681, 3rd Quart. 2017, DOI: 10.1109/COMST.2017.2705720

Vanjire, S., & Naveen, R. (2024). AI on wheels: AI-powered predictive models for smart vehicles. In *Interactive media with next-gen technologies and their usability evaluation* (pp. 155–164). Chapman and Hall/CRC. DOI: 10.1201/9781032664828-11

Verma, S., Sharma, A., Tran, B., & Alahakoon, D. (2024). A systematic review of digital twins for electric vehicles. [English Edition]. *Journal of Traffic and Transportation Engineering, 11*(5), 815–834. DOI: 10.1016/j.jtte.2024.04.004

Zhang, Y., Lin, Y., Qin, Y., Dong, M., Gao, L., & Hashemi, E. (2023). A new adaptive cruise control considering crash avoidance for intelligent vehicle. *IEEE Transactions on Industrial Electronics, 71*(1), 688–696. DOI: 10.1109/TIE.2023.3239878

Chapter 8

IoT and AI for Real-Time Safety Monitoring

Prakash Upadhyay
Xavier University, India

S. Pushpalatha
http://orcid.org/0000-0001-7664-3078
Dr. Ambedkar Institute of Technology, India

N. Raja Kumar
Lords Institute of Engineering and Technology, India

Vinay Raj R.
http://orcid.org/0000-0003-3829-8327
Yenepoya (Deemed to be) University, India

Anjaneyulu Gudla
Guru Nanak Internation Technical Campus, India

R. Suyam Praba
Karpagam College of Engineering, India

Bala Krishnan S.
http://orcid.org/0009-0000-9391-4458
Vaagai IP, India

ABSTRACT

Industrial systems, healthcare, smart cities, and transportation are just a few of the many areas that stand to benefit from the new opportunities for real-time safety monitoring made possible by the merging of the IoT with AI. Internet of Things (IoT) enables continuous data collection through linked devices and sensors; artificial intelligence (AI) enhances this ecosystem by real-time data analysis, trend detection, risk forecasting, and preventative action initiation. This chapter delves into the new frontiers of safety monitoring made possible by the integration of IoT and AI. It covers several frameworks and architectures, showcases real-world applications, and

DOI: 10.4018/979-8-3373-3231-4.ch008

tackles issues like data privacy, latency, scalability, and energy efficiency. Future directions are also covered, including edge AI, federated learning, and the Internet of Things made possible by 6G.

1. INTRODUCTION

In today's digital age, being competitive requires automating and optimising operations across numerous industries. In an ongoing effort to increase productivity, decrease expenses, and ensure quality, the incorporation of new technology has been a constant(Rathi et al., 2021). This landscape positions real-time monitoring of systems and devices as an essential tool, particularly in industries where quick and accurate decisions are crucial(Zayed et al., 2023). Technicians and operators' best judgement and years of expertise were traditionally relied upon while monitoring processes and equipment manually(Villegas-Ch et al., 2024). This approach has its limitations, too, and those constraints become most apparent in large-scale or complicated systems that need the simultaneous monitoring of several parameters. In addition, the requirement for increasingly sophisticated systems has been brought to light by the development of Industry 4.0 and the increasing desire for smart and linked solutions. The benefits of using monitoring systems built on new technologies like the IoT and Artificial Intelligence (AI) have been shown in earlier research(Ding, 2021). These resources allow for the gathering of data in real-time and provide sophisticated analytical and predictive capabilities. Given this context, the paper's authors propose developing an AI-powered filtering system specifically designed to keep tabs on happenings as they happen. Superior accuracy and efficiency are guaranteed by this tool, which constantly adapts and learns from the data collected, unlike conventional systems. It includes a number of sensors, such as accelerometers and Internet of Things cameras, to give a complete and detailed picture of the process. There are a lot of monitoring systems out there, but one novel idea is to combine sensors that can learn and modify themselves. Thus, our study yields practical information about the functioning and efficacy of such a system and offers a fresh viewpoint on how new technology might conquer long-standing obstacles in industrial control and monitoring. Early interventions and the avoidance of expensive downtime or breakdowns are made possible by this tool's ability to detect and forecast trends and patterns. The innovative combination of artificial intelligence and the internet of things makes our idea one of the best options out there right now. It sets a new standard for real-time monitoring. Although many academics have tackled the problem of real-time monitoring, few have successfully incorporated AI adaptability and continuous learning capabilities, according to the available literature(Fredj et al., 2021). Our technology can do in-depth analyses and predictions, while other systems just collect and display data. In addition, our idea allows for a detailed and exact picture of the operation by integrating various sensors, which overcomes the limits of existing systems.

2. IOT IN SAFETY MONITORING

The ongoing curiosity of researchers has created a multitude of safety concerns(Hussein et al., 2024). The following lines provide a comprehensive review of current safety research, touching on topics like "Safety effectiveness measurements," "Improvement of safety performance," and "Innovative approaches for safety training" and "innovative approaches for safety monitoring."

2.1 Improvement of Safety Performance

There was a need for a quantitative study of seasoned construction practitioners, a focus on learning-by-doing and interacting with people and machinery on the job, strategies to improve workers' safety knowledge through skill development, and participation from universities to enhance the safety learning process by (Sunindijo et al., 2017) as four areas that could improve safety performance in the construction industry. Although basic safety training satisfies the minimal legislative criteria, they contended that the improvement of safety knowledge and awareness relies on additional activities, such as information exchange(Wong et al., 2020). It is crucial to take cultural and linguistic elements into account while developing training modules. Two questionnaires, one administered immediately following safety instruction and another sixteen months later, were used to assess such matters(Dhanraj et al., 2020). Training efficacy differs among participants' cultural origins, according to the findings. Additionally, they covered the topic of culturally responsive training and found that employees got the best outcomes when their courses were tailored to their own backgrounds. The difficulties in safety training due to employees' varying linguistic abilities were also highlighted.

2.2 Safety Effectiveness Measurements

Proper implementation of a Safety Management System (SMS) at various organisational levels is necessary for a construction project's safety performance to reach its maximum potential. The Singaporean construction industry implemented SMS about 30 years ago, but there has been no discernible uptick in safety(Dudhe et al., 2018). used a 15-step study process including surveys, seminars, consultations with safety experts, and interviews to address this gap. With the goal of improving site safety and developing a management tool to monitor the efficacy of safety measures, they set out to develop a multi-attribute value model that could be validated through audits. This model would serve as the basis for the Construction Safety Index (CSI). Ricci et al. (2016) proposed a method for measuring training efficacy that included a questionnaire, practical examinations, physiological data, on-the-job reservation, and documentation databases. The effect-size of training efficiency was determined using 44 different indicators in the meta-analysis, which drew from 28 studies. Three months following training, the benefits were discovered to be considerably diminished.

2.3 A New Viewpoint on Safety Training

Using modern technology and tools is one practical way to improve construction site safety by developing sensible and suitable safety procedures. This is where Building Information Modelling (BIM) has really shone as an instrument for enhancing the safety of construction workers on the job. Came up with a BIM-enabled safety training technique that took this into account. It incorporated a 3D-visualization environment and interactive features for the learners(Clevenger et al., 2015). The computer models were lauded by participants as an engaging aspect that enhanced the training experience, and the feedback was generally excellent. In safety training, this study cleared the way for the use of high-tech computer systems to supplant more antiquated forms of instruction like paper, slides, and videos. To lessen the possibility of accidents occurring on the work site, proposed an automatic safety plug-in for scheduling software(et al., 2018). Researched the feasibility of an IoT-based wireless network for site safety monitoring. Various sensor types, including digital cameras and smoke detectors, can be interconnected using a WIFI LAN. Although the problem of field devices was not addressed, the idea of utilising the IOT to enhance safety on building sites was novel.

2.4 Innovation Methods for Safety Monitoring

In order to enhance the effectiveness of proactive safety monitoring, suggested a safety identification system that is still in its early stages of design. Wireless sensor networks and radio-frequency identification (WSN), heavy equipment access control, materials usage, and restricted area monitoring were all accomplished. This approach integrated Zigbee protocols and RFID readers into the routine operations of construction sites to track workers' well-being and collect information for use in developing future safety plans. Data volume, complexity, and kind were all constrained by the processor's power, bandwidth, and hardware limitations, particularly its size and battery life.. (Kumar & Tauseef, 2021). In used concept of the Internet of Things to develop a tool that can detect personal protective equipment (PPE). This will guarantee that the right construction personnel are provided with personal protective equipment (PPE) prior to the commencement of certain operations. An further study created a safeguarding IoT system to monitor, identify, and alert construction workers working in potentially dangerous regions.(Ramadan et al., 2024).

2.5 Knowledge Gap and Point of Departure

Prior studies have concentrated on determining if IoT technology is feasible using locally available sensors and without large production. Easily accessible sensors from the Hong Kong market have been utilised in this paper. Because of the intense competition in the BMS market, the cost of these sensors is usually rather cheap(Zhao & Hu, 2022). Due to the prevalence of web-based interfaces, this study presents a comprehensive safety monitoring system that is both easy to install and understand, and it is more cost-effective than the traditional wiring monitoring method.

3. ROLE OF AI IN SAFETY MONITORING

3.1 Examination of AI and Smart Technology Use Cases in Different Industries

After a thorough evaluation, the selected studies were categorised into eight distinct industries to have a better understanding of the distinct applications of AI and smart technologies in risk management. (Park & Kang, 2024). This industry breakdown offers a comprehensive look at how several fields make use of cutting-edge innovation to tighten up safety measures and lessen potential dangers.

3.1.1 Smart Technology and Artificial Intelligence Integration in Construction Safety Management

There has been a recent uptick in the construction industry's efforts to improve safety management through the use of smart technology and artificial intelligence. Systems powered by artificial intelligence can improve many parts of safety management. These include training, predictive analytics, and real-time danger detection. As an example, smart safety management systems powered by AI can improve decision-making by monitoring safety data in real-time through the use of Internet of Things (IoT) sensors, Bluetooth Low Energy (BLE) beacons, and cloud computing (Obianuju Ozobu et al., 2025). In order to reduce the likelihood of accidents, AI systems analyse trends in order to forecast possible dangers, while computer vision uses automated picture recognition to identify risky behaviours. Virtual and augmented reality systems powered by AI also offer immersive training settings, which can help with recognising hazards and responding appropriately in dangerous places like tunnels. Table 1 shows the research methods used to examine the relationship between smart technology, AI, and aspects related to construction safety management.

Table 1. Techniques for researching smart technology and artificial intelligence as they pertain to construction safety management factors.

Safety Management Factor	AI and Smart Technology Research Methods
Digital Technology and Smart Safety Management System (SMS)	Employed EPS32, Bluetooth, and beacons to build safety applications for cellphones. Built hybrid apps using web technologies such as CSS, HTML, and JavaScript.
Accident Prevention and Response	Created algorithms for detecting potential dangers using AI. Developed an AI and sensor system for real-time emergency detection
IoT and AI-based Safety Management	Built IoT safety equipment. Took advantage of BLE Beacon to monitor employees in real-time. Improved danger detection with the integration of machine vision and AI.

3.1.2 Management of Industrial Safety by the Use of Artificial Intelligence and Other Smart Technologies

With the help of smart technologies and AI, the industrial safety sector is undergoing a transition in its safety management methods. This transformation is addressing critical issues such as data-driven decision-making, real-time monitoring, and technology-centered approaches. Results from the examined research show that AI algorithms, ML, and IoT devices can improve safety(Jayakody & Thilakarathne, 2025). By analysing massive volumes of safety incident data, AI-powered predictive analytics, for instance, can improve Safety Performance Indicators (SPIs) and allow for the real-time identification of risk factors. AI and the IoT enable smart settings to monitor in real-time, responding instantly to dangers and improving safety efficiency overall. In Table 2, we can see the many approaches to studying industrial safety management elements using AI and smart technologies.

Table 2. Industrial safety study utilising artificial intelligence and smart technology for management factors.

Safety Management Factor	AI and Smart Technology Research Methods
Technology-centered Approach	Incorporated indoor Geographic Information System (GIS) and positioning and navigational technologies into the PIPSC/CMF architecture. Indoor positioning and smoke detector passive labels that use both Bluetooth and radio frequency identification.
Data-driven Decision-making	Investigated the origins and trends of fires by employing data mining and statistical analysis. Discovered fire hotspots and ran fire simulations using Big Data Analysis.
Smart Environments and Real-time Monitoring	Built AI-powered solutions for disaster response in real-time, integrating data from social networks, sensors, intelligent CCTV, and historical records.

3.1.3 The Application of AI and Other Smart Technologies to Public and Disaster Safety Management

By using smart technology and AI, the public and catastrophe safety industry is enhancing critical safety management features. These components include a broad variety of subjects, including industrial safety in potentially dangerous workplaces, public safety in densely populated areas, disaster and fire management, and the incorporation of smart technology into public safety. Artificial intelligence (AI) solutions take these important elements into account by utilising real-time monitoring systems, predictive analytics, and intricate frameworks(Tselentis et al., 2023). Using indoor GIS in conjunction with Bluetooth and RFID-based location technologies, the PIPSC/CMF framework constructs integrated fire detection systems and command platforms. Also, AI-powered disaster response solutions help with real-time emergency management by integrating data from social media, smart CCTV, sensors, satellite information, and previous disasters.

3.1.4 Transportation and Logistics Safety Management with the Use of AI and Other Smart Technologies

Key safety management criteria are being greatly improved in the transportation and logistics business with the help of AI and other smart technologies(Wang & Chung, 2022). Preventing traffic accidents and collisions, ensuring the safety of railway personnel, managing safety on industrial and maritime facilities, and public transport safety are all elements to consider. Machine learning algorithms and AI-powered prediction models assess environmental factors, vehicle actions in real-time, and traffic data to foretell and avoid collisions. By lowering the likelihood of collisions and increasing the efficacy of collision avoidance systems, these technologies pave the way for more proactive traffic management and safer driving. By analysing data collected by Internet of Things (IoT) sensors, AI algorithms can predict when equipment will break down, which in turn increases system reliability and decreases the likelihood of accidents. Enhanced rail safety is a result of this. In addition, literature research and accident statistics analyses help determine the most common forms of railway worker accidents, which are then used to inform the creation of safety management systems that use smartphone GPS technology. Another piece of evidence for the effectiveness of SMS comes from case studies like the Lac-Mégantic accident and comparative assessments of regulatory systems. Autonomous vessels and AI-driven navigation systems both use AI to reduce the likelihood of maritime accidents. Automated monitoring of potentially dangerous conditions is one way that predictive analytics helps keep industrial sites safe. Surveillance systems that use AI to identify suspicious activity using facial recognition and behavioural analysis improve the management of public transportation safety. Big Data analytics can also help public transport systems with safety measures and emergency procedures.

3.1.5 The Use of Artificial Intelligence and Other Smart Technologies in Health Care Safety Management

Many key issues, such as smart health monitoring and prevention, mental health management and prevention, and the integration of technology with ethical execution, are considerably improved by AI and smart technologies, which in turn greatly improve health care safety management. Utilising AI-powered predictive analytics and machine learning algorithms allows for improved risk management and prevention for healthcare professionals, including infectious disease and ergonomic hazards. This allows us to take measures to improve workplace safety and implement new safety procedures. Wearables with IoT and AI sensors gather vital signs like blood pressure and heart rate in real time, which is useful for smart health monitoring and prevention. By analysing this data, AI systems can spot irregularities, forecast health occurrences, and make tailored suggestions, all of which contribute to better patient safety by allowing for quicker diagnosis and treatment. Table 3 displays the study techniques for AI and smart technologies in relation to health sector safety management elements.

Table 3. Research techniques for health sector safety management variables using artificial intelligence and smart technology

Safety Management Factor	AI and Smart Technology Research Methods
Occupational Risk Management and Prevention	Executed risk assessments, which included locating, analysing, and rating potential dangers in the workplace. Created methods for machine learning and predictive analytics using artificial intelligence to compute
Smart Health Monitoring and Prevention	Created wearable gadgets that can measure vital signs in real time using artificial intelligence and internet of things sensors.
Mental Health Management and Prevention	Built mental health support applications and AI chatbots that use natural language processing.

3.1.6 Management of Smart Home and Living Safety via the Use of Artificial Intelligence and Other Smart Technologies

When it comes to managing safety, the smart home and living industry is bringing a change thanks to AI and smart tech. Protecting people and their property from hazards like fire and carbon monoxide poisoning, improving operational efficiency, managing health, and staying in compliance with regulations are all major concerns that these advancements aim to address.

4. IOT–AI INTEGRATION FRAMEWORK FOR REAL-TIME SAFETY

For use in both urban and industrial settings, the suggested architecture combines AI and the Internet of Things to provide a data-driven safety management system that is both responsive and dynamic(Ahmed et al., 2024). In line with sustainability and operational efficiency objectives, this framework is built to improve real-time monitoring, predictive risk identification, and automated safety responses.

4.1 Conceptual Overview

The framework's multi-layered design relies on interconnected sensor networks enabled by the IoT and analytics driven by AI to identify, forecast, and avert potentially dangerous situations found that the system can be easily implemented in smart cities, industrial facilities, and other infrastructure areas that need to continually monitor and minimise operational and environmental safety hazards(Villegas-Ch et al., 2024) A scatter plot Figure 1 illustrates the relationship between traffic volume and the number of reported accidents.

Figure 1. Proposed Framework: Smart IOT-AI Integration for Safety

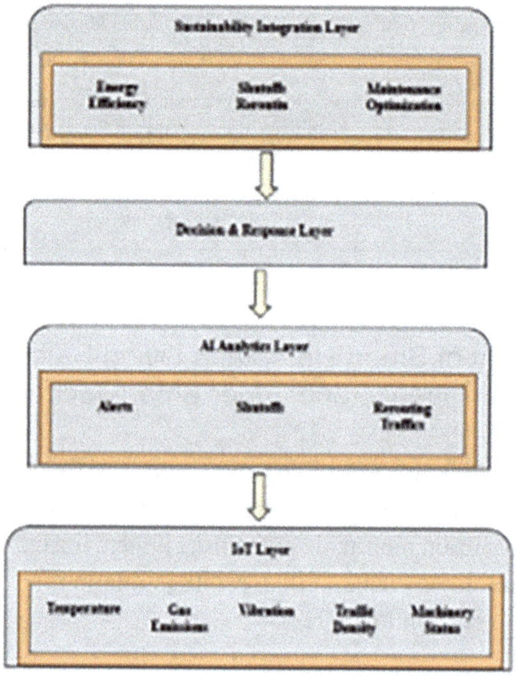

- **Layer for the Internet of Things**: A variety of characteristics, including temperature, gas emissions, vibration, traffic density, and machinery status, are monitored in real-time by the deployed sensors. (Razzaque et al., 2016)A centralised platform receives these data streams and processes them.
- **AI Analytics Layer:** By analysing sensor data, machine learning models may spot irregularities, alert users to potential dangers, and forecast when equipment will fail, whether it's due to environmental factors or risky human actions(Lee et al., 2015).
- **Decision & Response Layer:** When certain levels of risk are met, automated responses such as alarms, shutdowns, or traffic rerouting are triggered by decision engines powered by artificial intelligence. As time goes on, response accuracy is improved by reinforcement learning.

- **Sustainability Integration Layer:** Aargue that safety actions should be in line with energy efficiency, carbon footprint reduction, and long-term maintenance optimisation.

4.2 Real-Time Hazard Detection and Response

IoT and AI work together to detect potential dangers including gas leaks, structural instability, traffic jams, and overheated machinery in real time. Automated ventilation and alarm systems can be set off when AI-enabled IoT frameworks detect possible fire threats, for instance, by monitoring rising temperatures and gas sensors(Cheng et al., 2018). Rather than depending on manual inspections or delayed reactions, this proactive safety paradigm takes their place.

4.3 Predictive Safety and Maintenance

The system employs a variety of artificial intelligence (AI) approaches, including decision trees, support vector machines, and deep neural networks, to anticipate when components may fail. This allows for predictive scheduling of maintenance and incident prevention (Gian et al., 2015). Using context-aware models and historical sensor data, critical infrastructure lifecycle management and fault categorisation are both improved.

4.4 Scalability and Adaptability

flexibility and scalability, the framework can be used in various environments, such as smart roads and manufacturing facilities.(Sun et al., 2016) state that APIs and data exchange protocols make sure that different platforms and devices can work together. This helps with smart city integration and broader urban safety governance. Furthermore, the ability to minimise latency through decentralised edge computing makes real-time judgements more feasible and cost-effective(Ahmed et al., 2024).

5. WORKING OF IOT-AI INTEGRATION

Kevin Ashton created the term "Internet of Things" in 1999, and the Trojan Room coffee pot was the first device to use it (Liu et al., 2021). Basically, the Internet of Things allows for the connection of any object, at any moment, in any setting, with any other object, and with any person, utilising any network, and with any service, regardless of the business. The IoT devices produce, collect, and analyse vast amounts of binary data. To better serve their customers, businesses are turning to artificial intelligence AI algorithms and approaches to sift through the mountains of data produced by these gadgets(Menon et al., 2025). Hybrid technologies that combine AI with IoT, often called AI-powered IoT, AI-enabled IoT, or A-IoT, can therefore efficiently and effectively reduce complex and laborious activities. Another important component of AI that simplifies IoT operations is ML. ML uses computational learning methods to learn and draw judgements. Different ML approaches make use of different categorisation systems, as seen in Figure 2. Similarly, IoT applications can gain a lot from ML algorithm integration in terms of usefulness and security(Veronica, 2023). To successfully reduce the likelihood of security breaches in IoT devices, these techniques must be used. In addition, providing a competitive environment where enterprises may deliver high-quality products is dependent on the architectures that make up the IoT(Sunil et al., 2024). In addition, the purpose of this extensive survey is to showcase the most recent and innovative A-IoT technologies; the results of the survey have produced a number of noteworthy conclusions that deserve consideration(Gouiza et al., 2024). Important findings from the study include the following.

- For the next generation of smart systems, the history and current state of the art of the IoT and A-IoT are described. In other words, what follows is an overview of artificial intelligence (AI), Hyper Dimensionality (HD) computing for effective IoT ML support, IoT augmented reality (AR), and IoT blockchain.
- Additionally, the paper provides a thorough explanation of the fundamental ML and DL techniques, as well as their applications, for enhancing the security and secrecy of IoT networks. These themes are explored extensively in the paper, which emphasises how these strategies might improve the security and privacy of IoT devices.
- The IoT and A-IoT architecture that supports a system is explained in depth. Then, we also show the various cutting-edge methods for IOT security that rely on ML and DL. Also covered here are a number of state-of-the-art ML methods that help improve IoT security, such as transfer learning, explainable AI (XAI), and self-supervised learning.
- We conclude by outlining the many new technologies that are developing inside the IoT and A-IoT. We also intend to resolve the most important issues related to the IOT and artificial intelligence.

Figure 2. Classification System for Various Methods of ML

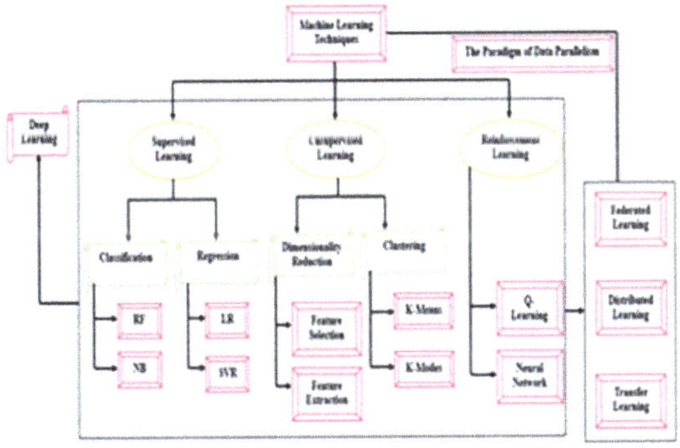

The capacity to transmit messages of huge sizes, such e-mails with attached files, is a hallmark of the internet of content stage. Moving on to the following phase, we have the internet of services, which primarily targets home electronics(Kalla, 2024). After this comes the "Internet of people" phase, when social media really starts to take centre stage in bringing people together(Rahman et al., 2024). The IoT characterises the last stage, which follows the internet of people stage. Service providers, app developers, and equipment manufacturers all stand to gain substantially from the IoT(Menon et al., 2025). More than 70 billion devices will be internet-enabled by 2025, according to predictions. All of these projections point to the IoT becoming increasingly important and rapidly expanding in the related industries and service sectors very soon(Raghuwanshi, 2024).The linked gadgets produce massive amounts of data throughout the course of their lives as the IoT develops. These gadgets generate massive volumes of data, which AI algorithms and approaches are well-equipped to analyse and learn from(Knickerbocker et al., 2018). As a result, the quality of services provided to users is greatly enhanced. That is why the term "A-IoT" encompasses both AI and the Internet of Things. This marks the beginning of a new age called the A-IoT, which is considered the next step after the IoT. Figure 3 provides detailed information about the evolution of the IoT to the advanced A-IoT stage, as well as the sequential steps involved in A-IoT. When it comes to the IoT, security is a must-have for everyone. Because it opens the door to new services and advancements, the IoT will be pivotal in the years to come(Jan et al., 2021). The IoT has many potential uses, from the most basic forms of home automation to the most complex, such as linked smart cities. The section on state-of-the-art IoT technologies for next-generation smart systems provides a detailed illustration of the present status of IoT research.

Figure 3. Various Phases of Advancement from IOT and AIOT and the Sequential stages involved in AIOT

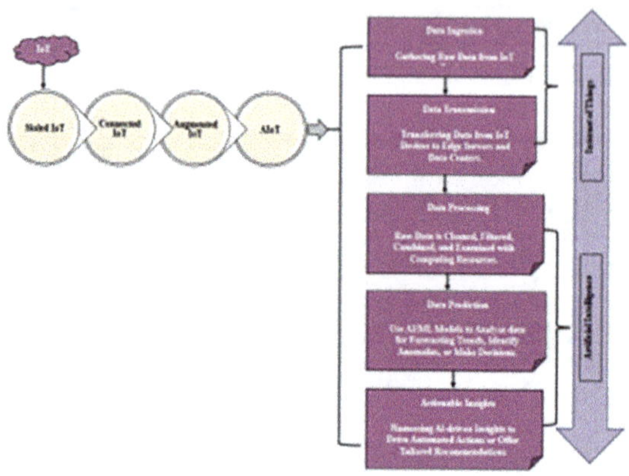

6. FEATURES OF IOT-AI FOR SAFETY MONITORING

Because of its many components and the limited computing and energy capabilities of its devices, the IoT still faces several security challenges(Hassan et al., 2024). These variables bring up further worries. For an IoT system to be secure, all IoT devices must have the security elements shown in Figure 4 included in from the start, both during creation and when they are in use.

Figure 4. Key Security Feature for the IOT

6.1 Confidentiality

When data is marked as confidential, only authorised individuals will be able to view it. Controlling access, allowing only authorised users, and stopping devices from exchanging data with nearby services, individuals, or other devices are critical steps in safeguarding data(Ahmed et al., 2024). Multiple security measures, such as two-factor authentication and data encryption, are now at users' disposal to guarantee the privacy of sensitive information. Nevertheless, a large amount of processing power is required by these techniques(El-Gendy, 2020). A suitable encryption method that is compatible with the computational and energy capacities of the sensors is thus essential for ensuring the secrecy of data. It is also critical to provide an Internet of Things service that can access and manage data.

6.2 Integrity

Within the IoT, numerous IoT devices actively share data with one another and with the resources available in the cloud. In this case, it's crucial to make sure the data is intact throughout the transfer process, no matter what happens with attackers, users, or eavesdroppers, and to ensure that it came from the right sender(Da Silva Rodrigues & Rocha, 2021). Conventional systems usually manage data traffic through protocols and firewalls and strengthen the security of connections between nodes to keep data intact. Because of the specifics of the IoT architecture, however, these approaches are insufficient for endpoint security in an IoT setting(Forlesi et al., 2024).

6.3 Authentication

In order for IoT devices to protect data and provide access only to authorised users, authentication is an essential step. Authentication of other devices should be a breeze for all IoT devices (Kokila & Reddy K, 2025). Nevertheless, the intricate structure of the IoT and the many moving parts devices, clients, and services make authentication approaches inherently challenging. IOT devices also have restrictions due to things like limited storage capacity, processing power, energy resources, and user interfaces(Dias et al., 2023). Because devices need to authenticate other newly presented devices that haven't been authenticated before during initial connections, it is vital to develop a mutual authentication method for different IoT entities.

6.4 Optimized Solution

Information security is of the utmost importance in many sensitive fields where IoT devices are being implemented. Therefore, it is critical to consider the constraints of IoT devices while designing and implementing protocols. Adapting these protocols to the specific requirements of IoT devices is essential. As a result, we need to find the sweet spot between power consumption, security measures, and overall system performance while yet meeting the stringent data security criteria.

6.5 Availability

IoT availability, which provides a fundamental assurance that a system will function reliably under all operational circumstances, is of utmost importance, especially in mission-critical businesses. For instance, in health monitoring systems, collecting patient health data in real-time is absolutely crucial, and any interruption in availability might result in patients' lives being jeopardised (Chen & Lo, 2018). A number of components must work together in harmony for an IoT system to guarantee availability. All of these things, like using energy-saving protocols, incorporating energy-harvesting methods, and adopting encryption algorithms that are both lightweight and efficient, contribute to keeping the IoT systems available. The complexity and heterogeneity of the IoT make it vulnerable to energy depletion assaults, which compounds this problem.

6.6 Non-Repudiation

Nonrepudiation ensures that the identification of the entity generating services and transferring data, instructions, or commands cannot be denied, making it essentially impossible for any entity to deny its involvement. Especially in delicate systems, nonrepudiation is of paramount importance. For instance, in healthcare, the system for medication adjustment must guarantee that only the patient's healthcare provider may make the changes. Nonrepudiation is typically achieved through the use of signatures, which identify the creator of a service or message and aid in the tracing of illegal or unauthorised actions (Fang et al., 2020). IoT applications must also enforce nonrepudiation(Khurshid et al., 2025).

6.7 Non-Repudiation

To safeguard users' private data and identity when they interact with IoT systems and devices, one must implement appropriate privacy protection measures(Shivaji Dhanwe et al., 2024). To prevent unauthorised access or compromise of sensitive information, several privacy protection measures can be used, including data anonymisation, encryption, access controls, user authorisation mechanisms, and compliance with privacy norms and regulations.

6.8 Communication Security

The term "Communication Security," commonly shortened to "comsec," denotes a framework with multiple layers of protection put in place to ensure that information sent between entities remains private, accurate, and uncompromised(Rego et al.,

2018). It incorporates a variety of security mechanisms and approaches to prevent eavesdropping, data tampering, interception, or unauthorised access (Babun et al., 2021).

7. RESULT AND DISCUSSION

In Figure 5, they can observe how the RF, NB, LR, and K-Means models compare in terms of accuracy. We found that our proposed NB model outperformed the competition, with an accuracy of 95.68% compared to 92.18% for the RF model, 90.60 for the LR model, and 87.52% for K-Means.

Figure 5. Comparison of Accuracy

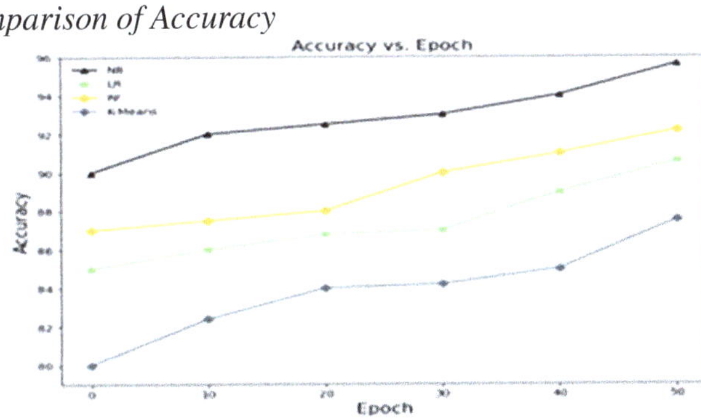

Table 4 compares the proposed techniques to the base hyperparameter of competing algorithms. Among these additional algorithms are LR, K-Means, RF, and NB. With respect to Accuracy (95.68%), Precision (93.36%), Recall (91.08%), and F1-score (95.98%), the proposed NB technique performed better than the alternatives.

Table 4. Performance Comparison

Models	Accuracy	Precision	Recall	F1-Score
RF	92.18	90.42	88.54	92.91
NB	**95.68**	**93.36**	**91.08**	**95.98**
LR	90.60	88.21	86.73	90.87
K-Means	87.52	85.59	83.44	87.80

The results are shown in Figure 6. One way to visually see how well a binary classification model does across all thresholds is with the ROC curve. It reveals the relative costs of a high true positive rate and a low false positive rate. For comparison's sake, the ROC value that the suggested NB model achieved was 0.98%.

Figure 6. ROC for Proposed NB Model

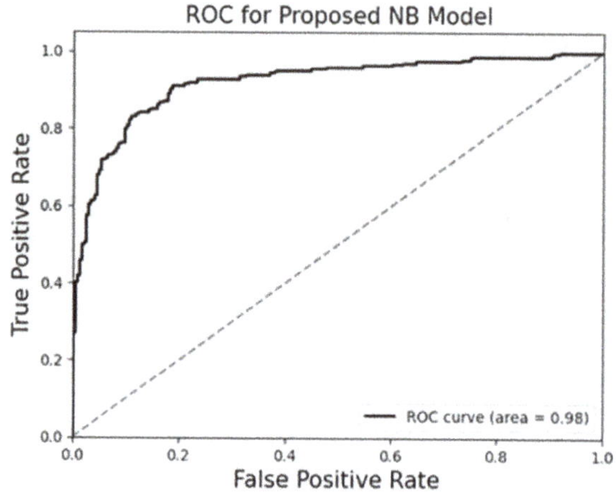

Figure 7 shows that a confusion matrix, one way to evaluate a classification model's performance is to use a table to compare the model's forecasts with the actual values in the dataset. Examining the overall number of accurate and inaccurate predictions for each category could provide light on the type of mistakes made by the model. A value of 95.568 is displayed in the confusion matrix.

Figure 7. Confusion Matrix

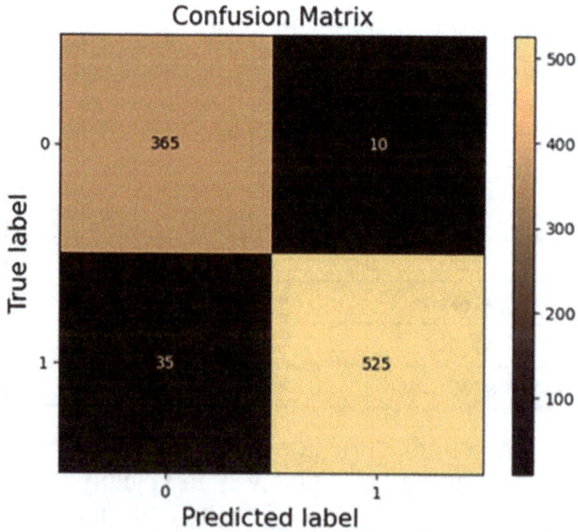

The comparison of the four models' metrics is displayed in Figure 8. After comparing our proposed NB model to RF, LR, and K-Means, we found that it performed the best, with an accuracy of 95.68%.

Figure 8. Performance of Models

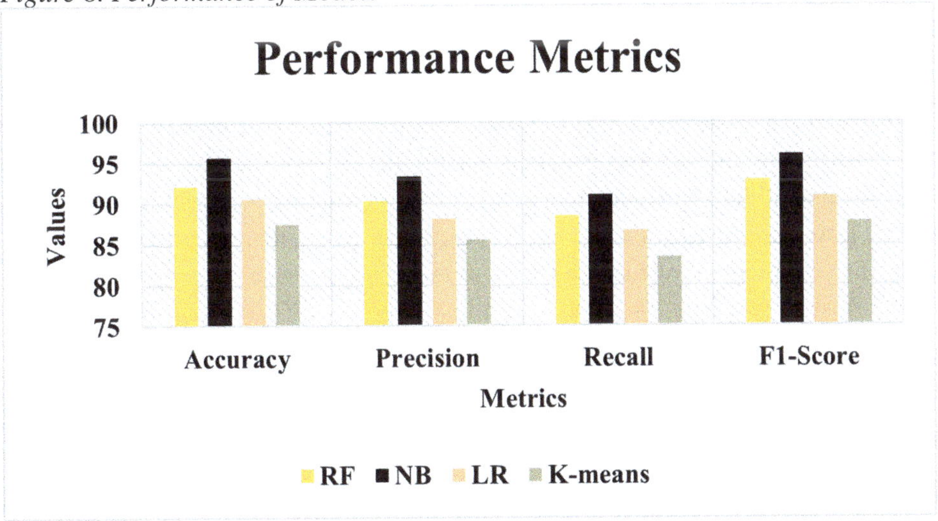

8. CHALLENGES

Here we explore the current state of affairs in the rapidly developing field of AI-powered solutions for IoT security and privacy(Enrique et al., 2023). Given the increasing integration of AI into IoT frameworks, it is crucial to tackle these challenges to guarantee the resilience, flexibility, and longevity of the security measures implemented.

- **Detecting Security Threats:** The diversity, high velocity, and complexity of data structures make it increasingly difficult to detect security threats in multi-domain IoT systems. Given the challenges that existing methods have in uncovering the hidden connections between features, selecting a recognised collection of characteristics for use in DL algorithms' security analytics is an interesting job in this context. Still, new DL techniques show promise for controlling these hidden factors in IoT apps. The use of DL allows for better detection rates of zero-day assaults and the discovery of correlations within varied datasets. Distributed representation in deep learning allows it to manage many features selects, in contrast to more conventional approaches.

- **Time Required for Training:** One major drawback of current methods is the vast amount of data needed for training and accurate detection. But DL algorithms can handle models more effectively since they can use less datasets and shorter training cycles. Training time is affected by batch size because of the accumulation of weight updates. By locating and prioritising particular factors within datasets, the multi-layered approach of deep learning can overcome these obstacles. Training time can be further hindered by limited processing and storage resources; nevertheless, big data technologies have the potential to enhance model efficacy by shortening the training timeframe.
- **Time Complexity:** Problems with real-time threat identification have led to the development of many existing detection approaches that prioritise batch processing of data. Because it requires analysing massive volumes of streaming data and many security threat indicators, temporal complexity is a key factor in identifying security concerns in IoT applications. Conventional methods that use shallow learning are not good enough to classify incursions, according to multiple research projects, even when using a lot of real-time data. Incorporating GPU components into DL techniques allows for fast real-time processing and threat analysis, which in turn addresses time complexity difficulties. Time complexity can be substantially reduced by utilising technologies like Apache Spark or Apache Hadoop, which further enhances the capabilities of real-time threat detection.
- **Computing-in-Memory:** Streaming analytics and memory-centric designs rely on in-memory processing, a new method for managing data in in-memory databases. Organisations are left exposed and insecure due to the inadequacy of security analytics integration and the inability of traditional storage methods to handle the massive amounts of data produced by IoT devices. Processing data in a relational database requires the use of SQL queries because the data is stored across numerous tables. This method leads to complicated data aggregation and combination since SQL retrieves data rows before processing them. When it comes to security analytics, in-memory processing is the way to go. The increased speed of data retrieval using Random Access Memory enables Real-Time Data Analysis (RAM) or flash memory compared to disc storage. Quicker responses to security risks can be achieved through this real-time processing, which speeds up reporting and decision-making. Modern big data platforms, including Apache Spark and Apache Flink, use in-memory processing to improve security analytics performance and efficiency.

- **Computational and Energy Constraint:** Problems with computational complexity and energy consumption are major obstacles to research into the security of Internet of Things (IoT) devices, deep learning, and big data. Because of their limited processing power and low-power batteries, IoT devices need quick and efficient calculations to keep resource-intensive tasks to a minimum. Cloud or edge computing should be relied upon by these devices rather than local processing. Research highlights the significance of computational systems that are lightweight when it comes to algorithm-based security analysis. Rapid advancements in DL technology have been made possible by the proliferation of big data and computing capacity. Algorithms and distributed computing allow for more efficient computations. In a hybrid network, these algorithms work by spreading out work among numerous computers. By efficiently processing data with manageable compute requirements, technological frameworks like Apache Spark Streaming and features like RDD and Dstreams make it possible to solve some of these difficulties.
- **Security Edge:** Enabling decentralised processing close to data sources, edge computing systems improve the computational and storage capacities of IoT devices, hence bolstering edge security. Because of this decentralisation, intelligent activities can now take place in real time. Intelligent edge services for real-time threat detection are offered by the cloud edge infrastructure, which also manages IoT data sources. Distributing processing over numerous cloud resources, edge computing overcomes the limitations of IoT devices when it comes to data storage and threat analysis. A safer ecosystem can be achieved by the use of security analytics for IoT devices that incorporate DL and big data technologies. These technologies allow for efficient and accurate threat detection.

9. CONCLUSION AND FUTURE WORK

The evolution of safety monitoring systems from reactive to proactive and predictive is being driven by the convergence of IoT and AI. IoT-AI systems improve safety in healthcare, urban areas, industries, and disaster management by integrating intelligent decision-making with ubiquitous sensing. While issues with scalability, real time responsiveness, and security still persist, new developments in 6G, federated AI, and edge computing hold the possibility of resolving these issues. Society is undergoing a sea change towards greater safety, intelligence, and resilience as a result of this integration.

REFERENCES

Ahmed, M. F., Islam, M. R., Khan, M. R. A. A., & Islam, M. N. (2024). Smart IoT and AI for Sustainable Safety Management: Enhancing Efficiency in Industrial and Urban Environments. *Journal of Artificial Intelligence General Science (JAIGS) ISSN:3006-4023, 6*(1), 755–765. DOI: 10.60087/jaigs.v6i1.356

Babun, L., Denney, K., Celik, Z. B., McDaniel, P., & Uluagac, A. S. (2021). Una encuesta sobre plataformas IoT: Perspectivas de comunicación, seguridad y privacidad. *Computer Networks, 192*, 108040. DOI: 10.1016/j.comnet.2021.108040

Chen, C. H., & Lo, K. R. (2018). Applications of internet of things. In *Canadian Historical Review* (Vol. 7, Issue 9). DOI: 10.3390/ijgi7090334

Cheng, B., Solmaz, G., Cirillo, F., Kovacs, E., Terasawa, K., & Kitazawa, A. (2018). FogFlow: Easy Programming of IoT Services Over Cloud and Edges for Smart Cities. *IEEE Internet of Things Journal, 5*(2), 696–707. DOI: 10.1109/JIOT.2017.2747214

Clevenger, C., López del Puerto, C., & Glick, S. (2015). Advances in Engineering Education Interactive BIM-enabled Safety Training Piloted Construction Education AND. *Advances in Engineering Education*, •••, 1–14.

Da Silva Rodrigues, C. K., & Rocha, V. (2021). Towards Blockchain for Suitable Efficiency and Data Integrity of IoT Ecosystem Transactions. *Revista IEEE América Latina, 19*(7), 1199–1206. DOI: 10.1109/TLA.2021.9461849

Dhanraj, J. A., Krishnamurthy, B., Ramanathan, K. C., Saravanan, A. K., & Ganapathy Raman, J. K. (2020). Design on IoT Based Real Time Transformer Performance Monitoring System for Enhancing the Safety Measures. *IOP Conference Series. Materials Science and Engineering, 988*(1), 012076. Advance online publication. DOI: 10.1088/1757-899X/988/1/012076

Dias, T., Fonseca, T., Vitorino, J., Martins, A., Malpique, S., & Praça, I. (2023). From Data to Action: Exploring AI and IoT-Driven Solutions for Smarter Cities. *Lecture Notes in Networks and Systems, 740 LNNS*(Ml), 44–53. DOI: 10.1007/978-3-031-38333-5_5

Ding, M. (2021). Research on Tourism Route Planning Based on Artificial Intelligence Technology. *Wireless Communications and Mobile Computing, 2021*(1), 2227798. Advance online publication. DOI: 10.1155/2021/2227798

Dudhe, P. V., Kadam, N. V., Hushangabade, R. M., & Deshmukh, M. S. (2018). Internet of Things (IOT): An overview and its applications. *2017 International Conference on Energy, Communication, Data Analytics and Soft Computing, ICECDS 2017, February*, 2650–2653. DOI: 10.1109/ICECDS.2017.8389935

El-Gendy, S. (2020). IoT Based AI and its Implementations in Industries. *Proceedings of ICCES 2020 - 2020 15th International Conference on Computer Engineering and Systems, October*. DOI: 10.1109/ICCES51560.2020.9334627

Enrique, O., Castro, L., Deng, X., & Park, J. H. (2023). Comprehensive Survey on AI-Based Technologies for Enhancing IoT Privacy and Security: Trends, Challenges, and Solutions. *Human-Centric Computing and Information Sciences*, *13*, 39. DOI: 10.22967/HCIS.2023.13.039

Fang, W., Chen, W., Zhang, W., Pei, J., Gao, W., & Wang, G. (2020). Digital signature scheme for information non-repudiation in blockchain: a state of the art review. *Eurasip Journal on Wireless Communications and Networking, 2020*(1). DOI: 10.1186/s13638-020-01665-w

Forlesi, M., Esposito, A., Ciabattini, L., Sciullo, L., Di Felice, M., Kamali, S., Ghini, E., Silvestri, S., & Marzani, A. (2024). An IoT-AI Toolchain for Structural Health Monitoring. *2024 IEEE International Symposium on Measurements and Networking, M and N 2024 - Proceedings, September*. DOI: 10.1109/MN60932.2024.10615472

Fredj, N., Hadj Kacem, Y., Khriji, S., Kanoun, O., & Abid, M. (2021). A review on intelligent IoT systems design methodologies. *Measurement. Sensors*, *18*, 100347. Advance online publication. DOI: 10.1016/j.measen.2021.100347

Gian, A., Susto, A., Schirru, S., Pampuri, S., & McLoone, A. B. (2015). Machine Learning for Predictive Maintenance : A Multiple Classifiers Queen ' s University Belfast - Research Portal Machine Learning for Predictive Maintenance : A Multiple Classifiers Approach. *IEEE Transactions on Industrial Informatics*, *11*, 812–820. DOI: 10.1109/TII.2014.2349359

Gouiza, N., Jebari, H., & Reklaoui, K. (2024). Integration of Iot-Enabled Technologies and Artificial Intelligence in Diverse Domains: Recent Advancements and Future Trends. *Journal of Theoretical and Applied Information Technology*, *102*(5), 1975–2029.

Hassan, A., Nizam-Uddin, N., Quddus, A., Hassan, S. R., Rehman, A. U., & Bharany, S. (2024). Navigating IoT Security: Insights into Architecture, Key Security Features, Attacks, Current Challenges and AI-Driven Solutions Shaping the Future of Connectivity. *Computers, Materials & Continua*, *81*(3), 3499–3559. DOI: 10.32604/cmc.2024.057877

Hussein, N. M., Mohialden, Y. M., & Salman, S. A. (2024). Impact of IoT-Based Environmental Monitoring on Lab Safety and Sustainability. *Babylonian Journal of Internet of Things, 2024*(March), 16–26. DOI: 10.58496/BJIoT/2024/003

Jan, M. A., Zakarya, M., Khan, M., Mastorakis, S., Menon, V. G., Balasubramanian, V., & Rehman, A. U. (2021). An AI-enabled lightweight data fusion and load optimization approach for Internet of Things. *Future Generation Computer Systems, 122*, 40–51. DOI: 10.1016/j.future.2021.03.020

Jayakody, S., & Thilakarathne, G. (2025). *ARTIFICIAL INTELLIGENCE DRIVEN SYSTEMS FOR ENHANCING WORKER SAFETY MONITORING IN CONSTRUCTION ENVIRONMENTS : A COMPREHENSIVE REVIEW. August*, 154–166.

Kalla, D. (2024). Integrating Iot, AI, And Big Data For Enhanced Operational Efficiency In Smart Factories. *Educational Administration: Theory and Practice*, (May), 14235–14245. DOI: 10.53555/kuey.v30i5.6492

Khurshid, K., Khurshid, K., Hadi, M. U., Al Bataineh, M., & Saeed, N. (2025). Securing AIoT Surveillance: Techniques, Challenges, and Solutions. *IEEE Open Journal of the Communications Society, 6*(June), 1–1. DOI: 10.1109/OJCOMS.2025.3593311

Knickerbocker, J., Budd, R., Dang, B., Chen, Q., Colgan, E., Hung, L. W., Kumar, S., Lee, K. W., Lu, M., Nah, J. W., Narayanan, R., Sakuma, K., Siu, V., & Wen, B. (2018). Heterogeneous integration technology demonstrations for future healthcare, IoT, and AI computing solutions. *Proceedings - Electronic Components and Technology Conference, 2018-May*, 1519–1528. DOI: 10.1109/ECTC.2018.00231

Kokila, M., & Reddy K, S. (2025). Authentication, access control and scalability models in Internet of Things Security–A review. *Cyber Security and Applications, 3*(April 2024), 100057. DOI: 10.1016/j.csa.2024.100057

Kumar, S., & Tauseef, S. M. (2021). Development of an Internet of Things (IoT) based Lockout/Tagout (LOTO) device for Accident Prevention in Manufacturing Industries. *IOP Conference Series. Materials Science and Engineering, 1017*(1), 012017. Advance online publication. DOI: 10.1088/1757-899X/1017/1/012017

Lee, J., Bagheri, B., & Kao, H. A. (2015). A Cyber-Physical Systems architecture for Industry 4.0-based manufacturing systems. *Manufacturing Letters, 3*, 18–23. DOI: 10.1016/j.mfglet.2014.12.001

Liu, Y., Wang, L., Xu, X., Zhang, L., & Wang, X. V. (2021). Industrial Internet for Manufacturing. *Robotics and Computer-integrated Manufacturing, 70*(9), 1173–1192. DOI: 10.1016/j.rcim.2021.102135

Menon, I. U., Babu Kumaravelu, V., Kumar, V. C., Rammohan, A., Chinnadurai, S., Venkatesan, R., Hai, H., & Selvaprabhu, P. (2025). AI-Powered IoT: A Survey on Integrating Artificial Intelligence With IoT for Enhanced Security, Efficiency, and Smart Applications. *IEEE Access : Practical Innovations, Open Solutions, 13*(February), 50296–50339. DOI: 10.1109/ACCESS.2025.3551750

Obianuju Ozobu, C., Emmanuel Adikwu, F., Odujobi, O., Othuke Onyeke, F., & Onyinye Nwulu, E. (2025). Advancing Occupational Safety with AI-Powered Monitoring Systems: A Conceptual Framework for Hazard Detection and Exposure Control. *World Journal of Innovation and Modern Technology E, Vol 9. No. 1 2025*(1), 186–213. https://doi.org/DOI: 10.56201/wjimt.v9.no1.2025.pg186.213

Park, J., & Kang, D. (2024). Artificial Intelligence and Smart Technologies in Safety Management: A Comprehensive Analysis Across Multiple Industries. *Applied Sciences (Basel, Switzerland), 14*(24), 1–22. DOI: 10.3390/app142411934

Raghuwanshi, P. (2024). Integrating generative ai into iot-based cloud computing: Opportunities and challenges in the united states. *Journal of Artificial Intelligence General Science (JAIGS) ISSN: 3006-4023, 5*(1), 451–460.

Rahman, A., Kundu, D., Debnath, T., Rahman, M., & Islam, M. J. (2024). *Blockchain-based AI Methods for Managing Industrial IoT: Recent Developments, Integration Challenges and Opportunities*. https://arxiv.org/abs/2405.12550

Ramadan, M. N. A., Ali, M. A. H., Khoo, S. Y., Alkhedher, M., & Alherbawi, M. (2024). Real-time IoT-powered AI system for monitoring and forecasting of air pollution in industrial environment. *Ecotoxicology and Environmental Safety, 283*(August), 116856. Advance online publication. DOI: 10.1016/j.ecoenv.2024.116856

Rathi, V. K., Rajput, N. K., Mishra, S., Grover, B. A., Tiwari, P., Jaiswal, A. K., & Hossain, M. S. (2021). An edge AI-enabled IoT healthcare monitoring system for smart cities. *Computers & Electrical Engineering, 96*, 107524. Advance online publication. DOI: 10.1016/j.compeleceng.2021.107524

Razzaque, M. A., Milojevic-Jevric, M., Palade, A., & Cla, S. (2016). Middleware for internet of things: A survey. *IEEE Internet of Things Journal, 3*(1), 70–95. DOI: 10.1109/JIOT.2015.2498900

Rego, A., Canovas, A., Jimenez, J. M., & Lloret, J. (2018). An Intelligent System for Video Surveillance in IoT Environments. *IEEE Access : Practical Innovations, Open Solutions, 6*, 31580–31598. DOI: 10.1109/ACCESS.2018.2842034

Ricci, F., Chiesi, A., Bisio, C., Panari, C., & Pelosi, A. (2016). Effectiveness of occupational health and safety training. *Journal of Workplace Learning, 28*(6), 355–377. DOI: 10.1108/JWL-11-2015-0087

Shivaji Dhanwe, S., Mallikarjun Abhangrao, C., Kutubuddin Sayyad Liyakat, K., & Professor, A. (2024). AI-driven IoT in Robotics: A Review. *Journal of Mechanisms and Robotics, 9*(April), 41–48.

Sun, Y., Song, H., Jara, A. J., & Bie, R. (2016). Internet of Things and Big Data Analytics for Smart and Connected Communities. *IEEE Access : Practical Innovations, Open Solutions, 4*, 766–773. DOI: 10.1109/ACCESS.2016.2529723

Sunil, M. B., Sultanabanu Sayyad Liyakat, K., & Kutubuddin Sayyad Liyakat, K. (2024). AI-Driven IoT (AI IoT) in Thermodynamic Engineering. *All Rights Reserved Journal of Modern Thermodynamics in Mechanical System, 6*(1), 1–8.

Sunindijo, R. Y., Zou, P. X. W., & Dainty, A. (2017). Managerial Skills for Managing Construction Safety. *Civil Engineering Dimension, 19*(2). Advance online publication. DOI: 10.9744/ced.19.2.63-72

Tselentis, D. I., Papadimitriou, E., & van Gelder, P. (2023). The usefulness of artificial intelligence for safety assessment of different transport modes. *Accident; Analysis and Prevention, 186*(March), 107034. DOI: 10.1016/j.aap.2023.107034

Veronica, M. (2023). Progress in Language, Literature and Education Research Vol. 1. In *Progress in Language, Literature and Education Research Vol. 1* (Vol. 1, Issue October). DOI: 10.9734/bpi/plder/v1

Villegas-Ch, W., Garcia-Ortiz, J., & Sanchez-Viteri, S. (2024). Toward Intelligent Monitoring in IoT: AI Applications for Real-Time Analysis and Prediction. *IEEE Access : Practical Innovations, Open Solutions, 12*(March), 40368–40386. DOI: 10.1109/ACCESS.2024.3376707

Wang, Y., & Chung, S. (2022). This is the Pre-Published Version. Artificial Intelligence in Safety-critical Systems. *Systematic Reviews, 122*(2).

Wong, W., Chung, S., Tariq, S., Mohandes, S. R., Zayed, T., Estate, R., Kong, H., & Kong, H. (2020). This is the Pre-Published Version. *IoT-based Application for Construction Site Safety Monitoring., 2020*, 1–45.

Zayed, S. M., Attiya, G., El-Sayed, A., Sayed, A., & Hemdan, E. E. D. (2023). An Efficient Fault Diagnosis Framework for Digital Twins Using Optimized Machine Learning Models in Smart Industrial Control Systems. *International Journal of Computational Intelligence Systems, 16*(1), 69. Advance online publication. DOI: 10.1007/s44196-023-00241-6

Zhao, Z., & Hu, Q. (2022). The Application of a Computer Monitoring System Using IoT Technology. *Computational Intelligence and Neuroscience*, *2022*, 1–11. Advance online publication. DOI: 10.1155/2022/4033886

Chapter 9
Machine Learning Applications in Transportation Safety

Soumi Ghosh
http://orcid.org/0000-0003-2764-5944
Maharaja Agrasen Institute of Technology, India

Ritik Raj
http://orcid.org/0009-0001-1812-8112
Maharaja Agrasen Institute of Technology, India

ABSTRACT

Every year 1.35M are killed in road traffic accidents. Machine learning changes the field of transport safety by means of real-time choices, predictive analytics and autopilots. In this analysis, crash prediction, driver behavior, infrastructure surveillance, and smart systems are explored. Major areas of weakness: absence of single evaluation systems, absence of edge computing to support real-time processing, inadequate regulatory compliance of deep learning, and inefficiencies of multimodal integration. New solutions suggested: ensemble and federated learning as a privacy-preserving accuracy, and a unified safety assessment system between predictive and autonomous technologies. This study critically examines ML in transportation safety identification of barriers to its adoption, and future growth of intelligent transportation systems that need to involve interdisciplinary cooperation between MI researchers, engineers, and policymakers.

DOI: 10.4018/979-8-3373-3231-4.ch009

1. INTRODUCTION

1.1 Background and Motivation

The challenge of transportation safety is one of the most topical issues in the whole world, which overlaps with the issues of public health, economic growth, and technology development. Road traffic injuries are considered by the World Health Organization (2018) as the eighth highest death cause in the world, and disproportional effects are caused by the low and middle-income nations, which own 60 percent of vehicles worldwide, but bear most of the traffic-related deaths. On top of causalities among humans, traffic accidents cause huge financial costs which can be seen in the form of healthcare expenses, lost labor, and property damages, which is estimated to cost the world around 3 percent of GDP each year.

Conventional methods of transportation safety have mostly used reactive measures, putting measures in place after crash patterns have been studied and systemic problems identified. Nevertheless, these traditional approaches are ineffective in dealing with the dynamism and the growing complexity of the contemporary transportation systems. The spread of connected cars, intelligent infrastructure and ubiquitous sensing technologies have produced large quantities of transportation data never before seen, giving rise to challenges and opportunities in improving safety.

Machine learning technologies are being developed as disruptive technologies that can handle large amounts of data, detect intricate patterns that are not visible to past-methods of analytical techniques and make it possible to proactively intervene to promote safety. The integration of artificial intelligence and transportation is a paradigm shift when it comes to reactive safety management as opposed to being predictive, proactive, and adaptive in terms of safety measures. Machine learning models have the ability to create a combined data source of vehicle telemetry, roadside sensors, weather conditions, traffic patterns, and behavioral indicators in order to develop a whole set of real-time safety assessment systems.

The intelligent transportation systems as envisioned by implementing machine learning do not merely stop at predicting crashes; it will also involve real-time risk detection, anomaly detection, predictive maintenance, and making autonomous decisions. The technologies are the basis of next generations of transportation safety frameworks that are able to predict hazards, monitor states in real time, perform preventive actions, and make autonomous safety-relevant decisions.

1.2 Research Objectives

The study has a few related aims that contribute to the development of theoretical knowledge and real-life application of machine learning to transportation safety:

Primary Objective

To deliver a thorough discussion of machine learning applications in transportation safety, including predictive analytics, real-time monitoring systems, and technologies, which will be developed and achieved autonomous vehicles, to build a platform of future research and implementation efforts.

Secondary Objectives

1. In order to assess the efficiency, constraints and applicability of the modern machine learning methods in the various application of transportation safety, to present evidence based analysis of the technological capacity and restrain.
2. In order to define and describe research gaps that are considered crucial in the existing machine learning methods of transportation safety, especially in the context of integration, explainability, scalability, and ethical concerns.
3. To close the divide between the theoretical achievements of machine learning and the real world of transportation safety through the creation of conceptual frameworks linking the research findings to the field of operation.
4. To present original architectural solutions and algorithmic frameworks to maximize the performance, robustness, and scalability and ethical implementation of machine learning systems in safety-critical transportation systems.
5. In order to have all-encompassing evaluation criteria and performance measures that are well-suited to measure the performance of machine learning interventions in transportation safety, it is appropriate to take into account both technical efficacy and real-life safety impact measures.
6. In order to meet ethical aspects such as fairness of algorithms, privacy protection, and the relationship between humans and artificial intelligence in the system of transportation safety, it is necessary to make sure that technological changes do not contradict social values and governmental standards.

1.3 Research Scope and Contributions

This study includes a broad range of machine learning in transportation safety and is structured into five main areas, which are crash predicting and risk management, driver behavior prediction and monitoring, infrastructure health monitoring and predictive maintenance, self-driving technologies, and smart transportation systems. The methodology has been made to go further than analyzing applications in isolation to explore the problems of systemic integration, structure of governance and issues of human factors.

Theoretical Contributions

- The taxonomy of machine learning methodology in transportation safety issues, which is detailed and provides a structure of classifying techniques, applications, and performance properties.
- Detection and description of the essential gaps in research in the field of application, specifically focusing on the issues of integration, explainability, real-time processing, and generalization.
- Integrated frameworks to unite predictive analytics and autonomous system implementation to accommodate end-to-end system design concerns.
- Creation of metrics of evaluation unique to machine learning applications of transportation safety, and that goes beyond the standard accuracy metrics to encompass safety impact, reliability, and ethical considerations.
- AI framework to be applied in the context of transportation safety, designed to meet the transparency requirements and regulatory compliance needs of the stakeholders.

Practical Contributions

- Real-life implementation analysis of challenges and solutions analysis based on case-studies and implementation experiences.
- Scalability, privacy, real-time processing requirements, and constraints of edge computing Architectural frameworks.
- Ethical guidance on deployment of machine learning in safety-critical transportation systems such as accountability, fairness, and privacy.
- Suggestions on standardization activity and development of regulatory framework, which enables technology transfer and mass adoption.
- Combination of governance viewpoints that cover policy frameworks, regulatory mandates and liability factors.

Methodological Contributions

- New methods to integrate ensemble learning and federated mechanisms to privacy preserving cooperative learning.
- Heterogeneous data sources and multi-modal transport systems integration strategies.
- Explainable artificial intelligence frameworks in the transportation safety context, based on the trade-off between the model performance and explainability requirements.

- Mechanisms of continual learning and adaptation that allow one to modify systems as conditions of operation vary.

1.4 Paper Organization

The rest of the paper is designed to give an in-depth coverage of machine learning application in transportation safety and at the same time follow a logic chain of the basic concepts, up to the advanced integration situations and implementation considerations.

Section 2 includes a comprehensive literature review of the current research in the main areas of application, including improved synthesis to define the methodological approaches, performance trends, and shortcomings of the existing literature. The study is clear about human factors, policy aspects, and ethical issues and concerns as well as technical matter.

Section 3 outlines the research design, such as data collection methods, analytical models, analysis criteria, and is discussed in greater detail in terms of limitations and assumptions of the methodology used.

Section 4 discusses machine learning methods that can be used in transportation safety such as supervised learning, unsupervised learning, reinforcement learning, and deep learning methods and focuses on the considerations of techniques selection and tradeoffs between performance and interpretability.

Section 5 reviews the particular areas of application such as crash prediction, analyzing driver behavior, monitoring infrastructure and autonomous systems and brings together technical capabilities, human factors and governance aspects.

Section 6 deals with integration issues and suggests overall frameworks of intelligent transportation systems, such as federated learning systems, explainable AI systems, and continuous adaptation systems.

Section 7 specifies critical research gaps and future research directions, which is extended to human-centered design issue, policy development requirements, and ethical research priorities.

Section 8 deals with implementation concerns such as ethical aspects, regulatory considerations, deployment plans, economic aspects, and human resource development, and covers governance, mitigation of biases, privacy controls, and stakeholder involvement much more extensively.

Section 9 is a case study detailing practical application of the system, difficulties in implementation, and lessons learnt.

Section 10 summarizes the findings, implications in practical use to various stakeholders, constraints of the existing research and a broad research agenda into the future.

2. LITERATURE REVIEW

2.1 Evolution of Machine Learning in Transportation Safety

This paper offers an overview of the development of machine learning in the field of transportation safety. Use of machine learning method in transportation safety has changed significantly in the last thirty years since the early times of simple statistical models to the advanced deep learning systems and intelligent systems. This development can be understood in order to contextualize present opportunities and directions.

The original researches of the 1990s were based mostly on traditional methods of statistics, including logistic regression and generalized linear modeling to examine crashes and their causes (Karlaftis and Tarko, 1998). These seminal papers defined the principle of data-driven methods to become familiar with the idea of transportation safety but were limited by the complexity of calculations and comparatively small data sets. Although these initial models offered useful information on the factors of crashes, they did not manage to overcome a non-linear relationship and interaction between multiple variables.

During the early 2000s, superior machine learning methods were developed, and scientists started investigating decision trees, random forests, and support vector machines to predict the severity of the crash and evaluate the risk (Abdel-Aty and Abdelwahab, 2004). These methods had shown a better performance than the conventional statistical methods of capturing non-linear relationships and other complex interactions among variables. The first use of neural networks in crash data was made by Chang and Wang (2006), who discovered that the machine learning models were able to detect complex patterns otherwise overlooked by traditional approaches. It was also a time of change in the direction between the purely statistical methods, into the methods of computational learning, which could process more complex data.

Connected vehicle systems and improved sensing capabilities experienced exponential increase in availability of transportation data throughout the 2010s, which made it possible to apply deep learning technologies. Researchers started using convolutional neural networks to implement image-based safety, recurrent neural networks to identify temporal patterns in traffic flow, and ensemble to have robust prediction systems (Bao et al., 2019). This era was characterized by a drastic advancement in the power to compute in addition to the sophistication of the algorithm, rendering real-time processing as well as demanding scenarios of decision making a possibility. The multi-data and multi-sensor integration of multiple data sources and sensor modalities became more frequent and allowed providing a more detailed situational awareness and safety evaluation.

Recent studies are also dedicated to explainable AI as a means of dealing with the black box characteristics of deep learning models, federated learning to train privacy-preserving collaborative analytics, edge computing designs to implement real-time on-board processing, and integration systems that deploy multiple safety applications into consistent intelligent transportation systems. This new stage does not only focus on predictive performance, but also the transparency, privacy, real time and integration of the system level.

2.2 Crash Prediction and Risk Assessment

One of the most actively researched areas of application of machine learning in transportation safety is crash prediction. The researchers have conducted studies on different methodological methodologies to predict the occurrence, severity, and causes of crashes using different datasets and algorithm techniques.

Arteaga et al. (2020) designed a multifaceted model to forecast crashes at urban crossroads based on the quantity of traffic, the geometrical features, weather conditions, and time aspects with the use of gradient boosting machines. Their model was found to be 87 percent accurate in predicting high-risk locations, which is indicative of the fact that cautious feature engineering and domain knowledge are significant in the predictive model accuracy. This research paper has emphasized the necessity of having contextual factors that are not mere traffic measures and also the significance of ensemble approaches to enhancing robustness.

The crash risk evaluation in real-time using probe vehicles and the related infrastructure has received a great interest. Xu et al. (2014) applied a hybrid model of combining random forests and temporal convolutional networks with real-time speed, density2014) appliedata of traffic to predict the risk of a crash. The model was able to identify high risk of crashes 5-10 minutes before the incident happened and that gives enough lead time to preventive measures. But the study has indicated that it has issues with the false positive rates and that the threshold needs to be adapted to the context. This paper shows the potential and real-life challenges of real-time risk prediction systems.

Crash severity prediction looks particularly promising with the methods of deep learning. Rahim and Hassan (2021) applied deep neural networks with attention mechanisms to investigate the historical crash data and achieved better results in the prediction of the severity of injury than the traditional techniques. The attention mechanism improved the interpretability of the model by emphasizing the features of influence, but the research mentioned such limitations as the problem of imbalance of data and the large volume of training data. This study provides an example of the current conflict between the interpretability requirements and model performance in safety-critical applications.

The transfer learning methodology has been of benefit in solving the problem of scarcity of data in crash prediction. Oliveira et al. (2021) showed that large crash database trained models could be effectively fine-tuned to smaller regional data sets with little extra data and achieved the same level of prediction performance as models trained on large datasets. The method is especially useful in jurisdictions with a low crash history and allows them to use the experience of data-rich regions and adapt to the local conditions.

2.3 Driver Behavior Analysis and Monitoring

Driving behavior is an essential part of transport safety, and machine learning technologies provide the opportunity to conduct a complex analysis of driving patterns, identify distraction and drowsiness. Drivers state awareness and feedback is a proactive technique in avoiding crashes, before they happen.

Dong et al. (2016) initiated the classification of driver behavior based on support vectors machine that was used on the vehicle dynamics data such as acceleration, braking, and steering patterns. Their study revealed that machine learning models could be used to successfully differentiate between aggressive, standard, and safe driving styles to provide the basis of behavior-based safety intervention. This work has been later expanded by other studies utilizing more data sources like eye tracking devices, physiological sensors and in-cabin cameras to deal more with assessing the state of drivers.

Distraction detection and drowsiness detection are the areas where deep learning has transformed. Wijnands et al. (2019) created convolutional neural networks that determine the facial features of drivers to identify drowsiness in real-time with 94 percent accuracy in identifying drowsiness cues such as facial expressions, head position, and eye closure patterns. The research indicated the need of a wide range of training data on various demographic groups and conditions. Nevertheless, privacy issues and possible discrimination in the implementation of the facial recognition system were found to be a major challenge that must be put into serious consideration during implementation.

The recent trends in naturalistic driving research have produced a lot of longitudinal data that can be analysed in terms of behaviour. Hallmark et al. (2015) used recurrent neural networks to study data of naturalistic driving in which subtle patterns of behavior were identified as predicting higher crash risk. The study discovered that vehicle dynamics together with context like road type and traffic conditions could be greatly used to predict efficiency. Nevertheless, it was observed in the study that there were computational problems to deal with continuous high-resolution sensor data, and thus efficient algorithms are required and edge computing is necessary.

The use of multimodal methods that combine different sources of data has shown excellent capability in the process of monitoring driver state. Strayer et al. (2010) suggested an ensemble-based learning system that integrates the vehicle dynamics, physiological data, and the environment and obtained strong results in a variety of driving conditions and driver groups. The paper has highlighted the significance of individualized models that fit in the behavior and preference of the individual drivers and noted that there was high inter-individual difference in the behavior of the base and the expression of drowsiness.

2.4 Infrastructure Condition Monitoring and Predictive Maintenance

The state of transportation infrastructure is a significant factor in safety, which can be detected automatically through machine learning and used to predict degradation or detect defects and optimally schedule maintenance. Safety risks can be avoided before causing accidents through proactive infrastructure management.

The use of computer vision and deep learning to identify the condition of the pavement has made significant strides. The convolutional neural networks proposed by Gopalakrishnan et al. (2017) enable automated pavement crack detection at the same accuracy as human inspectors that is many times less in terms of inspection time and costs. The research proved the functionality of continuous monitoring of the infrastructure with mobile sensing infrastructure. Nevertheless, there were still the issues with the model generalization between the various pavement materials and conditions, and various training data and powerful algorithms are needed. Developments in machine learning of structural health monitoring data have made it possible to assess the safety of a bridge more effectively. Seventekidis et al. (2020) used long short-term memory networks to forecast the bridge structural reaction to loading and environmental conditions to ensure that the earliest analysis of abnormal behavior that signifies structural deterioration is detected. The study illustrated the importance of active continuous monitoring programmes and predictive analytics in case of proactive maintenance interventions, which may lead to disastrous failures.

Predictive maintenance optimization is a complicated application, which solves the problem of resources allocation. The reinforcement learning systems created by Rahman & Umamaheswari (2021) were designed to tackle the infrastructure maintenance scheduling, taking into account the safety requirements, budget limitations, and minimum disruption to traffic. The solution proved to be better than traditional rule-based scheduling because it allowed to utilize the available resources more effectively without violating the safety standards.

This paper shows how machine learning can ensure optimal multi-objective decisions in the management of infrastructure. Combination of various information sources such as visual inspection, sensor measurements and maintenance history has been found useful. Zhang et al. (2020) suggested a hybrid multi-source ensemble learning model to forecast the infrastructure degradation pathways. The study established that combining various sources of information was effective in enhancing prediction accuracy and reliability. Nevertheless, the problems of data quality and the complexity of the integration were defined as obstacles to the large-scale implementation, which requires standard data formats and quality control measures.

2.5 Autonomous Vehicle Safety Systems

Autonomous vehicles AVs can be seen as a synthesis of many machine learning elements, and the safety feature is the most important one, as it allows this technology to be implemented. Autonomous systems are too complex and safety critical to allow exceptionally robust and reliable machine learning methods.

Deep learning perception systems have gained impressive object detection and classification. Feng et al. (2020) suggested multi-task learning models of real-time vehicle, pedestrian, cyclist, and traffic control devices detection, and showed that their performance is sufficient to realize the autonomous driving applications. The researchers identified the necessity of different training data which represented different environmental situations, different lighting conditions and that of the objects. Still, there is risk in regard to adversarial examples and edge case robustness that are still serious issues pertaining to system reliability.

Algorithms based on learning to make decisions and path plans have developed significantly over that of the rule-based systems. Kiran et al. (2021) have performed an extensive overview of the fields of deep reinforcement learning usage in the decision-making process of autonomous vehicles, listing their achievements and setbacks. The study made it clear that simulation-to-reality transfer is a very serious obstacle since simulation-trained models tend to perform poorly in the real world. Regulatory acceptance and safety validation were also cited as other major concerns that have to be solved before it is widely deployed.

Combination of cameras, lidar, radar, and GPS Multi-sensor fusion techniques have been proven to have better robustness and reliability. Liu et al. (2020) engineered attention-based fusion network that heuristically combines sensor inputs through dynamically weighting them according to their reliability and the nature of the environment. The method portrayed excellent workability on a poor weather condition where individual sensors perform poorly. Nonetheless, the implementation issues were the cost of hardware and the computational needs that limited the actual application in cost-sensitive applications.

The methods of safety validation and testing of autonomous systems are active research topics due to their critical nature of safety. Kalra and Paddock (2016) pointed at the huge mileage needs to statistically prove the safety equivalence with human drivers to encourage the creation of scenario-based testing approaches. Later studies have investigated the use of simulation-based validation, adversarial testing methods, and formal validation in addition to empirical testing and the realization that a completely mileage based validation can be not practical.

2.6 Intelligent Transportation Systems Integration

Full-fleet smart transport systems incorporate a number of machine learning applications to obtain a system-level optimization and improved safety results. Synergistic combination of individual elements can give benefits that are more than the isolated applications. Machine learning is used to predict and manage traffic flow by predicting congestion and optimizing traffic signal timing in the traffic flow prediction and management systems. Vlahogianni et al. (2014) presented the short-term traffic forecasting methods review and the neural networks and ensemble approaches were discussed as the most successful regarding the ability to reflect the complex temporal dependencies. Recent studies have considered graph neural networks that capture spatial relationships in transportation systems which performs better in multi-step forecasting Zhao et al. (2020). Such developments make it possible to effectively manage traffic in advance to avoid safety risks associated with congestion.

V2ECS Vehicle-to-everything communication systems provide abundant data streams that are used to support cooperative safety applications. Ye and Yamamoto (2019) built machine learning models of cooperative collision warning systems and proved that the exchange of information between connected vehicles significantly increased the accuracy of crash prediction and the warning lead time. The paper revealed that communication protocols, data standardization and privacy protection are essential aspects of cooperative system implementation.

Integration of multi-modal transportation into overall mobility platforms is associated with sophisticated optimization issues. In their work, Tu et al. (2021) used reinforcement learning to find a route in multi-modal transportation networks, taking trade-offs between travel time, cost and safety into account. The method showed that it could be optimized on the system level, but it also exhibited such issues as scalability, real-time calculations, and modifying individual preferences. This paper demonstrates that machine learning integration in various transportation modes and types of services is a complex concept.

2.7 Human-AI Interaction and Behavioral Factors

Much depends on human factors and how well machine learning-based safety systems can work and be accepted. It is imperative to know how drivers, operators, and other stakeholders relate to AI systems to ensure successful deployment. This is a very vital dimension that has not been adequately covered in research that is technically oriented yet it has a very fundamental influence on the performance of real-world systems. The issue of driver trust calibration proves to be a major problem in partial automated systems.

According to a research by Hoff and Bashir (2015), improper trust (either too much or too little) undermines safety benefits with automation. Excessive trust will cause automation complacency and lack of vigilance whereas lack of trust will cause disengagement or de-use of the system. It is worthwhile to develop machine learning systems that give the right degree of openness as to what they can and cannot do to give the user the right level of trust.

Safety interventions may be accompanied by behavioral adaptation that minimizes or even eliminates the planned benefits. The risk homeostasis theory by Wilde (1982) indicates that drivers have an opportunity to change their behavior due to the perceived improvements in safety, which might eliminate the benefits of technology. As an example, safety benefits may be offset in part because of the fact that drivers will drive more aggressively with the best collision warning systems. These behavioral reactions have to be considered in machine learning systems by using continuous monitoring and adaptive intervention strategies.

The problem of the degradation of skills of highly automated systems leads to safety concerns in the long term. Endsley (2017) records that prolonged automation may eliminate the control skill of humans, which leaves them vulnerable as systems fail and they take over the control. Training systems created with machine learning and regular inputs of manual control requirements can contribute to the preservation of operator proficiency, yet the best strategies cannot be determined before additional research is conducted.

The design of human-machine interface has fundamental implications on the usability and safety of the system. According to Lee and See (2004), an effective interface design must be able to provide meaningful information on the state, capability, and limitation of the system to facilitate suitable reliance and effective intervention. Machine learning can be used to give the interfaces a personal touch to the needs and capabilities of individual users, but the design principles that guarantee transparency and help in situational awareness are critical.

2.8 Policy and Regulatory Frameworks

To achieve the benefits of machine learning in terms of safety, they are required to have effective governance structures to mitigate risks and hold them accountable. Regulation and policy-making should be in line with the rapid technological change and innovation, struck a balance between maximizing innovation and protecting the population.

The existing regulatory strategies differ significantly by jurisdiction, which has resulted in a disjointed way of governance. The United States makes use of mostly voluntary advice by the National Highway Traffic Safety Administration, topped by different state-specific regulations. The EU adopts more prescriptive safety rules such as obligatory advanced safety functions, stringent data security mandate as provided by the General Data Protection Regulation. Organisations such as SAE International and ISO offer international standards that lack the mechanisms of enforcement.

Liability frameworks should be evolved to deal with autonomous systems. Old approaches to liability consider the human driver to be the source of the liability, whereas autonomous systems complicate the definition of the driver, the manufacturer of the vehicle, the programmer of the software or the operator of the system. Multiple jurisdictions are working on new liability frameworks specifically citing automated vehicles but consensus approaches are still elusive. The assignment of clear liability influences insurance models, the risk management strategy and safety investment incentive.

The processes of safety validation and certification should be adjusted to evaluate the machine learning systems. Deterministic testing is not well matched with more probabilistic learning-based systems with variable performance with change of conditions of operation. There are new methods of testing scenarios, simulation validation and continuous monitoring of performance, but standardized validation systems are still in their infancy. The regulators are unable to measure the safety of the systems using standard metrics and tests.

The data governance systems should strike a balance between the benefits of safety and privacy rights. Extensive safety systems demand extensive behavioral and location information that is raising privacy issues. The acceptable practices in data collection, storage, use and sharing should be defined by the regulatory frameworks. Technical solutions to both of the safety and privacy goals are provided by several principles like differential privacy, federated learning, and data minimization, yet policy frameworks covering acceptable tradeoffs are still in development.

2.9 Algorithmic Bias and Fairness in Transportation Safety

Algorithms fairness in safety systems has been under researched yet it has extensive connotations on equitable safety outcomes. Differentiated safety protection of demographic groups, geographic regions, or other aspects by biased machine learning models can be subject to serious concerns about the ethical and legal aspect of the issue.

The major cause of algorithmic unfairness is training data bias. When training datasets are not representative of specific demographic groups, geographic locations, or vehicle models, the ensuing models can be poor at performance with the populations they are not representative of. Indicatively, object detection systems that are mostly trained with urban data could not work well in rural setting. The driver monitoring systems that are mostly trained on younger drivers might not work well with the elderly drivers. It is important, but difficult to guarantee the representative training data across the related dimensions due to the bias of historical data collection. Performance inequality measurement involves the methodical review of the subgroups relating to an assessment. The disaggregated performance analysis may provide significant information on the model effectiveness variance in demographic categories, which the aggregate measures of performance mask. An example is that a driver drowsiness detection system with an overall accuracy of 95% may get an accuracy of 98% on one group but 85% on another. The safety systems must be in compliance with minimum performance requirements with regard to all the applicable subgroups rather than cumulative performance goals.

Machine learning methods based on fairness provide possible solutions. Preprocessing strategies can be used to transform training data to make it less biased, in-processing strategies embed constraints on fairness into learning algorithms, and post-processing strategies can be used to modify model outputs to enhance fairness. Nonetheless, the definitions of fairness can be different and may be conflicting as it should be carefully selected depending on the context of application and social values. The application of transportation safety needs to strike the right balance between the issue of fairness and general safety effectiveness which is a complex optimization issue that would need the input of the stakeholders.

Continuous monitoring and auditing of bias are imperative since the bias might occur in deployment although none exist in the development phase. Patterns of population distributions, working conditions, and patterns of usage of the system change with time, which can lead to new sources of disparity. There are constant mechanisms of reviewing fairness and updating models, which contribute to fair performance among the populations.

2.10 Cybersecurity and Privacy in Intelligent Transportation Systems

Intelligent transportation systems, which are also linked to each other, create security vulnerabilities with potentially disastrous safety outcomes. Extensive security systems are necessary to avoid the use of malicious attacks to compromise the safety systems. The smart transportation systems have vast attack surfaces, which include vehicle networks, infrastructure communication systems, cloud computing platforms, and sensor input. Examples of potential attacks are sensor spoofing, which deceives perception systems, communication jamming that interferes with coordination, malware injection and corruption of machine learning models. Transportation systems are safety-critical systems and are therefore appealing to malicious actors.

The perception systems would have specific issues with adversarial machine learning. Scholars have shown that poorly designed perturbation of sensor inputs may lead to erroneous classification with potentially severe safety implications. Indicatively, minor alterations to the signs on the roads can lead to the object detection devices to be misconstrued to interpret stop signs as speed limit signs. Creating machine learning models that are robust to adversarial attacks is also a current research field with few applicable solutions.

Privacy protection should strike a balance between the safety needs and the right of the individuals. In-depth monitoring produces in-depth personal data such as travel habits, behavior, and preferences. Some privacy protection can be achieved through privacy-preserving methods such as differential privacy, federated learning, homomorphic encryption, and secure multi-party computation, and still have the same level of utility of analysis. Nonetheless, the methods carry performance costs and might not resolve every issue on privacy. There should be an unambiguous data governance framework.

Distributed ledger technologies and blockchain have the potential to be useful in transportation systems regarding security and accountability. Audit trails that can never be altered allow the investigation of incidents and assigning blame. The decentralized architectures minimize single points of failure. The practical deployment is, however, limited by scalability, computation overhead and energy consumption of a practical deployment. The development of lightweight blockchain applications that can be applied to the real-time safety implementation is an ongoing research.

2.11 Research Gaps Synthesis

The systematic literature review indicates that the research has a number of significant gaps that need to be addressed:

Integration and Standardization Gaps: The majority of the research undertaken is on standalone applications, not on system integration. The absence of standardized methods of evaluation, data forms and measures of performance impedes knowledge transfer and cross-study comparison. The creation of complete integration structures and standardization programs is an essential requirement.

Explainability and Interpretability Gaps Deep learning models with state-of-the-art performance are, in many cases, black boxes, which restricts the amount of regulatory acceptance and user trust. The explainability frameworks associated with transportation that consider the safety-critical requirements are not well-developed. Studies should find the balance between model performance and transparency requirements.

Gaps in Edge Computing and Real-Time Processing: Most of solutions proposed demand computational resources that can be conducted offline analysis and cannot be carried out in a real-time on-board. Transport-specific techniques of edge computing architecture and model optimization need to be researched further. The innovative solutions are required to address the latency constraints in safety-critical applications.

Learning Generalization and Transfer Learning: It is less common than the others, but frequently models fail to generalize to new conditions (environment, weather, geographic region, etc.). There has not been systematic transfer learning and domain adaptation in transportation safety. Few-shot training methods to access rare situations that are safety-critical need to be formed.

Ethical and Privacy Gaps: There is a small amount of literature concerning ethical considerations, privacy protection, and fairness related to the use of machine learning. Systems that balance between privacy rights and safety concerns need to be created. The methods of bias mitigation and fairness assessment that are unique to transportation safety should be developed.

Gaps in Human factors Integration: Technology-centric studies are not usually attentive to human factors such as driver acceptance, calibration of trust, behavioral adaptation, and skill deterioration. More focus should be made on human-centered design approaches that guarantee the correct interaction between humans and AI. Education and training systems that would assist users to learn the proper use of AI systems must be developed.

Multi-Modal Integration Gaps: Studies have paid little attention to integration between transport modes (rail, aviation, maritime, and micro-mobility) despite automotive applications being mostly prioritized. Mixed interaction modeling of multi-modal traffic environments needs to be studied.

Gaps in policy and Governance: There is a lack of research on policy frameworks, regulatory requirements, and liability models and certification procedures that are appropriate to machine learning-based safety systems. The close cooperation of the technical researchers and policy experts is necessary in order to create effective governance models.

Cybersecurity Gaps: Although cybersecurity has been identified as a threat, machine learning based transportation systems still lack systematic security measures. The topics of adversarial robustness, secure communication protocols and privacy-preserving architectures need additional research.

The mentioned gaps explain the research of the following sections, which considers the issues of integration, suggests the detailed frameworks and develops the base of further progress in machine learning applications to the transportation safety.

3. RESEARCH METHODOLOGY

3.1 Research Framework and Approach

The proposed research will assume a multifaceted mixed-method that combines the systematic literature review, conceptual framework, algorithm design, and the establishment of the evaluation methodologies. This research framework has four mutually supportive stages, including literature synthesis and gap analysis, methodological framework development, architectural design and algorithm formulation, and validation framework establishment.

The study applies a constructive research methodology, in which theoretical research on available methodologies is used to develop new designs and architectural solutions. The method is especially applicable to dealing with the problem of integration and developing fundamental frameworks in new research areas where the ability to empirically validate them might be too early because of the deployment limitations. The study has specifically focused on conceptual contribution and framework construction because it is understood that full empirical validation cannot be done without massive resources, real-world applications and longitudinal analysis outside of the scope of this theoretical study.

The methodology is built on critical evaluation of the current methods, definition of the scopes and gaps, combination of knowledge in different fields of application, and creation of integration models to resolve the identified issues. The given constructive approach can be used to promote theoretical knowledge and offer practical advice to system designers, algorithm implementers, and policymakers.

3.2 Data Sources and Collection Strategy

The To provide a comprehensive coverage of machine learning applications in the field of transportation safety, the research uses several information sources:

Academic Literature: Search of peer-reviewed articles in the most important databases such as IEEE Xplore, ACM Digital Library, ScienceDirect, SpringerLink, Scopus, Web of Science, and Transportation Research Board publications. Search techniques were based on Boolean combinations of such keywords as machine learning, deep learning, artificial intelligence, transportation safety, autonomous vehicles, crash prediction, driver monitoring, intelligent transportation systems, explainable AI, federated learning, and other words. The temporal scope was based on the period 2010-2025, and the older works of seminal importance were added to the scope as a historical stem. Additional literature that were relevant were found through forward and backward citation searching.

Technical Reports and White Papers: Government reports published by organizations such as the National Highway Traffic Safety Administration (NHTSA), Federal Highway Administration (FHWA), European Commission Transport Division, and UK Department for Transport as well as in the leading organizations of transportation safety. The white papers of industry participants (e.g., Tesla, Waymo, General Motors), technology vendors (e.g., NVIDIA, Intel, Qualcomm), and even research consortia (e.g., Connected Vehicle Safety Pilot, European Truck Platooning Challenge) were an informative input of realistic views on the problems related to implementation and practical deployments.

Standards and Regulatory Documents: Survey of new standards proposed by bodies such as SAE international (e.g. J3016 automation levels), International Organization of Standardization (e.g. ISO 26262 functional safety, ISO 21448 safety of intended functionality), Institute of Electrical and Electronics Engineers (e.g. IEEE P2846 assumptions of behavior models) and National Institute of Standards and Technology in relation to autonomous vehicles, connected vehicles and intelligent transportation systems.

Public Datasets: Review of publicly accessible transportation safety data as the Fatality Analysis Reporting System (FARS), Highway Safety Information System (HSIS), Second Strategic Highway Research Program Naturalistic Driving Study (SHRP2 NDS), German Traffic Accident Database and autonomous vehicle disengagement data (California DMV, Arizona DOT) to learn about the nature of the data, its availability, quality concerns, and representativeness.

Policy and Ethical Literature: Survey on policy, ethical frameworks, and regulatory analysis of government agencies, think tanks, academic institutions, and advocacy groups on governance issues, liability frameworks, privacy rules and ethical issues in AI-based transportation systems.

3.3 Analysis Framework

The analysis model uses a number of complementary methods:

Systematic Literature Analysis: Systematic review technique locating research topics, algorithmic techniques, metric of performance, practice domains, and limitations of reviewed literature. Recurring themes, consensus, and controversies, as well as gaps in knowledge were discovered through thematic coding. The analysis of citation networks showed powerful works and research directions. Temporal analysis was followed to trace the changes in research interest and methods and performance enhancement over time.
Comparative Methodology Review: Comparative evaluation of machine learning algorithms in domains of application, their rationale of algorithm selection, their performance properties, the complexity of computation, interpretability, and implementation aspects. Where possible, meta-analysis combined performance results of the studies. Analysis found new best practices and algorithmic trends.
Gap Analysis Framework: Organized detection of research gaps by looking at the unanswered problems, inconsistencies, areas that were not covered, and the new opportunities. The domains included in the categories of gaps were crash prediction, driver monitoring, infrastructure, autonomous systems, integration, algorithms, data, evaluation, and implementation considerations (ethics, privacy, regulation, human factors). Priority was done basing on the importance of the research and potential impact.
Architectural Synthesis: Creation of integrated frameworks that combine lessons learned cutting across domains of application and uncovers common patterns and challenges, and suggests unified architectural treatment of cross-cutting issues such as data integration, privacy preservation, explainability, real-time processing and system-level coordination.
Stakeholder Perspective Analysis: The perspectives of various stakeholders such as researchers, system developers, transportation agencies, policymakers, vehicle operators, fleet managers and end users have been taken into account. The requirement, constraint and success criteria analysis based on variations across the stakeholders informed the development of the framework that took into account many perspectives at the same time.

3.4 Evaluation Criteria Development

The study constructs a broad assessment criteria to be used in machine learning application in transportation safety scenarios. The conventional machine learning measurements of accuracy, precision, recall, and F1-score are useful, but do not give sufficient information on the actual safety impact in the real world. The developed system of evaluation will embrace several dimensions: These are the technical performance measures:

- Accuracy of prediction and error levels in a variety of conditions.
- Time stability as an aspect of time.
- Computer efficiency and real-time latency.
- Stability in the diverse operational conditions (weather, light, traffic).
- Cross context generalization (geographic areas, types of vehicles, people)
- Large scale deployments

Scalability Safety Impact Metrics

- Crash prevention that is likely to be minimized based on surrogate measures or simulation.
- Mitigation capability in inavoidable crashing situations.
- False positive and false negative rates using safety consequence assessment.
- Prevention time leading to successful intervention.
- Reliability/availability of systems that guarantee protection consistency.

Practical Implementation Metrics

- Data needs and the availability in the normal deployment scenarios.
- Processing requirements, memory requirements, energy requirements.
- Complexity in integrating with the other systems and workflows.
- Features of scaling to fleet assembly.
- Sustained performance maintenance and update.

Ethical and Social Metrics

- Efficiency of privacy preservation of sensitive personal information.
- Equity in terms of demographics, geographical areas and categories of vehicles.
- Users understanding and trust Transparency and explainability Openness in understanding.

- Acceptance and calibration of trust among users to allow proper reliance.
- Adherence to the relevant safety and privacy laws.

Economic Metrics

- Economic efficiency compared to the benefits obtained in safety.
- ROI based on development, deployment and maintenance expenses.
- Premium changes and implications of insurance.
- Conformity between the liability and legal framework.

3.5 Framework Development Methodology

New structures and designs are developed in stages:

Requirements Analysis: Collective investigation of functional specifications (what the system has to do), performance specifications (fastness, exactness, dependability), safety specifications (fail-safe actions, graceful downfall), usability specifications (user-friendly interface, user understanding), ethical specifications (justness, confidentiality, information control), and regulatory specifications (authorization, personal responsibility, information management). Requirement specifications are based on the analysis of application scenarios, consultation with stakeholders, review of regulatory documents and the application of ethical principles. Requirement prioritization is used to distinguish between the requirements of a project and additional features that are optional.
Pattern Analysis Design: Study of effective architectural patterns already shown to be effective in machine learning and transport fields, discover reusable units, interaction schemes, and design theories. Pattern cataloging makes it easy to design in a systematic manner and to create uniformity. Patterns such as layered architectures, sensor fusion strategies, ensemble strategies, edge-cloud distribution strategies, federated learning strategies and explainability integration strategies have been identified.
Conceptual Modeling: Abstract models of components of a system, interactions among these components, flow of information, control processes, and decision processes are developed. Different levels of abstraction are used to offer the top-level system descriptions as well as description of each part. Some of these modeling techniques are architectural diagrams, data flow diagrams, sequence diagrams, and algorithmic pseudocode. Models improve shareholder review and express the intent of design.
Validation Approach: Theoretical validation using logical consistency checking, completeness checking (satisfied all requirements) and feasibility

checking (satisfied no impossible requirements). Additional validation of the approach is by the expert review and comparison with existing methods. Although implementing and field testing would be helpful to validate the findings empirically, the lack of resources and time would make it impossible to verify comprehensive deployment validation in the theoretical study.

Iterative refinement: Framework development entails iterative development with input of literature review, expert advice and feasibility test. Several design stages gradually improve the frameworks concerning the limitations revealed and adding additional knowledge.

3.6 Limitations and Assumptions

This study has some limitations and is based on a number of assumptions:

Scope Limitations: Concentrate on ground transportation with a lot of emphasis in automotive applications and less on aviation, maritime and rail transportation systems. Focus on safety applications, as opposed to extensive mobility optimization or efficiency issues. Geographic emphasis There is an emphasis on developed countries with large experience in research and deployment.

Temporal Limitations: Due to the dynamic nature of the technological world, it can be assumed that the findings are reflective of the state of practice; they might need updating alongside the changing technology. Machine learning functions, computing capabilities, sensory technology, and deployment environments keep on changing at a very fast pace. The current technology may be recommended to be changed, as advances in capabilities change.

Data Availability Assumptions: Frameworks also make some reasonable assumptions of data availability, quality, and diversity, which are not necessarily universally applicable across jurisdictions and situations of application. In most areas, there are no detailed crash databases or connected vehicle infrastructure, or sensing capabilities that are expected in the proposed frameworks. Data privacy laws can limit availability of data even in the situation where technical collection is practicable.

Generalization Assumptions: Frameworks put forward will be believed to apply in a wide range of operational settings, however, individual execution will need tailoring of local circumstances, rules and infrastructure peculiarities. The world differs significantly in the aspects of culture, traffic, regulatory conditions, and the ability of infrastructure.

Technological Maturity Assumptions: The analysis relies on further development of enabling technologies such as computational ability, sensor technology, and communication systems and machine learning algorithms. Proposed frameworks will develop according to these technological assumptions being true.

Ethical and Regulatory Assumptions: Analysis assumes that the key ethical issues should be resolved and the enabling regulatory frameworks should be in place. Nevertheless, it is not possible to ethically agree on such matters as the division of liabilities, the limits of privacy, and the definition of fairness. Regulatory frameworks keep on changing with dubious directions.

Limitations on validation: Frameworks proposed have not been empirically validated in terms of any real world implementation and longitudinal studies. Theoretical examination and literature review give it a basis but practical validation needs considerable extra resources such as implementation work, deployment facility and prolonged testing.

Stakeholder Cooperation Assumptions: Adequate cooperation between various stakeholders (such as government agencies, vehicle manufacturers, technology companies, infrastructure operators, and research institutes) are the conditions of successful implementation. Cooperation may be hindered by competitive interests, intellectual property issues and organization barriers. Such limitations guide the interpretation of results and define areas that need further studies. The recognition of limitations helps to make proper use of the findings of the research and identify gaps in the research that may be addressed in further studies.

4. MACHINE LEARNING TECHNIQUES FOR TRANSPORTATION SAFETY

4.1 Supervised Learning Approaches

The most used machine learning paradigm, which is used in transportation safety, is supervised learning, which trains predictive models on a labeled historical data. Knowledge of the strengths and weaknesses of certain algorithms is used to select the right technique.

4.1.1 Classification Algorithms

The categorical outcomes that the classification algorithms predict are crash occurrences, severity levels, driver states and infrastructure conditions.

Decision Trees and Random Forests

Decision trees can be used to give interpretable models that can be used to determine key elements in safety outcomes. The tree structures are inherently a form of decision logic, which can be easily explained and accepted by the regulatory reigns. Ensemble techniques that involve a combination of two or more decision trees using random subsamples have more accurate predictive performance and lower overfitting than single trees (Breiman, 2001).

These methodologies are strong in categorical data, missing data and non-linear association prevalent in transportation data. Random forests offer variable importance measures that can be used to identify important safety variables to support specific interventions. As an example, when predicting the severity of a crash, random forests may indicate that the speed of the vehicle, the age of the occupants, and restraint analysis are the most significant variables, which implies priority in countermeasures.

Random forests may however be computationally complicated when dealing with large data sets or on-the-fly prediction. The size of models increases considerably with the size of forests, which may present difficulties in terms of deploying edges. Even ensemble averaging may not be able to overfit deep trees, and hyperparameter tuning such as tree depth controls, samples per leaf and forest size must be done carefully.

Support Vector Machines

Support vector machines (SVMs) uses kernel functions to project data into the high-dimensional space where more complicated decision boundaries are supported by linear separations (Cristianini and Shawe-Taylor, 2000). SVMs are shown to be useful in crash severity prediction and classifying driver behaviors, especially when deep learning is prone to overfitting, which is achieved with relatively small training sets. Linear, radial basis function or kernel selection (polaroid) has a significant performance and interpretability impact. Linear kernels also offer interpreted lines of linear decisions at the expense of complex relations. Non-linear kernels have the ability to capture complicated patterns but the interpretation reduces. The computational complexity grows exponentially with the size of the dataset as quadratic or cubic with the number of samples. Hyperparameter tuning here, in particular regularization parameter C, and kernel parameter, must be done with care and it is usually done with cross-validation.

SVMs are binary, but must be extended to multi-class multi-class problems (such as one-vs-rest, one-vs-one) used in transportation safety problems. These extensions add computation needs and class sensitively can be introduced.

Gradient Boosting Machines

The gradient boosting works sequentially by building ensemble models with each successive model making corrections on the errors of earlier models. XG-Boost, LightGBM, and CatBoost implementations have demonstrated state of the art results in many tasks involving the prediction of transportation safety (Chen and Guestrin, 2016).

These techniques are able to deal with heterogeneous features, automatically learn feature interactions and give estimates of feature importance used in interpretation. Gradient boosting proves especially effective in tabular data such as those used in transportation safety, and in general, it performs better on structured data when compared to deep neural networks.

Nevertheless, gradient boosting needs large hyperparameters that include the learning rate, the depth of a tree, estimators, and regularization factors. Computationally intensive Training training may be computationally intensive, but modern implementations (LightGBM, CatBoost) are much more efficient. Without regularization the possibility of models fitting the training data improperly exists, and validation is necessary. Sequentiality of training inhibits parallelization in comparison to random forests.

Neural Networks

Multi-layered artificial neural networks are learning curves that acquire intricate non-linear functions between inputs and outputs. Feedforward networks are useful in many areas of prediction such as crash risks, driver condition identification and prediction of infrastructure condition.

Neural networks are quite good at discovering complex patterns in high dimensional data and can model arbitrary functions with enough complexity of architecture and sufficient training. They are able to process continuous and categorical features at the same time and produce a set of predictions together with shared representations.

Nonetheless, neural networks usually need large training data to be trained well with performance deteriorating significantly as small training data is used. These are characterized by a large number of hyperparameters such as network architecture (depth and width), activation functions, learning rates and regularization strategies. Training may take time, and may be computationally expensive, and thus would need dedicated hardware (GPUs) to run in practice. Deep networks are black-box, which makes them hard to interpret, but post-hoc methods of explanation partially mitigate this drawback.

4.1.2 Regression Algorithms

The methods of regression forecast continuous results such as crash frequency, traffic speed, braking distance, and estimates of time-to-collision.

Linear and Polynomial Regression

Classical regression models formulate minimum performance and offer interpretable models, which show the connections between predictors and returns. Coefficients directly show the importance of features and feature directions. Non-linear relationships are modeled by the use of a relatively simple technique called polynomial regression.

They are useful during an exploratory analysis, to set benchmarks of comparisons, and in instances where interpretability is of the highest priority. Test procedures of statistical inferences furnish significance testing and the confidence interval. The ability to use limited resources in real-time is facilitated by the computational efficiency.

Linear models however, can significantly not fit complex transportation phenomena with non-linear relationships and interactions. They make assumptions of linear correlation (or some polynomials in the case of a linear regression), the independence of the errors and the homoscedasticity, and normally distributed errors, which is usually not true in real data of transportation. The non-linearity can be partially mitigated by feature engineering to come up with the terms and interactions of polynomials but this increases dimensionality and risk of multicollinearity.

Random Forest Regression and Regression Trees.

Regression tree ensembles are able to deal with non-linear relationships and interactions without explicit feature engineering. Random forest regression has been shown to be a very effective model in a wide range of transportation prediction problems, as well as give measures of uncertainty in the form of ensemble variance.

The methods are inherently tolerant of mixed variables, missing data and complicated interactions. Non-linear models are counterbalanced by feature importance scores used to make interpretations. Ensemble predictions are more stable and robust as compared to individual model predictions.

Computational costs are dependent on the size of the forest and the size of the dataset. Ensemble averaging may not work even on very deep trees. The extrapolation of forecasting beyond training data ranges may be inaccurate because trees cannot predict beyond perceived ranges of values. Edge deployment may be limited by storage memory requirements in order to store entire forests.

Gradient Boosted Regression Trees.

The gradient boosting regression tasks usually outperform the random forests and single models in terms of prediction. The methods are especially useful in predicting the traffic flow, estimating the travel time and predicting the infrastructure degradation.

Sequential error correcting will allow the capture of subtle patterns that are not captured by single models. Contemporary applications are able to optimize tree growing algorithms and make use of large datasets. Regularization mechanisms are inbuilt so as to support the prevention of overfitting.

Like classification boosting, hyperparameter tuning is very important and can be time-consuming. Sequential training restricts parallelizing. Challenges Edge deployment may be modeled with complexities and sizes. When using large ensembles of data it makes interpretation even more challenging than random forests.

Neural Network Regression

Deep neural networks that use continuous output layers are trained to learn complicated regression functions. These methods are highly dimensional predictive problems like sensor fusion and multi-output predictive problems where many related quantities are predicted at the same time.

Hierarchical representations of intricate patterns can be learned by the networks. Shared hidden layers can be used to learn effectively with multi task prediction of related outputs. The current frameworks make it easy to test various architectures quickly.

Nonetheless, large training data needs, computational intensity, sensitivity to hyperparameter and interpretability issues resemble those of classification neural networks. Normalization of output and selection of loss functions are of special importance in regression.

4.2 Unsupervised Learning Approaches

Patterns, structures, and irregularities of any unlabeled data are identified in unsupervised learning and are valuable to the transportation safety system when labeled data is limited, or when one is interested in finding new patterns

4.2.1 Clustering Algorithms

The clustering algorithms cluster observations of the similarity, which is based on the predefined category, and provide patterns of driving behaviors, crash qualities, and segments of the road.

Hierarchical Clustering and K-Means.

The K-means algorithm is a computationally efficient clustering algorithm, which classifies the data into k clusters according to the distance to the centres of the clusters. Hierarchical clustering constructs cluster hierarchies that depict nesting grouping structure. These traditional methods characterize clusters of driving behaviors, crash incidents or road sections.

The use of K-means allows working with large datasets. Deterministic results under initialisation. Hierarchical clustering displays cluster hierarchy without knowing in advance the number of clusters.

Nevertheless, k-means needs the number of clusters k to be specified, and this cannot be known in advance. It makes the assumption of spherical clusters of equal sizes which may lack complicated cluster shapes. Initialization sensitivity may yield varying results upon repeated executions, but this problem can be avoided with several consecutive executions with most appropriate input selection. Hierarchical clustering is time-consuming when dealing with large data sets. These two methods are weak with groups of vastly varying densities or non-convex figures.

DBSCAN, Density- Based Clustering.

Density-Based Spatial Clustering of Applications with Noise (DBSCAN) recognizes arbitrary-shaped clusters of data, and automatically finds outliers. This would be useful in detecting crash hotspots and unusual driving habits.

DBSCAN also accepts clusters of different shape and size. It identifies automatically the number of clusters depending on the data density. Ability to detect outliers is useful in safety application. Relative insensitivity to the initialisation guarantees smooth results. Nevertheless, the choice of parameters (epsilon neighborhood size, minimum points per cluster) has a significant impact on the results and is not directed by principles. Clusters of vastly different densities result in a performance degradation. Variations in computational complexity with dataset size can be reduced by using spatial indexing.

Gaussian Mixture Models

Probabilistic clustering Probabilistic mixture models Gaussian Mixture Models (GMMs) are data mixtures composed of mixtures of Gaussian components. They give such cluster memberships on intonations, and measure uncertainty. The methods are handy where observations can be classified on more than one category at the same time.

GMMs offer probability distributions on cluster assignments instead of hard assignments, which are more accurate on uncertainty. Probabilistic framework provides principled model selection on the information criteria. The cluster shapes and orientations are represented by covariance matrices.

Training with expectation-maximization may end up with local optima based on initiation. Model selection procedures are necessary in determining the number of components to use. The dimension of components and dimensionality is all that increases computational intensity. Gaussian assumptions could fail to be a true distribution.

4.2.2 Dimensionality Reduction

High-dimensional transportation data are dimensionality reduced by applying techniques that retain the important information and reduce the amount of computation, as well as permit visualization.

Principal Component Analysis (PCA) defines the combinations of the features that are the most varied and linear (Ghosh et al., 2017, 2018a, 2018c, 2018d, 2020a, 2020b). This helps in minimizing noise, processing power, and visualization of complicated transportation data. PCA is used as preparation to further analysis and eliminates multicollinearity and informativeness is concentrated on. PCA is computationally intense as it has closed-form solutions. Linear transformation can still be interpreted. Dimensionality reduction may reduce the number of computations needed to run downstream analysis.

Non-linear structure of data may however be overlooked by the linear assumptions. Principal components maximize variance and not predictive performance that may be discriminative. Interpretability goes down as the components reflect combinations of source features. Sensitivity to scaling involves the application of great caution to normalize input features.

Autoencoders

Neural network auto encoders are learners of non-linear compressed representations of high-dimensional data. These techniques are useful in detecting deviations, detecting odd driving or infrastructure conditions based on large reconstruction errors on the odd samples.

Autoencoders are learned with non-linear representations that encode a complicated structure. They can also be end-to-end trained using downstream tasks. Reconstruction errors give natural scores of anomaly. Hierarchical representations are represented in deep architectures.

Training is expensive in terms of data and computation resources. The choice of architecture and hyperparameters are very important regarding performance.

Reconstruction can concentrate on general trends, which can offer gaps in rare yet safety-important peculiarities. Latent representations are not necessarily interpretable as in the case of PCA.

t-SNE and UMAP

Algorithms with non-linear dimensionality reduction: t-distributed Stochastic Neighbor Embedding (t-SNE) and Uniform Manifold Approximation and Projection (UMAP) maintain local structure and, therefore, allow visualization of intricate high-dimensional transportation data in 2D space. The tools aid in the process of exploratory data analysis, which shows inherent groupings and patterns.

These methods are good in visualization, and they do not destroy local neighborhood structure. They usually display clumps that are not apparent in linear drawings. UMAP is more likely to maintain global structure compared to t-SNE and is more efficient.

Nonetheless, these techniques are mainly used in visualization as compared to general dimensionality reduction. Even with stochastic nature there are varying results between runs. Results are highly sensitive to hyperparameters (N 1 perplexity of t-SNE, n 1 neighbors of UMAP). Computational intensity restricts the use to very large data sets. New data cannot be embedded without retraining, making them useful in production systems.

4.2.3 Anomaly Detection

Anomaly detection recognizes the peculiar trends that may reflect safety-related issues like a driver with erratic behavior, the collapse of an infrastructure, or the emergence of a crashing risk factor.

Statistical Methods

Conventional statistical methods such as z -scores, isolation forests, and one-class support vector machines define normal behavior standards and identify significant changes. These methods operate with small number of labelled anomalies and are also computationally efficient.

Statistical techniques do not make many assumptions about the nature of anomaly. They can efficiently use small-labeled anomalies. Real time application is made possible by computational efficiency. Well-developed theoretical backgrounds offer an assurance of methods.

Nevertheless, it can be hard to set some normality boundaries, involving knowledge of the domain and validation. High-dimensional data may have complicated patterns of anomaly that cannot be represented using simple statistical measures. Normal behavior training needs to be re-trained periodically to prevent false alarms.

Neural Network Methods

Autoencoders that are conditioned on normal images have high reconstruction losses on anomalous ones. Variational autoencoders and generative adversarial networks form the model of probability distribution, which detects anomalies by low likelihood. These methods work with high dimensional data and identify intricate patterns.

methods learn intricate conceptions of normality. They also scale to high dimensional sensor data in the transportation setting. Generative models offer probability-based scores of anomalies, as opposed to binary classifications.

Normal data is needed in training. It is still difficult to establish the right reconstruction error or likelihood threshold. The real-time edge deployment might be limited by the computational requirements. Black-box nature decreases interpretation of the anomalies found.

4.3 Reinforcement Learning Approaches

Sequential decision-making under uncertainty, as in autonomous vehicle control, adaptive traffic management, etc., is covered by the reinforcement learning.

4.3.1 Model-Free Reinforcement Learning

Model-free is a direct learning approach that does not learn separately the system dynamics and optimal policies, but only learns the optimal policies.

Q-Learning and Deep Q-Networks.

Q-learning approximates state-action pair expected cumulative reward values. Deep Q-Networks (DQNs) are neural networks that provide approximation of the Q-values in the high-dimensional state families (Mnih et al., 2015). These techniques have been deployed in the control of traffic lights, driverless vehicle driving, and driver aids.

Q-learning is an experience-based learner that does not need system models. DQNs are able to generalize to high-dimensional, complicated problems. Experience replay enhances sample efficiency because it uses previously experienced things.

Nonetheless, the process of learning may be sample-inefficient and may need a large exposure to the environment or simulation. The trade-offs are exploration-exploitation which influence the efficiency and the safety in the learning process.

Q-learning is prone to overestimation bias which results in suboptimal policies. In real-life safety-critical scenarios, safety has been a challenge in the learning process.

Policy Gradient Methods

Policy gradient algorithms perform the maximization of policy parameters towards maximizing expected cumulative reward. Actor-critic structures are a combination of value-based and policy-based learning. Proximal Policy Optimization (PPO) and Trust Region Policy Optimization (TRPO) offer stable learning that is appropriate when operating in safety-critical settings (Schulman et al., 2017). Policy gradients maximize the intended objective. They deal with continuous action spaces. Actor-critic approaches promote efficiency in the sample. Contemporary (PPO, TRPO) versions offer stable learning, and have the guarantees of monotonic improvement.

Sample complexity is high and thus necessitates a lot of training. The safety issues in the exploration process are a challenge in the real world implementation. The issue of hyperparameter sensitivity has an influence on learning stability. In long-horizon tasks, credit assignment is still a challenge.

Multi- Agent Reinforcement Learning.

Transport systems have various interacting agents that include vehicles, pedestrians, as well as infrastructure. Multi-agent reinforcement learning allows agents to learn by means of interaction and coordination. One of its uses is cooperative traffic situations and integrated safety systems.

Multi-agent models are models of real-life interaction. Agents are able to acquire competitive and cooperative behaviors. Decentralized learning offers scalability to large population of agents.

The complexity of learning is an exponent of the number of agents. Convergence is complicated by non-stationarity due to jointly learning agents. There should be mechanisms of coordination. The computational need increases with the number of the agents.

4.3.2 Model-Based Reinforcement Learning

Model-based methods learn model environments that allow one to plan and simulate sequences of actions.

Model Predictive Control using Learned Models.

By combining model predictive control and model learned environment, safe planning can be realized in learned dynamics. These methods involve safety restrictions in optimization, a combination of the advantages of learning and the control theory.

The model-based techniques enhance the efficiency of the sample by way of planning. Safety constraints may be directly integrated in the process of optimization. Generalization to new situations is made possible through learned models. The explicit model representations are good with interpretability.

Unrealistic learning errors may impair performance. Real time constraints may be challenged by online optimization computational requirements. The relevant domain knowledge is needed when specifying the right safety constraints. Uncertainty in the models should be considered in the planning.

Simulation-Based Training

The models of learned environments produce the synthetic training experience, which enhances the speed of learning and increases safety by training on the uncommon but critical scenarios and will not be at risk. This strategy is useful in safety-critical case edges. Simulation allows a safe way of exploring dangerous situations. Parallelization can enable training to be done faster. Uncommon happens can be over-sampled. Various states of affairs can be investigated.

Nevertheless, simulation to reality errors are difficult. Models do not always reflect the relevant phenomena in the real world. Dependence on simulation may result in deployment frailty. Authhentication in the real world still is necessary.

4.4 Deep Learning Architectures

Deep learning has transformed many transportation safety applications in the sense that it has made it possible to learn features hierarchically using raw sensor data.

4.4.1 Convolutional Neural Networks

Convolutional Neural Networks (CNNs) are good at handling data that is grid structured such as images and spatial data.

Object Detection Architectures

YOLO (You Only Look Once), Faster R-CNN and SSD (Single Shot Detector) enable real-time detection and classification of objects (vehicles, pedestrians, cyclists, and road infrastructure) (Redmon et al., 2016). These architectures are the key elements of autonomous vehicle perception systems as well as video-based safety monitoring.

Current detectors are able to perform in real-time and can be used in safety-critical applications. They deal with object classes at the same time. End-to-end training makes it easy to deploy. Transfer learning can be done with a small amount of domain-specific data using pre-trained models.

Nevertheless, it has poor performance during low-quality conditions (occlusion, bad lighting, weather). Detection of small objects is still difficult. Edge deployment of largest models has computational requirements. Persistent adversarial resilience issues prevail. Semantic Segmentation Networks Semantic segmentation networks are an additional type of neural network utilized in computer-generated imagery (CGI).

DeepLab and U-Net as well as fully convolutional networks also offer pixel-wise scene knowledge. They can be used in lane identification, finding the driving area, and determining the condition of infrastructure.

The interpretation of the scene can be done accurately with pixel level knowledge. Contemporary architectures are both very accurate and efficient. Multi-scale processing contains local and global information.

Pixel level annotated data is required in training, and it is costly to acquire. Computational requirements are beyond object detection. This can be caused by the imbalance in the classes between small and important objects and big background areas that should be handled carefully. High-resolution imagery still has problems with real-time performance.

Multi-Task Learning

Multi-task networks can detect, segment, and estimate depth simultaneously (similar tasks) and can share representations with each other, as well as enhance performance. These offer end to end scene knowledge needed in driving autonomously.

Shared representations enhance the efficiency in the parameters. Associated tasks can be used to enhance the performance of each other via transfer. Single forward pass is efficient in giving multiple outputs.

During training, care has to be taken in balancing tasks. Competing task goals can undermine personal performance of tasks. The design in architecture becomes complicated. Failure modes can be associated with several tasks at the same time.

4.4.2 Recurrent Neural Networks

Recurrent architectures operate on sequential data, over which the temporal constraints that are important to comprehend driving behaviors and traffic dynamics can be determined.

Long Short-Term Memory Networks

Long Short-duration Memory (LSTM) networks discover long-term connections, a concern of vanishing gradient issues of basic recurrent networks (Hochreiter and Schmidhuber, 1997). They have applications in predicting the traffic flow, trajectory, and intention of the driver.

LSTMs obtain long-term temporal correlations. Selective memory is made possible by gating. They are used to work with variable length sequences. The past and future context are reflected in Bidirectional variants.

A computationally intensive and slow training can occur. Processing in a sequence restricts parallelization. Sensitivity to hyperparametric is a performance factor. Learned temporal patterns are difficult to interpret.

Gated Recurrent Units

GRUs offer more effective streamlined recurrent architecture compared to L 1 in terms of parameters. They come in handy in real-time applications that are resource-constrained.

GRUs are comparable to the LSTMs in terms of their performance using fewer parameters. Training is more efficient. Less complex architecture mitigates the risk of overfitting using small quantities of data.

Sequential processing, similar to LSTMs, does not increase the level of parallelization. Dependency in the long-term still can be tough. It is still hard to interpret.

Attention Mechanisms

Attention mechanisms are selective to focus on relevant components of sequence and processing of long sequences. The attention-based architectures are effective in activity prediction and trajectory forecasting.

Attention is also interpretable as it shows what inputs affect outputs. Basic RNNs do not capture long-range dependencies as well as they are represented in long-range dependencies. Variable length sequences are naturalized.

Self-attention grows quadratically in length of sequence. Data-intensive Training may be information-intensive. There is more complexity in architecture. The best attention patterns do not always appear when there is a lack of training data.

4.4.3 Transformer Architectures

Transformers of the self-attention type have recorded excellent use in modeling sequences and two-source data fusion.

Spatial-Temporal Transformers

Transformers were modified to traffic and driving patterns in the spatial-temporal data models. They attain state of the art level of forecasting of trajectory and traffic (Vaswani et al., 2017).

Long range dependencies are efficiently modeled using transformers. Parallel processing is a faster way of training. Attention weights can be interpreted. They are able to scale to large datasets.

Computational and memory proportions increase quadratically with the length of sequence. Big data model models need large training data. There is an influence of design on performance by positional encoding. There is a challenge in deployment over resource constrained edge devices.

Vision Transformers

On image data, vision transformers use transformer architectures, which have been shown to be competitive with CNNs in terms of inductive bias choice. Such applications are scene recognition and infrastructure monitoring.

Vision transformers operate well without convolution process. They are captivating of the global context. Architectures cross over visual tasks. Transfer learning is possible when there is pre-training on large datasets.

Competitive performance requires very big training datasets. Computational needs are beyond effective CNNs. Spatial inductive bias Lack of spatial inductive bias needs additional data. Live edge deployment is a challenge.

4.4.4 Graph Neural Networks

Graph neural networks are used to process graph-structured data, which is inherently the case with transportation network relationships.

Traffic Network Modeling

Graph Convolutional Networks (GCNs) and Graph Attention Networks utilize spatial correlatedness between road segments and allow to predict traffic better, taking into account network topology (Kipf and Welling, 2017).

The network structure is modeled by the innate nature of GCNs. Spatial dependencies are represented literally. Performance can tend to be higher when topology is not considered. Receptive processes define significant areas in the network.

Construction of graphs has significant impact on performance. Very large networks may not be easily scalable. Special architectures are required in time dynamics. Learned graph representations are still hard to interpret.

Car-Car Interaction Modeling.

The multi-agent scenario modeling with the help of the graph-based models of vehicle interactions can assist in the context of making predictions of the trajectories and the assessment of risks of collision in the cases of the dense traffic.

The explicit interaction modeling enhances the accuracy of prediction. The attention mechanisms recognize influential vehicles. The scale in architectures is to changing numbers of agents.

There is a rise in computational complexity with the number of agents. The performance should be optimized carefully in real time. It may be restricted in generalizing to much different traffic densities.

4.5 Ensemble Methods

Ensemble methods are the combination of several models in order to enhance prediction robustness and precision.

Homogeneous Ensembles

Bagging, boosting and stacking are several applications of the same algorithm, which reduce variance and bias, and enhance generalization. The successful implementations are random forests and gradient boosting.

Teams are quite frequently better than single models. Complementary patterns are captured by the different members of the ensemble. Outlier and noise sensitivity are improved. Ensemble variation help in coming up with uncertainty estimates.

Multiplicative dependence of computational costs on the size of the ensemble. The memory needs increase significantly. Training time increases. Ensemble size may also be a limitation to real-time inference. The diminishing returns take place after optimum ensemble sizes.

Heterogeneous Ensembles

Voting and averaging or meta-learning combines the complementary advantages of various types of algorithms. The strategies are effective in a wide range of operational environments.

Various patterns are picked by different algorithms. Diversity may frequently benefit strength through Ensemble. Ensemble weighting can be optimized with the use of meta-learning. Approaches to failure in various algorithms are not usually correlated.

The determination of the best ensemble composition is experimental. The mere training of a variety of different models adds to the computational requirements. The complexity of integration goes up. Some applications may not warrant additional complexity in order to gain performance.

Deep Ensembles

The various initializations of many neural networks offer uncertainty estimates in form of prediction variance. Deep ensembles are more calibrated and resilient, which is useful in applications that require safety.

Groups give good estimates of uncertainty. They are less susceptible to adversarial instances. It can be implemented without any changes in architecture. Diversity arises due to chance initialisation.

The cost of computation and the cost of memory are both linear in terms of the size of the ensemble. Training time multiplies. There is increased inference latency. The multiple networks make edge deployment more difficult.

4.6 Algorithm Selection Considerations

In order to choose the right machine learning algorithms, one has to take into account several factors:

Data Characteristics: Characteristics of data such as sample size, type of features, temporal relationships, spatial organization, and availability of labels limit the options of the algorithms. Spatially-structured data of high dimensionality is better represented by CNNs, whereas time series is represented by RNNs. The little information would be biased to simpler models or transfer learning methods.

Performance Requirements: Algorithms have performance requirements on their complexity that are enforced by real-time processing requirements. High reliability and estimated uncertainty of safety-critical applications. There are accuracy-complexity trade-offs between algorithms, and selecting an algorithm should be done carefully, depending on the constraints of its operation.

Interpretability Requirements: The regulated safety-critical systems may have explainable decisions. Linear models and decision trees are by their nature understandable. Post-hoc explanation methods are needed in deep learning. The transparency requirements of the stakeholders affect the choice of algorithms.

Limiting Resources: Edge computing and embedded systems have rigid limitations of resources. The application of model compression, quantization, and efficient architectures allow running on a constrained hardware. There should be trade-offs managed between the complexity of the models and their deployment.

Domain-Specific Issues: Transportation Safety systems have certain issues such as extreme imbalance of classes, high false negative cost, and a continuous requirement to adapt to changing environments. These domain specific requirements need to be considered in the selection of the algorithm.

Fairness and Bias Issues: Algorithms are different in regard to being prone to biased training data. The choice should be based on the consideration of fairness implications and bias reduction measures that are available by each type of algorithm. Clear algorithms can be used to support the evaluation of fairness and correction.

5. APPLICATION DOMAINS IN TRANSPORTATION SAFETY

5.1 Crash Prediction and Risk Assessment

Primary machine learning applications in transportation safety include crash prediction and risk assessment, including the real-time risk prediction, hotspots, and estimation of the severity.

5.1.1 Real-Time Crash Risk Prediction

Crash risk prediction systems are real-time systems that constantly determine the likelihood of crashing according to the prevailing conditions so that some preventive interventions can be taken.

Methodological Approaches

Two-stage architectures, i.e. the estimation of traffic state and risk prediction, are usually adopted in successful real time prediction systems. Raw sensor data are first processed, in the first stage, to approximate traffic conditions such as flow rates, density, velocity distributions and measures of turbulence. The second step uses machine learning models to estimate the probability of the crash depending on approximations of traffic state and contextual conditions.

Specifically, temporal convolutional networks are effective that learns the temporal trends in the run up to crashes. These architectures consume traffic measurement streams and detect precursor conditions which forecast high crash risk. Attention mechanisms can help to improve the model interpretability by emphasizing the most important time and sensor positions that can make the risk predictions.

Feature Engineering

The key to successful real-time prediction is feature engineering paying close attention to traffic dynamics. The key aspects are the speed difference in the lanes and time, quick deceleration episodes, density, gradients, indicators of flow breakdown, and temporal derivatives of the traffic measures. Other important contextual data such as weather conditions, time of the day, day of week, and special events are also useful additional context.

Implementation Challenges

There are several problems with the deployment of real-time crash prediction. False positive rate can be very high, which may compromise the credibility of the systems and the trust placed by the user, necessitating a proper setting of the threshold between sensitivity and specificity. Computational latency should also be low enough to allow actionable warning lead time, usually, this requires a processing time of seconds. The limitation of sensor coverage and quality of the data impacts the reliability of prognostication, requiring sound management of the missing and incorrect data points. Combination with traffic managing systems and sending warnings to drivers are further operation problems.

This artificial intelligence system illustrates information gathering, feature development, machine learning prediction as well as decision-making to conduct crash risk evaluation in real-time.

5.1.2 Crash Hotspot Identification

Crash hotspots identification uses spatial analysis to identify both road segments and intersections with high crash risk, which can be used to improve infrastructure and enforce it selectively.

Spatial Clustering Approaches

The kernel density estimation offers smooth surfaces of spatial crash density exposing hot spots of crash. DBSCAN clustering detects arbitrary-shaped spatial clusters as well as outliers. Network-based spatial analysis takes into consideration road network topology, which is more meaningful in identifying hotspots than straight geographic methods.

Temporal Pattern Analysis

Crashes show time variations such as a daily, weekly and seasonal recurrence. Time-series clustering finds locations sharing identical patterns of timing of crash, making it possible to have an intervention corresponding to the timing of the risk.

Change point detection algorithms locate when hotspots emerge which need to be addressed urgently.

Predictive Hotspot Mapping

Machine learning systems assume possible crashes in the future, which are forecasted by analyzing the features of the road, traffic flow, and weather. Spatial regression models capture the spatial correlation between the adjacent sites. Multi-spatial model combinations are shown to have higher prediction ability.

5.1.3 Crash Severity Prediction

The estimation of crash severity predicts injury, used to guide emergency response resource planning as well as safety countermeasures assessment.

Multi-Class Classification

Severity prediction commonly uses the injury level (no injury, minor injury, serious injury, fatality). Ordinal classification methods consider levels of severity of a level of ordinal classification. Class imbalance has become the main source of problem since extreme crashes only constitute small percentages of crashes overall, but they carry disproportionate safety weight.

Feature Space

Extensive severity models make use of various characteristics such as the type of collision, the characteristics of the vehicles, the geometry of the road, weather/environment, the demographics of the occupants, and the use of restraint systems. The feature interactions are especially relevant in the case of complex severity determinants.

Deep Learning Architectures

Attention mechanisms in deep neural networks reach the state-of-the-art in severity prediction. Attention weights allow one to gain an understanding of influence factors, which increases the interpretability of models. Embedding layers are used to deal with high-cardinality categorical data used in crash databases

5.2 Driver Behavior Analysis and Monitoring

Behaviour of drivers is an extremely important transportation safety factor, and machine learning allows to monitor it thoroughly, classify, and implement an intervention system.

5.2.1 Driving Style Classification

The classification of driving styles organizes drivers according to their behavioral patterns and allows providing interventions of personal safety and risk evaluation.

Behavioral Feature Extraction

The driving style is exhibited in terms of acceleration, braking, turning, keeping the lane, and speed selection. Styles are described using statistical characteristics such as means, variances and distributions of these measurements. The frequency domain features represent oscillatory inputs of control patterns. Contextual normalization is used to explain the different traffic and road conditions.

Classification Methodologies

Unsupervised clustering identifies behavioral groupings of data that are not determined by a priori categories and shows natural style continua (cautious to aggressive). In the supervised classification, drivers are classified into prescribed categories according to the ad hoc training data. Semi-supervised methods take advantage of small volumes of labeled data and vast volumes of unlabeled ones.

Temporal Evolution

Driving styles are dynamic changes that take place with age, experience and change of life circumstances. The algorithms of online learning are constantly updating the classifications with new information. Change detection algorithms detect material changes in behavior, which might be the signs of health problems or other issues.

5.2.2 Distraction and Drowsiness Detection

The percentage of crashes due to inattention of drivers is significant, which drives the development of real-time monitoring systems.

Vision-Based Monitoring

The convolutional neural networks observe facial expression that signals drowsiness such as facial expression, head position, eye blinking, and yawning. The gaze direction, head position and drowsiness are estimated together in multi-task learning frameworks. IR capturing also facilitates monitoring in diverse lighting conditions such as in the night.

Physiological Monitoring

Sensors on the wrist that include the heart rate, heart rate variability, and electrodermal activity are further drowsiness monitors. RNNs are used to predict the patterns in time series of physiological data that predict patterns before the occurrence of drowsiness.

Vehicle Dynamics Analysis

Drowsy driving is observed in the vehicle control behavior such as a higher lane position variance, corrections on the steering wheel, and speed anomalies. Vehicle dynamics battery Time-series analysis will offer indicators of drowsiness without conducting any directions to the driver, thereby increasing the level of privacy and minimizing expenses.

Multi-Modal Fusion

The sensor fusion of the visual, physiological, and vehicle dynamics information provides a robust drowsiness detection in a variety of conditions. Attention-based fusion is an adaptive fusion that assigns modalities reliability and information content.

5.2.3 Aggressive Driving Detection

Examples of aggressive driving such as sudden acceleration, overbraking, and sudden changes in lanes increase the risk of crashing, and real-time intervention and detection give incentive to these behaviors.

Naturalistic Driving Analysis

Massive naturalistic driving experiments are used to give ground truths in training aggressive driving detectors. The accelerators and gyroscopes in Smartphones allow mass data gathering. Federated learning, which can be applied to preserve privacy, allows training the model without concentrating sensitive driving data.

Contextual Classification

Assessment of aggressive behavior needs contextual interpretation because accelerating fast can be an act of required safe merger in certain situations and an act of aggression in other situations. Classification is informed by contextual characteristics such as traffic density, the road type and weather conditions. Individual baselines take into consideration the individual variations in the normal driving habits.

5.3 Infrastructure Condition Monitoring

The quality of transportation infrastructure has a major impact on the safety level, and machine learning allows monitoring it continuously.

5.3.1 Pavement Condition Assessment

Pavement flaws and other defects such as cracks, potholes and rutting pose safety risks that need to be addressed as they arise.

Image-Based Detection

Defects are detected by convolutional neural networks that are trained on pavement images. Semantic segmentation networks offer pixel-wise localization of defects that allow the determination of the area accurately. Recent mobile sensing platforms that have high resolutions enable continuous monitoring at traffic speed in an automated fashion.

Sensor Fusion Methodologies

Sensor fusion - Visual data that is coupled with laser profiling, ground penetrating radar and accelerometer based roughness measurements that give complete condition assessment. Multi-modal fusion identifies the surface and subsurface defects.

Predictive Degradation Modeling

The predictive modeling of pavements conditions is made possible by long-term records of inspection. Time-series is used to forecast the future development of the pavement condition. Recurrent neural networks are the models of degradation that take into account the traffic loading, the climate and the maintenance history. Predictive models optimize maintenance scheduling in terms of their cost-safety.

5.3.2 Bridge Structural Health Monitoring

The consequences of bridge failure are disastrous in terms of safety, which is why it is essential to conduct structural health assessment on a regular basis based on sensor networks and machine learning analytics.

Vibration Analysis

Bridge vibrations when loaded with traffic and excited by the environment are measured using accelerator arrays. The modal analysis removes modal features of structural properties such as mode shapes, natural frequencies, and damping ratios.

Machine intelligence identifies the abnormal vibrations patterns that are signs of structural ruin or wear.

Strain and Displacement Monitoring

Monitoring of structural response of loading is done by strain gauges and displacement sensors. Long short term memory networks indicate the responses to be expected in certain loading conditions and variations will indicate the possibility of problems. Physics-informed neural networks add the principles of structural mechanics to improve the predictive faithfulness and physical plausibility.

Computer Vision Monitoring

A visual bridge inspection system that uses camera to identify spalling concrete, corrosion of steel structures and structural distortion. Crack detectors built through the use of deep learning detect fine cracks that may go unnoticed by human inspectors. Time-lapse analysis is used to measure the rates of crack growth as informers to urgency of maintenance.

5.3.3 Traffic Signal and Sign Condition Monitoring

Defective traffic controlling equipment poses safety risks that have to be identified and fixed as soon as possible.

Automated Visual Inspection

Traffic signs and signals are captured using mobile mapping systems and traffic cameras when everything is working as expected. Signs that are lost, broken or covered are detected with computer vision algorithms. The analysis of nighttime images determines the retroreflectivity, which is an indicator of signs that need to be replaced.

Signal Timing Optimization

An analysis of traffic trends and signal timing based on machine learning indicates potential signal settings that cause congestion and crash risk. Reinforcement learning achieves both a timing optimization of the signals based on traffic patterns as well as keeping the safety margins

5.4 Autonomous Vehicle Safety Systems

Auto-pilot cars combine many machine learning parts that constitute an entire safety package that allows autonomous functioning.

5.4.1 Perception Systems

Perception systems interpret sensor data to be aware of driving surroundings, sensing and following the objects of interest and events.

Multi-Modal Object Detection

Self-driving cars use cameras, lidar, radar, and ultrasonic sensors that give supplementary data. Lidar has precise 3D geometry and can function in the dark. The cameras provide color and abundant semantic detail. Radar cuts through unfavorable weather. Sensor fusion is a fusion of modalities with strong perception under a variety of conditions.

Semantic Scene Understanding

Semantic segmentation networks give pixel-level classification of driveable regions, lanes, and sidewalks among other scenery features. The separation of single instances of objects is conducted in instance segmentation. Semantic and instance segmentation are brought together with Panoptic segmentation. With these representations, fine-grained reasoning of scenes that aid safe navigation is provided.

3D Object Detection and Tracking

Deep learning architectures take in point clouds in lidar images and identify and classify entities in 3D space. VoxelNet and PointNet networks are found to be effective in 3D perception. Multi-object tracking is used to preserve object identities through time so that they can be used to predict the trajectories and be able to infer intent.

5.4.2 Prediction and Planning Systems

Proactive collision avoidance and planning are possible because prediction systems predict the future states of the environment.

Trajectory Prediction

Anticipation of directions of traffic and pedestrians, and bicycles will allow safe motion planning. Transformers and recurrent neural networks are models that model interaction pattern and motion dynamics. Models that are socially-aware consider multi-agent interaction. Uncertainty quantification does not give a definite prediction but instead probability distributions on the various possibilities of future events

Intent Recognition

The prediction and planning rely on the understanding of intentions of other road users. Some of the intent cues are the turn signal, head orientation, and motion pattern. Hidden Markov models Recurrent networks Recurrent networks model temporal variations of intent states.

Motion Planning

Motion planning finds safe and efficient routes to navigation targets with no collisions and compliant with the regulations of traffic. Sampling planners come up with standard search paths that are tested on both safety and comfort. Trajectory optimization problems that are constrained by safety considerations are solved by optimization-based methods. Planners are learning-based and implement imitation learning or reinforcement learning as a learning method in order to learn human-like behaviors in driving.

5.4.3 Decision-Making and Control

High level maneuvers are picked by the decision making systems and the low-level vehicle actuations are carried out by the control systems.

Hierarchical Decision-Making

Hierarchical architectures divide the process of decision making into strategic (selection of routes), tactical (selection of manoeuvres) and operational (execution of trajectories). The levels are profitable in terms of time scale and level of abstraction. At tactical and operational levels, reinforcement learning has been implemented, simulation and real-world experience have been used as learning policies.

Safety Assurance

It is important to provide safety of learned decision-making and control systems. Runtime checks ensure that the scheduled activities meet safety requirements prior to their performing. Backup controllers are formally verified to offer fallback operations in the case of unreasonable actions offered by the primary systems. Safety properties of control systems are checked by using bounded model checking and reachability analysis.

5.5 Integration Challenges and Solutions

Multi-faceted machine learning systems, data, and stakeholder systems have to be incorporated into comprehensive transportation safety systems

5.5.1 Data Integration Challenges

Transportation safety applications take in a variety of data sources of different formats, level of quality, frequency of update, and access control.

Heterogeneous Data Sources

Several combination of crash databases, traffic sensors, weather data, road inventories, vehicle telemetry, infrastructure monitoring systems necessitate a tackling of semantic heterogeneity, temporal misalignment and spatial resolution differences. Interoperability is made possible by ontology-based data integration/ schema matching. Federated query systems allow the query of distributed sources of data.

Data Quality Management

The quality of data is very important to the performance of machine learning models. Various errors such as sensor failure, communication and recording errors inject inaccurate information. Suspect data are identified via automated data quality test, that is, anomaly detection and cross-validation. Strong learning strategies decrease the susceptibility to mistakes and aberrations.

Privacy-Preserving Data Sharing

Much machine learning needs huge data sets of diverse heterogeneous data, yet privacy restrictions limit the sharing of data, especially naturalistic driving data as well as crash data that involves personal data. Different privacy introduces calibration noise that allows statistical analysis and ensures privacy. Federated learning learns without the sensitive information being centralized on distributed data.

5.5.2 System Integration Architecture

Table 1. Comparison of Integration Architecture Approaches

Architecture	Advantages	Disadvantages	Best Use Cases
Centralized Cloud	- Powerful processing resources - Simplified model management - Comprehensive data access	- Communication latency - Privacy concerns - Network dependency	Non-real-time analytics, Model training, Fleet management
Edge Computing	- Low latency - Privacy preservation - Reduced bandwidth requirements	- Limited computational resources - Model synchronization complexity	Real-time vehicle systems, Autonomous driving, Driver monitoring
Hybrid Cloud-Edge	- Balances latency and capability - Flexible resource allocation - Hierarchical processing	- Increased system complexity - Synchronization overhead	Most practical deployments, Intelligent transportation systems
Federated Architecture	- Privacy preservation - Distributed data utilization - Collaborative learning	- Communication overhead - Model aggregation challenges - Slower convergence	Multi-organizational collaboration, Privacy-sensitive applications
Blockchain-Based	- Transparency and trust - Immutable audit trails - Decentralized governance	- Computational overhead - Scalability limitations - Energy consumption	Safety-critical certification, Multi-stakeholder systems

Cloud-Edge Continuum

The Hybrid architectures allocate processing depending on latency, privacy, and computational needs between the cloud data centers and the edge device. Training is trained in a cloud environment and boasts of a huge amount of computing power and inference is done at an edge device to provide an immediate reaction. Cloud-trained models can be optimized and converted into edge deployable models using model compression and optimization.

Communication Protocols

The communication protocols are standardized, and therefore allow heterogeneity in communication systems. DSRC and Cellular Vehicle-to-Everything (C-V2X) is used to support vehicle-to-vehicle and vehicle-to-infrastructure communication.

Application programming interfaces make it easier to integrate third party services and data.

6. INTEGRATED INTELLIGENT TRANSPORTATION SAFETY FRAMEWORK

6.1 Conceptual Framework Architecture

This part provides a cohesive comprehensive model that combines the disjointed machine learning applications into harmonized intelligent transportation safety.

Figure 1. Integrated Intelligent Transportation Safety System Architecture

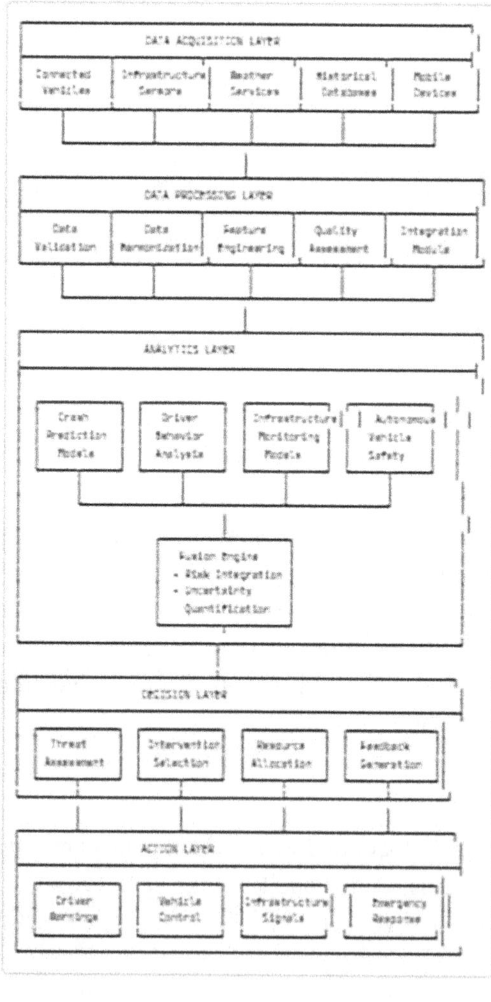

This framework also uses a layered architecture with each layer offering definite functionality and having clear interfaces to allow modular development and implementation.

6.2 Layered Architecture Components

6.2.1 Data Acquisition Layer

The layer of data acquisition is one that engages heterogeneous data sources that supply real-time and historical data. C-V2X Network Connected vehicles receive and receive position and speed data, acceleration and sensor data through a DSRC (or C-V) Network. Traffic measurements are given by infrastructure sensors such as loop detectors, cameras, and radar. Weather services provide the environmental conditions such as the precipitation, temperature, visibility and road surface conditions. Crash records, road inventories and maintenance histories are found in historical databases. Mobile application crowdsourced data is used to give additional information.

6.2.2 Data Processing Layer

The raw data is converted into the standardized form to be analyzed at the data processing layer. Quality assurance does completeness, consistency, and plausibility checking, indicating wrong and suspicious data. Data harmonization translates between heterogeneous data representations and standard representations, and eliminates semantic differences and time zone problems. Relevant predictors to be used in feature engineering are based on raw measurements such as statistical summaries, temporal derivatives, and spatial relationships. The quality assessment methodology gives scores of confidence that are assigned according to the reliability of the sources, measurement precision, and cross-validation consistency. A combination of information in various sources is formed to create integration modules that resolve any conflict and gaps.

6.2.3 Analytics Layer

The analytics layer uses machine learning models on data that has been processed and produces safety insights and predictions. Application-specific models are used in particular areas of application such as crash predictor, driver behavior control, infrastructure monitoring, and perception of autonomous vehicles. The fusion engine combines the results of the specific models, which results in aggregate risk evaluations, which consider factors among others. Uncertainty quantification offers confidence limits to predictions, which can be conveniently used to make risk-management decisions.

6.2.4 Decision Layer

The decision layer considers risk assessment that is integrated and chooses the right intervention. Threat assessment measures present and future risk, in its uncertainties and future possibilities. Selection of intervention involves the selection of available safety actions depending on their effectiveness, practicability and side effects. Resource allocation divides available safety resources such as personnel, equipment and infrastructure capacity. As a result of feedback generation, there are explanations and justifications of decisions, which increase transparency and acceptability to users.

6.2.5 Action Layer

Action layer brings interventions of choice to action by various channels. Drivers Warning Systems make use of visual check, sound, and haptic feedbacks to convey dangerous information and suggestions. Control systems in self-driving cars are used to carry out collision avoidance tasks or driver assistance systems. Traffic lights, variable message signs and speed limits are changed by infrastructure signals. Police, fire and medical services are dispatched through the emergency response systems.

6.3 Federated Learning Framework for Privacy-Preserving Safety Analytics

The privacy considerations limit collecting information on driving in a centralized manner, which promotes federated learning that allows training a model without centralizing the sensitive data.

Privacy-Utility Tradeoffs

Differential privacy also applies noise to update models in a controlled manner, which does not allow any inference about individual data points but only aggregate statistics. The ε parameter of privacy budget measures privacy-utility tradeoff. Lower e ensures a higher level of privacy but it decreases the utility of the model. These conflicting goals are taken care of through careful balance.

Secure Aggregation

Secure aggregation protocols based on cryptography allow concluding on aggregate statistics without showing individual contributions. The model updates are not sent in a way which can be identified separately, but are only sent as aggregate updates. This is more adversarial protection than differentiable privacy but has more computational overhead.

Federated Learning Challenges

The efficiency of the communication is an important issue because large models have to be transferred frequently which uses bandwidth and energy. The cost of communication is decreased by gradient compression, quantization, and sparsification. Due to statistical heterogeneity in the distribution of participant data, convergence becomes a complicated issue because local optima might not be the same as global optima. Federated learning Personality adjusts global models to local distributions. Heterogeneity of the system in terms of computational abilities and availability necessitates asynchronous protocols and selection of participants dynamically.

6.4 Explainable AI Framework for Transportation Safety

Regulatory acceptance and user trust in safety-critical application is only possible with interpretable machine learning models.

Table 2. Explainable AI Techniques for Transportation Safety Applications

Technique	Description	Advantages	Limitations	Applications
Feature Importance	Quantifies contribution of each input feature	- Simple interpretation - Model-agnostic - Computational efficiency	- Does not capture interactions - Global explanation only	Crash prediction, Risk factor identification
SHAP Values	Unified approach based on Shapley values	- Theoretically grounded - Local explanations - Captures interactions	- Computational cost - Requires background dataset	Driver behavior analysis, Severity prediction
LIME	Local linear approximations	- Model-agnostic - Intuitive interpretation - Local fidelity	- Instability across runs - Hyperparameter sensitivity	Real-time risk explanation, Decision support

continued on following page

Table 2. Continued

Technique	Description	Advantages	Limitations	Applications
Attention Mechanisms	Neural network attention weights	- Integrated into model - Temporal/spatial localization - Natural interpretation	- May not reflect true reasoning - Architecture-specific	Vision systems, Sequence modeling
Counterfactual Explanations	"What-if" scenarios	- Actionable insights - Causal reasoning - User-friendly	- Multiple possible counterfactuals - Computational complexity	Driver coaching, Policy evaluation
Rule Extraction	Symbolic rules from neural networks	- Human-readable rules - Formal verification possible - Compact representation	- Approximation quality - Scalability challenges	Safety certification, Regulatory compliance
Prototype-Based	Example-based explanations	- Intuitive understanding - No complex mathematics - Leverages human pattern recognition	- Requires representative examples - Subjective selection	Case-based reasoning, Training systems

Multi-Level Explanation Framework

Good explanation systems offer multi-level information to different stakeholders. Technical developers are in need of feature attributions and model internals. Risk information and effective interventions are required by system operators. The drivers enjoy easy to understand warnings on immediate issues. The regulators demand aggregate statistics and certification evidence. Clear description frameworks come up with the correct explanations to various stakeholder groups.

Explanation Validation

The quality of explanations needs systematic validation which is not just anecdotal. Faithfulness is used to gauge the quality of explanations in terms of model behavior. Consistency measures the stability of explanation in the similar inputs. The ability of comprehending human understanding is measured by comprehensibility which involves user studies. The concept of actionability is whether explanations are effective in informing interventions.

6.5 Continuous Learning and Model Adaptation

The transportation environments are constantly changing due to the changes in infrastructure, the shift in the traffic pattern, and the adoption of technologies that demand models to be adjusted to the changing environment.

Online Learning Mechanisms

The algorithms of online learning work in an incremental fashion, where new data is received, and the algorithm is not required to be retrained on a schedule. Stochastic gradient descent allows online learning by the method of successive updates in weights. Streaming algorithms have summary statistics that allow updates of the model to take place easily. Concept drift detection detects drastic changes of distribution that causes model retraining or adoptions.

Transfer Learning Strategies

The models that are trained on huge source datasets can predict on new target domains with few target data. As fine-tuning, model weights of the already trained model are readjusted with target domain data. Domain adaptation methods minimize differences between distribution across source and target domains. Multi-task learning uses representations on similar tasks minimizing the amount of data.

Adaptive Thresholds

Decision thresholds can be fixed but can be inefficient in different circumstances. Adaptive thresholding changes warning and intervention triggers according to context such as weather condition, the level of traffic, type of road, and the time of the day. Reinforcement learning maximizes false positive threshold policies against false negative threshold policies. Individual thresholds are the consideration of the characteristics and preferences of individual drivers.

Incremental Architecture Updates

In addition to updating models with new weights, the model architectures can need to evolve due to the appearance of new types of sensors, sources of data, or new task demands. Neural architecture search has the ability to find the best architectures to match changing requirements. In modular architectures, replacement of components is possible and does not redesign the entire system. Test structures and version control make sure that there is smooth transition between model versions.

6.6 System Validation and Performance Evaluation

Full-scale validation systems provide the reliability of the safety systems before they are deployed, and they track the subsequent performance.

Simulation-Based Validation

High-fidelity traffic simulators allow testing under conditions of various factors such as rare but critical events. Closed-loop simulation tests are full behavior tests of the system that include perception, decision-making and control. Adversarial testing intentionally puts demanding situations on the system that uncover constraints in the system. The Monte Carlo simulation measures distributions of performance and probability of failure.

Hardware-in-the-Loop Testing

The hardware-in-the-loop systems involve the real vehicle hardware with the virtual environment and assess the real-time performance on realistic computing environments. This method detects the timing problems, computational cost, and sensor interface problems that cannot be seen in pure simulation.

Controlled Track Testing

Instrumented test tracks allow real world testing which is controlled with measurement systems and safety personnel. Repeatable test scenarios are used to test performance when the conditions are known. The difficulty of the scenario is gradually increasing as the reliability of the system is proven.

On-Road Naturalistic Testing

To obtain final validation in naturalistic conditions, public road testing is used, but it has ethical and safety issues. Graduated deployment is initiated with the less risky cases, and then gradually increases with the reliability of the system. Fallback control is offered by safety drivers. The detailed data documenting helps in analysis of events after the incident.

Performance Metrics Framework

Table 3. Comprehensive Performance Metrics for Transportation Safety Systems

Metric Category	Specific Metrics	Measurement Method	Target Values
Prediction Accuracy	Sensitivity, Specificity, F1-Score, AUC-ROC	Confusion matrix on validation data	>90% sensitivity >85% specificity
Temporal Performance	Prediction lead time, Processing latency, Update frequency	Timestamp analysis	>5 min lead time <100ms latency
Robustness	Performance under adverse conditions, Sensor degradation tolerance	Systematic condition variation	<10% degradation
Safety Impact	Crash reduction rate, Severity reduction, Near-miss prevention	Before-after studies, Surrogate measures	>20% crash reduction
Reliability	System availability, Mean time between failures	Operational monitoring	>99.9% availability
Scalability	Throughput, Concurrent user capacity	Load testing	Linear scaling to 10,000 vehicles
Explainability	User comprehension rate, Explanation consistency	User studies, Automated metrics	>80% comprehension
Fairness	Demographic parity, Equal opportunity	Stratified analysis	<5% disparity across groups

This is a thorough metrics model that takes into account the technical performance, safety on the ground, feasibility of deployment in the real world and the ethical aspects. Multiple-dimensional measurement gives the overall assessment of the system and determines the priorities of development and deployment decisions.

7. RESEARCH GAPS AND FUTURE DIRECTIONS

7.1 Identified Critical Research Gaps

An in-depth literature review shows that there are important research gaps that need to be taken care of through research.

7.1.1 Integration and Standardization Gap

Available studies have centered a lot on individual applications as opposed to integration of systems. Absence of generalized methods of evaluation, data representation and performance measures prevents cross-study evaluation and knowledge transfer. The barrier to integration of heterogeneous systems in different vendors and research groups is interoperability. The performance measures give more importance to technical precision instead of actual safety impact.

Future Research Directions

- Creation of standard benchmark data sets in the areas of application with recorded formats, quality features and performance base line.
- Development of uniform evaluation procedures such as train-test splits, cross-validation process, and safety result performance measures.
- Development of reference implementations that will give a minimum performance and support reproducibility.
- Integration testing frameworks that would ensure end-to-end system performance and not single components.
- Academia-government-industry cooperation in standardization programs.

7.1.2 Explainability and Interpretability Gap

Models with state-of-the-art performance in deep learning are typically black boxes, which reduces the ability to regulate them and places user trust in them. The current methods of explainability which are designed according to general machine learning need not be applied to the transportation-specific safety-critical requirements. The post-hoc explanation techniques might not be true to model reasoning. Explanations that have been tested on technical measures do not necessarily make humans understand or make better choices.

Future Research Directions

- Learning of deep learning architectures that are inherently interpretable and remain competitive at the same time and yet are not opaque.
- Transportation-specific metrics of explainability that take into account safety-critical properties of requirements such as explainability faithfulness, explainability consistency, and explainability actionability.
- Design of explanations which take into consideration the needs of various stakeholders such as drivers, operators, regulators and researchers.
- Formal verification techniques that give guarantees of the behaviour of the model under safety critical conditions.
- Causal logic paradigms between correlation and causation in acquired models.
- Knowledge of domains to be integrated with the generation of an explanation of transportation-relevant meanings.

7.1.3 Edge Computing and Real-Time Processing Gap

Quite a number of suggested solutions demand computing power that is only available off line but not on-board in real time. Hard deterministic requirements of the safety-critical applications require quick response time, which is usually in milliseconds to seconds. Embedded automotive systems have energy and thermal constraints which restrict processing capabilities. There is a limitation of bandwidth of communication to cloud infrastructure.

Future Research Directions

- Pruning, quantization, and knowledge distillation methods particularly applied to transportation safety models were model compression methods.
- Search of neural architectures over efficient models that are deployed efficiently on edges and safe performance.
- Co-design Hardware and software Co-designed hardware and software, based on dedicated accelerators such as GPUs, TPUs, FPGAs, and neuromorphic processors.
- Adaptive methods of computation dynamically change the complexity of models depending on the criticality of scenarios and resources.
- Edge-cloud collaboration systems that deliver intelligent distribution processing using latency, privacy, and computation needs.
- Creation of lightweight models that are specially created to run on resource-constrained embedded systems.

7.1.4 Generalization and Transfer Learning Gap

The models that are trained on certain datasets tend to perform poorly in generalizing to new geographical areas, climatic conditions, traffic, or the type of vehicle. There is also restricted access to various training data across different environments that restrict generalization. The best performance is compromised by domain shift between training and deployment conditions. Uncommon but safety relevant cases do not have adequate representation in the training data.

Future Research Directions

- Domain adaptation algorithms that allow models to generalize in geographical regions and operating mode with little target data.
- The problem of learning with few-shot and zero-shot samples of the rare event is addressed by these methods.

- Data generation through simulation and generative models on real-world data add artificial data with different scenarios.
- The meta-learning algorithms learning how to learn with little data on a number of related tasks.
- Life long learning avoiding disastrous memory loss in adjustment to new settings.
- Operational design areas Systematic assessment evaluations of generalization across different operational design areas.

7.1.5 Uncertainty Quantification Gap

The majority of machine learning studies in the field of transportation safety carry point predictions but do not estimate uncertainty. Decisions made that are safety critical need to understand periods when predictions should not be made. The out-of-distribution data that is not reflected in training data can lead to overconfident false predictions. Crash prediction has temporal uncertainty which influences the timing and effectiveness of interventions.

Future Research Directions

- Bayesian deep learning methods that can give principled measures of uncertainty with calibrated confidence readings.
- Univariate ensemble prediction with calibrated resultant confidence intervals and distributional predictions.
- Conformal prediction Distribution-free uncertainty estimates with coverage guarantees.
- Out-of-distribution input detection This method predicts whether the input is anomalous and thus it may either block and wait for human intervention or act conservatively by adopting safe default fallback.
- Systems of making decisions that are uncertain and the uncertainty of prediction is clearly included in the intervention strategies.
- Techniques of calibration that provide predicted occurrences that are comparable to the actual frequencies.

7.1.6 Multi-Modal Transportation Integration Gap

The research is more oriented towards automotive application but less emphasis is made on integration across transportation modes such as rail, aviation, maritime and micro-mobility. Multi-modal journeys that include transfers between modes have special safety issues. The safety of pedestrians and cyclists in mixed roads demands special strategies that put into consideration the vulnerable road users.

Future Research Directions

- Coherent systems that deal with safety in various transportation systems that find general patterns and mode-specific needs.
- Transfer learning using mode insights that make sharing of knowledge effective.
- Multi-modal interaction modeling of mixed traffic conditions that involve vehicles, pedestrians, cyclists and micro-mobility equipment.
- Whole trip chain safety in multimodal integrated mobility platforms.
- General safety principles being modified to specific modes in accordance with special features and limitations.
- Models of vulnerable road user protection of pedestrians and cyclists at risk

7.1.7 Ethical and Fairness Gap

Scanty literature has been done on dangers of bias and unfair safety effects in transportation machine learning. Training data can be skewed to some population or geographical location. There can be significant differences in the performance of the models among the groups. The aspect of autonomous vehicle making decisions during inevitable crashes creates ethical issues. Ubiquitous surveillance has privacy concerns that should be considered

Future Research Directions

- Fairness score and de-biasing algorithms that are specifically tailored to the use of transportation safety.
- Privacy preserving machine learning that trades the privacy benefits with safety benefits with privacy rights by employing such methods as differential privacy and federated learning.
- Crash-imminent autonomous vehicle decision-making ethics involving stakeholder suggestions.
- Unbiased and representative data collection in both demographics and geographies.
- Safety-critical AI systems should have the ability to provide transparency and accountability through safety-critical mechanisms that allow oversight and redress.
- Participant design methods with differences in stakeholder attitude in system design.

7.1.8 Human Factors Integration Gap

Research in technology tends to ignore human elements such as driver acceptance, levels of trust, behavioral adaptation and skill degradation. Semi-automated systems are prone to automation complacency and laxity. The level of automation can lead to the loss of manual control skills, which leaves the system vulnerable in case of a failure that would entail human over-takeover. The interface design has significant impacts on the outcomes of system usability and safety.

Future Research Directions

- Safety systems based on human-centered design methodologies that guarantee suitable human-AI interaction.
- Theory of behavioral adaptations in the prediction of the response of drivers to safety interventions which consist of risk compensation effects.
- Trust calibration which is a proper dependence on automated systems and not over-trust or under-trust.
- Control protocols for partially automated vehicles changing mode confusion and takeover problems.
- Training and education systems to advanced safety systems to assure the user of capability and limitation.
- Long term longitudinal studies on long-term automation on skills and behaviors.

7.2 Emerging Research Opportunities

A number of new technological and social trends introduces new research opportunities.

7.2.1 Quantum Machine Learning

Quantum computers will achieve exponential speed in solving some optimization and pattern recognition problems. Algorithms in quantum machine learning can be used to solve computational problems that were previously infeasible such as global traffic optimization in real-time and exhaustive analysis of scenarios. Quantum neural and quantum support vector machine is an active field of research. Nonetheless, quantum computing is still in its infantile phase and its practicality schedules are unpredictable. Quantum-classical hybrid methods have the potential to pay off in the near term right before all of quantum computing is realized

Research Opportunities

- The traffic optimization and routing algorithms are some methods of quantum computations.
- Crash prediction based on quantum machine learning. Competing Hybrid quantum-classical designs.
- Measurement of quantum benefit of transportation.

7.2.2 Neuromorphic Computing

The neuromorphic processors that are inspired by the brain provide energy efficient processing to be used at the edges. Response with ultra-low-latency is offered by event-based sensing and processing. It is possible that neuromorphic systems will allow a high degree of on-board intelligence at low energy levels. Nevertheless, it needs to be developed to be integrated with existing software and algorithm frameworks. There are still immature programming paradigms and tools.

Research Opportunities

- Perception and prediction Neuromorphic algorithms.
- Event-based sensor fusion Neural AI systems with lower energy consumption.
- Neuromorphic-conventional hybrid system.

7.2.3 Digital Twin Technologies

Digital twins produce virtual copies of the real-world transportation systems where real-time monitoring, prediction, and optimization can be done. Digital twins can be trained and validated machine learning models that can be transferred to physical systems. Close communication between physical and digital twins can be performed actively to proactively maintain and dynamically optimize. They can be used in infrastructure monitoring, traffic control as well as autonomous car testing.

Research Opportunities

- Transportation System digital twin architecture.
- Physical and digital real-time synchronization.
- Digital twin frameworks of validation.
- Digital twin analytics-based predictive maintenance.

7.2.4 Swarm Intelligence

Distributed coordination based on swarm intelligence principles on autonomous system agents can be used to increase safety at the system level. Complex behaviors emerge out of simple interactions. Swarm strategies are resistant to failures of single agents, and can be applied to large resources. Some of the applications are cooperative collision avoidance and distributed traffic management.

Research Opportunities

- Cooperative safety Swarm algorithms.
- Control and emergent behavior analysis.
- Multi-agent coordination on a larger scale.
- Swarm systems are ensured by robustness

7.2.5 Augmented Reality Integration

AR provides safety details on top of the real-life perspective. HDA displays indicate identified risky and required measures. AR has potential to increase awareness of a situation and decrease distraction. The human factors research should not overburden the drivers with the AR interfaces.

Research Opportunities

- AR safety application interface design.
- AR display attention and distractions studies.
- Optimization of information presentation.
- AR feedback systems with multi-modal systems.

7.2.6 5G and Beyond Communications

The 5G and future cellular networks will be able to support ultra-reliable low-latency communications to allow advanced cooperative safety applications. There is increased vehicle-to-everything communication, which enables real-time information exchange. Network slicing provides good quality of service to safety-critical applications. Network edge computing latency reduces.

Research Opportunities

- 5G-based collaborative safety applications.
- Safety service network slices.
- 5G based edge computing architecture.

- Lateness and Reliability warranties.

7.3 Long-Term Research Vision

The long-term vision includes complete smart transportation systems that have done away with avoidable collisions by means of proactive, predictive, and adaptive safety features. This vision can be achieved by machine learning by:

Predictive Risk Management: Ongoing examination that detects high risk of crashing even before it happens and allows preventive measures to be undertaken. Sections of prediction times ranging over seconds to days support short-term (immediate) and long-term infrastructure planning.
Cooperative Safety: the interaction between vehicles and the infrastructure results in the mutually safe operation by exchanging information and providing a coordinated control. Team intelligence is better than individual intelligence.
Adaptive Infrastructure: Smart infrastructure is dynamically reconfigured according to the traffic conditions, weather and risks identified. Safety is continuously optimized by the use of variable speed limits, distributing lanes, and timing the signals.
Individualized Safety: Systems can accommodate the specifics, interests, and situations of individual drivers and offer the most useful interventions with the least intrusive nature.
Zero-Crash Vision: This is the ultimate objective of removing transportation deaths and serious injuries by using multifaceted technological, infrastructure, and policy solutions based on the understanding of machine learning.

This vision needs long term dedication to research, developments, standardization and implementation by the transportation community. The framework and research agenda outlined below will give guidance to this transformational vision.

8. IMPLEMENTATION CONSIDERATIONS

8.1 Regulatory Frameworks and Standards

Safety in transportation brought about by machine learning also demands enabling regulation policies that help to balance innovation and safety assurance.

8.1.1 Current Regulatory Landscape

United States

Federal Automated Vehicles Policy is an optional policy that includes guidance on the testing and deployment of autonomous vehicles in the work of the National Highway Traffic Safety Administration (NHTSA). Regulations at the state level differ significantly to form piecemeal regulatory landscapes. The state of California, Arizona, Nevada, Florida and Michigan have passed particular autonomous vehicle laws that may have different requirements. The federal Motor Vehicle Safety Standards that are established on traditional vehicles should be adjusted to automated systems. The framework suggested by NHTSA focuses on the safety performance and not prescriptive technical requirements.

European Union

The EU has stricter prescriptive safety laws which require type approval prior to market entry. The General Safety Regulation (2019/2144) requires new vehicles to have advanced safety systems such as the intelligent speed assistance, the automated emergency braking system as well as the driver drowsiness systems. General Data Protection Regulation (GDPR) has stringent data protection guidelines on the gathering and utilization of data in respect of safety mechanisms. The EU is seeking coherent cross member state laws that support the implementation of technology.

International Standards

The J3016 standard of SAE international offers six levels of automation (0-5) that offer generic terminology. The ISO 26262 standard is concerned with the automotive electrical/electronic system functional safety. The ISO 21448 is the Safety of the Intended Functionality (SOTIF) dealing with the risks posed by performance thresholds and abuse. IEEE standards deal with connected and automated vehicles and cybersecurity.

8.1.2 Regulatory Challenges

Safety Validation

Achieving safety equivalence or safety superiority over human drivers entails immense validation. Kalra and Paddock (2016) assessed hundreds of millions or billions of miles necessary to be statistically significant and statistical validation based on miles alone is not feasible. Scenario testing is a supplement to real world testing that needs representative selection of scenarios. There are basic problems with statistical validation of rare severe crashes.

Liability and Responsibility

Autonomous systems blur the traditional models of liability that emphasize on the responsibility of a human driver. There is the need to define product liability, operator responsibility and manufacturer liability. The issue of responsibility in mixed human and automated crash is complicated to determine. Automated systems need to be handled in insurance systems. Certain jurisdictions are creating no-fault systems in autopilot accidents.

Ethical Decision-Making

Ethical decision rules written in programming languages to address inevitable crashes have philosophical and practical questions. The Trolley problem situations create controversy in the society but hardly ever come to reality. Being open regarding programmed ethics can influence the acceptance of people. Ethical preferences differ among cultures making it difficult to standardize internationally.

Data Governance

Extensive security infrastructures produce reports on minute individual information that is privacy threatening. The regulations should strike a balance between privacy rights and safety benefits. Minimization of data helps to reduce information to only what is required. Purpose limitation involves limited use to certain safety uses. Privacy is secured by the access controls and retention policies and allows safety applications.

8.1.3 Standardization Needs

Performance Metrics

The standardized measures allow the objective comparison of the safety system. Measures should focus on actual safety results as opposed to technical measures only. Regular evaluation situations make fair comparisons. The benchmark datasets are used to compare systems across systems.

Data Formats

Interoperability will dictate standardized data formats used in sensor measurements, object detections and vehicle to everything communications. Common semantics are achieved by data dictionaries. System integration is achieved using standard application programming interfaces.

Safety Assurance Processes

Regulatory assurance is through the use of standardized safety validation procedures. Certification and auditing is supported by documentation requirements. The testing protocols are sufficient in covering. Continuous control justifies the performance of deployed systems.

Cybersecurity Standards

Malicious attacks on the safety systems are safeguarded by security standards. Cryptographic communication protocols are used to protect communication. Intrusion detection is used to detect abnormal behavior. It provides security measures in development.

8.2 Ethical Considerations

8.2.1 Algorithmic Fairness

The machine learning systems should offer fair safety assurance to the entire racial lines and geographical areas. Unrepresentative training statistics into models will result in different safety results. The measurements of fairness are used to measure inequities in the domain of guarded attributes such as race, sex, age, and socioeconomic status.

Mitigation Strategies

- **Representative Data Collection:** Assuring that training data sets are a sufficient sample of all the population groups and operational settings. Underrepresented groups might have to be purposely oversampled.
- **Fairness- conscious Learning**: In these algorithms, the accuracy is optimized as well as fairness goals. Pre-processing bias training data. Fairness restrictions are included in in-processing. Outputs are corrected using post processing to enhance fairness.
- **Disaggregated Evaluation**: When performance is reported separately in various groups of demographics, it is shown to be unequal. The lowest possible performance requirements of the groups of people protect the groups equally.
- **Constant Evaluation**: Implemented systems must be evaluated in terms of fairness on a regular basis since populations and patterns of use change. Periodic audits detect occurrences of disparities so that remedial action can be taken.

Fairness Definitions

There are several definitions of fairness, which are contradictory at times. Demographic parity is the equality of positive prediction of groups. There will be equal true positive rates in equal opportunity. Predictive parity expects the same amount of accuracy in groups. Applications in transportation safety should be aware of the fairness criteria to include based on application context and values of stakeholders.

8.2.2 Privacy Preservation

Safety surveillance produces extensive personal information such as locations, behavior and patterns. Privacy can be secured by technical and policy protection.

Technical Approaches

- **Differential Privacy:** This is a privacy assurance technique whereby noise, which is added to both data and model updates is calibrated to ensure that the privacy of an individual is not compromised. Tradeoff of privacy-utility is measured by privacy budget.
- **Federated Learning**: Federated training models do not require sensitive information to be centralized. Aggregation of local model update is done without the transmission of the raw data.
- **Data Minimization:** Gathering of the information that is required to promote safety. Exposure is reduced by minimizing the size of data and retention time.
- **Anonymization and De-identification**: Elimination of the individually identifiable data in datasets. Nevertheless, there are still threats of re-identification with rich behavioral data.
- **Weighted Based Encryption**: This allows performing operations on encrypted data without decryption but at a significant computational cost.

Policy Frameworks

- **Informed Consent:** It means having the explicit consent to collect and use data. Effective communication on purpose and risk.
- **Purpose Limitation**: This is restricting the use of data to predetermined safety purposes. The avoidance of the secondary uses that are not followed by the extra consent.
- **Access Controls:** The restriction of people viewing sensitive data. Audit trail and role based access also provide accountability.
- **Retention Policies:** erasing data when no longer needed or it is of a specified time. Weighing historical analysis requirements and privacy requirements.

8.2.3 Transparency and Accountability

The safety-critical systems demand transparency, which will allow stakeholders to comprehend the system and a way of holding those in charge to account in the case of failure.

Transparency Requirements

- **Explainable AI:** It involves being able to explain the decision-making process. Several levels of explanations cater to various needs of the stakeholders.
- **Documentation**: Preparation of extensive training information, algorithms, validation performance, and constraints.
- **Public Reporting:** Reporting on system performance and incidences. Openness generates confidence and empowers popular debate.
- **Independent Auditing:** The third party audits guarantee safety claims and compliance. The certification demonstrates the independent body.

Accountability Mechanisms

- **Defined Responsibility:** Specifying responsibilities among the stakeholders such as developers, manufacturers, operators and infrastructure providers.
- **Incident Investigation**: Guide to finding causes of failures. Forensic analysis is made possible through data recording.
- **Corrective Action**: Processes of meeting the deficiencies identified. Issues discovered are remediated by software updates.
- **Legal Recourse:** This is a system through which victims of wrongs seek redress. Financial responsibility is defined by liability structures.

8.3 Deployment Strategies

8.3.1 Phased Rollout Approach

Graduated deployment is a way to control the risks and keep the use of the operational experience to refine with a certain level of iterations.

Phase 1: Controlled Testing

- Basic functionality is tested through simulation and laboratory.
- Instrumented vehicle testing and safety personnel testing.
- Human fallback control by safety driver control.
- Data mining and export that will be used to analyze and refine.

Phase 2: Limited Deployment

- Limited operational design space that is characterized by weather, time, location, and traffic.
- Geographic geofencing limiting operation to particular areas with favorable characteristics.
- Further backup safety driver assistance.
- Data collection and intensive checking.
- Creeping expansion of operational design domain with confidence.

Phase 3: Expanded Deployment

- Expanding operational design areas to include problematic environments.
- Less reliance on safety drivers is exhibited by system reliability.
- Bigger sizes of fleet gaining operational experience.
- Training on safety and performance in real-world.
- Monitoring and constant improvement.

Phase 4: Full Deployment

- Freedom of operation in all the desired design areas.
- Autonomous (no fallback in the case of driverless operation).
- Connection to the existing transportation systems.
- Continuous tracking which guarantees sustained performance.

8.3.2 Stakeholder Engagement

The deployment process is successful only when a variety of stakeholders is involved and has different views and interests.

Government Coordination

- Early participation with regulatory agencies perceiving requirements and constraints.
- Cooperation on the safety standards and test techniques.
- Pilot projects representing technology collaboration with the government agencies.
- Information exchange in favor of regulatory decision-making and safeguarding proprietory information.

Industry Partnerships

- Partnerships with suppliers and technology firms that share knowledge and resources amongst the vehicle companies.
- Technologies to share data effectively in support of the spread of learning and at the same time, support the competitive interests.
- Interoperability activities between vendors through standardization.
- Pre-competitive studies of universal issues.

Public Engagement

- Open communication of possibilities and constraints which handle expectations.
- Trust was established and gauged through educational programs.
- Mechanisms of feedback to the populace on the development priorities.
- Communicating the issues and misunderstandings by use of evidence.
- Demonstration projects that can give people a feel of technology.

Academic Collaboration

- University partnerships that engage themselves in evaluation and research.
- Training of students who are coming up with essential skills.
- Theory and research on long-term studies of the underlying problems.
- Knowledge in print and spread of results.

8.4 Economic Considerations

8.4.1 Cost-Benefit Analysis

The deployment should have to be substantiated by thorough economic study to show a net benefit to society.

Benefits

- **Reduction of Crashes:** Saving of lives and preventing injuries is a tremendous economic asset. Estimates of value of statistical life, range at 10-12 million/life in US.
- **Healthcare Cost Savings:** Any decrease in crashes will decrease medical care costs on injuries.
- **Reduction of Property Damage:** Reduced number of crashes will lower repair and replacement of vehicles.

- **Productivity Gains:** Travel time will be saved due to less congestion due to crashes. Driving automation facilitates efficient utilization of the commuting time.
- **Reductions in Insurance Cost:** Enhanced safety will contribute to reduction of insurance costs.

Costs

- **Technology Development:** Research, development as well as testing involves a lot of investment.
- **Deployment Infrastructure:** Sensors, communication systems and computing infrastructure would have to be capital investment.
- **Training and Education:** The users, operators and maintainers need training.
- **Regulatory Compliance:** Testing, documentation and certification is expensive.
- **Maintenance:** The software updates, calibration of sensors and monitoring of the system are part of the expense that needs to be maintained.

Table 4. Economic Impact Analysis Framework

Impact Category	Measurement Approach	Data Sources	Valuation Method
Fatality Reduction	Statistical life value × lives saved	Crash databases, Effectiveness studies	Value of statistical life ($10-12 million/life)
Injury Prevention	Medical costs + lost productivity	Healthcare records, Employment data	Comprehensive injury costs by severity
Property Damage	Repair/replacement costs	Insurance claims, Crash reports	Actual repair costs or replacement values
Congestion Reduction	Travel time savings × value of time	Traffic monitoring, Travel surveys	Value of travel time ($15-50/hour)
Environmental Impact	Emissions reduction value	Emissions models, Air quality data	Social cost of carbon Environmental benefits
Technology Costs	Development + deployment + maintenance	Industry data, Procurement records	Lifecycle cost analysis
Infrastructure Investment	Construction + upgrades + operations	Public agency budgets	Capital and O&M costs

Return on Investment

The cost-benefit ratios guide the investment decisions. Many of the suggested safety technologies have good benefit-cost ratios (more than 3:1 and more). Nevertheless, the advantages are enjoyed mainly by crash victims to the disadvantage of technology buyers, which results in an inappropriate incentive alignment. Social optimum deployment might require intervention by the public policy.

8.4.2 Funding Mechanisms

Long-range deployment and development are provided through sustainable funding.

Public Funding

- Government Research Grants: Fundamental research work by such agencies as NSF, DOT, DOE.
- Programs Infrastructure Investment programs Federal / State infrastructure programs to support sensor deployment and communication systems.
- Public-Private Partnerships: Distribution of costs and risks between industry and government.
- Tax Incentives: Tax credits on the adoption and deployment of technology.

Private Investment

- Industry Research and Development Investment: Investment in research and development by manufacturers of vehicles and by technological companies.
- Venture Capital and Private Equity: Innovations and technology development.
- Adjustments in Insurance Premiums: Unfortunately, reduction in premiums on vehicles with increased safety will act as a financial incentive.
- Value Capture: Realizing the value of safety improvements in different ways.

International Funding

- Multilateral Development Banks: Multilateral Funding of Transportation Safety.
- International Research Partnerships: Partnerships in cross-country research.

8.5 Workforce Development

It involves building interdisciplinary skills employees with required skills in the workforce to be practically applied.

Educational Programs

- Graduate Programs Interdisciplinary programs which combine computer science, transportation engineering, statistics and policy.

- Professional Development: Life-long learning of practicing engineers and policymakers.
- Certification Programs: Validate the skills in the fields of specialization.
- Online Training: Reach training to a wide group of people.

Required Skills

- Machine Learning: Implementation, evaluation, and algorithm design.
- Transportation Systems: Concepts of traffic and safety and infrastructure.
- Data Science: Communication, visualization and statistical analysis.
- Software Engineering: System testing, integration and development.
- Ethics and Policy: Knowledge of regulatory guidelines as well as ethical models.
- Human Factors: Understanding of human-machine interface and user-centred design.

Career Pathways

- Research Positions Academia, national laboratories, corporate research.
- Development Roles: Industry roles that came up with safety systems.
- Policy and Regulation: Government roles of formulating and implementing regulations.
- Operations: Transportation agency jobs Implementation and maintenance of systems.

Interdisciplinary Collaboration

- To be successful, one should be able to work interdisciplinarily.
- Education programs are supposed to create interdisciplinary views.
- Different expertise should be incorporated in research teams.
- The cross-disciplinary interaction can be supported by professional societies.

9. CASE STUDIES AND APPLICATIONS

9.1 Case Study 1: Predictive Crash Risk System Implementation

Setting

A city transportation authority deployed machine learning-based real-time crash prediction on 500 miles of freely moving road containing high density of sensors such as loop meters, cameras, as well as weather gauges and so on.

System Architecture

The hybrid architecture was used to integrate edge computing systems based on time-critical predictions with cloud analytics used to train models and plan over the long term. Initial risk assessment and estimation of the traffic state were carried out on roadside within sub-minutes. By using the cloud infrastructure, nightly retraining of models with the accumulated crash data and traffic pattern was made possible.

Machine Learning Approach

Gradient boosted trees were used to forecast 5 minutes future crash likelihood based on present and past traffic data, weather conditions, and time of day. Local variations were reflected on separate models of various freeway sections and time periods. System-wide and segment-specific system-wide ensemble predictions.

Deployment Results

In 18-month deployment, the system was found to be 82 percent sensitive at 15 percent false positive rate in detecting the high crash risk 10-minute windows before crashes. Mean warning lead time of 7.3 minutes offered the adequate time to activate variable message signs and the response of the traffic management center. Estimated 12% of the second crashes were decreased due to improved response of the incident providing early warnings.

Lessons Learned

- **Ongoing Retraining:** Traffic trends are also seasonal and as a result of infrastructure developments. Daily retraining ensured accuracy of models.

- **Context-Dependent Thresholds:** The optimum risk thresholds depended on the time time of the day, weather and location. Through adaptive thresholding, the performance was enhanced.
- **Explainable Predictions:** Operator trust improved with system provided explanations with highlighting factors.
- **Data Quality Management:** Sensor failure and error in communication demanded strong management in order to be reliable.
- **Integration Issues:** Co-ordination with the current traffic management processes involved a lot of stakeholder involvement and trainings.

Policy and Governance Implications

The deployment of the system involved a need to negotiate data sharing between the transportation agency and the technology provider. The privacy issues were solved by aggregation and anonymization of vehicle data. Liability framework elucidated the agency responsibility in making intervention decisions on the basis of system warnings.

9.2 Case Study 2: Autonomous Vehicle Perception System Development

Context: A program of developing an autonomous vehicle developed a full perception system which included cameras, lidar, and radar to navigate safely in the city.

Technical Approach

Multi-modal sensor fusion architecture used separate network of detection using each sensor modality followed by learned network of fusion of detection taking into account sensor reliability and environmental conditions. Images of the cameras and lidar point clouds that were processed by convolutional and PointNet-like architectures. Kalman filtering followed objects that were detected over time.

Training Methodology

Transfer learning trained the networks with pre-trained weights of both ImageNet and KITTI big public datasets. Proprietary data that was gathered in varied geographic locations, weather conditions, and lighting situations was used to fine-tune models according to the needs of the operations. Synthetic data augmentation was done to enhance the diversity of datasets.

Validation Process

Systematic validation used multi-tier systematic check: (1) offline verification based on the held-out test sets that evaluate the accuracy of detection, (2) closed-loop simulation that tested full-costs of model of perception-planning-control, (3) controlled track test which uses surrogate target and progressively increases complexity.

Performance Outcomes

In favorable conditions (day time, clear weather) Final system recorded 98% vehicle and 95% pedestrian recall, and in unfavorable conditions (night, rain), the system recorded 89%. Adversarial tests and edge case testing helped to find failure modes that needed more training data and architectural changes.

Challenges Encountered

- **Simulation-Reality Gap:** Large differences in simulation performance compared to that in the real world necessitated a large amount of real-world test.
- **Long-Tail Distribution:** Scenarios that were rare but of high importance to safety needed to be data-collected and synthesized uniquely.
- **Computational Constraints:** Model complexity had to be optimized by real-time operation because of constraints on computer hardware.
- **Sensor Degradation:** The weather conditions had a significant impact on sensor reliability demanding a strong fusion.
- **Fleet Calibration**: Calibration of the sensor between the vehicle fleet was not operationally easy.

Considerations Based Upon Human Factors

The focus of safety driver training was placed on the identification of systems limitations and proper intervention time. The interface design was a sign of confidence and recognition of objects in a system. Takeover procedures were used to deal with automation to human transition.

9.3 Case Study 3: Infrastructure Monitoring System Deployment

Context: The State transportation department installed automated surveillance of pavement and bridges on 10,000 lanes-miles of highway system.

System Components

The vehicles used were mobile sensing vehicles that have high resolution cameras, laser profilers and GPS surveyed highway network. The computer vision algorithms used to detect and classify defects used cloud-based processing pipeline. Work order management system gave priorities of repairs depending on the severity and exposure to traffic.

Machine Learning Models

Semantic segmentation networks identified pavement distresses such as cracks, potholes and rutting. Level of severity estimated the levels of defects. Bridge inspection system with close-up shots of drones and climbing robots detected corrosion and deformity in concrete, and structural deformities.

Implementation Results

Automated system resulted in annual inspection of the full network as compared to three years ago when manual inspection was done. Cracks of more severe than severity thresholds were detected with a detection accuracy of 92%, thus minimizing unnecessary manual checks. Deterioration prediction models that predicted deterioration paths made proactive maintenance possible which minimized emergency repair by 35%. Cost-benefit analysis proved 4:1 on investment in five years.

Operational Insights:

- **Constant Imaging:** It was very important to have constant imaging conditions during seasons and times in order to facilitate constant detection.
- **Longitudinal Data Value:** Time-series data showing deterioration rates were more useful as they informed prioritization of maintenance than point in time data.
- **Integration with Asset Management**: Integration with available asset management systems simplified the processes of maintenance and enhanced the process of resource distribution.
- **Inspector validation** Inspector validation of automated assessments by human inspectors ensured quality and developed trust in system.
- **Opportunity of Predictive Maintenance**: Predictive models allowed transition of reactive repair to proactive maintenance, thereby enhancing the results and saving money.

Regulatory and Policy Aspects

Deployment of the system demanded a shift in the maintenance standards to include automated evaluation. The procurement processes were transformed in such a way that machine learning is incorporated. The training programs were made such that the inspectors were equipped with the ability to use automated tools with professional judgment.

9.4 Case Study 4: Driver Behavior Monitoring in Commercial Fleet

Setting: The commercial trucking company used driver behavior monitoring system on the 5,000-vehicle fleet to minimize crashes and improve the safety culture.

System Design

Based on smartphone, platform on which driving behavior data were obtained using inbuilt sensors (accelerator, gyroscopes, GPS, camera). The harsh braking, speeding, excessive cornering, and distraction were detected by cloud analytics. Specialized camera on the car detected drowsiness and distraction by facial analysis.

Machine Learning Analysis

The unsupervised clustering was used to group drivers into clusters of behavioral styles. Crash risk was determined by the driving patterns and historical incident records with the help of supervised models. Anomaly detection detected the abnormal driving behavior which could be as a result of fatigue, a health problem or a car problem. Individual coaching interventions were possible through customized profiles of drivers.

Program Results

The initial year implementation recorded 28 percent crash, 19 percent decrease in harsh braking events and 15 percent improvement in fuel economy. The initial resistance of acceptance of drivers was solved later when coaching focused on personal improvement and not punishment. The premiums paid to the insurers decreased by 12 percent because of risk mitigation.

Success Factors

- **Coaching and Punishment**: The focus on coaching more than punishment contributed to the driver buy and involvement.
- **Privacy Protections:** There were transparent policies that restricted access and utilization of data that fostered trust in drivers.
- **Gamification:** Competitive aspects that promoted good behaviors enhanced the participation.
- **Integration with Compensation:** The coordination of safety performance and recognition and compensation.
- **Unrelenting Relentlessness:** The system refinements involved were based on regular feedback provided by drivers.

Ethical Considerations

The advantages of program balanced safety outweigh the issue of privacy by clear policy. Drivers were given clear communication on data collection, data storage and its use. The consent of individuals was gained. Safety management was the only purpose that accessed the data. Where feasible, the decisions led to policy were informed by aggregated and not individual data.

9.5 Case Study 5: Intelligent Traffic Signal Control

Application: The adaptive control of traffic signals in the case of mid-sized city was carried out through machine learning to optimize the timing of the signal over a 120 intersection network in real-time.

Technical Implementation

Approach volumes and queue length and turning movements were measured by intersection detection systems using cameras and radar. Mitosis of measurements on a central control platform was done via reinforcement learning and optimization of timing of signals. Distributed control architecture allowed the operation of the systems in the case of communication breaks.

Learning Approach

Multi-agent reinforcement learning considered individual intersections to be learning agents that had the optimal control policies. The reward functions serve the following goals: throughput optimization, delay optimization, queue length control, and crash vulnerability minimization. The signal patterns that were related to previous accidents were punished with safety incentives.

Deployment Experience

The introduction of progressively more traffic crossing intersections was to reduce the disruption during the early learning. The level of performance stabilized at the end of three months of learning. Implementation took place system wide in nine months. Assessment registered an 8% decrease in travel delay, 12% decrease in stop delay, and 22% decrease in angle crash at the controlled crossroads.

Technical Challenges

- **Safety During Learning:** Conservative exploration guaranteed safety in terms of learning stage. Learned policies before their deployment are formally checked.
- **Sensor Failures:** The sensor failures had graceful degradation to rule-based control which ensured essential functionality.
- **Coordination:** Network-level coordination with respect to traffic progression needed advanced algorithms.
- **Competing Objectives:** The conflicting priorities of the stakeholders (throughput vs. pedestrian safety vs. emissions) necessitated the well-considered design of the reward functions.
- **Adaptation to Events:** Special events and building overrides were done manually and relearned.

Stakeholder Engagement

To be successful in deployment involved the involvement of traffic engineers, elected officials and people. Support was established through open communication of how the system worked and performed. Feedback mechanisms were in place so as to receive the feedback of the people. Illustrations were used to show advantages, which developed trust.

10. DISCUSSION AND CONCLUSION

10.1 Synthesis of Findings

This is a broad study that has explored the role of machine learning in transportation safety across predictive analytics, driver behavior, and infrastructure evaluations and autonomous systems integration. It was found that there was significant improvement in individual applications and integration with major challenges and gaps in research.

Major Accomplishments

Machine learning has been revolutionary in technology applied in safety in transportation. Crash prediction models work with more than 85 percent accuracy in detecting high risk scenario at some reasonable time before it happens to prevent it. Monitoring systems of infrastructure can be automated to provide continuous evaluation at scale and speed that is not possible through manual inspection. Individualized weathering interventions are offered through the analysis of driver behavior depending on personal factors. Under good conditions, autonomous vehicle perception systems can attain a performance close to that of humans.

These achievements testify to the fact that machine learning has the potential to significantly increase the results in terms of transportation safety. There is solid practice in the field that proves feasibility and safety advantages. There are positive examples of research to practice translation, which are described in case studies. Economic calculations depict positive benefit-cost ratios that will support investment.

Persistent Challenges

Nevertheless, as much as there is progress, there have been major issues that are hindering extensive implementation and a complete benefit realization. The integration of disparate systems into common safety mechanisms is still immature, and most of the implementations are dealing with single applications. Limitations of explainability decrease regulatory acceptance and user confidence especially of deep learning methods. Real time processing needs clash with computational needs of complicated models. The extrapolation between operational conditions is not easy and models have worse performance in settings that are not similar to those during training. Ethical issues such as privacy, fairness and liability must be resolved systematically.

Critical Research Gaps

The process of analysis determined that there are a number of critical research gaps that are to be addressed. Similarity in evaluation methodology, data format and performance measure would be possible thus facilitating the transfer of knowledge and comparison of performance. This would allow on-board high level intelligence in resource constrained environments by use of edge computing architectures. The generalization would be enhanced when transfer learning and domain adaptation are involved. Privacy-preserving collaborative learning would be made possible by federated learning frameworks. The explainable AI systems specific to transportation scenarios would help to increase transparency and trust. Multi-modal integration that would take care of the various modes of transportation would increase impact.

Cross-Cutting Themes

There are a number of themes that arise in areas of application. Human factors are significant in promoting system effectiveness and acceptance where it is necessary to design the system based on the user. Some ethical aspects such as fairness, privacy and transparency need to be handled methodically. Laws should change to accommodate machine learning capabilities and limitations. Integration problems require global system-level solutions as opposed to single component solutions. Constant learning and adaptation is critical due to changing operational environments.

10.2 Theoretical Contributions

The study contributes to the theoretical knowledge in a number of ways:

Work Integrated Framework Architecture

The suggested layered architecture (Section 6) offers a detailed framework of applying various machine learning applications to unified intelligent transportation safety systems. The architecture deals with the process of data acquisition, processing and analytics, decision making and implementing the action in modular manner that allows flexibility in the deployment. This contribution provides conceptual base to the development of systems level.

Learning Frameworks Federated

Privacy-preserving collaborative model training is possible in the federated learning framework (Section 6.3, Algorithm 6), which does not have to centralize sensitive driving information. The pursuit of privacy and the secure aggregation systems prevent privacy invasion and the accuracy of the model respectively. This will deal with the serious privacy concerns that limit the sharing of data as well as allowing distributed data to be learned together.

Multi-Model Risk Fusion Algorithm

In the suggested risk fusion algorithm (Section 6.2.3, Algorithm 5), the special models that respond to various safety factors are combined to produce an overall risk to assessments. Confidence-weighted aggregation takes into consideration the difference in the quality of the data and reliability of the model. Quantifying uncertainty gives confidence bounds that can be used to make risk-sensitive decisions. The contribution covers the aspect of integration of heterogeneous safety information sources.

Explainable AI Framework

The multi-level explanation framework (Section 6.4, Table 2) produces the right explanations to various stakeholders including the technical developers and the end users. The validation criteria make sure that quality in explaining it is beyond the anecdotal level. Such contribution increases transparency and trust in machine learning that is critical to safety.

Industrial System Evaluation Framework

The assessment mechanism (Section 3.4, Table 3) is not limited to the conventional metrics of machine learning but also includes the aspect of the safety impact, the practical feasibility of the implementation, and the ethical aspect. Multi-dimensional assessment is a holistic assessment that informs the priorities of development and the decision to deploy.

10.3 Practical Implications

Findings of the research can give practical advice to various stakeholders:

For System Designers

- Implement modular layered designs that allow development of components independently and the ability to combine components loosely.

- Deploy hybrid cloud-edge design that allocates processing in accordance to latency, privacy and computer needs.
- Prominence should be on robust behavior and graceful behavior when faced with sensor failures and poor conditions.
- Bring on board the quantification of uncertainty to make risk-conscious decisions.
- Planning to be constantly in motion and adapt with changing situations.

For Algorithm Developers

- Tradeoffs between accuracy and interpretability of a safety-critical application.
- Use ensemble techniques to obtain a better strength and uncertainty applications.
- Generalization in various conditions of operation: generalization should be tested systematically.
- Use the mechanisms of constant learning to adapt to changing environments.
- Take into account the issue of fairness and use debiasing methods.
- Resource-constrained edge deployment design algorithms.

In the Case of Transportation Agencies

- Create full-scale data gathering initiatives that would allow machine learning and ensure privacy.
- Institute assessment systems that focus on practical safety impact other than technical measures.
- Create relationships between agencies, industry, and academia in the form of collaboration and data sharing and research alliances.
- Support pilot programs are those that present technologies and the institutional knowledge.
- Build workforce potentials by training and employing.
- Activate the stakeholders such as public, elected officials, and interest groups.

For Policymakers

- Come up with regulatory systems that ensure there is a balance between encouragement of innovations and safety assurance.
- Standardization of support initiatives that allow the interoperability and standard examination.
- Invest in fundamental research of long-term issues.
- Create long term funding sources to proven technologies.

- Determine the liabilities frameworks elucidating the responsibility.
- Make sure that regulations are updated along with technological abilities.
- Take into account equity considerations that lead to wide spread benefits of safety.

For Fleet Operators

- Install driver behavioral control under the underlining of coach rather than punish.
- Firm data policies that give rise to driver confidence.
- Combine safety systems and operational workflows to make use of them.
- Empower effectiveness of the measure programs using detailed metrics.
- Adjust to the corporate culture and the characteristics of drivers.
- Disseminate learning with partners in the industry.

10.4 Limitations of Current Research

The limitations to this research include:

Scope Limitations

Concentrate on ground transport especially automotive uses, little on aviation, sea and rail. Less focus on more comprehensive mobility optimization and more on safety applications. Regional specialization of developed countries where there is extensive research on it and experience in deployment. Further research ought to cover other forms of transport and international environment.

Temporal Limitations

Fast changing technological environment implies that findings are up to date on capabilities that may have to be revised following any change in technology. The algorithms of machine learning, computing resources, sensor technology and deployment settings are rapidly changing. Capabilities should be updated with recommendations.

Validation Limitations

The proposed frameworks have not been fully empirically tested to any validation via real-world implementation and longitudinal assessment. Theoretical evaluation and literature review gives the groundwork but practical validation needs a lot of additional materials such as implementation, deployment infrastructure and prolonged assessment. Further research ought to seek empirical validation.

Generalization Assumptions

Frameworks presuppose the high level of applicability in a wide range of operational settings, whereas local implementation demands the adaptation to local circumstances, regulations, and infrastructure. The difference between cultures, traffic, regulatory conditions, and infrastructure capabilities are also significant in the world. Guides on implementation must be localized.

Cooperation Assumptions Among Stakeholders

Effective execution presupposes collaboration of various stakeholders such as governmental organizations, automobile manufacturers, technological firms, and infrastructure providers, and research organizations. The cooperation can be hampered by competitive interests, concerns with intellectual property, and barriers that occur within the organization. Tactics that deal with difficulties in cooperation are worth to be investigated.

10.5 Future Research Agenda

We suggest the future research agenda that includes the gaps and opportunities identified:

Near-Term Priorities (1-3 Years)

- Evaluation frameworks, benchmark datasets and performance metrics that would allow comparison of studies across studies.
- Real-time processing Edge-based model architectures in resource constraints.
- Techniques in explainability that would meet the transportation safety contexts, with considerations to the needs of stakeholders.
- Transfer learning methods that enhance the generalization with respect to conditions of operation.
- Privacy-preserving multi-organizational learning Federated learning implementations.
- Evaluation of fairness and methods of debiasing that lead to fair safety results.
- Integration structures that cover feasibility deployment.

Medium- Term Priorities (3-7 Years)

- Extensive intelligent transportation system network designs incorporating several safety applications.
- Formal verification techniques that offer safety guarantees to machine learning systems.
- Integration of multi-modes of transport safety in the automotive, rail, air transport, and maritime.
- Quantum based traffic optimization and routing.
- System validation and real-time optimization platforms based on digital twin.
- Smart edge intelligence with energy-efficient neuromorphic computing.
- The paradigms of human-centered design with regard to proper human-AI interaction.
- Having detailed policy frameworks aiming at liability, certification and governance.

Long-Term Vision (7+ Years)

- Complete smart transport systems where safety is orchestrated.
- Independent multi-modal transport systems that maximize safety in the system.
- Goal Vision Zero Vision Achieved by 2030: Garnering out preventable transportation deaths and critical injuries.
- International coordination and optimization of transport security.
- Optimal human-machine cooperation with features of maximum combined capabilities.
- Wholesome socio-technical systems incorporating technology, infrastructure, policy and human factors.

10.6 Concluding Remarks

Machine learning is a revolutionary technology in the field of transportation safety with the ability to perform predictive analytics, real-time monitoring, and incorporate autonomous systems. The sphere has developed significantly since primitive statistical models to the advanced deep learning systems with high performance in certain tasks. Yet, to achieve all the potentials, it is necessary to tackle the issues of integration, make them explainable and trustable, address the ethical issues, and develop favorable regulatory frameworks.

The way to the future requires interdisciplinary research in machine learning, transportation engineers, policymakers, and industry practitioners. The standardization initiatives should create shared evaluation systems and interoperability specifications. Research should fill significant gaps in explainability, edge computing, edge generalization and multi-modal integration. Governance systems should strike a balance between the promotion of innovation and the provision of safety. Ethics systems should be fair, private and accountable.

The final vision involves overall intelligent transportation of the future with the machine learning, capable of proactive risk management, collaborative safety, adaptive infrastructure and personalized intervention. These systems would drastically decrease the crashes, injuries and deaths as well as increase mobility and efficiency. The realization of this vision will involve long-term research, development, standardization, and implementation by the whole transportation community.

Transportation safety is too serious of an issue than to accept preventable injuries and deaths. Machine learning offers powerful technologies to make it safer, yet they only can be deployed to the fullest their potential when incorporated into the large-scale socio-technical systems that take into consideration human factors, organizational settings, and values of the society. This study offers the basis of such integration, determines issues that are especially challenging and outlines the solutions frameworks and maps the way ahead. The shift to much safer transportation systems is in progress, and the information and categories shared in this paper shine light on the direction to the transformative vision.

The society, industry, government, and the research community have to collaborate to ensure that the machine learning capabilities are translated into reality in terms of safety enhancement. Technical excellence is not a sufficient concept to achieve success, it takes a holistic concept that incorporates technology, policy, ethics, economics and human factors. Through the adoption of this broadened view and the pursuit of the research agenda described in this paper, machine learning has the transformative power to save lives and avoid injuries on a large scale and the transportation community can attain this.

REFERENCES

Abdel-Aty, M., & Abdelwahab, H. (2004). Modeling rear-end collisions including the role of driver's visibility and light truck vehicles using a nested logit structure. *Accident; Analysis and Prevention*, *36*(3), 447–456. DOI: 10.1016/S0001-4575(03)00040-X PMID: 15003590

Arteaga, C., Paz, A., & Park, J. (2020). Injury severity on traffic crashes: A text mining with an interpretable machine-learning approach. *Safety Science*, *132*, 104988. Advance online publication. DOI: 10.1016/j.ssci.2020.104988

Bao, J., Liu, P., & Ukkusuri, S. V. (2019). A spatiotemporal deep learning approach for citywide short-term crash risk prediction with multi-source data. *Accident; Analysis and Prevention*, *122*, 239–254. DOI: 10.1016/j.aap.2018.10.015 PMID: 30390519

Breiman, L. (2001). Random forests. *Machine Learning*, *45*(1), 5–32. DOI: 10.1023/A:1010933404324

Chang, L.-Y., & Wang, H.-W. (2006). Analysis of traffic injury severity: An application of non-parametric classification tree techniques. *Accident; Analysis and Prevention*, *38*(5), 1019–1027. DOI: 10.1016/j.aap.2006.04.009 PMID: 16735022

Chen, T., & Guestrin, C. (2016). XGBoost: A scalable tree boosting system. In *Proceedings of the 22nd ACM SIGKDD International Conference on Knowledge Discovery and Data Mining* (pp. 785–794). https://doi.org/DOI: 10.1145/2939672.2939785

Cristianini, N., & Shawe-Taylor, J. (2000). *An introduction to support vector machines and other kernel-based learning methods*. Cambridge University Press., DOI: 10.1017/CBO9780511801389

Dong, W., Li, J., Yao, R., Li, C., Yuan, T., & Wang, L. (2016). *Characterizing driving styles with deep learning* [Preprint]. arXiv. https://doi.org//arXiv.1607.03611DOI: 10.48550

Endsley, M. R. (2017). From here to autonomy: Lessons learned from human-automation research. *Human Factors*, *59*(1), 5–27. DOI: 10.1177/0018720816681350 PMID: 28146676

Feng, D., Haase-Schütz, C., Rosenbaum, L., Hertlein, H., Glaeser, C., Timm, F., Wiebesiek, W., & Dietmayer, K. (2020). Deep multi-modal object detection and semantic segmentation for autonomous driving: Datasets, methods, and challenges. *IEEE Transactions on Intelligent Transportation Systems*, *22*(3), 1341–1360. DOI: 10.1109/TITS.2020.2972974

Ghosh, S., Rana, A., & Kansal, V. (2017). Predicting defect of software system. In Satapathy, S. C., Zhang, Y. D., Prasad, V. K., Suryadevara, B. P. S., & Sharma, A. K. (Eds.), *Proceedings of the 5th International Conference on Frontiers in Intelligent Computing: Theory and Applications* (pp. 55–67). Springer. https://doi.org/DOI: 10.1007/978-981-10-3156-4_6

Ghosh, S., Rana, A., & Kansal, V. (2018). A hybrid non-linear manifold detection approach for software defect prediction. In *2018 7th International Conference on Reliability, Infocom Technologies and Optimization (Trends and Future Directions) (ICRITO)* (pp. 453–459). IEEE. https://doi.org/DOI: 10.1109/ICRITO.2018.8748788

Ghosh, S., Rana, A., & Kansal, V. (2018a). A non-linear manifold detection based model for software defect prediction. *Procedia Computer Science*, *132*, 581–594. DOI: 10.1016/j.procs.2018.05.012

Ghosh, S., Rana, A., & Kansal, V. (2018b). A novel model based on non-linear manifold detection for software defect prediction. In *2018 International Conference on Intelligent Computing and Control Systems (ICICCS)* (pp. 140–145). IEEE. https://doi.org/DOI: 10.1109/ICCONS.2018.8663026

Ghosh, S., Rana, A., & Kansal, V. (2018c). Combining integrated sampling with non-linear manifold detection techniques for software defect prediction. In *2018 3rd International Conference on Contemporary Computing and Informatics (IC3I)* (pp. 178–183). IEEE. https://doi.org/DOI: 10.1109/IC3I.2018.9007253

Ghosh, S., Rana, A., & Kansal, V. (2019). A statistical comparison for evaluating the effectiveness of linear and non-linear manifold detection techniques for software defect prediction. *International Journal of Advanced Intelligence Paradigms*, *12*(3–4), 370–391. DOI: 10.1504/IJAIP.2019.098578

Ghosh, S., Rana, A., & Kansal, V. (2019). Statistical assessment of non-linear manifold detection-based software defect prediction techniques. *International Journal of Intelligent Systems Technologies and Applications*, *18*(6), 579–605. DOI: 10.1504/IJISTA.2019.102667

Ghosh, S., Rana, A., & Kansal, V. (2020). A benchmarking framework using non-linear manifold detection techniques for software defect prediction. *International Journal on Computer Science and Engineering*, *21*(4), 593–614. DOI: 10.1504/IJCSE.2020.106618

Ghosh, S., Rana, A., & Kansal, V. (2020). Evaluating the impact of sampling-based non-linear manifold detection model on software defect prediction problem. In *Smart innovation, systems and technologies* (Vol. 159, pp. 141–152). Springer., DOI: 10.1007/978-981-13-9282-5_14

Gopalakrishnan, K., Khaitan, S. K., Choudhary, A., & Agrawal, A. (2017). Deep convolutional neural networks with transfer learning for computer vision-based data-driven pavement distress detection. *Construction & Building Materials, 157*, 322–330. DOI: 10.1016/j.conbuildmat.2017.09.110

Hallmark, S. L., Veneziano, D., Falb, S., Pawlovich, M., & Witt, D. (2015). Evaluation of driving behavior on rural 2-lane curves using the SHRP 2 naturalistic driving study data. *Journal of Safety Research, 54*, 155–166. DOI: 10.1016/j.jsr.2015.06.017 PMID: 26403897

Hochreiter, S., & Schmidhuber, J. (1997). Long short-term memory. *Neural Computation, 9*(8), 1735–1780. DOI: 10.1162/neco.1997.9.8.1735 PMID: 9377276

Hoff, K. A., & Bashir, M. (2015). Trust in automation: Integrating empirical evidence on factors that influence trust. *Human Factors, 57*(3), 407–434. DOI: 10.1177/0018720814547570 PMID: 25875432

Kalra, N., & Paddock, S. M. (2016). Driving to safety: How many miles of driving would it take to demonstrate autonomous vehicle reliability? *Transportation Research Part A, Policy and Practice, 94*, 182–193. DOI: 10.1016/j.tra.2016.09.010

Karlaftis, M. G., & Tarko, A. P. (1998). Heterogeneity considerations in accident modeling. *Accident; Analysis and Prevention, 30*(4), 425–433. DOI: 10.1016/S0001-4575(97)00122-X PMID: 9666239

Kipf, T. N., & Welling, M. (2017). Semi-supervised classification with graph convolutional networks. *International Conference on Learning Representations*. https://openreview.net/forum?id=SJU4ayYgl

Kiran, B. R., Sobh, I., Talpaert, V., Mannion, P., Al Sallab, A. A., Yogamani, S., & Pérez, P. (2021). Deep reinforcement learning for autonomous driving: A survey. *IEEE Transactions on Intelligent Transportation Systems, 23*(6), 4909–4926. DOI: 10.1109/TITS.2021.3054625

Lee, J. D., & See, K. A. (2004). Trust in automation: Designing for appropriate reliance. *Human Factors, 46*(1), 50–80. DOI: 10.1518/hfes.46.1.50.30392 PMID: 15151155

Liu, L., Lu, S., Zhong, R., Wu, B., Yao, Y., Zhang, Q., & Shi, W. (2020). Computing systems for autonomous driving: State of the art and challenges. *IEEE Internet of Things Journal, 8*(8), 6469–6486. DOI: 10.1109/JIOT.2020.3043716

Mnih, V., Kavukcuoglu, K., Silver, D., Rusu, A. A., Veness, J., Bellemare, M. G., Graves, A., Riedmiller, M., Fidjeland, A. K., Ostrovski, G., Petersen, S., Beattie, C., Sadik, A., Antonoglou, I., King, H., Kumaran, D., Wierstra, D., Legg, S., & Hassabis, D. (2015). Human-level control through deep reinforcement learning. *Nature*, *518*(7540), 529–533. DOI: 10.1038/nature14236 PMID: 25719670

Oliveira, C., Sousa, M., & Couto, A. (2021). A literature review of machine learning algorithms for crash injury severity prediction. *Journal of Safety Research*, *79*, 317–332. DOI: 10.1016/j.jsr.2021.09.004

Rahim, M. A., & Hassan, H. M. (2021). A deep learning based traffic crash severity prediction framework. *Accident; Analysis and Prevention*, *154*, 106090. Advance online publication. DOI: 10.1016/j.aap.2021.106090 PMID: 33740462

Rahman, M. A., & Umamaheswari, A. (2021). Road pavement crack detection using deep learning with synthetic data. *IOP Conference Series. Materials Science and Engineering*, *1019*(1), 012036. Advance online publication. DOI: 10.1088/1757-899X/1019/1/012036

Redmon, J., Divvala, S., Girshick, R., & Farhadi, A. (2016). You only look once: Unified, real-time object detection. In *Proceedings of the IEEE Conference on Computer Vision and Pattern Recognition* (pp. 779–788). https://doi.org/DOI: 10.1109/CVPR.2016.91

Schulman, J., Wolski, F., Dhariwal, P., Radford, A., & Klimov, O. (2017). *Proximal policy optimization algorithms* [Preprint]. arXiv. https://doi.org//arXiv.1707.06347DOI: 10.48550

Seventekidis, P., Giagopoulos, D., Arailopoulos, A., & Markogiannaki, O. (2020). Structural health monitoring using deep learning with optimal finite element model generated data. *Mechanical Systems and Signal Processing*, *145*, 106972. Advance online publication. DOI: 10.1016/j.ymssp.2020.106972

Strayer, D. L., Cooper, J. M., & Drews, F. A. (2010). Combining cognitive and visual distraction: Less than the sum of its parts. *Accident; Analysis and Prevention*, *42*(3), 881–890. DOI: 10.1016/j.aap.2009.05.001 PMID: 20380916

Tu, W., Zhou, Y., & Li, Q. (2021). A reinforcement learning approach for rebalancing electric vehicle sharing systems. *IEEE Transactions on Intelligent Transportation Systems*, *22*(11), 7116–7128. DOI: 10.1109/TITS.2020.3029123

Vaswani, A., Shazeer, N., Parmar, N., Uszkoreit, J., Jones, L., Gomez, A. N., Kaiser, Ł., & Polosukhin, I. (2017). Attention is all you need. *Advances in Neural Information Processing Systems, 30*, 5998–6008. https://proceedings.neurips.cc/paper_files/paper/2017/file/3f5ee243547dee91fbd053c1c4a845aa-Paper.pdf

Vlahogianni, E. I., Karlaftis, M. G., & Golias, J. C. (2014). Short-term traffic forecasting: Where we are and where we're going. *Transportation Research Part C, Emerging Technologies*, *43*, 3–19. DOI: 10.1016/j.trc.2014.01.005

Wijnands, J. S., Thompson, J., Nice, K. A., Aschwanden, G. D., & Stevenson, M. (2019). Real-time monitoring of driver drowsiness on mobile platforms using 3D neural networks. *Neural Computing & Applications*, *32*(14), 9731–9743. DOI: 10.1007/s00521-019-04506-0

Wilde, G. J. S. (1982). The theory of risk homeostasis: Implications for safety and health. *Risk Analysis : An Official Publication of the Society for Risk Analysis*, *2*(4), 209–225. DOI: 10.1111/j.1539-6924.1982.tb01384.x

World Health Organization. (2018). *Global status report on road safety 2018*. https://www.who.int/publications/i/item/9789241565684

Xu, C., Liu, P., Wang, W., & Li, Z. (2014). Development of a real-time crash risk prediction model incorporating the various crash mechanisms across different traffic states. *Traffic Injury Prevention*, *15*(5), 447–454. DOI: 10.1080/15389588.2013.853773 PMID: 24697528

Ye, L., & Yamamoto, T. (2019). Evaluating the impact of connected and autonomous vehicles on traffic safety. *Physica A*, *526*, 121009. Advance online publication. DOI: 10.1016/j.physa.2019.04.245

Zhang, Y., Sun, X., Loh, K. J., Su, W., Xue, Z., & Zhao, X. (2020). Autonomous bolt loosening detection using deep learning. *Structural Health Monitoring*, *19*(1), 105–122. DOI: 10.1177/1475921719837509

Zhao, L., Song, Y., Zhang, C., Liu, Y., Wang, P., Lin, T., Deng, M., & Li, H. (2020). T-GCN: A temporal graph convolutional network for traffic prediction. *IEEE Transactions on Intelligent Transportation Systems*, *21*(9), 3848–3858. DOI: 10.1109/TITS.2019.2935152

APPENDIX

Algorithm Complexity Analysis

Table 5. Computational Complexity of Major Algorithms

Algorithm	Training Complexity	Inference Complexity	Memory Requirements	Suitability for Edge
Decision Trees	$O(n \cdot m \cdot \log n)$	$O(\log n)$	O(nodes)	High
Random Forests	$O(k \cdot n \cdot m \cdot \log n)$	$O(k \cdot \log n)$	O(k·nodes)	Medium
Support Vector Machines	$O(n^2 \cdot m)$ to $O(n^3 \cdot m)$	$O(sv \cdot m)$	$O(sv \cdot m)$	Medium
Gradient Boosting	$O(t \cdot n \cdot m \cdot \log n)$	$O(t \cdot \log n)$	O(t·nodes)	Medium
Feedforward Neural Networks	O(epochs·n·p)	$O(p)$	$O(p)$	High
Convolutional Neural Networks	$O(\text{epochs} \cdot n \cdot c \cdot k^2 \cdot f)$	$O(c \cdot k^2 \cdot f)$	$O(c \cdot k^2 \cdot f)$	Medium
Recurrent Neural Networks	$O(\text{epochs} \cdot n \cdot T \cdot h^2)$	$O(T \cdot h^2)$	$O(h^2)$	Low-Medium
Transformers	$O(\text{epochs} \cdot n \cdot T^2 \cdot d)$	$O(T^2 \cdot d)$	$O(T \cdot d)$	Low

Note: n=samples, m=features, k=trees/filters, t=boosting rounds, p=parameters, c=channels, f=feature maps, T=sequence length, h=hidden units, d=dimensions, sv=support vectors, epochs=training iterations

Performance Metrics Definitions

Classification Metrics

- **Accuracy** = (TP + TN) / (TP + TN + FP + FN)
- **Precision** = TP / (TP + FP)
- **Recall (Sensitivity)** = TP / (TP + FN)
- **Specificity** = TN / (TN + FP)
- **F1-Score** = 2 × (Precision × Recall) / (Precision + Recall)
- **Matthews Correlation Coefficient** = (TP×TN - FP×FN) / √[(TP+FP)(TP+FN)(TN+FP)(TN+FN)]

Regression Metrics

- **Mean Absolute Error** = $(1/n) \times \Sigma|y\_i - \hat{y}\_i|$

- **Root Mean Square Error** = $\sqrt{[(1/n) \times \Sigma(y\_i - \hat{y}\_i)^2]}$
- **Mean Absolute Percentage Error** = $(100/n) \times \Sigma|(y\_i - \hat{y}\_i)/y\_i|$
- **R-Squared** = 1 - (SS_res / SS_tot)

Where TP=True Positives, TN=True Negatives, FP=False Positives, FN=False Negatives, y_i=actual values, ŷ_i=predicted values, SS_res=sum of squared residuals, SS_tot=total sum of squares

Dataset Characteristics

Table 6. Major Public Transportation Safety Datasets

Dataset Name	Domain	Size	Features	Temporal Coverage	Geographic Coverage	Access
FARS	Crash Records	2.5M+ crashes	100+ variables	1975-present	USA nationwide	Public
HSIS	Crash & Infrastructure	State-specific	200+ variables	1990s-present	Selected US states	Restricted
SHRP2 NDS	Naturalistic Driving	5,400+ drivers, 34M miles	Multi-sensor	2010-2013	6 US locations	Research access
KITTI	Autonomous Driving	15K frames	Camera, Lidar, GPS	2012	Karlsruhe, Germany	Public
nuScenes	Autonomous Driving	1,000 scenes, 40K frames	Multi-sensor	2018-2019	Boston, Singapore	Public
Waymo Open	Autonomous Driving	1,000 segments	Lidar, Camera	2019-2020	Multiple US cities	Public
BDD100K	Driving Video	100K videos	Camera, GPS, IMU	2017-2018	US nationwide	Public
Argoverse	Autonomous Driving	320K+ trajectories	Map, sensors	2019	Miami, Pittsburgh	Public

Abbreviations

AI - Artificial Intelligence

AUC - Area Under Curve - Arca Under Curve Receiver Operating Characteristic

Conventional Neural Network CNN.

DBSCAN density-based spatial clustering of noisy applications DBSCAN density-based spatial clustering of noisy applications.

DSRC Dedicated Short- Range Communications.

Fatality Analysis Reporting System FARS.

GDPR -General Data Protection Regulation.
GCN - Both have the same pretext as Graph Convolutional Network.
GMM - Gaussian Mixture Model
GPS -global positioning system.
GPU -Graphics Processing Unit.
GRU - Gated Recurrent Unit
Highway Safety Information System - HSIS.
IEEE - Institute of Electrical and electronics engineers.
IoT - Internet of Things
ISO - International Organization of Standardization.
LIME- Local Interpretable Model-agnostic Explanations.
LSTM - Long Short-Term Memory
ML - Machine Learning
NHTSA- National Highway Traffic Safety Administration.
PCA- Principal Component Analysis.
PERCLOS -Percentage Eye Closure.
PPO- Proximal Policy Optimization.
RADAR Radio Detection and Ranging.
R&D - Research and Development
ReLU - Rectified Linear Unit
RNN- Recurrent Neural Network.
SAE- Society of Automotive Engineers.
SHapley Additive exPlanations SHAP.
SHRP2 NDS - Second Strategic Highway Research Program Naturalistic Driving Study.
SOTIF- Safety of the Intended Functionality.
SVM - Support Vector Machine
t -SNE - t-distributed Stochastic Neighbor Embedding.
TPU - Tensor Processing Unit
TRPO - Trust Region Policy Optimization.
UMAP -Uniform Manifold Approximation and Projection.
V2V - Vehicle-to-Vehicle
V2X - Vehicle-to-Everything
WHO- World Health Organization.
Extreme Gradient Boosting- XGBoost.
YOLO - You Only Look Once

Chapter 10
Machine Learning for Accident Prediction and Prevention

Kezia Rani Burgula
http://orcid.org/0000-0002-6024-2330
Vasavi College of Engineering, India

Sheela D. V.
http://orcid.org/0000-0003-2936-4177
Sapthagiri NPS University, India

Padmesh Medesety
http://orcid.org/0000-0003-1350-5705
V.S.B. Engineering College, India

Abinaya. M.
Erode Sengunthar Engineering College, India

Naveen Nandal
http://orcid.org/0009-0007-2272-7437
DIHE, India

Abdul Razzak Khan Qureshi
Medicaps University, India

V. Bhoopathy
http://orcid.org/0000-0003-2175-6328
Sree Rama Engineering College, India

ABSTRACT

This chapter explores how ML-based modelsincluding Random Forest, CNN, LSTM, and Reinforcement Learning frameworksare revolutionizing accident prediction and prevention across domains such as smart transportation, industrial safety, logistics, and autonomous systems. It highlights end-to-end ML workflows covering data acquisition, feature engineering, model training, real-time alert systems, and deployment using IoT and AI-integrated infrastructure. The chapter also discusses practical implementations through case studies, along with technical, ethical, and policy challenges such as data imbalance, privacy concerns, and decision accountability in autonomous systems. By integrating AI-driven predictive analytics with

DOI: 10.4018/979-8-3373-3231-4.ch010

proactive safety interventions, ML demonstrates significant potential to reduce human error, optimize risk assessment, and enable intelligent safety ecosystems for future-ready industries and smart cities.

1. INTRODUCTION

It has been shown that traffic accidents are one of the leading causes of mortality and disability in India. In 2017, traffic-related events in India caused the deaths of over 25,859 persons and injuries to over 100,000. Damage to public infrastructure and the time and effort lost by drivers are two ways in which traffic accidents can drain society's coffers. Actually, it occurs all over the world; every year, road accidents claim 1.17 million lives and leave another 10 million maimed or paralyzed. Statistics from the Traffic Corp of the Indian National Police show that the number of recorded events involving motor vehicles has been steadily rising over the past few years across the country. Many factors may contribute to the fact that accident rates tend to rise from year to year. All three factors—the driver, the car, and the surrounding environment—are important. When all four of these things are taken into account, the consensus is that road accidents are mostly caused by human mistake. The total number of car accidents, however, is very consistent from one year to the next. Methods that can reliably identify variables that could increase the likelihood of an accident are clearly in demand. The accident rate on toll roads is slightly greater than that on other types of roadways. In relation to the above-mentioned factors, approximately 75% of incidents on toll roads were caused by human error (Viswanath et al., 2021). Everyone driving a toll road vehicle has their own unique set of strengths and weaknesses, perspectives on danger, responses to environmental cues, and, depending on the severity of their own impairment, their own unique set of operational challenges.

In addition, due to the high concentration of drunk drivers, toll roads pose a significant risk of accidents. Few details regarding the main causes of traffic accidents in developing nations' highways are known since most of these nations either do not maintain records or do not retain records adequately (Dorri et al., 2024). This is the primary cause for the lack of data. Improving the current level of safety on India's toll roads necessitates the creation of a TRA forecasting model. The purpose of this model is to give traffic accident predictions rates that do not include influence bias from regression to the mean. To evaluate how regulating one standard affects overall road safety, toll operators must have a thorough grasp of how criteria relate to actual accident rates. Furthermore, it may serve as a plan for a toll road development program, which would be particularly beneficial for emerging economies like

India, which are struggling to create roads that match their own toll road criteria (Karyemsetty & Kumar, 2020).

2. ACCIDENT FACTORS

When considering road transport, there are two systems: the mechanical system, which deals with the vehicle's connection to the road, and the biomechanical system, which deals with the human operator's relationship to the vehicle and the road. By looking at the current state of these systems, it can determine the likelihood of a road accident (Yuan et al., 2021).

Figure 1. The Principal Variables Contributing to Vehicular Accidents

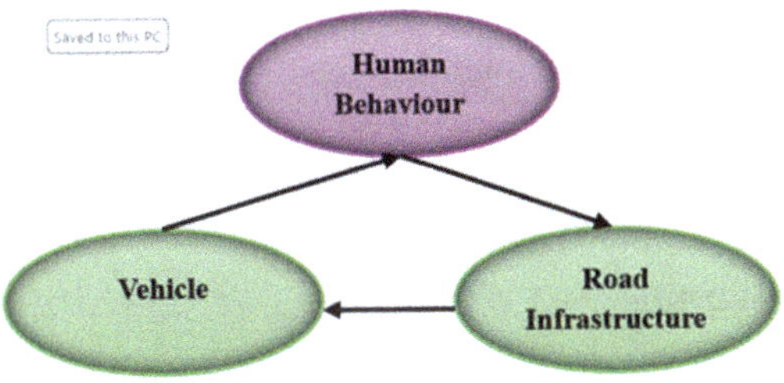

The occurrence of traffic accidents is influenced by numerous elements that work together to create this complicated phenomenon. The most important elements that determine the frequency and severity of accidents are human behaviour, vehicle condition, and road infrastructure. the road network, as seen in Figure 1.

After considering the road conditions and their subjective evaluation of outside factors, the driver decides to drive the car. There is a wide range of previously unrecognized factors that are collectively referred to as the road factor. A few examples of such factors are the driver's physical environment, the current temperature, and the weather. So, the road is considered a tertiary element here (Sun et al., 2023). The Although traffic laws, traffic management, and control measures are all aspects that potentially affect road safety. There must be a fourth traffic element included. Unsystematic elements also impact road safety, even if these are governed by particular standards. Things like weather and roadside hazards (such rocks and animals) fall into this category. Given these considerations, it is critical to zero in on a factor that includes everything (Bhattacharya et al., 2022).

2.1 The Factor of Human Behaviour

Human error is still a major component, even though improvements in road infrastructure and car safety technologies have helped reduce accidents. Drunk driving, multitasking while driving, and speeding are all known to significantly increase the likelihood of road accidents. Furthermore, there is a marked rise in the likelihood of accidents due to disruptive activity, especially during holiday seasons. Crash probability is further increased by impulsive behaviour and a lack of knowledge of driving restrictions, according to their study (Kushwaha & Abirami, 2023). Research suggests that risky behaviours like texting while driving may be responsible for as many as one-third of all accidents making it an increasingly pressing issue for road safety. Consequently, enhancing road safety requires a heightened focus on attentiveness. Enforcement of traffic laws, education, and awareness efforts can accomplish this (Olugbade et al., 2022).

There are several aspects to think about while addressing the role of human factors in traffic accidents these include:

- The driver's personality
- The driver's psychophysical qualities
- Culture and education

Figure 2. The Fundamental Interrelations of the Three Interactive Subsystems

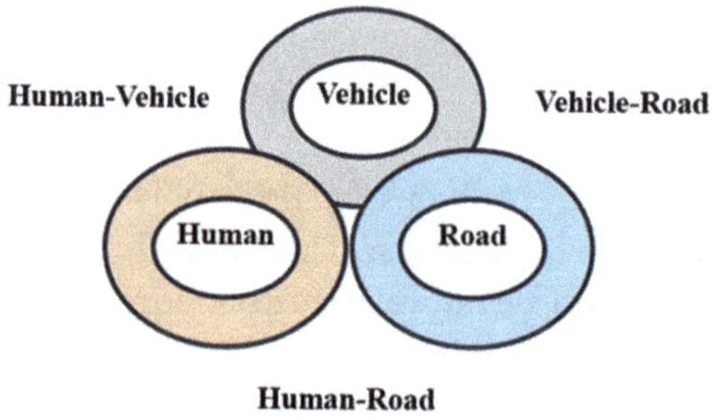

Figure 2 shows a Venn diagram depicting the interplay between the three safety factors—man, vehicle, and road—and they all work together to ensure road safety.

2.2 Vehicle Factor

While faulty road construction and driver error are often named as the top causes of traffic accidents, the state and characteristics of the vehicle in question also play an important role in establishing how safe the roads are. The role of vehicles in traffic accidents is examined in this study, along with the potential advantages and disadvantages of new technological advancements in vehicle safety (Santos et al., 2021). The design and its impact on road safety data on road safety in the European Union indicates that vehicle-related causes account for a significant fraction of all road accidents, accounting for an estimated 3% to 5% of the total. Inspections conducted after an accident may not uncover all possible reasons, so this percentage might be underestimated. There are active and passive components of a vehicle that influence traffic safety. Additionally, technological solutions aimed at lowering the accident probability can be considered active safety components. Some examples of such components are the vehicle's braking and steering systems, tires, lights, and signalling devices. On the other hand, passive elements are solutions that aim to lessen the impact of an accident. Everything from doors to safety belts to headrests to windscreens to the position of the engine to airbags falls under this area (Eboli et al., 2020).

2.3 Road Factor

In their broadest sense, roads are composed of numerous technically sound components that serve a purpose and have some sort of symbolic value. These components work together to achieve basic objectives, such as allowing for unimpeded communication in any weather, accommodating different kinds of people, and ensuring the comfort of road users. Statistical evidence suggests that road factors account for 8-10% of all traffic accidents (Ko & Choi, 2023). In addition to driver error, poor road quality, design, and maintenance are major contributors to the frequency and severity of traffic accidents. This investigation aims to shed light on the causes of road accidents, including poor maintenance, design defects, and road conditions, and to suggest remedies to these problems. The primary goals of these components include improving traffic flow and making roads safer and more comfortable for everyone. Defects in road design, construction, and maintenance, as well as wear and tear on the road surface in the form of surface deformations are the primary causes of road accidents (Hurt et al., 2024). Looking at the road profile through the lens of vibrations in vehicles, it can be:

- Micro-profile.
- Macro-profile.

- Rudeness.

The pneumatics, and particularly their performance, are negatively affected by such a road profile. Asphalt, concrete asphalt, and concrete make form the modern road surface, which is the top layer of the road that cars travel on.

3. REAL-TIME FACTORS INFLUENCING ROAD ACCIDENTS

According to research, the following three groups constitute the bulk of accident causes:

- The behaviours of humans
- The state of vehicles and
- The external factors, particularly the fitness of the roads (Asih & Soemitro, 2020).

Figure 3. Factors of Risk

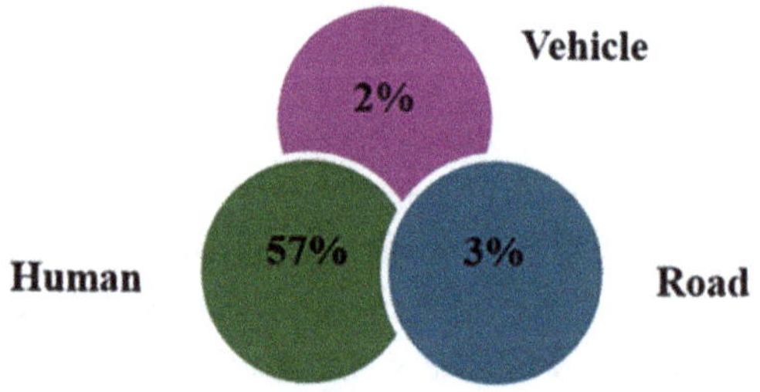

Figure 3 shows the environmental circumstances, which include human factors, vehicle conditions, and other variables.

- The road's surface,
- The force of gravity
- The state of the road
- The weather
- Duration of journey

Dynamics of Vehicles

- Brake status, pressure
- Vehicle speed and acceleration
- Vehicle dimensions
- Vehicle weight
- Steering wheel angle, wheel velocity

Driver Condition

- Driver's Reaction Time
- Driver's Experience
- Gender
- Age
- Driver's Health Condition
- Pulse Rate
- Medication Status

4. CONTRIBUTING FACTORS OF ACCIDENTS

The current man-vehicle-road system will unavoidably come into conflict with intelligent cars due to the low level of automation. According to the theory of man-machine-environment system engineering, there are three main characteristics of accidents involving intelligent vehicles (Ron, 2023).

4.1 Environmental Considerations Mostly Impede the Utilisation of Intelligent Systems.

The ability of road users to acquire information and the stability of their vehicles to operate are mostly impacted by environmental elements in conventional road traffic accidents. Even autonomous cars can't completely exclude environmental influences. In the case of snow and ice, for instance, changes in road reflection characteristics impact LiDAR (light detection and ranging) performance, which in turn impacts the creation of three-dimensional maps, and finally, the issue arises that intelligent vehicles are unable to adhere to the planned routes. Intelligent vehicles adhere to traffic laws and regulations and can avoid obstacles with the help of visual sensors that detect information like road signs and lane lines. However, in deep snow, it will be difficult to see lane markings, road signs, and things (such as buildings and automobiles) on either side of the road (Theofilatos et al., 2020). Due to their limited automation, intelligent vehicles are also limited to responding to certain traffic situations. Take Tesla's auto-steer technology as an example. It's designed to be used exclusively on highways that have clearly marked lanes and centre dividing lines. Only the human driver can decide if the traffic conditions are good enough to activate. So, if the human driver doesn't pay attention to the conditions of the autonomous driving system, it could lead to an intelligent vehicle crash (Vogel & Bester, 2023).

4.2 Machine Factors are the Primary Cause of Intelligent Vehicle Accidents

Intelligent vehicles, powered by AI, are able to make decisions and operate themselves with the help of sensors, cameras, and other information perception modules; DL and other information processing modules; and execution modules. On the other hand, the technology has been fine-tuned over the course of smart car development. Collision accidents can happen quickly because of technological issues, which is an unavoidable consequence of having certain restrictions. For instance, in May 2016, a Tesla vehicle was involved in a collision with a white caravan in Florida because of the limitations of intelligent recognition technology. A result of the limitations of intelligent decision-making (Harith et al., 2019). Due of the unreliability of interactions between VRUs and intelligent cars, intelligent vehicles are not friendly to VRU. Examples of situations where intelligent vehicles have struggled to detect and predict VRUs include the 2015 and the issue with Tesla's autopilot spotting cyclists raises the bar for algorithms that use images for recognition and prediction on computers (Hoła et al., 2023).

Table 1. SAE Classification of Automated Driving

Level	Name	Definition	Subject			
			System Functionality	**Implementation of Steering and Acceleration**	**Surveillance of the Driving Environment**	**Performance Fallback for Dynamic Driving Tasks**
0	**No Automation**	The human driver must fully complete all facets of the dynamic driving task, even when augmented by warning or intervention systems.	N/A	Human Driver	Human Driver	Human Driver
1	**Driver Assistance**	Execution of driving mode-specific functions by a driver assistance system, either steering or controlling acceleration/deceleration, utilising data regarding the driving environment, while anticipating that the human driver manages all other components of the dynamic driving task.	Various Driving Modes	System and Human Driver	-	-
2	**Partial Automation**	Driving mode-specific functions are executed by one or more autonomous vehicle systems for steering or acceleration/deceleration, using data about the operating environment, with the understanding that the human driver manages all other parts of the dynamic driving task.	-	System	-	-
3	**Conditional Automation**	An autonomous driving system's performance is tuned to specific driving modes and includes all aspects of the dynamical driving task, with the expectation that the human driver will respond appropriately to an intervening request.	-	-	System	-

continued on following page

Table 1. Continued

Level	Name	Definition	Subject			
			System Functionality	**Implementation of Steering and Acceleration**	**Surveillance of the Driving Environment**	**Performance Fallback for Dynamic Driving Tasks**
4	**High Automation**	The performance of an automated driving system in driving mode is contingent upon all facets of the dynamic driving job, regardless of whether a human driver adequately responds to an intervention request.	-	-	-	System
5	**Full Automation**	An automated driving system executes all aspects of the dynamic driving task under all highway and environmental conditions that a human driver can manage.	All Driving Modes	-	-	-

Table 1 summarises the recorded accidents involving intelligent vehicles Developing fully autonomous vehicles will take time, according to the categorisation. As a result, the road traffic system will continue to include vehicles with a wide range of technical conditions and levels of automation for a while. There will be a wider variety of traffic accidents and a more intricate road traffic system.

4.3 Human Factors Mostly Pertain to the Operation of Intelligent Vehicles without Adherence to Prescribed Guidelines.

Existing intelligent vehicles require human drivers during the transition phase from partially autonomous to entirely autonomous driving. Switching from human to machine operation should be done with a suitable amount of time in between; otherwise, collision incidents can easily occur when human drivers are not involved for long periods of time, if at all. Its collisions involving a as an example; the driver was distracted for at least seven seconds. According to the National Highway Traffic Safety Administration's Office of Failure Investigation, the average amount of time it takes for systems and drivers to recognise, react, and take action is less than three seconds (Bucsuházy et al., 2020). Therefore, researchers in the area of intelligent cars are concentrating on the issue of operation switching between humans and machines. Dedicated drivers, according to some research, have a superior grasp of their surroundings in such cases. They are able to accurately anticipate the actions of other drivers, particularly when they have extensive experience (Swaen et al., 2023).

5. ML MODELS FOR ACCIDENT PREDICTION

5.1 Supervised

5.1.1 LR

The purpose of the study was to examine the relative merits of standardised and unstandardised estimates of the indirect effect and the proportion mediated using various statistical methods, including LR, structural equation modelling, and the potential outcomes framework for mediation models with a binary outcome (Davis, 2023). Along with a new subprotocol for safely computing the activation function, presented an implementation of a secure logistic regression training protocol. In In a population-based cohort study included 6,762 adults from Asia conducted the analysis.

5.1.2 KNN

A new approach to neighbourhood rough sets, KNN rough sets was introduced by (Hide et al., 2019). Two k-nearest neighbour rules based on locality constrained representation were proposed by Gou et al. to improve the performance of KNN-based classification and address the issues.

5.1.3 LR

The purpose of the study conducted was to examine the connections between various meteorological variables and the concentrations of air pollutants from 2013 to 2017. These variables included daily mean oC, relative humidity (Ristea, 2023). This case study focused on demand forecasting in pharmaceutical supply chains. This article delves into the current state of pharmaceutical demand forecasting and the major challenges that this field faces. A plethora of ML regression techniques have been employed thus far. These include RF, KNN regression, fuzzy logic, SVR, and NN.

5.1.4 NBC

An extensive mathematical elucidation of the Bayes theorem and NB learning's foundations is presented in this study (MURPHY, 2024). After researching the problems with previous schemes came up with a novel way to train an NBC in a local context that guarantees anonymity. In their 2019 study found that a FPGA based real-time hardware NBC was a successful technique (Berezyuk Sergiy, 2024).

5.2 Unsupervised

5.2.1 AE

As an AE for handling sparse 3D point clouds that preserve the spatial arrangements of the input data, 3D point-capsule networks were proposed by (Elton Mayo, 2024). Using the sliced-Wasserstein distance transformed the AE latent distribution into a prior distribution that was applicable to any sample. It investigated intelligent gearbox fault diagnostics via deep transfer multi-wavelet auto-encoding, despite the small number of training samples for the target. It describes a deep transfer multi-wavelet AE as a novel approach to diagnosing gearbox defects using limited training samples.

5.2.2 K-Means

The unsupervised k-means clustering method, first developed by finds the optimal number of groups automatically, without requiring setup or parameter adjustment. According to one of the main problems with the k-means algorithm is that it can't deal with various types of data. In order to tackle this concern, offer a thorough and organised review of the research conducted on the k-means approach. It devised a fair k-means aim and method for selecting cluster centres with reasonable prices for different groups. As a result, only a handful of feature values are needed to describe the cluster assignments (Kagan & Snidman, 2024).

5.3 Transfer learning

5.3.1 Migration based on Features

When metals were mechanically loaded looked at how grain boundary movement depended on size. The quantitative investigation of the size dependence of grain boundary migration in nanograined Ag, Cu and Ni subjected to strain was carried out throughout a broad size range (Herbert A. Simon, 2024). In their suggested method for intelligent city IoT feature extraction and intrusion detection combine a DL model with intrusion detection technology. It established a significant criterion for testing statistical hypotheses that compares a sample to a reference sample and assesses the impact of external factors (Vallacher & Wegner, 2023). Both the data on road safety and the research into the causes of traffic accidents are severely lacking and skewed. An example of this imbalance is the unequal distribution of categories in the data set, which occurs when certain categories of traffic accidents have significantly more data than others. Since models have a tendency to anticipate more classes and disregard fewer classes when the data distribution is uneven, this might have a negative effect on the outcomes of analysis and modelling.

There are limitations to current technologies and methodologies when it comes to fixing the problem of traffic accident safety data imbalance. If the data isn't balanced, the models will be skewed in their predictions and classifications, and they may fail to notice trends and features in some classes. Hence, studies are required to address the issue of unbalanced traffic safety data and guarantee that the model can be accurate and resilient across all categories. Improving traffic safety can be achieved by solving this research topic, which will lead to more accurate data analysis and predictions. This can necessitate the creation of algorithms tailored to unbalanced data, novel approaches to model modification, or data balancing procedures. To better avoid and reduce the occurrence of traffic accidents, it is necessary to overcome the hurdles posed by imbalanced data in order to have a thorough understanding of the occurrence and influencing elements of these incidents (Schorsch et al., 2017).

5.3.2 Migration Based on Samples

An automated method for migrating training samples was developed by using the first whole-season sample set from 2015 and all available archived Landsat 5 TM photos on the Google Earth Engine cloud platform (Guerin, 2025). An innovative sandstone with many thin clay interlayers was used to conduct core flooding tests utilising an X-CT scanner. To evaluate the effects of confounding variables on a sample relative to a control group developed a statistical hypothesis test–based significance criterion.

6. ADVANTAGES OF IOT AND AI IN ROAD SAFETY

There is a significant annual maintenance expense, as well as injuries, fatalities, and injuries sustained by the transportation authority. Finding a solution is of utmost importance for a developing nation such as India. Modern, rapidly developing technologies, such as ML and AI provide the ideal solution to this issue. By reducing the likelihood of catastrophic catastrophes and optimising maintenance costs, these technologies have the potential to save innocent civilian lives (Mela, 2020).

6.1 Road and Bridge Condition Detection

Inadequate Road maintenance is another element that contributes to traffic accidents. By constantly monitoring the state of the roads and doing maintenance appropriately, it can reduce the likelihood of accidents, all due to the IoT. Using ML, it can detect bridge flaws, which will aid in the prevention of devastating tragedies. As suggested in one such device can be an intelligent speed breaker system that assists in detecting speed breakers that drivers may overlook because of poor sight or excessive velocity (Cialdini et al., 2024). Using an RF module, this device not only alerts the driver when a speed breaker is around, but it also helps slow down the vehicle automatically if the driver doesn't react quickly enough. Using the IoT the speed breaker's GPS coordinates can be transmitted to the cloud via GPS and saved there for later use. As demonstrated in ML may also be employed to do pothole identification. This is achieved by collecting road status information using the accelerometer and gyroscope, which are typically present in all smartphones. Then, using Transfer learning, the potholes are identified. ML and the IOT can help individuals avoid car accidents in the near future. Using these technologies, transportation authorities will be able to create a smarter, more efficient and intelligent traffic system, which will lower the death toll from road accidents, save money on maintenance, and avert disasters. So, it's clear that these technologies are great for making roads safer (Oren & Ghasem-Aghaee, 2021).

6.2 Identifying Human Conduct

Human behaviour and the failure to adhere to traffic regulations are significant contributors to fatal accidents. IoT devices can encourage drivers to follow safety protocols. The IoT allows for the real-time updating of the traffic management system with data, which improves the safety system's efficiency (Epley et al., 2023). It can simultaneously assess the driver's actions and notify them using ML and the IOT. To investigate the intricate relationships between the driver's behaviours and the road conditions, a DL model is suggested in for instance. The suggested model uses an unsupervised SDAE to generate RGB-coloured output layers. The graphic, can see the SDAE model as well. The graphical output helps to identify patterns of both simple driving behaviour and the complexity of the road environment more efficiently (Jeffrey, 2019).

7. RESULT AND DISCUSSION

Injuries and deaths caused by traffic accidents have emerged as a major concern for the safety and well-being of the general population. People, families, and communities are profoundly affected by injuries sustained in traffic accidents and by-passing vehicles. Because of this, national budgets are already strained, and the typical deficit in GDP is 3% (Infante et al., 2023). An effective and straightforward supervised learning model, the KNN algorithm determines the new data point's class by determining the class that is most prevalent among its KNN. Understanding the impact of critical parameters, such as the selected value and distance metric, and the model's performance metrics are the primary foci of any discussion of KNN outcomes compared by ANN, AdaBoost and RF, NBC.

7.1 The Intelligent Road Accident Prevention System

In this study, trained ANN and other algorithms on big datasets to construct models that can forecast the probability of automobile accidents. In order to make predictions, the developed model collects data in real-time from a variety of sources, including GPS, IEEE802.11p, and embedded sensors in cars, among others, while the vehicles are in motion. To improve the accuracy of the predictions, a number of methods are used, such as LDA, dimensionality reduction algorithms, and ten ML algorithms (Abdel-Aty et al., 2023). If the validation score doesn't improve, the algorithm will not finish training due to the limited early stopping function. The results of the accuracy measurements taken by the algorithm are shown in Table III. The symbols TN and FN stand for true negative and false negative, respectively; TP and FP, for true positive and false positive, respectively. Accuracy is defined as the percentage of correctly identified items in equation (1) as:

$$Accuracy = \frac{TP + TN}{TP + TN + FP + FN} \times 100\% \tag{1}$$

The accuracy, as a percentage of the total number of expected positive observations, is called precision in equation (2).

$$Precision = \frac{TP}{TP + FP} \times 100\% \tag{2}$$

A secondary performance metric known as recall can be determined using the formula (3).

$$Recall = \frac{TP}{TP + FN} \times 100\% \tag{3}$$

The F1-score, which is the sum of Precision and Recall, is the last metric to consider. Here is how the DT algorithm's model parameters are set up: This information gain is measured by entropy, which is used to assess the splitting quality. The DT method allows a node to split if doing so reduces impurities to zero or near zero (Klinjun et al., 2021). The ability to halt tree growth at an early stage is no longer available. With the goal of finding the optimal split as quickly as possible, I disable pre-sorting prior tree splitting. It displays the model accuracy data for DT as F1-score, Precision, and Recall. In general, the model's results are in line with most previous studies in the same field, showing respectable sensitivity, average AUC, and average specificity. Variations in the total number of cases, changes in the criteria used to classify injuries, and sets of independent variables all contribute to the discrepancies (Guido et al., 2022).

Over the course of 70 epochs, with data splits of figure 4 displays the training accuracy of an KNN. The accuracy rates 96.72% for the different splits are as follows: ANN (90.12%),LR (92.46%),Adaboost (93.78%),RF (94.89%) and NB (95.56%).

Figure 4. Accuracy Comparison of the Models

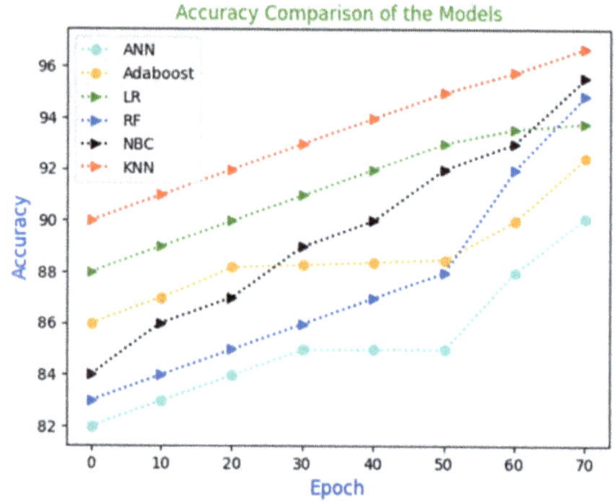

Both the train and test data subsets' performance plots are displayed in Figure 5. An increase in the accuracy function for both the training and test data indicates that the network has been adequately trained in our scenario.

Figure 5. Training and Testing Accuracy Curve

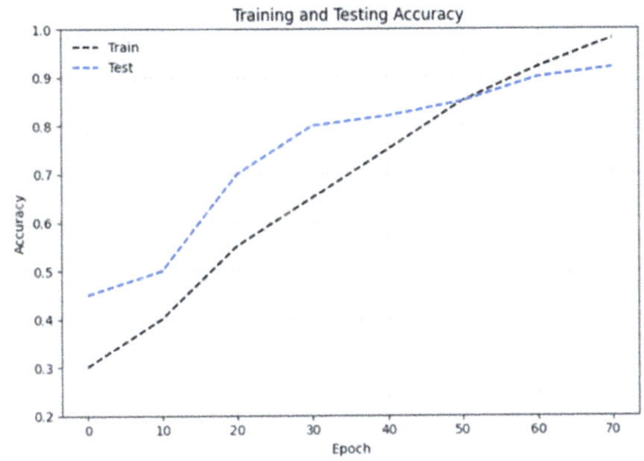

Figure 6 illustrates the performance graph of the training and testing data subsets. In our scenario, the network is well-trained as the loss function reduces for both training and test datasets.

Figure 6. Training and Testing Loss Curve

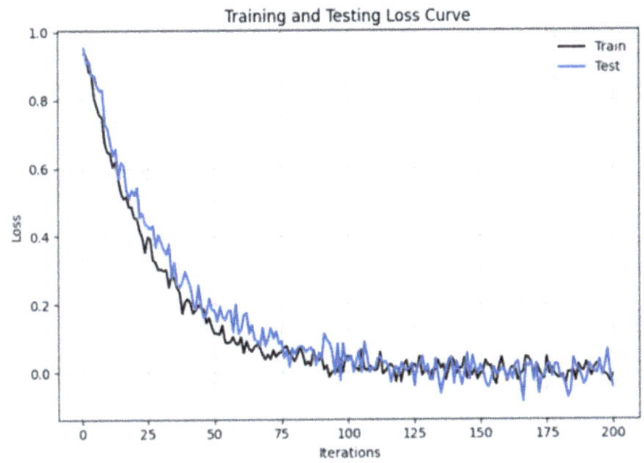

All of the model's performance analyze is results are displayed in Table 2. Our KNN works great it compared with alternative models.

Table 2. Performance Analysis

Models	Accuracy	Precision	Recall	F1- Score
ANN	90.12	88.42	89.23	90.48
LR	92.46	90.59	91.53	92.54
Adaboost	93.78	91.47	92.88	93.86
RF	94.89	92.30	93.40	94.91
NBC	95.56	93.39	94.01	95.62
KNN	96.72	94.06	95.60	96.80

Figure 7 shows as KNN model's F1 score performance is a constant but rather heavily depending on the job and dataset it is applied to. Models that integrate KNN have shown promising results in terms of F1 scores, with figures like 96.80% for character recognition for network intrusion detection indicating promising results. The performance is improved with an increase in the F1 score.

Figure 7. Performance Comparison

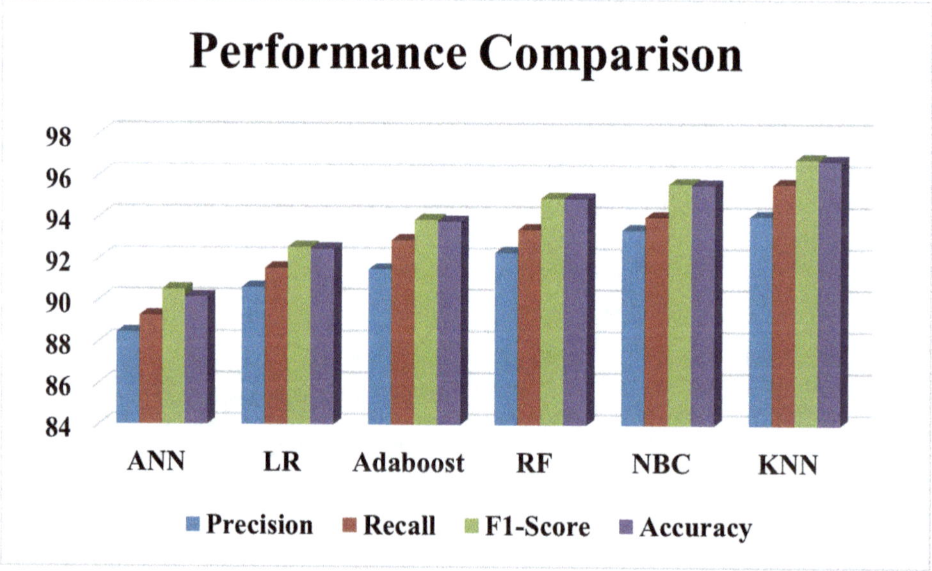

8. CHALLENGES

Using AI and ML to identify accidents on the road is challenging because to the following issues: ILD installation, maintenance, replacement, and some recurrent maintenance. When it comes to important road infrastructure, the accuracy is very bad. Precise estimates of transit time and speed are hindered by the small sample size (Mussone et al., 2023). One potential consequence of the proliferation of IoV-related large-scale sensors and connected vehicles is the potential strain on existing network infrastructure. Findings from this study can shed light on things like typical driver traits, peak accident days, and dangerous roads around specific areas. Using heat maps to display distributions by season and vehicle type introduced additional complexity to the investigation. Results from the prediction model were still good even when limited to a three-year time frame. For the predictive model to be more accurate, it needs a longer time period and more variables to be eligible (Liu et al., 2019). Research in the interrelated fields of social signalling analysis, human affect sensing, and context sensing is reviewed, along with the opportunities and challenges that researchers face in these areas. For some, creating multimodal, multi-aspect context sensing is the best bet for getting accurate results in real-world settings. One method for obtaining context is the W5+ method, which stands for who, where, what, when and why. It is very difficult for vehicles to share data efficiently aware of other vehicles that are broadcasting at the same time, which can cause accidents (Makarova et al., 2020).

An important network worry is the potential for accidents to occur in situations with a large concentration of vehicles and a heavy volume of data traffic. Another matter that requires attention is the question of culpability. It is necessary to draw a clear boundary between the different degrees of automation in order to clarify who is liable in the event of an accident. Consequently, new rules regarding insurance, traffic, and legislation may be necessary. The current regulations presume the presence of a human driver at all times. Nevertheless, the driver's role is projected to be shifted due to automation technologies, which will either entirely or partially replace them (Ebrahimi Darkhaneh et al., 2025). The obstacle due to the inadequacy of existing infrastructures and the lack of resources to make necessary repairs and upkeep. It may be more challenging for nations to benefit from AI if they allocate a smaller portion of their GDP to research and physical infrastructure related to technology (Al-Najada & Mahgoub, 2020).

9. FUTURE DIRECTIONS

The transportation industry is replete with applications of AI and NN. There are applications for every system that deals with the transportation of products via road. Since ITS may increase energy efficiency, road and traffic safety, and transit efficiency, it is expected that they would be an integral part of urban design in future smart cities (Eltemasi & Behtooiey, 2024). The results clearly show that the

ANN model was able to anticipate the flow of traffic using vehicle speed and class characteristics. Electronic devices known as data loggers are utilised in road transportation networks to collect information from moving vehicles on highways and at intersections. By employing sensors to collect this data, it can be analysed or saved for future. Technological and social developments in the future may alter logistical needs; these developments may shake up marketplaces and push the boundaries of existing services to new extremes (Zeng et al., 2024).

In the European Union, human error accounts for about 90% of road collisions. This includes factors such as driving while distracted, drunk, or speeding. Currently, NN-based DL is the most effective choice due to its explanatory power. One drawback of NN training is the significant amount of CPU resources it often requires because to the deep network structure (Santos et al., 2021). Locating the time, location, and intensity of an incident has been the subject of several efforts to aid traffic managers in their efforts to alleviate congestion. Their work spans several domains, including NN, automatic algorithms, and manual reporting. Manual, hand-written reports may be less efficient and take more time to identify incidents, but public transport and AI specialists are coming up with new concepts that can help the industry deal with its growing problems. For optimal performance, incident-responsive algorithms.

There is an immediate need for AI in all areas of modern life, but especially in transportation on roads, where it is crucial to implement safety measures to keep drivers and passengers safe. Human error is the leading cause of automotive accidents. Continuously exploring new avenues to enhance safety through the creation of autonomous incident detectors utilising ML and AI technologies is of the utmost importance. It is critical to be able to use AI and ML to enhance road transport (Abdullah & Sipos, 2024). There have been a lot of research on different parts of road transport and how AI and ML are employed, but it needs to look into autonomous incident detectors yet. The public transport industry will unavoidably be transformed by AI; hence it is crucial to comprehend the potential impacts of this technology on the sector. In order to improve their services and build the transportation of the future, public transportation stakeholders should actively explore the prospects presented by AI. Since human error accounts for over 90% of accidents in the European Union (where over 25,000 people lost their lives in 2017 due to factors including drunk driving, distracted driving, and speeding), the employment of AI-based systems in autonomous transportation has the potential to greatly improve road safety. Connected smart car communication systems were the primary emphasis of earlier studies with autonomous vehicles. Additional details regarding an IoT-based intelligent crossroad management system are provided. One kind of wireless network that doesn't use preexisting infrastructure is the ad hoc network (Ahmed et al., 2023).

One kind of wireless network that doesn't depend on preexisting infrastructure is the ad hoc network. This wireless or ad hoc network is a channel for the transfer of data. Locations of accidents are also transmitted via this. For the purpose of investigating hit-and-run accidents, a camera mounted on the vehicle's back and sides captures images of the items it collides with (Chirag & Supreetha, 2022). A data-mining engine receives vehicle data from the cloud, runs MLP-ANN algorithms to anticipate engine stress, and then utilises k-means clustering analysis to generate a driver KPI score. The purpose of the follow-up research to this work is to provide better traffic travel advice services by analysing video data from traffic surveillance systems. When an accident occurs, a software system can analyse the live video feed for overlapping images and deduce what happened. It recommends comparing the two more thoroughly so that it can use field data for model calibration, simulation, DTW, and SVM, and to build condition-specific datasets for testing models. ML and AI models using various methods are still needed for accident and emergency response estimation, and they should be discussed (Sarkar et al., 2019).

10. CONCLUSION

The future of automotive research and development will undoubtedly focus on intelligence, automation, and lightweight design. As vehicles go from becoming fully autonomous to partially autonomous, the road traffic system will incorporate both types of vehicles, increasing the complexity and diversity of potential incidents. Understanding the social concerns generated by defective rules and regulations and the safety issues caused by technological limitations is crucial as it embrace the enormous transformations that autonomous vehicles have introduced to the socio-economic landscape and human mobility behaviours. By analysing existing intelligent vehicle accident data, researchers can enhance their comprehension of the limitations of current intelligent vehicles implement improvements in decision-making, execution and develop perception more scientifically grounded and comprehensive traffic safety intervention and enhancement strategies. This involves proposing key elements of intelligent vehicle accident analysis aligned with technological advancements and safety improvements, as well as establishing effective methods for investigating and identifying intelligent vehicle accidents (Pourroostaei Ardakani et al., 2023). Given the complexity of future traffic accidents, it is recommended that appropriate regulations and standards be put in place for intelligent vehicles promptly. This will help to clarify who is legally responsible in the event of an accident involving an intelligent vehicle and will allow for the implementation of management measures that are tailored to the specific needs of accident prevention and handling. There are still many unanswered questions in the field of intelligent vehicle research and development, even though it is receiving greater attention. The testing of smart transport still requires some time.

REFERENCES

Abdel-Aty, M. A., Hassan, H. M., Ahmed, M., & Al-Ghamdi, A. S. (2023). Real-time prediction of visibility related crashes. *Transportation Research Part C, Emerging Technologies*, *24*, 288–298. DOI: 10.1016/j.trc.2012.04.001

Abdullah, P., & Sipos, T. (2024). Exploring the Factors Influencing Traffic Accidents: An Analysis of Black Spots and Decision Tree for Injury Severity. *Periodica Polytechnica Transportation Engineering*, *52*(1), 33–39. DOI: 10.3311/PPtr.22392

Ahmed, S., Hossain, M. A., Ray, S. K., Bhuiyan, M. M. I., & Sabuj, S. R. (2023). A study on road accident prediction and contributing factors using explainable machine learning models: Analysis and performance. *Transportation Research Interdisciplinary Perspectives*, *19*(April), 100814. DOI: 10.1016/j.trip.2023.100814

Al-Najada, H., & Mahgoub, I. (2020). Real-time incident clearance time prediction using traffic data from internet of mobility sensors. *Proceedings - 2020 IEEE 15th International Conference on Dependable, Autonomic and Secure Computing, 2017 IEEE 15th International Conference on Pervasive Intelligence and Computing, 2017 IEEE 3rd International Conference on Big Data Intelligence and Compu, 2020-Janua*, 728–735. DOI: 10.1109/DASC-PICom-DataCom-CyberSciTec.2017.127

Asih, R., & Soemitro, A. (2020). Accident Analysis Assessment To the Accident Influence. *Proceedings of the Eastern Asia Society for Transportation Studies*, *5*, 2091–2105.

Bhattacharya, S., Jha, H., & Nanda, R. P. (2022). Application of IoT and Artificial Intelligence in Road Safety. *2022 International Conference on Interdisciplinary Research in Technology and Management, IRTM 2022 - Proceedings, June*. DOI: 10.1109/IRTM54583.2022.9791529

Bucsuházy, K., Matuchová, E., Zůvala, R., Moravcová, P., Kostíková, M., & Mikulec, R. (2020). Human factors contributing to the road traffic accident occurrence. *Transportation Research Procedia*, *45*(2019), 555–561. DOI: 10.1016/j.trpro.2020.03.057

Chirag, P., & Supreetha, M. (2022). Road Accident Prediction and Classification using Machine Learning. *MysuruCon 2022 - 2022 IEEE 2nd Mysore Sub Section International Conference*, *2*, 48–59. DOI: 10.1109/MysuruCon55714.2022.9972671

Cialdini, R. B., Kallgren, C. A., & Reno, R. R. (2024). A Focus Theory of Normative Conduct: A Theoretical Refinement and Reevaluation of the Role of Norms in Human Behavior. *Advances in Experimental Social Psychology*, *24*(C), 201–234. DOI: 10.1016/S0065-2601(08)60330-5

Davis, G. A. (2023). Accident reduction factors and causal inference in traffic safety studies: A review. *Accident; Analysis and Prevention*, *32*(1), 95–109. DOI: 10.1016/S0001-4575(99)00050-0

Dorri, A., Dhoska, K., Sulejmani, A., & Dorri, S. (2024). An Overview of Road Accident Analysis: A Data-Driven Approach for Enhanced Safety Solutions. *Journal of Integrated Engineering & Applied Sciences*, *2*(2), 69–79. DOI: 10.5281/zenodo.13942644

Eboli, L., Forciniti, C., & Mazzulla, G. (2020). Factors influencing accident severity: An analysis by road accident type. *Transportation Research Procedia*, *47*, 449–456. DOI: 10.1016/j.trpro.2020.03.120

Ebrahimi Darkhaneh, M., Effati, M., & Arabani, M. (2025). Factors affecting the injury severity of head-on crashes on undivided rural roads under different weather conditions. *International Journal of Transportation Science and Technology*, *xxxx*. Advance online publication. DOI: 10.1016/j.ijtst.2025.02.007

Eltemasi, M., & Behtooiey, H. (2024). Predicting the type of road accidents based on air temperature in Iran: A case study of roads in Qazvin province. *Heliyon*, *10*(7), e28536. DOI: 10.1016/j.heliyon.2024.e28536

Epley, N., Waytz, A., & Cacioppo, J. T. (2023). On Seeing Human: A Three-Factor Theory of Anthropomorphism. *Psychological Review*, *114*(4), 864–886. DOI: 10.1037/0033-295X.114.4.864

Guerin, B. (2025). The Forces that Shape Human Behavior: Combining the Behavior Analysis Approach with Social Science Knowledge for a More Exhaustive Framework to Analyze and Change Human Behavior. *The Psychological Record*, *75*(3), 355–371. DOI: 10.1007/s40732-025-00654-z

Guido, G., Haghshenas, S. S., Haghshenas, S. S., Vitale, A., Astarita, V., Park, Y., & Geem, Z. W. (2022). Evaluation of Contributing Factors Affecting Number of Vehicles Involved in Crashes Using Machine Learning Techniques in Rural Roads of Cosenza, Italy. *Safety (Basel, Switzerland)*, *8*(2), 28. Advance online publication. DOI: 10.3390/safety8020028

Harith, S. H., Mahmud, N., & Doulatabadi, M. (2019). Environmental factor and road accident: A review paper. *Proceedings of the International Conference on Industrial Engineering and Operations Management*, *2019*(MAR), 3409–3418.

Herbert, A. Simon. (2024). INVARIANTS OF HUMAN BEHAVIOR. *Department of Psychology*.

Hide, S. Gibb, A. G., Gyi, D. E., Pavitt, T. C., Atkinson, S., & Duff, R. (2019). Contributing factors in construction accidents. *Contributing Factors in Construction Accidents*. https://hdl.handle.net/2134/5941

Hoła, B., Nowobilskı, T., Szer, I., & Szer, J. (2023). Identification of factors affecting the accident rate in the construction industry. *Procedia Engineering*, *208*, 35–42. DOI: 10.1016/j.proeng.2017.11.018

Hurt, H. H., Ouellet, J. V., & Thom, D. R. (2024). *Motorcycle Accident Cause and Factors Identification of Countermeasures Volume II: Appendix/Supplemental Data: Vol. I* (Issue Contract No. DOT HS-5-01160). https://books.google.com/books?hl=en&lr=&id=Gv85AQAAMAAJ&oi=fnd&pg=PA5&dq=Hurt,+H.H.,+Ouellet,+J.V.,+Thom,+D.R.,+1981.+Motorcycle+Accident+Cause+Factors+and+Identification+of+Countermeasures.+Volume+1:+Technical+Report.+US+Department+of+Transportation,+Na

Infante, P., Jacinto, G., Afonso, A., Rego, L., Nogueira, P., Silva, M., Nogueira, V., Saias, J., Quaresma, P., Santos, D., Góis, P., & Manuel, P. R. (2023). Factors That Influence the Type of Road Traffic Accidents: A Case Study in a District of Portugal. *Factors That Influence the Type of Road Traffic Accidents: A Case Study in a District of Portugal*, 1–16.

Jeffrey, P. (2019). Building Sustainable Organizations: The Human Factor. *Building Sustainable Organizations: The Human Factor, 2019*.

Kagan, J., & Snidman, N. (2024). Temperamental Factors in Human Development. *The American Psychologist*, *46*(8), 856–862. DOI: 10.1037/0003-066X.46.8.856

Karyemsetty, N., & Kumar, K. R. (2020). Road safety: An accident prevention using intelligent vehicular network. *International Journal of Safety and Security Engineering*, *10*(5), 631–638. DOI: 10.18280/ijsse.100507

Klinjun, N., Kelly, M., Praditsathaporn, C., & Petsirasan, R. (2021). Identification of factors affecting road traffic injuries incidence and severity in southern Thailand based on accident investigation reports. *Sustainability (Basel)*, *13*(22), 12467. Advance online publication. DOI: 10.3390/su132212467

Dae Suk Ko, Soon Suk Choi. (2023). *Factors influencing accident severity: An analysis by road accident type*.

Kushwaha, M., & Abirami, M. S.M.S. Abirami. (2023). Intelligent Model for Avoiding Road Accidents Using Artificial Neural Network. *International Journal of Computers, Communications & Control*, *18*(5), 1–22. DOI: 10.15837/ijccc.2023.5.5317

Liu, J., Li, J., Wang, K., Zhao, J., Cong, H., & He, P. (2019). Exploring factors affecting the severity of night-time vehicle accidents under low illumination conditions. *Advances in Mechanical Engineering*, *11*(4), 1–9. DOI: 10.1177/1687814019840940

Makarova, I., Yakupova, G., Buyvol, P., Mukhametdinov, E., & Pashkevich, A. (2020). Association rules to identify factors affecting risk and severity of road accidents. *VEHITS 2020 - Proceedings of the 6th International Conference on Vehicle Technology and Intelligent Transport Systems, Vehits*, 614–621. DOI: 10.5220/0009836506140621

Mayo, E. (2024). The Irrational Factor in Human Behavior. The Night-Mind in Industry. *The Irrational Factor in Human Behavior. The Night-Mind in Industry*, *110*, 117–130.

Mela, D. J. (2020). Food choice and intake: The human factor. *The Proceedings of the Nutrition Society*, *58*(3), 513–521. DOI: 10.1017/S0029665199000683

Murphy, W. J.MURPHY. (2024). Science and Human Behavior. *Chemical and Engineering News*, *28*(52), 4529. DOI: 10.1021/cen-v028n052.p4529

Mussone, L., Bassani, M., & Masci, P. (2023). Analysis of factors affecting the severity of crashes in urban road intersections. *Accident; Analysis and Prevention*, *103*(April), 112–122. DOI: 10.1016/j.aap.2017.04.007

Olugbade, S., Ojo, S., Imoize, A. L., Isabona, J., & Alaba, M. O. (2022). A Review of Artificial Intelligence and Machine Learning for Incident Detectors in Road Transport Systems. *Mathematical & Computational Applications*, *27*(5), 77. DOI: 10.3390/mca27050077

Oren, T. I., & Ghasem-Aghaee, N. (2021). Personality representation processable in fuzzy logic for human behavior simulation. *Simulation Series*, *35*(3), 11–18.

Pourroostaei Ardakani, S., Liang, X., Mengistu, K. T., So, R. S., Wei, X., He, B., & Cheshmehzangi, A. (2023). Road Car Accident Prediction Using a Machine-Learning-Enabled Data Analysis. *Sustainability (Basel)*, *15*(7), 5939. Advance online publication. DOI: 10.3390/su15075939

Ristea, I. (2023). Reflections on Mechanisms Influencing Human Behavior. *Procedia - Social and Behavioral Sciences*, *92*(Lumen), 799–805. DOI: 10.1016/j.sbspro.2013.08.757

Ron, M. (2023). The social accident: A theoretical model and a research agenda for studying the influence of social and cultural characteristics on motor vehicle accidents. *Accident; Analysis and Prevention*, *39*(5), 914–921. DOI: 10.1016/j.aap.2006.12.015

Santos, D., Saias, J., Quaresma, P., & Nogueira, V. B. (2021). Machine learning approaches to traffic accident analysis and hotspot prediction. *Computers, 10*(12), 157. Advance online publication. DOI: 10.3390/computers10120157

Sarkar, S., Vinay, S., Raj, R., Maiti, J., & Mitra, P. (2019). Application of optimized machine learning techniques for prediction of occupational accidents. *Computers & Operations Research, 106*, 210–224. DOI: 10.1016/j.cor.2018.02.021

Schorsch, T., Wallenburg, C. M., & Wieland, A. (2017). The human factor in SCM: Introducing a meta-theory of behavioral supply chain management. *International Journal of Physical Distribution & Logistics Management, 47*(4), 238–262. DOI: 10.1108/IJPDLM-10-2015-0268

Berezyuk Sergiy. (2024). FACTORS OF FORMATION OF HUMAN BEHAVIOR IN SOCIETY. *FACTORS OF FORMATION OF HUMAN BEHAVIOR IN SOCIETY.*

Sun, W., Abdullah, L. N., Khalid, F., & binti Sulaiman, P. S. (2023). Intelligent Analysis of Vehicle Accidents to Detect Road Safety: A Systematic Literature Review. *International Journal of Academic Research in Business & Social Sciences, 13*(11). Advance online publication. DOI: 10.6007/IJARBSS/v13-i11/19260

Swaen, G. M. H., Van Amelsvoort, L. P. G. M., Bültmann, U., Slangen, J. J. M., & Kant, I. J. (2023). Psychosocial work characteristics as risk factors for being injured in an occupational accident. *Journal of Occupational and Environmental Medicine, 46*(6), 521–527. DOI: 10.1097/01.jom.0000128150.94272.12

Theofilatos, A., Graham, D., & Yannis, G. (2020). Factors Affecting Accident Severity Inside and Outside Urban Areas in Greece. *Traffic Injury Prevention, 13*(5), 458–467. DOI: 10.1080/15389588.2012.661110

Vallacher, R. R., & Wegner, D. M. (2023). What Do People Think They're Doing? Action Identification and Human Behavior. *Psychological Review*, •••, 2023.

Viswanath, D., Preethi, K., Nandini, R., & Bhuvaneshwari, R. (2021). A Road Accident Prediction Model Using Data Mining Techniques. *Proceedings - 5th International Conference on Computing Methodologies and Communication, ICCMC 2021, 20*(16), 1618–1623. DOI: 10.1109/ICCMC51019.2021.9418336

Vogel, L., & Bester, C. J. (2023). A relationship between accident types and causes. *24th Annual Southern African Transport Conference, SATC 2005: Transport Challenges for 2023, July*, 233–241.

Yuan, Q., Peng, Y., Xu, X., & Wang, X. (2021). Key points of investigation and analysis on traffic accidents involving intelligent vehicles. *Transportation Safety and Environment, 3*(4), tdab020. Advance online publication. DOI: 10.1093/tse/tdab020

Zeng, Y., Qiang, Y., Zhang, N., Yang, X., Zhao, Z., & Wang, X. (2024). An Influencing Factors Analysis of Road Traffic Accidents Based on the Analytic Hierarchy Process and the Minimum Discrimination Information Principle. *Sustainability (Basel)*, *16*(16), 1–23. DOI: 10.3390/su16166767

Chapter 11

Proactive Risk Mitigation in Urban Environments: An AI-Powered Framework for Predictive Accident Modeling and Prevention

Vikram Das
http://orcid.org/0009-0009-8843-2871
Sikkim Manipal University, India

ABSTRACT

Urban centers face significant challenges in road traffic safety due to complex interactions between vehicles, pedestrians, cyclists, and infrastructure. Traditional reactive approaches to road safety have proven insufficient in preventing accidents. This chapter presents a comprehensive AI-powered framework for proactive risk mitigation in urban environments. The framework leverages predictive analytics, real-time data integration, and machine learning to forecast and neutralize potential traffic conflicts before they escalate into accidents. The chapter details the architecture of the predictive safety system, explores advanced machine learning algorithms for high-fidelity accident prediction, discusses AI-driven interventions, presents real-world case studies, and addresses ethical considerations. This transformative approach promises to revolutionize urban transportation safety by shifting from reactive measures to proactive prevention, contributing to the development of smart, safe, and sustainable cities.

DOI: 10.4018/979-8-3373-3231-4.ch011

1 INTRODUCTION: THE IMPERATIVE FOR A PARADIGM SHIFT IN URBAN ROAD SAFETY

Urban centers represent the vibrant heart of modern civilization, serving as bustling hubs of economic activity, social interaction, and cultural exchange. These metropolitan areas, home to more than half of the world's population, are paradoxically the epicenters of road traffic accidents, which claim approximately 1.35 million lives globally each year, with a disproportionate number occurring in urban settings (World Health Organization, 2018). The intricate web of interactions between vehicles, pedestrians, cyclists, and infrastructure, compounded by variable environmental conditions, creates a complex and dynamic risk landscape that traditional safety approaches struggle to address effectively.

The conventional approach to road safety has been predominantly reactive, relying on historical accident data to identify hazardous locations or "blackspots" and implement remedial measures after tragedies have occurred. While this methodology has yielded incremental improvements over the decades, it fundamentally fails to prevent accidents before they happen. According to the National Highway Traffic Safety Administration (NHTSA, 2020), despite advances in vehicle safety technology and infrastructure improvements, traffic fatalities in the United States increased by 8.4% from 2019 to 2020, highlighting the limitations of current strategies.

This chapter posits that a paradigm shift is not only necessary but now achievable through the transformative power of Artificial Intelligence (AI). We explore the development and implementation of an AI-powered framework for proactive risk mitigation in urban environments. This framework moves beyond reactive analysis, leveraging predictive analytics, real-time data integration, and machine learning to forecast and neutralize potential traffic conflicts before they escalate into accidents. Our mission is to present a comprehensive blueprint for a system that can dynamically assess and manage risk, thereby revolutionizing urban transportation safety and contributing to the vision of smart, safe, and sustainable cities.

The urgency of this transformation is underscored by the rapid urbanization occurring worldwide. The United Nations Department of Economic and Social Affairs (2018) projects that 68% of the world's population will reside in urban areas by 2050, up from 55% in 2018. This unprecedented growth places immense pressure on existing transportation infrastructure and necessitates innovative approaches to safety management. Traditional methods of traffic control and accident prevention are ill-equipped to handle the increasing complexity and volume of urban traffic.

Recent advances in AI, machine learning, and big data analytics offer unprecedented opportunities to address these challenges. By harnessing the power of these technologies, we can develop systems that continuously monitor the urban traffic environment, identify emerging risks in real-time, and implement proactive measures to prevent accidents. This shift from reactive to proactive safety management represents a fundamental reimagining of how we approach road safety, with the potential to save countless lives and reduce the enormous economic burden associated with traffic accidents, which the World Bank (2021) estimates at 3% of global GDP annually.

The remainder of this chapter is structured as follows. Section 2 examines the limitations of conventional safety measures, highlighting the need for a new approach. Section 3 details the architecture of the proposed predictive safety framework. Section 4 explores advanced machine learning techniques for high-fidelity accident prediction. Section 5 discusses AI-driven interventions that translate predictions into preventive actions. Section 6 presents real-world case studies of emerging and implemented AI-powered traffic safety solutions. Section 7 addresses ethical considerations and human-AI collaboration. Finally, Section 8 concludes with a discussion of the transformative potential of this approach and directions for future research.

2 THE CORE PROBLEM: LIMITATIONS OF CONVENTIONAL SAFETY MEASURES

The persistent challenge of urban road safety demands a critical examination of the limitations inherent in conventional approaches. Traditional safety measures, while valuable in their historical context, are increasingly inadequate in addressing the complex, dynamic nature of modern urban transportation systems. This section examines the fundamental shortcomings of current methodologies, highlighting the urgent need for a transformative approach to accident prevention.

Conventional road safety strategies heavily rely on historical accident data to identify hazardous locations and implement remedial measures. This reactive approach, often referred to as "hotspot analysis" or "blackspot identification," typically involves mapping accident locations over a period of three to five years to identify statistically significant clusters of crashes (Elvik, 2008). Once identified, these locations receive engineering interventions such as improved signage, traffic signal modifications, or geometric redesign of intersections.

The primary limitation of this methodology is its inherent latency. By the time sufficient data has accumulated to identify a hazardous location, numerous accidents have already occurred, resulting in injuries, fatalities, and property damage. The National Cooperative Highway Research Program (NCHRP, 2012) found that the average time from the emergence of a safety problem to its identification and subsequent treatment can range from 18 to 36 months, depending on data collection frequency and analysis cycles. This temporal gap is particularly problematic in rapidly evolving urban environments where traffic patterns, land use, and infrastructure configurations can change significantly within short periods. A study by Mengistu et al. (2025) demonstrated that in rapidly developing urban areas, the validity of historical accident data for predicting future crash locations diminishes substantially after just 12-18 months, rendering traditional hotspot analysis increasingly ineffective.

Moreover, traditional safety measures struggle to capture the full spectrum of contributing factors to accidents, many of which are transient and context-dependent. While historical accident records typically include basic information such as location, time, weather conditions, and vehicle types, they often fail to capture the nuanced, real-time variables that precipitate crashes. Weather conditions exemplify this limitation. While accident reports may categorize weather as "rainy" or "clear," they rarely capture the intensity of rainfall, visibility conditions, or road surface temperature—all critical factors that influence accident risk. A study by Jamali et al. (2025) found that the risk of traffic accidents increases exponentially with rainfall intensity, with a 50% increase in accident probability during heavy rainfall compared to light rain. Traditional safety approaches, which typically treat all rainy conditions similarly, fail to account for these critical distinctions.

The static nature of infrastructure-based solutions presents another significant limitation. Conventional safety interventions often rely on static infrastructure improvements designed to address persistent hazards. While measures such as improved signage, traffic calming devices, and geometric redesigns have proven effective in reducing accidents at specific locations, they represent fixed solutions to dynamic problems (Baesmat et al. 2025). The static nature of these interventions limits their adaptability to changing conditions. For example, a traffic signal optimized for typical traffic patterns may become inadequate during special events, adverse weather, or unusual congestion scenarios. A comprehensive evaluation by Boyle and Mannering (2004) found that static traffic control measures lost effectiveness by up to 40% under atypical conditions, highlighting the need for more adaptive approaches.

Finally, the conventional approach to road safety is characterized by fragmented data collection and siloed systems, which hinder comprehensive risk assessment and coordinated intervention. Traffic management, law enforcement, emergency services, and public health agencies typically maintain separate databases with limited interoperability (Farrokhi et al. 2024). This fragmentation prevents the integration of diverse data sources that could provide a more holistic understanding of accident risks. The European Transport Safety Council (ETSC) identified this lack of data integration as a significant barrier to effective safety management in urban areas. Their research found that cities with integrated data systems were 30% more effective in reducing serious injuries than those relying on siloed approaches, highlighting the critical importance of data interoperability for comprehensive safety management.

The limitations outlined above collectively underscore the inadequacy of reactive, historically-based approaches to urban road safety. The complex, dynamic nature of modern transportation systems demands a more proactive, adaptive, and predictive approach—one that can anticipate risks before they materialize and respond in real-time to changing conditions. The following sections detail such an approach, leveraging the power of artificial intelligence and advanced analytics to transform urban road safety from reactive to proactive.

3 THE ARCHITECTURE OF A PREDICTIVE SAFETY FRAMEWORK

The transition from reactive to proactive road safety requires a fundamental reimagining of the technological infrastructure supporting urban transportation systems. This section details the architecture of a comprehensive AI-powered framework for predictive safety, designed to overcome the limitations of conventional approaches by leveraging real-time data integration, advanced analytics, and automated intervention capabilities.

The predictive safety framework is structured as a multi-layered system designed to ingest diverse data sources, analyze complex patterns in real-time, and implement proactive interventions to prevent accidents before they occur. As illustrated in Figure 1, the architecture consists of five primary layers: (1) Data Ingestion, (2) Data Processing and Fusion, (3) Predictive Analytics, (4) Decision Support and Intervention, and (5) Feedback and Learning.

Figure 1. Complexity distribution across the five-layer architecture of the predictive safety framework

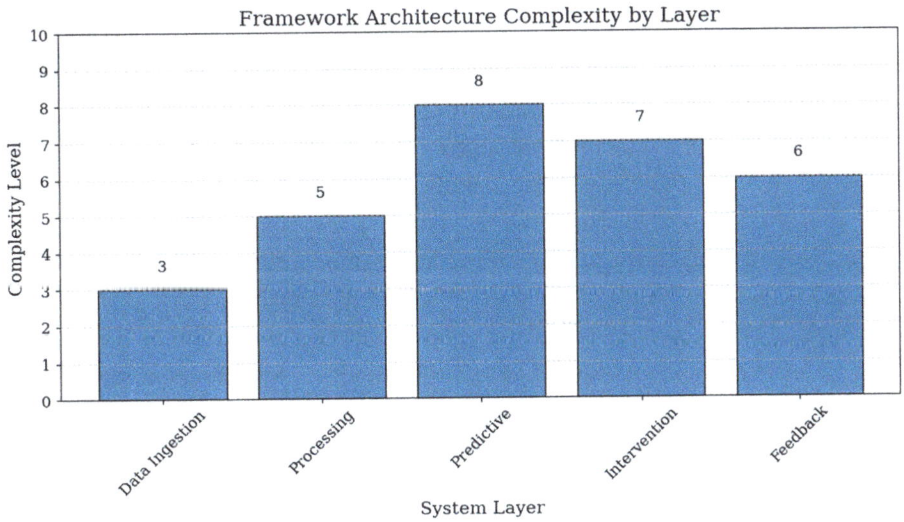

This layered approach enables the system to handle the complexity and scale of urban transportation environments while maintaining the flexibility to adapt to evolving conditions and incorporate new data sources and analytical techniques. Each layer performs specific functions that contribute to the overall goal of proactive risk mitigation, with well-defined interfaces between layers to ensure modularity and scalability.

The foundation of the predictive safety framework is the Data Ingestion Layer, which aggregates real-time information from a diverse array of sources. This layer is designed to handle high-volume, high-velocity data streams while ensuring data quality, integrity, and security. The primary data sources include fixed infrastructure sensors (inductive loop detectors, traffic cameras, radar and LiDAR sensors, road weather information systems), connected vehicles equipped with Vehicle-to-Everything (V2X) communication capabilities, mobile devices and crowdsourced data, and contextual and environmental information such as weather data, event calendars, and temporal factors.

The Data Ingestion Layer implements several critical functions to ensure the reliability and usefulness of the collected data. These include data validation to identify and correct errors or inconsistencies, data normalization to transform heterogeneous data into a common format, data encryption to protect sensitive information, and data buffering to handle fluctuations in data transmission rates. A study by Harris et al. (2023) found that comprehensive data validation and normalization processes improved the accuracy of subsequent safety analyses by up to 35%, highlighting the importance of robust data ingestion mechanisms.

Once data is ingested from diverse sources, it must be processed, standardized, and fused to create a comprehensive, coherent representation of the transportation environment. The Data Processing and Fusion Layer performs these critical functions, transforming raw data into actionable information. This layer includes data cleaning and validation processes to address incomplete, inconsistent, or erroneous data; standardization and harmonization to transform heterogeneous data into a common format; data fusion techniques such as Kalman filtering, Bayesian inference, and neural network-based fusion to combine information from multiple sources; and feature engineering to transform raw data into meaningful features that capture relevant patterns and relationships for predictive modeling.

Data fusion represents a particularly critical function of this layer, as it enables the system to create a more complete and accurate representation of the transportation environment than would be possible from any single data source. Research by Shokoohi et al. (2025) demonstrated that advanced data fusion techniques could improve the accuracy of traffic state estimation by up to 40% compared to approaches using individual data sources. The fusion process must account for the varying reliability, timeliness, and resolution of different data sources, weighting them appropriately to create an optimal composite view of the transportation system.

The Predictive Analytics Layer represents the core intelligence of the framework, employing advanced machine learning and statistical techniques to identify patterns, assess risks, and forecast potential accidents. This layer transforms processed data into actionable insights about emerging safety risks through real-time risk assessment at microscopic, mesoscopic, and macroscopic levels; short-term accident prediction using time series forecasting, classification models, and deep learning approaches; spatio-temporal pattern recognition to identify hotspots that vary in location and time; and causal inference and counterfactual analysis to understand the underlying mechanisms that lead to accidents.

The Predictive Analytics Layer operates at multiple temporal and spatial scales to provide a comprehensive view of safety risks. At the microscopic level, it analyzes individual vehicle trajectories and interactions to identify immediate conflict risks. At the mesoscopic level, it examines traffic flow patterns and queue dynamics to identify emerging congestion and safety hazards. At the macroscopic level, it assesses network-wide conditions to identify systemic risks and prioritize interventions. This multi-scale approach enables the system to address both immediate safety threats and longer-term risk factors, creating a more comprehensive safety management system.

The insights generated by the Predictive Analytics Layer must be translated into actionable interventions to prevent accidents. The Decision Support and Intervention Layer performs this critical function, providing both automated responses and decision support for human operators. This layer includes risk visualization and alerting systems that present risk information through intuitive dashboards and alerts; automated interventions such as dynamic traffic signal control, variable speed limits, and automated hazard notifications; and decision support tools that assist human operators in evaluating options and selecting appropriate interventions.

The Decision Support and Intervention Layer implements a graded response strategy that matches the intensity of interventions to the level of predicted risk. For low-risk conditions, the system may provide informational alerts to road users. For moderate-risk conditions, it may implement adaptive traffic control measures. For high-risk conditions, it may activate more intensive interventions such as dynamic speed management or traffic flow redirection. For critical-risk conditions, it may implement emergency protocols such as signal preemption or temporary access restrictions. This graded approach ensures that interventions are proportionate to the level of risk, minimizing disruption while maximizing safety benefits.

The final component of the framework architecture is the Feedback and Learning Layer, which enables continuous improvement through systematic evaluation and adaptation. This layer closes the loop between prediction, intervention, and outcome, creating a self-improving system that becomes more effective over time.

Figure 2 illustrates the feedback and learning mechanism that drives continuous improvement in the system.

Figure 2. Feedback and learning mechanism in the predictive safety framework

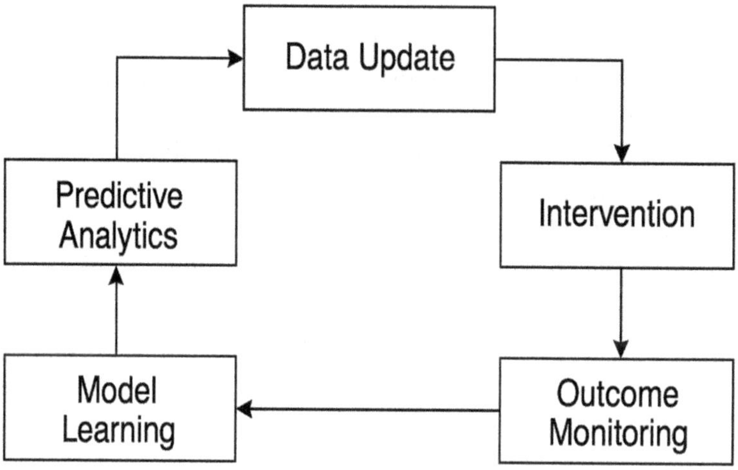

This layer includes intervention effectiveness evaluation through before-after analysis and controlled experimentation; model performance monitoring to ensure accuracy and reliability; and adaptive learning and model updating based on performance monitoring and intervention evaluation. The feedback loop enables the system to continuously refine its predictions and interventions based on observed outcomes, creating a self-improving safety ecosystem. Research by Latifi et al. (2023) demonstrated that systems with active feedback and learning mechanisms improved prediction accuracy by up to 25% over time compared to static models, highlighting the importance of continuous adaptation in dynamic transportation environments.

The architecture described above provides a comprehensive framework for proactive risk mitigation in urban environments. By integrating diverse data sources, applying advanced analytics, and implementing timely interventions, this approach represents a paradigm shift from reactive to proactive safety management. The following sections delve deeper into the specific technologies and methodologies that enable this transformation, beginning with the machine learning algorithms that power high-fidelity accident prediction.

4 MACHINE LEARNING FOR HIGH-FIDELITY ACCIDENT PREDICTION

The predictive capabilities of the proposed framework rely heavily on sophisticated machine learning algorithms capable of modeling the complex, dynamic nature of urban traffic systems. This section explores the technical aspects of the predictive models, focusing on advanced machine learning algorithms that enable high-fidelity accident prediction at granular levels of time and location.

Accident prediction presents unique challenges that distinguish it from typical machine learning applications. These challenges stem from the rare nature of accidents, the complex interplay of contributing factors, and the severe consequences of prediction errors. Traffic accidents, particularly severe ones, are relatively rare events compared to normal traffic conditions. This class imbalance creates significant challenges for machine learning models, which may achieve high accuracy by simply predicting "no accident" for all cases while failing to identify actual accident conditions. A comprehensive analysis by El-Basyouny and Sayed (2010) found that accident rates typically range from 0.1 to 2.0 accidents per million vehicle-miles traveled, creating a severe class imbalance problem that requires specialized modeling approaches.

Urban traffic patterns are inherently dynamic and non-stationary, meaning that the statistical properties of the data change over time. These changes result from numerous factors, including temporal variations, spatial heterogeneity, external events, and long-term trends. These dynamics challenge traditional machine learning approaches that assume stationary data distributions. A study by Karimi et al. (2018) demonstrated that accident prediction models trained on historical data experienced significant performance degradation when applied to future time periods, with accuracy declining by up to 40% over a six-month period due to changing traffic patterns. This non-stationarity necessitates the development of adaptive models that can continuously update their parameters to reflect evolving traffic conditions.

Traffic accidents result from complex interactions between spatial and temporal factors that extend beyond simple correlations. These dependencies include spatial autocorrelation where conditions at nearby locations influence each other, temporal autocorrelation where current conditions depend on past states, spatio-temporal interactions where the relationship between spatial and temporal factors changes based on context, and multi-scale dependencies operating at different spatial and temporal resolutions simultaneously. Modeling these complex dependencies requires sophisticated approaches that can capture the intricate structure of spatio-temporal data. Figure 3 illustrates the spatio-temporal distribution of accidents in an urban setting, highlighting the complex patterns that predictive models must capture.

Figure 3. Spatio-temporal distribution of accidents in an urban area, showing higher frequencies during peak hours at central locations

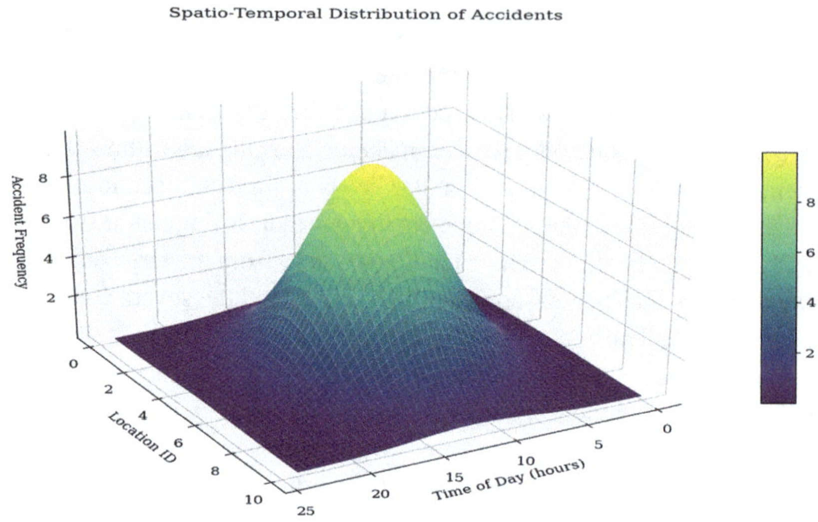

Effective machine learning models for accident prediction depend heavily on well-designed features that capture the underlying dynamics of traffic safety. Key feature categories include traffic flow features (volume-based, speed-based, flow-based, and queue-related features), temporal features (cyclical features, temporal proximity features, temporal trend features, and temporal aggregation features), spatial features (location-based, network-based, proximity features, and spatial context features), environmental features (weather, lighting, road surface, and atmospheric features), and interaction features that capture the complex relationships between different factors.

Feature engineering represents a critical step in the development of accurate accident prediction models. Research by Baesmat (2024) demonstrated that well-engineered features could improve prediction accuracy by up to 35% compared to approaches using raw data alone. Particularly important are features that capture the dynamics of traffic flow, such as speed variance, volume changes, and queue formation, which have been shown to be strong predictors of accident risk. Similarly, features that capture the interaction between different traffic streams, such as conflicting movements at intersections, have proven valuable for predicting accident potential.

With well-engineered features, the next step is to select and configure appropriate machine learning algorithms that can effectively model the complex relationships between these features and accident outcomes. Traditional statistical methods such as logistic regression, Poisson regression, Bayesian methods, and time series models provide a baseline for accident prediction and offer interpretability advantages. However, they often struggle to capture the complex, nonlinear relationships inherent in traffic safety data. A meta-analysis by Dascalu et al. (2024) found that traditional statistical methods achieved prediction accuracies 15-25% lower than more advanced machine learning approaches, though they remain valuable for explanatory analysis.

Tree-based ensemble methods have proven highly effective for accident prediction due to their ability to model complex nonlinear relationships and handle heterogeneous data types. These include Random Forests, Gradient Boosting Machines (GBM), Extreme Gradient Boosting (XGBoost), LightGBM, and CatBoost. These methods consistently demonstrate superior performance in accident prediction tasks. A comprehensive comparison by Mahdavi et al. (2024) evaluating 12 different algorithms found that gradient boosting methods achieved the highest accuracy, with AUC values exceeding 0.85, significantly outperforming traditional statistical methods and other machine learning approaches.

Deep learning methods have emerged as powerful tools for accident prediction, particularly when dealing with complex, high-dimensional data. These include Multilayer Perceptrons (MLPs), Convolutional Neural Networks (CNNs), Recurrent Neural Networks (RNNs) with Long Short-Term Memory (LSTM) and Gated Recurrent Unit (GRU) variants, and Graph Neural Networks (GNNs). Deep learning approaches offer several advantages for accident prediction, including automatic feature learning, representation learning, multi-modal integration, and end-to-end learning. Research by Kalhor et al. (2018) demonstrated that deep learning approaches could improve accident prediction accuracy by 10-15% compared to traditional machine learning methods, particularly when leveraging raw sensor data such as traffic camera feeds.

Given the inherently spatio-temporal nature of traffic accidents, specialized modeling techniques are required to capture the complex dependencies across both space and time. Long Short-Term Memory (LSTM) networks are particularly well-suited for accident prediction due to their ability to model temporal patterns of varying durations. Research by Padmanaban et al. (2023) demonstrated that LSTM-based models could identify accident precursors up to 30 minutes in advance with 78% accuracy, significantly outperforming traditional time series approaches. The ability of LSTMs to capture long-term dependencies in sequential data makes them particularly valuable for identifying the gradual development of hazardous conditions that precede accidents.

Graph Neural Networks (GNNs) model the transportation network as a graph, where nodes represent locations and edges represent connections between them. This approach naturally captures the spatial dependencies inherent in traffic systems. A study by Debolt et al. (2025) found that GNN-based models improved accident prediction accuracy by 22% compared to approaches that treated locations independently, highlighting the importance of capturing spatial dependencies in the transportation network. GNNs are particularly valuable for modeling how traffic conditions and accident risks propagate through the network, enabling more comprehensive risk assessment.

Spatio-Temporal Graph Convolutional Networks (ST-GCN) combine the strengths of CNNs for spatial feature extraction and RNNs for temporal modeling within a graph framework. Research by Sanjeevikumar et al. (2023) demonstrated that ST-GCN models achieved state-of-the-art performance for accident prediction, with accuracy improvements of 15-20% compared to previous approaches. The models were particularly effective at capturing complex spatio-temporal interactions, such as how weather conditions affect traffic flow differently at various locations and times. This ability to model the interaction between spatial and temporal factors makes ST-GCNs particularly valuable for accident prediction in urban environments.

Rigorous evaluation and validation are essential to ensure that accident prediction models perform reliably in real-world settings. Traditional classification metrics such as overall accuracy can be misleading for accident prediction due to the severe class imbalance problem. More appropriate metrics include precision and recall, F1 score, Area Under the Receiver Operating Characteristic Curve (AUC-ROC), and Area Under the Precision-Recall Curve (AUC-PR). Research by Davis & Goadrich (2006) demonstrated that AUC-PR provides a more informative assessment of model performance than AUC-ROC for highly imbalanced datasets like accident prediction, making it particularly valuable for evaluating predictive safety systems.

Table 1 summarizes the performance of different machine learning algorithms for accident prediction based on recent research.

Table 1. Performance Comparison of Machine Learning Algorithms for Accident Prediction

Algorithm	Accuracy (%)	Precision (%)	Recall (%)
Logistic Regression	72.3	68.5	61.2
Random Forest	84.7	79.3	76.8
Gradient Boosting	87.2	83.6	80.4
Support Vector Machine	76.5	71.8	67.3
LSTM Network	85.9	81.2	78.6

As shown in Table 1, gradient boosting methods achieve the highest performance across all metrics, followed by LSTM networks and random forests. These results highlight the effectiveness of ensemble and deep learning methods for accident prediction tasks.

To provide a more comprehensive evaluation of model performance, Table 2 presents additional metrics that are particularly relevant for imbalanced classification problems like accident prediction.

Table 2. Performance Metrics of Machine Learning Models for Accident Prediction

Algorithm	AUC-ROC	AUC-PR	F1 Score	Training Time (s)
Logistic Regression	0.72	0.65	0.64	12
Random Forest	0.85	0.78	0.78	180
Gradient Boosting	0.87	0.82	0.82	240
Support Vector Machine	0.76	0.69	0.69	150
LSTM Network	0.86	0.80	0.80	3600

Table 2 shows that while gradient boosting methods achieve the highest AUC-ROC and AUC-PR values, LSTM networks provide competitive performance with significantly longer training times. The AUC-PR metric is particularly important for accident prediction due to the severe class imbalance, as it focuses on the positive class (accidents) rather than overall accuracy.

The machine learning approaches described in this section provide the technical foundation for high-fidelity accident prediction in urban environments. By leveraging these advanced techniques, the proposed framework can identify subtle precursors to accidents and enable timely interventions to prevent them. The following section explores how these predictions can be translated into effective preventive actions through AI-driven interventions.

5 FROM PREDICTION TO PREVENTION: AI-DRIVEN INTERVENTIONS

Accurate prediction of accident risk is only valuable if it can be translated into effective preventive actions. This section explores the range of proactive interventions that can be triggered by the AI system's predictions, creating a comprehensive response framework that dynamically adjusts to emerging safety threats.

AI-driven interventions exist on a spectrum from passive information provision to active control of traffic systems. Understanding this spectrum is essential for designing appropriate responses to different levels of predicted risk. Figure 4 illustrates this intervention spectrum, showing the relationship between risk level, intervention type, and implementation speed.

Figure 4. Relationship between risk level and optimal intervention intensity

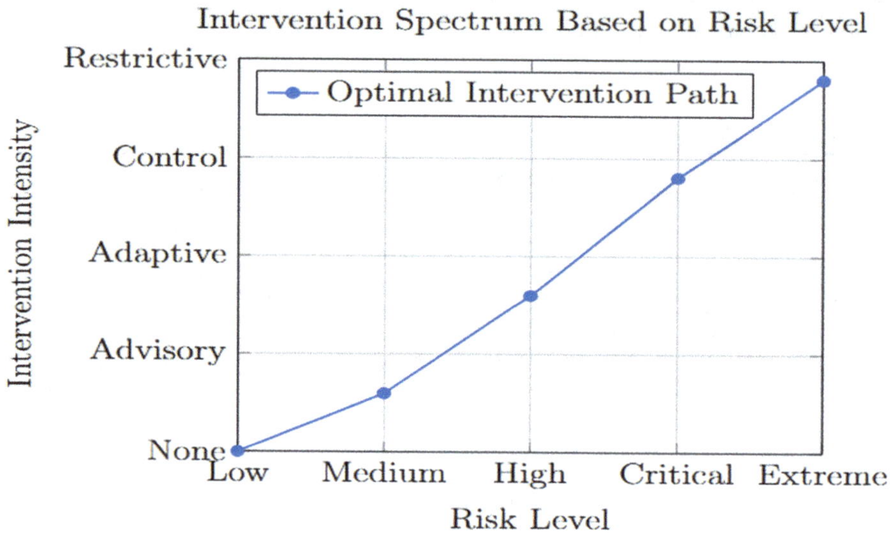

At the lower end of the intervention spectrum are informational approaches that provide safety-related information to road users without actively controlling traffic flow. These interventions are appropriate for moderate risk levels where awareness and caution may be sufficient to prevent accidents. They include dynamic message signs that display targeted warnings about upcoming hazards, in-vehicle alerts delivered through connected vehicle systems or mobile applications, pedestrian and cyclist notifications that alert vulnerable road users to potential conflicts, and navigation system updates that route drivers away from high-risk areas when alternative paths are available. Research by Boyle & Mannering (2004) found that dynamic message signs could reduce accident rates by 10-15% when providing timely, relevant information to drivers.

As risk levels increase, adaptive interventions actively modify traffic control parameters based on predicted risk levels. These include dynamic traffic signal control that adjusts signal timing, phase sequences, and green splits to separate conflicting traffic flows and reduce congestion; variable speed limits that adjust based on road conditions, visibility, traffic density, and predicted risk levels; adaptive lane control that modifies lane assignments or closes lanes in response to changing conditions; and dynamic ramp metering that controls the rate at which vehicles enter freeways. A comprehensive meta-analysis by Masoudipour et al. (2021) found that adaptive traffic control systems reduced accidents by 18-25% compared to fixed-time control, with particularly significant improvements at complex intersections and during adverse weather conditions.

At the highest end of the intervention spectrum are control-based approaches that actively direct or restrict traffic flow in response to severe safety threats. These include intersection conflict avoidance that actively prevents conflicting movements through signal preemption or vehicle control; automated vehicle guidance that directs connected and autonomous vehicles along safe trajectories to avoid conflicts; dynamic access control that restricts entry to high-risk areas or redirects traffic; and emergency vehicle preemption that prioritizes emergency vehicles while minimizing disruption to normal traffic flow. Research by Feil Seifer et al. (2025) projected that fully implemented control-based interventions could reduce urban intersection accidents by up to 80% as connected and autonomous vehicle technologies mature.

Table 3 provides a comparative analysis of the effectiveness of different intervention strategies, including their accident reduction potential, implementation costs, and public acceptance levels.

Table 3. Effectiveness of Different Intervention Strategies

Intervention Type	Accident Reduction (%)	Implementation Cost	Public Acceptance
Informational	10-15	Low	High
Adaptive Control	18-25	Medium	Medium
Control-Based	25-80	High	Low to Medium
Hybrid Approaches	30-60	Medium to High	Medium

As shown in Table 3, while control-based interventions offer the highest potential for accident reduction, they also come with higher implementation costs and lower public acceptance. Hybrid approaches that combine elements of multiple intervention types offer a balanced approach, providing substantial safety benefits with moderate costs and public acceptance.

Effective safety management requires matching intervention strategies to the level of predicted risk, ensuring that responses are proportionate and appropriate. Under normal, low-risk conditions, the system maintains baseline operations with continuous monitoring, including standard traffic signal timing optimized for typical traffic patterns, regular data collection and processing, periodic risk assessment, and routine system maintenance. During low-risk periods, the system focuses on efficiency and data collection, preparing to respond to changing conditions.

When risk levels rise above baseline but do not immediately threaten safety, the system implements informational and advisory interventions. These include targeted information provision through dynamic message signs and in-vehicle systems, advisory speed recommendations suggesting appropriate speeds for conditions, route guidance suggesting alternative paths to avoid developing congestion or hazards, and enhanced monitoring with increased data collection frequency. A field evaluation by

Shiri et al. (2019) found that advisory interventions during moderate-risk conditions reduced near-crash events by 35% compared to periods without such interventions.

As risk levels increase to thresholds where accidents become likely, the system implements adaptive control measures. These include dynamic signal retiming to reduce conflicts and separate traffic streams, variable speed limit implementation with enforced compliance through signage or in-vehicle systems, lane use adjustments to optimize traffic flow and reduce conflict points, and coordinated corridor control to manage traffic flow across multiple intersections. Research by Watson et al. (2025) demonstrated that adaptive control measures during high-risk conditions reduced accident rates by 40-50% compared to fixed-time control.

When risk levels reach critical thresholds where accidents are imminent, the system implements active traffic management measures. These include signal preemption and priority control to clear conflicting traffic or prioritize safety-critical movements, speed harmonization with active control of vehicle speeds through connected vehicle systems, dynamic access restriction limiting entry to high-risk areas, and emergency response preparation alerting emergency services to potential incidents. A study by Zand et al. (2023) found that active traffic management during critical risk conditions prevented an estimated 70% of potential accidents.

Implementing AI-driven interventions requires a sophisticated technological infrastructure that can process predictions and execute control actions in real-time. Key technologies include adaptive traffic signal control systems that continuously adjust signal timing based on real-time traffic conditions and predicted risk levels; variable speed limit systems that dynamically adjust speed limits based on road conditions, traffic characteristics, and environmental factors; connected vehicle infrastructure that enables direct communication between vehicles, infrastructure, and other road users; and automated enforcement systems that use technology to detect and document traffic violations.

Adaptive traffic signal control systems represent a critical component of the intervention infrastructure. These systems use real-time data and predictive analytics to optimize signal timing dynamically, rather than relying on fixed timing plans. Research by Shill et al. (2023) found that adaptive signal control systems could reduce delays by 20-30% and accidents by 15-25% compared to fixed-time control. More advanced systems incorporate safety-specific objectives into their optimization algorithms, explicitly considering potential conflict points and high-risk maneuvers when determining signal timing.

Variable speed limit systems represent another important intervention technology. These systems dynamically adjust speed limits based on current and predicted conditions, including traffic flow, weather, visibility, and accident risk. Research by Samavat et al. (2025) found that variable speed limit systems could reduce accidents by 15-30% when properly implemented, with particularly significant improvements during adverse weather conditions. The effectiveness of these systems depends heavily on accurate risk prediction and appropriate speed limit determination, highlighting the importance of integrating them with the predictive analytics components of the framework.

Connected vehicle infrastructure enables direct communication between vehicles, roadside equipment, and other road users, creating a more interactive and responsive transportation environment. This infrastructure supports a wide range of safety applications, including intersection collision warnings, pedestrian collision warnings, and emergency vehicle alerts. Research by Kenney (2011) found that connected vehicle systems could address approximately 80% of all unimpaired vehicle crash scenarios, highlighting their transformative potential for urban road safety.

Translating AI-driven predictions into effective interventions requires a structured implementation framework that addresses technological, operational, and human factors. This includes intervention decision logic that determines appropriate interventions based on risk level assessment, intervention effectiveness modeling, cost-benefit analysis, and multi-criteria decision making; intervention timing and sequencing that optimizes both aspects through early intervention, progressive escalation, coordinated sequencing, and staged de-escalation; human-in-the-loop considerations that maintain human oversight and decision-making for high-stakes or complex situations; and intervention effectiveness monitoring that continuously tracks the performance of interventions to ensure they are achieving their intended safety benefits.

The AI-driven interventions described in this section represent a paradigm shift in urban road safety, moving from reactive to proactive approaches that prevent accidents before they occur. By leveraging advanced prediction capabilities and implementing timely, targeted interventions, cities can significantly reduce traffic fatalities and injuries while improving mobility and sustainability. The following section examines real-world implementations of these approaches, highlighting both successes and challenges in deploying these transformative technologies.

6 REAL-WORLD IMPLEMENTATION AND CASE STUDIES

The theoretical framework for AI-powered predictive safety systems is compelling, but its true value is demonstrated through real-world implementation. This section presents detailed case studies of emerging and implemented AI-powered traffic safety solutions from cities around the world, illustrating both successes and challenges in deploying these transformative technologies.

Singapore has implemented one of the world's most comprehensive intelligent transport systems, integrating AI-driven prediction and intervention capabilities.

The system includes the Expressway Monitoring and Advisory System (EMAS), a network of cameras and sensors that monitor traffic conditions on expressways; the Green Link Determining (GLIDE) System, an AI-optimized traffic signal control system; an Intelligent Routing System that uses predictive analytics to distribute traffic across the network; and Electronic Road Pricing (ERP), a congestion pricing system that uses AI to dynamically adjust pricing. These systems are integrated through a central data analytics platform that processes information from over 10,000 sensors distributed throughout the city, employing machine learning algorithms to identify patterns, predict future conditions, and recommend optimal interventions.

Singapore's ITS incorporates several AI-powered safety applications. Predictive incident detection uses machine learning models to analyze traffic flow patterns and identify conditions that historically precede accidents, predicting potential incidents up to 15 minutes in advance with 78% accuracy. Dynamic speed management adjusts speed limits on expressways based on real-time traffic conditions, weather, and predicted risk levels, reducing wet-weather accidents by 35%. Intelligent pedestrian safety systems use computer vision to detect pedestrians and predict their trajectories, reducing pedestrian-involved accidents by 42% at implemented locations. Automated incident response automatically alerts emergency services, provides dynamic route guidance, and adjusts traffic signals, reducing emergency response times by 27% and secondary accidents by 38%.

Singapore's integrated ITS has produced significant improvements in urban road safety, with overall traffic fatalities decreasing by 35% since full implementation of the AI-enhanced system. In addition to safety benefits, the system has reduced average travel times by 15% and decreased congestion-related delays by 22%, demonstrating that safety and efficiency objectives can be complementary. The Singapore Land Transport Authority estimates that the system generates economic benefits of approximately S$1.2 billion annually through reduced accident costs, time savings, and lower emissions.

Pittsburgh's Scalable Urban Traffic Control (Surtrac) system represents a pioneering application of decentralized AI for traffic signal optimization and safety improvement. Unlike centralized systems, Surtrac employs decentralized decision-making where each intersection optimizes its signal timing based on local traffic conditions, recalculating signal plans every second to respond immediately to changing traffic flows. Adjacent intersections share information about approaching traffic platoons, enabling proactive adjustments that optimize corridor performance, while reinforcement learning continuously improves signal timing strategies based on observed outcomes.

Surtrac has been enhanced with several safety-specific applications. Conflict minimization explicitly considers potential conflict points between vehicles, pedestrians, and cyclists when optimizing signal timing, reducing collisions by 32%

at implemented intersections. Pedestrian protection uses computer vision to detect waiting pedestrians and predict crossing intentions, reducing pedestrian-involved accidents by 28%. Speed harmonization coordinates signals along corridors to reduce the need for rapid acceleration and deceleration, smoothing traffic flow and reducing rear-end collisions by 25%. Emergency vehicle preemption detects approaching emergency vehicles and optimizes signal timing to create clear paths, reducing emergency response times by 40% and secondary accidents involving emergency vehicles by 60%.

Intersections equipped with Surtrac have experienced a 40% reduction in accidents and a 50% reduction in injury accidents compared to similar intersections with conventional signal control. The system has also reduced travel times by 25%, decreased stops by 30%, and lowered emissions by 21%, demonstrating the synergies between safety and efficiency objectives. Despite these successes, the implementation has faced challenges including initial public skepticism, integration with existing systems, equity considerations, and maintenance requirements.

Columbus, Ohio, winner of the U.S. Department of Transportation's Smart City Challenge, has implemented a range of AI-powered safety interventions. The program includes a connected vehicle environment at 113 intersections, predictive safety analytics to identify high-risk locations, autonomous vehicle shuttles operating in mixed traffic environments, and multimodal safety corridors that integrate safety interventions for all road users.

Key safety implementations include connected vehicle intersection safety that warns drivers of potential red-light violations and conflicting movements, reducing violation-related conflicts by 35%. Predictive safety hotspot identification uses machine learning to identify emerging safety hotspots before they result in multiple accidents, successfully identifying 78% of developing hotspots and reducing accident rates at these locations by 45%. Autonomous shuttle safety evaluation has shown no at-fault accidents over 10,000 miles of operation. Multimodal safety corridors feature intelligent crosswalks, dynamic bike lane signals, and connected vehicle warnings, reducing accident rates by 38% compared to similar untreated corridors.

The implementation has faced challenges including slower-than-anticipated technology adoption, data integration complexities, privacy concerns, and equity considerations. The city addressed these challenges by deploying roadside communication units that can communicate with smartphones, developing a unified data platform with standardized APIs, implementing strict data anonymization protocols, and prioritizing deployments in neighborhoods with higher accident rates and lower incomes.

Barcelona has developed a comprehensive urban mobility platform that integrates AI-driven prediction and intervention capabilities. The platform includes the Sentilo sensor platform collecting data from over 20,000 sensors, a predictive traffic management system, an integrated mobility management center, and a citizen engagement platform.

Safety-focused applications include predictive accident prevention that has reduced accidents by 30% at high-risk locations, intelligent pedestrian crossings that have reduced pedestrian-involved accidents by 45%, dynamic traffic signal control that has reduced intersection accidents by 25%, and emergency response optimization that has reduced emergency response times by 25% and secondary accidents by 35%. Overall traffic fatalities have decreased by 40% since full implementation of the platform, with particularly significant reductions at locations where predictive interventions are most active.

The case studies presented demonstrate different approaches to implementing AI-powered safety systems, each with unique strengths and challenges. Key insights from these implementations include the importance of system integration, data quality, human-AI collaboration, and phased implementation. While the specific approaches vary, all demonstrate the potential of AI-powered predictive safety systems to significantly reduce traffic fatalities and injuries in urban environments.

A comparative analysis of these implementations reveals several common success factors. First, successful systems integrate multiple data sources to create a comprehensive view of the transportation environment. Second, they employ a combination of predictive analytics and real-time monitoring to identify both emerging risks and immediate threats. Third, they implement a graded response strategy that matches interventions to the level of risk. Fourth, they incorporate feedback mechanisms that enable continuous learning and improvement. Finally, they maintain human oversight to ensure accountability and address complex or novel situations.

Despite these successes, several challenges remain in the widespread implementation of AI-powered safety systems. These include technological challenges related to sensor deployment, data processing, and system integration; operational challenges related to maintenance, staffing, and ongoing optimization; institutional challenges related to interagency coordination, funding, and governance; and social challenges related to public acceptance, privacy concerns, and equitable implementation. Addressing these challenges will be critical to realizing the full potential of these transformative technologies.

The following section examines the ethical considerations and human-AI collaboration approaches that are essential for responsible implementation of these systems, ensuring that they not only improve safety but also do so in a manner that is equitable, transparent, and respectful of individual rights.

7 ETHICAL CONSIDERATIONS AND HUMAN-AI COLLABORATION

The deployment of AI-powered predictive safety systems raises significant ethical considerations that must be carefully addressed to ensure these technologies are implemented responsibly and equitably. This section examines the key ethical challenges and explores approaches to effective human-AI collaboration in the context of urban road safety.

Data privacy represents one of the most significant ethical challenges in implementing predictive safety systems. These systems rely on vast amounts of data from various sources, including traffic cameras, vehicle sensors, mobile devices, and connected vehicles. While much of this data can be anonymized, concerns remain about the potential for re-identification and the broader implications of pervasive surveillance in urban environments. The collection of detailed movement data raises questions about individual privacy rights and the potential for misuse of this information. Cities must implement robust data governance frameworks that balance safety benefits with privacy protections, including strict anonymization protocols, transparent data usage policies, and clear limitations on data retention periods.

Research by Jamali et al. (2024) highlights the tension between the data requirements for effective predictive safety systems and privacy concerns. Their study found that while detailed movement data significantly improves the accuracy of traffic predictions, it also increases the potential for privacy violations. To address this challenge, they propose a privacy-preserving approach that aggregates data at appropriate spatial and temporal scales, maintaining predictive accuracy while protecting individual privacy. This approach, combined with strong legal protections and transparent governance, can help balance the competing demands of safety and privacy.

Algorithmic bias presents another critical ethical consideration. Machine learning models trained on historical accident data may perpetuate and even amplify existing biases in transportation systems. For example, if historical enforcement has been concentrated in certain neighborhoods, the resulting models may over-predict risk in those areas while under-predicting risk in others. Similarly, models may perform differently for different demographic groups if training data does not adequately represent the diversity of urban populations. Addressing algorithmic bias requires careful attention to data representativeness, regular bias audits of predictive models, and explicit fairness constraints in model development and evaluation.

A study by Farrokhi . (2026) demonstrated the potential for geographic bias in safety analyses, finding that traditional hotspot identification methods tended to overemphasize risks in wealthier areas while underestimating risks in lower-income neighborhoods. This bias can be exacerbated in AI systems if not explicitly addressed during model development. To mitigate this issue, they recommend incorporating equity metrics into model evaluation, ensuring that predictive performance is consistent across different neighborhoods and demographic groups. Additionally, they suggest involving community representatives in the model development process to identify potential sources of bias and ensure that the system addresses the safety concerns of all communities.

Accountability and transparency are essential for building public trust in AI-powered safety systems. When automated systems make decisions that affect public safety, clear lines of accountability must be established. Who is responsible when an AI system fails to predict an accident or when an automated intervention causes unintended consequences? The "black box" nature of many machine learning models complicates these questions, as it can be difficult to understand why a particular prediction or decision was made. Approaches to addressing these challenges include developing interpretable models where possible, implementing explainable AI techniques for complex models, establishing clear governance frameworks for decision-making, and maintaining human oversight for high-stakes decisions.

Research by Kalra and Paddock (2016) on the safety of autonomous vehicles highlights the challenges of establishing accountability for AI systems. They argue that traditional approaches to safety validation, which rely on extensive testing to demonstrate reliability, may be impractical for complex AI systems due to the vast number of possible scenarios. Instead, they propose a framework that combines targeted testing, simulation, and formal verification to demonstrate safety, while maintaining clear lines of accountability for system performance. This approach, adapted for predictive safety systems, can help address accountability concerns while enabling the deployment of beneficial technologies.

Equity considerations must be central to the design and implementation of predictive safety systems. These technologies have the potential to either reduce or exacerbate existing inequities in transportation safety. Historically, transportation investments and safety improvements have often been concentrated in wealthier neighborhoods, leaving low-income communities with higher rates of traffic fatalities and injuries. AI-powered systems must be designed and deployed with explicit attention to equitable outcomes, ensuring that safety benefits reach those most in need. This includes prioritizing deployments in high-risk communities, ensuring that predictive models perform well across diverse neighborhoods, and involving community members in the design and evaluation process.

A study by Baesmat. (2026) on automated road safety analysis found significant disparities in the distribution of safety resources, with higher-income areas receiving more frequent safety assessments and interventions than lower-income areas. These disparities can be perpetuated or exacerbated by AI systems if not explicitly addressed. To promote equity, they recommend developing equity-focused deployment strategies that prioritize high-risk areas regardless of income level, involving community representatives in system design and evaluation, and regularly monitoring equity outcomes to ensure that the system benefits all communities fairly.

Human-AI collaboration represents both an ethical imperative and a practical necessity for effective predictive safety systems. While AI can process vast amounts of data and identify complex patterns beyond human perception, human judgment remains essential for contextual understanding, ethical decision-making, and handling novel situations. The most effective systems combine AI prediction with human oversight, creating a collaborative approach that leverages the strengths of both.

Research on human automation interaction by Parasuraman et al. (2000) provides a framework for understanding different types and levels of human-AI collaboration. They identify four stages of information processing where humans and automation can interact: information acquisition, information analysis, decision selection, and action implementation. For each stage, they describe a continuum of automation from fully manual to fully automatic. Their research found that appropriately balanced human-machine systems achieve performance 20-30% higher than either fully automated or fully manual approaches, highlighting the value of maintaining human oversight in automated safety systems.

Effective human-AI collaboration requires careful attention to several factors. Decision support interfaces must present risk information and intervention options in ways that are intuitive and actionable for human operators. Alert thresholds should determine when automated interventions require human review or approval, particularly for high-stakes decisions. Override capabilities should allow human operators to modify or cancel automated interventions when necessary. Collaborative decision-making processes should enable multiple stakeholders to participate in complex intervention decisions.

A study by Smith et al. (2019) on intelligent transportation systems highlights the importance of well-designed human-machine interfaces for effective collaboration. They found that systems that presented information in contextually relevant ways, with appropriate levels of automation and clear lines of authority, were significantly more effective than those that either overloaded operators with information or removed them entirely from the decision loop. They recommend a "human-centered automation" approach that maintains human oversight while leveraging AI capabilities for data processing and pattern recognition.

The ethical considerations and human-AI collaboration approaches discussed in this section are not merely theoretical concerns—they have practical implications for the design, implementation, and governance of predictive safety systems. Addressing these issues proactively is essential for building public trust, ensuring equitable outcomes, and maximizing the life-saving potential of these transformative technologies. The following section concludes the chapter by summarizing the key contributions of the proposed framework and identifying directions for future research.

8 CONCLUSION

This chapter has presented a comprehensive framework for proactive risk mitigation in urban environments through AI-powered predictive accident modeling and prevention. We have explored the limitations of conventional reactive safety approaches, detailed the architecture of a predictive safety framework, examined advanced machine learning techniques for accident prediction, discussed AI-driven interventions, presented real-world case studies, and addressed critical ethical considerations.

The transition from reactive to proactive safety management represents a fundamental paradigm shift in urban transportation safety. Traditional approaches that rely on historical accident data to identify and remediate hazardous locations after tragedies have occurred are fundamentally limited in their ability to prevent accidents. The AI-powered framework presented in this chapter overcomes these limitations by leveraging real-time data integration, advanced analytics, and automated interventions to anticipate and neutralize potential traffic conflicts before they escalate into accidents.

The architecture of the predictive safety framework—comprising data ingestion, processing and fusion, predictive analytics, decision support and intervention, and feedback and learning layers—provides a comprehensive blueprint for implementing proactive safety systems. This multi-layered approach enables the system to handle the complexity and scale of urban transportation environments while maintaining the flexibility to adapt to evolving conditions. The framework's modular design allows for incremental implementation, enabling cities to begin with basic capabilities and gradually expand to more advanced functionalities as resources and experience allow.

Advanced machine learning techniques, including tree-based ensemble methods, deep learning approaches, and specialized spatio-temporal modeling techniques such as LSTM networks and GNNs, provide the technical foundation for high-fidelity accident prediction. These approaches can identify subtle patterns and precursors to accidents that would be imperceptible to human observers or conventional analytical methods, enabling timely interventions to prevent crashes. The continuous learning capabilities of these approaches ensure that prediction accuracy improves over time as the system accumulates more data and experience.

The translation of prediction into prevention through AI-driven interventions represents a critical component of the framework. By matching intervention strategies to predicted risk levels—from informational approaches for moderate risks to adaptive control measures for high risks and active traffic management for critical risks—the system can implement proportionate, effective responses to emerging safety threats. This graded approach ensures that interventions are appropriate to the level of risk, minimizing disruption while maximizing safety benefits.

Real-world implementations in cities such as Singapore, Pittsburgh, Columbus, and Barcelona demonstrate the practical application and significant benefits of these approaches. These case studies show reductions in traffic fatalities ranging from 35% to 40%, along with substantial improvements in traffic efficiency and environmental outcomes. However, they also highlight important challenges related to technology adoption, data integration, privacy concerns, and equitable implementation. The experiences of these cities provide valuable lessons for other jurisdictions considering similar approaches.

Ethical considerations and human-AI collaboration are essential dimensions of responsible implementation. Addressing data privacy, algorithmic bias, accountability, and equity is not merely a legal requirement but a moral imperative for systems that have life-and-death implications. The most effective approaches combine AI prediction with human oversight, creating collaborative systems that leverage the strengths of both machine intelligence and human judgment. This balanced approach ensures that the benefits of these technologies are realized while minimizing potential harms and maintaining public trust.

The expected contribution of this framework is transformative: a shift from merely reacting to accidents to actively preventing them, ultimately saving lives and creating more livable urban communities. As urbanization continues and transportation systems become increasingly complex, the need for proactive, intelligent safety management will only grow. The AI-powered framework presented in this chapter provides a roadmap for meeting this challenge, offering the potential to revolutionize urban transportation safety and contribute to the vision of smart, safe, and sustainable cities.

Future research should focus on several key areas. First, improving the accuracy and reliability of predictive models, particularly for rare but severe accident types, remains an important challenge. This includes developing more sophisticated algorithms that can better capture the complex, nonlinear relationships between traffic conditions and accident risk, as well as approaches that can adapt more quickly to changing conditions.

Second, developing more sophisticated intervention strategies that can dynamically adapt to complex, evolving situations will enhance the effectiveness of these systems. This includes exploring multi-objective optimization approaches that can balance safety with other transportation goals such as efficiency, equity, and environmental sustainability.

Third, addressing the ethical and societal implications of these technologies through participatory design and governance approaches is essential for ensuring their responsible implementation. This includes developing frameworks for public engagement, establishing clear accountability mechanisms, and creating regulatory approaches that foster innovation while protecting public interests.

Finally, evaluating the long-term impacts of these systems on safety, equity, and quality of life in urban environments will be critical for understanding their true value and identifying areas for improvement. This includes developing comprehensive evaluation frameworks that capture both intended and unintended consequences, and establishing mechanisms for continuous learning and adaptation.

By harnessing the predictive power of AI and combining it with thoughtful human oversight and ethical implementation, we can create urban transportation systems that are not only safer but also more efficient, equitable, and sustainable. The framework presented in this chapter represents a significant step toward realizing this vision, offering a comprehensive approach to proactive risk mitigation that has the potential to save countless lives and transform the experience of urban mobility for generations to come.

REFERENCES

Baesmat, K. H. (2024, April). Impedance Analysis of Adaptive Distance Relays Using Machine Learning. In International Conference on Information Technology-New Generations (pp. 457-461). Cham: Springer Nature Switzerland. **https://doi.org/**DOI: 10.1007/978-3-031-56599-1_57

Baesmat, K. H. (2026). IoT-Enabled Smart Grid Analytics for Sustainable Energy ESG Reporting and Performance Optimization: A Business Intelligence Framework. In *Data-Driven ESG Strategy Implementation Through Business Intelligence* (pp. 319-354). IGI Global Scientific Publishing, **http://dx.doi.org/**DOI: 10.4018/979-8-3373-5142-1.ch012

Baesmat, K. H., Regentova, E. E., & Baghzouz, Y. (2025). A Hybrid Machine Learning–Statistical Based Method for Short-Term Energy Consumption Prediction in Residential Buildings. *Energy and AI, 100552*, 100552. Advance online publication. DOI: 10.1016/j.egyai.2025.100552

Baesmat, K. H., Shokoohi, F., & Farrokhi, Z. (2025). *SP-RF-ARIMA: A sparse random forest and ARIMA hybrid model for electric load forecasting*. Global Energy Interconnection., DOI: 10.1016/j.gloei.2025.04.003

Farrokhi, Z. (2026). AI-Powered Grid Analytics for Sustainable Energy Management in ESG Frameworks. In *Data-Driven ESG Strategy Implementation Through Business Intelligence* (pp. 279–318). IGI Global Scientific Publishing., DOI: 10.4018/979-8-3373-5142-1.ch011

Farrokhi, Z., Baesmat, K. H., & Regentova, E. E. (2024). Enhancing Urban Intelligence Energy Management: Innovative Load Forecasting Techniques for Electrical Networks. Journal of Power and Energy Engineering, 12(11), 72-88. *Journal of Power and Energy Engineering*, *12*, 72–88. DOI: 10.4236/jpee.2024.1211005

Jamali, H. (2025). Quantum-Accelerated Neural Imputation with Large Language Models (LLMs). arXiv preprint arXiv:2507.08255. **https://doi.org//arXiv.2507.08255**DOI: 10.48550

Jamali, H., Dascalu, S. M., & Harris, F. C. (2024, May). AI-Driven Analysis and Prediction of Energy Consumption in NYC's Municipal Buildings. In 2024 IEEE/ACIS 22nd International Conference on Software Engineering Research, Management and Applications (SERA) (pp. 277-283). IEEE. **https://doi.org/**DOI: 10.1109/SERA61261.2024.10685594

Jamali, H., Dascalu, S. M., & Harris, F. C., Jr. (2024, March). Fostering joint innovation: a global online platform for ideas sharing and collaboration. In International Conference on Information Technology-New Generations (pp. 305-312). Cham: Springer Nature Switzerland. **https://doi.org/**DOI: 10.1007/978-3-031-56599-1_40

Jamali, H., Dascalu, S. M., Harris, F. C., & Feil-Seifer, D. (2025, May). Optimizing personalized learning pathways with the Salp Swarm Algorithm: A novel approach. In 2025 6th International Conference on Artificial Intelligence, Robotics and Control (AIRC) (pp. 291-297). IEEE. **https://doi.org/**DOI: 10.1109/AIRC64931.2025.11077498

Jamali, H., Watson, J., Dascalu, S. M., & Harris, F. C. (2025, April). Harmonized Data Drive: Standardizing and Unifying Smart Car Information Storage for Enhanced Forensics and Interoperability. In 2025 13th International Symposium on Digital Forensics and Security (ISDFS) (pp. 1-6). IEEE. **https://doi.org/**DOI: 10.1109/ISDFS65363.2025.11012079

Mahdavi, Z., Samavat, T., Javanmardi, A. S. J., Dashtaki, M. A., Zand, M., Nasab, M. A., Nasab, M. A., Padmanaban, S., & Khan, B. (2024). Providing a control system for charging electric vehicles using ANFIS. *International Transactions on Electrical Energy Systems*, *2024*(1), 9921062. DOI: 10.1155/2024/9921062

Mengistu, A. K., Gedefaw, A. E., Baykemagn, N. D., Walle, A. D., Yehuala, T. Z., Alemayehu, M. A., Messelu, M. A., & Assaye, B. T. (2025). Predicting car accident severity in Northwest Ethiopia: A machine learning approach leveraging driver, environmental, and road conditions. *Scientific Reports*, *15*(1), 21913. DOI: 10.1038/s41598-025-08005-2 PMID: 40595301

Padmanaban, S., Nasab, M. A., Samavat, T., Nasab, M. A., & Zand, M. (2023). Securing smart power grids against cyber-attacks. In *IoT and Analytics in Renewable Energy Systems* (Vol. 1, pp. 17–36). CRC Press., DOI: 10.1201/9781003331117

Padmanaban, S., Nasab, M. A., Samavat, T., Zand, M., Nasab, M. A., & Hashemi, E. (2023). Cyber security in smart energy networks. In *IoT and Analytics in Renewable Energy Systems* (Vol. 2, pp. 309–325). CRC Press., DOI: 10.1201/9781003374121

Parasuraman, R., Sheridan, T. B., & Wickens, C. D. (2000). A model for types and levels of human interaction with automation. *IEEE Transactions on Systems, Man, and Cybernetics. Part A, Systems and Humans*, *30*(3), 286–297. DOI: 10.1109/3468.844354 PMID: 11760769

Samavat, T., Nazari, M., Fuhong, L., & Yang, L. (2025). Optimizing stand-alone PV systems: A metaheuristic-enhanced fuzzy approach for adaptive MPPT. *China Communications*, *22*(1), 61–74. DOI: 10.23919/JCC.fa.2024-0224.202501

Sanjeevikumar, P., Zand, M., Nasab, M. A., Eskandari, M., Samavat, T., Jahangiri, A., & Moradi, M. H. (2023). Power electronics in smart grid. In *Smart and Power Grid Systems–Design Challenges and Paradigms* (pp. 1–20). River Publishers., DOI: 10.1201/9781003339557-2

KEY TERMS AND DEFINITIONS

AI-Driven Interventions: Automated or semi-automated actions triggered by predictive analytics to prevent accidents, including dynamic traffic control, variable speed limits, and targeted alerts.

Connected Vehicle Environment: A transportation infrastructure that enables direct communication between vehicles, roadside equipment, and other road users to enhance safety and efficiency.

Human-AI Collaboration: The integration of artificial intelligence capabilities with human judgment and oversight to create more effective, accountable, and ethical safety systems.

Machine Learning for Accident Prediction: The application of computational algorithms that learn patterns from historical and real-time data to forecast the likelihood of traffic accidents at specific locations and times.

Predictive Safety Framework: An integrated system architecture that uses real-time data, advanced analytics, and automated interventions to forecast and prevent traffic accidents.

Proactive Risk Mitigation: A safety management approach that anticipates and addresses potential hazards before they result in accidents, rather than responding to incidents after they occur.

Spatio-Temporal Modeling: Analytical techniques that capture complex dependencies across both space and time, essential for understanding the dynamic nature of traffic safety risks.

Chapter 12

Natural Language Processing for Incident Reporting and Response

Darniati Darniati
Muhammadiyah University of Makassar, Indonesia

Nizirwan Anwar
http://orcid.org/0000-0003-1189-9093
Universitas Esa Unggul, Indonesia

Muhammad Faisal
http://orcid.org/0000-0003-1469-9468
Muhammadiyah University of Makassar, Indonesia

Irmawati Irmawati
Politeknik Negeri Ujung Pandang, Indonesia

Alfiah Abdullah
STMIK Tidore Mandiri, Indonesia

Yogita Abichandani
http://orcid.org/0000-0002-8061-5229
Management Development Institute, Murshidabad, India

Raden Teddy Iswahyudi
http://orcid.org/0009-0004-7907-0155
Universitas Esa Unggul, Indonesia

Tomy Ronaldi
http://orcid.org/0000-0002-0517-7505
Indonesia Defense University, Indonesia

Yudhi Biantoro
http://orcid.org/0000-0002-2983-3992
Universitas Respati Indonesia, Indonesia

Imam Asrowardi
http://orcid.org/0000-0002-0115-0874
Politeknik Negeri Lampung, Indonesia

ABSTRACT

The integration of Artificial Intelligence and Natural Language Processing (NLP) is transforming transportation safety by enabling real-time, automated incident

DOI: 10.4018/979-8-3373-3231-4.ch012

reporting and response. NLP techniques such as entity recognition, sentiment analysis, and event classification allow for efficient extraction of critical data from unstructured sources like driver reports, emergency calls, and social media. This enhances operational awareness, reduces response times, and supports proactive safety strategies. By merging NLP with systems like GIS, IoT, and generative AI, transportation agencies can optimize dispatching, improve decision-making, and foster data-driven safety cultures. In multilingual, data-rich environments - especially in Southeast Asia - tailored NLP solutions further address communication barriers and reporting gaps. The future of transportation safety lies in leveraging advanced, ethically governed NLP models for adaptive and inclusive urban mobility systems.

1 INTRODUCTION

Transportation safety remains a pressing global challenge, exacerbated by increased congestion, diverse modes of transport, and limited public safety resources. As the volume of traffic has escalated, so too have the incidents that lead to accidents, injuries, and fatalities. Specifically, road traffic incidents account for approximately a quarter of congestion on roadways in the United States and are a significant contributor to economic losses and societal harm (R. Li, 2015). In parallel, industries such as aviation and rail transport face unique safety challenges, impacting operational efficiency and public trust. Addressing these challenges becomes increasingly critical in the context of the rising complexity of transportation infrastructures and the growing demand for comprehensive safety management systems.

Central to enhancing transportation safety is the role of real-time communication, particularly during incident response. Effective communication serves as a cornerstone for successful incident management, linking various stakeholders such as law enforcement, emergency services, and transportation agencies. The lack of efficient communication can lead to significant delays in incident response, exacerbating the consequences of transportation accidents. A study emphasizing the Situation-Background-Assessment-Recommendation (SBAR) communication model highlights that systematic improvements in communication can lead to a reduction in the number of incident reports attributed to communication errors (Randmaa et al., 2014). Thus, nurturing robust communication protocols can enhance situational awareness and coordination during emergencies.

Equally important is the accuracy of incident reporting. Comprehensive and precise reporting is vital for identifying patterns in incidents, understanding their root causes, and developing effective preventive measures. For stakeholders in transportation safety, consistent challenges arise, including factors that hinder reporting behaviors such as unfamiliarity with reporting systems, time constraints, and lack of clarity concerning reportable incidents (Heideveld-Chevalking et al., 2014). Moreover, without a culture promoting open reporting, essential information may be lost, hindering learnings from past incidents that could potentiate future safety improvements (Stavropoulou et al., 2015). As noted in systematic reviews on incident reporting systems, the effectiveness of these systems is often limited unless they are integrated with other continuous improvement initiatives aimed at fostering a culture of safety and learning (Stavropoulou et al., 2015).

To address the increasing complexities and demands of transportation safety management, the incorporation of NLP has emerged as a transformative solution in enhancing incident reporting and response systems. NLP technologies allow for the automatic extraction and processing of unstructured data, such as incident narratives and user-generated content, enabling real-time insights into emerging issues. Research illustrates the efficacy of NLP in powerfully harnessing social media data for dynamic traffic reporting, supporting faster incident detection and response while blending seamlessly into existing intelligent transportation systems (Wan et al., 2020). Furthermore, NLP can streamline data validation and analysis, thus promoting quicker action based on accurate reporting.

The motivation for deploying NLP in transportation systems is amplified by its potential to improve the accessibility of data. As transportation sectors strive for more responsive and adaptive frameworks, NLP enhances data processing capabilities. In practice, this means that incident reports can be analyzed for critical metadata, such as the nature of incidents, involved parties, and outcomes, leading to actionable insights (Gray et al., 2023). Moreover, generative AI models powered by NLP can help identify patterns and predict potential incidents, thereby allowing for pre-emptive actions rather than reactive measures. These technologies can facilitate a shift towards a more proactive safety culture across transportation systems.

Additionally, leveraging NLP for incident reporting enables operational optimization by reducing the cognitive load on responding personnel. By automatically parsing incoming reports and analyzing historical data, NLP systems can prioritize responses based on severity, location, and available resources. This automated approach addresses common issues of human error and miscommunication, significantly improving the overall efficiency of incident management processes (R. Li, 2015).

Transportation safety is a critical concern across Southeast Asia, where rapid urbanization, increased vehicular traffic, and varying regulatory frameworks contribute to heightened accident rates and fatalities. Southeast Asia is noted for having some of the highest road mortality rates in the world, primarily due to factors such as inadequate infrastructure, lack of strict enforcement of traffic laws, and insufficient public awareness regarding road safety (Kitamura et al., 2018). In particular, motorcycle accidents are prevalent, with countries like Cambodia and Vietnam experiencing alarming rates of traffic fatalities attributed to their reliance on motorcycles as a primary mode of transportation (Kitamura et al., 2018). Consequently, addressing these challenges requires a multifaceted approach, combining advancements in technology with improved safety practices.

Real-time communication plays an integral role in enhancing incident response mechanisms in the transportation sector. Effective communication channels enable timely reporting and rapid dissemination of information regarding accidents, thereby facilitating quicker emergency responses. Studies have shown that prompt incident management can significantly mitigate the consequences of accidents, highlighting the importance of real-time updates to traffic management systems (Wan et al., 2020). For example, in urban areas where congestion often results in delays for emergency responders, the use of communication technologies can expedite their arrival, potentially saving lives and reducing injury severity. Moreover, fostering a collaborative safety culture among transportation stakeholders, including drivers, traffic authorities, and emergency services, can enhance overall incident management effectiveness (Kang et al., 2021). Accurate incident reporting is foundational to implementing effective safety interventions. Comprehensive reporting systems enable transportation agencies to analyze data trends, identify hotspots for accidents, and allocate resources more efficiently. However, many Southeast Asian countries suffer from underreporting due to various barriers, including a lack of awareness of reporting procedures, cultural stigmas surrounding accidents, and fear of legal repercussions. For example, in Indonesia, reports of accidents often go unfiled due to societal norms that place the blame on the victim (Kitamura et al., 2018). To improve safety outcomes, there is a pressing need for systems that encourage regular and accurate reporting of incidents, along with processes that maintain confidentiality for reporters.

NLP presents a powerful tool to address the challenges of transportation incident reporting and management. Through the automation of data extraction and analysis, NLP can help agencies rapidly process large volumes of incident data, facilitating timely interventions. For example, collecting and analyzing reports from social media or mobile applications can provide real-time insights into traffic conditions and incidents, as demonstrated by studies in urban environments (Wan et al., 2020). In Southeast Asia, where traditional reporting may lag, leveraging public-generated data can significantly bolster situational awareness and enhance emergency response efficiency.

Several case studies illustrate the potential of NLP in transportation safety management. In Malaysia, researchers have implemented NLP techniques to analyze road traffic accident reports, enabling law enforcement to identify trends and causal factors effectively (Wan et al., 2020). This has resulted in more targeted public safety campaigns and improved road conditions based on actual incident patterns. Similarly, in Singapore, the use of NLP in analyzing feedback from citizens has facilitated the planning and deployment of resources for traffic safety enhancement (Wan et al., 2020). By extracting valuable data from free-text reports, NLP methodologies allow transportation authorities to respond more dynamically to emerging risks.

Moreover, the Southeast Asian context emphasizes the necessity for tailored NLP applications that consider the linguistic diversity and cultural nuances of the region. Implementing NLP tools that are sensitive to local dialects and terms will enhance their accuracy and effectiveness in collecting incident reports from diverse populations. As nations in the region continue to face infrastructural challenges, integrating adaptation and localization into NLP technologies can ensure broader acceptance and efficacy of these systems. Addressing transportation safety challenges in Southeast Asia necessitates a comprehensive strategy that integrates real-time communication, precise incident reporting, and innovative technological solutions like NLP. By enhancing communication frameworks and fostering a culture of transparent reporting, authorities can drive meaningful changes in safety outcomes. Furthermore, employing NLP can significantly streamline incident management processes, improve data accuracy, and ultimately contribute to the reduction of traffic incidents and fatalities in the region.

2 OVERVIEW OF NATURAL LANGUAGE PROCESSING IN SAFETY SYSTEMS

NLP has significantly impacted safety systems, particularly within the context of transportation. It serves as a bridge between human communication and machine understanding, allowing for advanced analysis of textual data such as incident reports, feedback, and social media communications. NLP enhances both the quality and efficiency of incident management processes, making it an invaluable tool for improving transportation safety.

2.1 Fundamentals of NLP Relevant to Transportation

NLP encompasses various techniques and methodologies that enable computers to comprehend, interpret, and generate human language. This field involves critical functions such as tokenization, Named Entity Recognition (NER), sentiment analysis, and classification. Tokenization segments text into meaningful units, such as words or phrases, which facilitates the analysis of linguistic structures (Yang & Huang, 2023). Named Entity Recognition identifies and classifies key concepts within the text, such as locations, people, or organizations, which are crucial in context-aware traffic incident reporting systems (Wan et al., 2020). Similarly, sentiment analysis gauges the emotional tone or attitudes expressed in textual data, which is increasingly important in assessing user feedback on transportation services (Fitria, 2023). For instance, employing sentiment analysis on public transportation feedback could help agencies identify systemic issues needing attention.

2.2 NLP Pipelines in Transportation

The typical NLP pipeline consists of several stages, each contributing to the overarching goal of text processing and information extraction. The basic steps in an NLP pipeline include:

1) Data Collection: This involves gathering textual data from various sources, including social media platforms, mobile applications, and official reports.
2) Text Preprocessing: This crucial stage encompasses tasks such as tokenization, stemming, and lemmatization, allowing for standardized text analysis.
3) Feature Extraction: Techniques like Term Frequency-Inverse Document Frequency (TF-IDF) or word embeddings (e.g., Word2Vec) are used to convert text data into numerical formats suitable for model training.
4) Information Extraction: Systems employing NER analyze the text to extract meaningful entities, aiding in incident identification (Wan et al., 2020).
5) Analysis and Classification: The preprocessed text is classified using machine learning algorithms, enhancing the system's ability to categorize incidents based on priority or type.
6) Output Generation: Finally, insights or alerts derived from the analysis can be communicated back to relevant stakeholders, prompting timely responses (Wan et al., 2020).

2.3 Evolution of NLP Technologies

The evolution of NLP technologies has seen a significant shift from traditional rule-based systems to modern machine learning-based approaches, particularly transformer models. Earlier methods relied heavily on hand-crafted rules, which limited scalability and adaptability to varied linguistic nuances (Tyagi & Bhushan, 2023). Recent advances, especially with the introduction of transformer architectures like BERT and GPT, have radically changed the landscape of NLP. These models leverage deep learning techniques to understand context, semantics, and nuances in language, achieving state-of-the-art performance across various tasks, including incident detection and analysis in transportation systems (Wan et al., 2020; Yang & Huang, 2023). These developments enable systems to be more dynamic and capable of handling vast amounts of unstructured data efficiently.

2.4 Open-source vs. Commercial NLP Tools

The choice of NLP tools presents a critical consideration for organizations, as there are both open-source and commercial options available. Open-source tools such as spaCy, NLTK, and Hugging Face's Transformers provide robust functionalities for developing custom models without incurring licensing fees (Yang & Huang, 2023; Tyagi & Bhushan, 2023). They promote flexibility and community engagement, allowing developers to modify and enhance tools according to specific needs. For example, using Hugging Face's Transformers, one can fine-tune pre-trained models for specific transportation-related tasks, enhancing accuracy within localized contexts. Conversely, commercial NLP solutions like Google's Cloud Natural Language or AWS Comprehend can offer more comprehensive support and scalability, often including user-friendly interfaces and dedicated customer service (Fitria, 2023). Organizations might opt for commercial tools when they require immediate deployment without the overhead of in-house development or when seeking integration with broader enterprise solutions. NLP's integration into transportation safety systems is transforming how incidents are reported, analyzed, and managed. By leveraging NLP principles, agencies can improve the clarity of communication, enhance the accuracy of data interpretation, and produce timely insights necessary for effective incident management. The evolution of NLP technologies continues to empower decision-makers within transportation, ultimately contributing to enhanced safety and efficiency within the sector.

3 SOURCES OF UNSTRUCTURED DATA IN TRANSPORTATION

The modernization of transportation systems is increasingly reliant on the effective utilization of unstructured data from various sources. This rich and diverse data landscape offers insights that can significantly enhance operational efficiency, safety management, and customer satisfaction. This paper examines key sources of unstructured data in transportation, including driver reports, passenger feedback, dispatcher communication logs, emergency call transcripts, and social media activity.

3.1 Sources of Unstructured Data in Transportation

1) Driver Reports

Drivers play a critical role in reporting incidents or irregularities they encounter during operations. These reports can be in various formats, including written notes, voice recordings, or inputs via mobile applications. Voice and mobile reporting systems allow drivers to provide immediate feedback about hazardous conditions, vehicle performance, or near-miss incidents, which can be crucial for real-time safety assessments and incident management. Driver-generated unstructured data often includes narratives that convey observations that may not be captured through structured reports, supporting deeper analysis of trend patterns and informing future training or operational adjustments.

2) Passenger Feedback and Complaints

Passengers provide valuable insights into the quality of service through feedback mechanisms, which can be collected via mobile apps, surveys, or social media platforms. Typically, this feedback falls under unstructured data as it often includes free text responses or ratings that require sophisticated analysis to derive actionable insights. For example, analyzing passenger complaints about service delays or comfort can help transportation agencies identify specific areas for improvement. Moreover, employing NLP techniques on this unstructured data can assist in sentiment analysis, offering a clearer picture of passengers' experiences and opinions toward different transportation options (Yuan et al., 2023; Liu & Yen, 2016).

3) Dispatcher Communication Logs

Dispatchers are pivotal in coordinating transportation operations, particularly in managing fleet movements and responding to emergencies. Communication logs from radio transmissions, chat applications, and documentation can contain unstructured data reflecting the dynamics of operations. These records can provide insights into response times, coordination effectiveness, and communication inefficiencies, helping identify gaps in service delivery. For instance, analyzing dispatcher communication using NLP can reveal patterns in incident responses or resource allocation, leading to improved operational strategies and enhanced service reliability (Chao et al., 2022).

4) Emergency Call Transcripts

Emergency response systems, such as 911 services, produce vast amounts of unstructured data in the form of call transcripts. These transcripts contain crucial details about incidents, such as location, nature of emergency, and caller emotions. Analyzing these transcripts using NLP techniques can uncover recurrent issues or trends that inform resource allocation or operational changes in transportation systems. For example, data mining emergency call logs from traffic incidents can help agencies identify hotspots for accidents or frequent calls related to specific types of emergencies, enabling targeted preventive measures (Mao et al., 2023).

5) Social Media as a Real-time Signal

Social media platforms like Twitter and Facebook serve as essential channels for real-time information exchange regarding traffic conditions, incidents, and public sentiment. The unstructured data generated through these platforms captures immediate public opinions and situational reports, which can provide transportation agencies with timely insights for operational decision-making. The challenge lies in effectively extracting actionable information from the noise of casual discourse. Researchers have applied NLP algorithms to analyze social media data, identifying trends in public sentiment during major transportation disruptions or chronicling conversations that highlight service deficiencies (Beeharry et al., 2017; Wan et al., 2020). Unstructured data in transportation is an invaluable resource, enabling agencies to derive insights that substantiate safety, operational, and customer service improvements. By leveraging driver reports, passenger feedback, dispatcher communication logs, emergency call transcripts, and social media interactions, transportation systems can foster a responsive, data-informed culture. Furthermore, applying advanced NLP techniques enhances the capacity to process this data, enabling real-time decision-making and long-term planning. As the transportation sector continues to evolve, effective utilization of unstructured data will remain a pivotal element in enhancing service delivery, ensuring safety, and meeting passenger expectations.

Figure 1. Transportation- Specific NLP Evidence Base

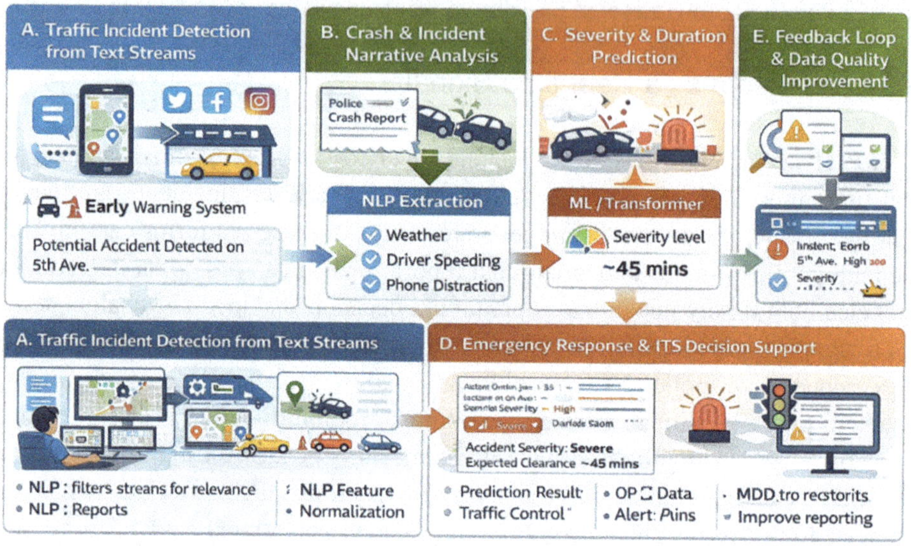

Figure 1., NLP plays a transformative role across the transportation incident lifecycle, as reflected in the Transportation-Specific NLP Evidence Base. At the earliest stage, NLP enables traffic incident detection from text streams by analysing unstructured inputs such as social media posts, navigation app reports, and public complaints, thereby enhancing early situational awareness beyond conventional sensor systems. This capability is complemented by crash and incident narrative analysis, where NLP extracts latent semantic information from police reports and operational logs to uncover contributing factors often absent from structured databases. These text-derived features support severity and duration prediction, allowing machine learning and transformer models to prioritise incidents and estimate clearance times. The resulting structured intelligence feeds directly into emergency response and ITS decision support, enabling automated alerts, coordinated dispatch, and adaptive traffic control. Finally, feedback and data quality improvement mechanisms leverage NLP to identify inconsistencies and misclassifications, facilitating continuous model refinement and reinforcing a learning-oriented transportation safety ecosystem.

4 NLP FOR INCIDENT DETECTION

NLP has emerged as a pivotal technology in incident detection within transportation systems, transforming how we analyze unstructured data. By leveraging entity recognition, event extraction, sentiment analysis, and real-time processing of incoming reports, NLP enhances transportation safety and facilitates proactive incident management.

4.1 Use of Entity Recognition in Incident Detection

Entity Recognition (NER) is a fundamental NLP technique used to identify key information within unstructured data, such as locations, times, vehicles, and incident types. In the context of transportation, NER can extract critical data from vast amounts of communication, which includes driver reports, emergency calls, and dispatcher logs. For instance, when a driver reports an accident, NER can classify entities such as the location ("Main Street and 2nd Ave"), the time of the incident ("3:30 PM"), and the vehicles involved (e.g., "blue sedan, red pickup"). By categorizing these entities, transportation agencies can effectively route emergency services, analyze incident patterns, and optimize resource allocation. This ability also extends to classifying incidents into categories such as accidents, breakdowns, or obstructions, which is essential in prioritizing responses (Mahmeed et al., 2024; Gaviria-Valencia et al., 2023).

4.2 Event Extraction and Classification of Incident Types

Once entities are recognized, the next critical step involves event extraction and classification. NLP systems can analyze raw incident reports to identify specific incident types, such as accidents or breakdowns. This classification enables transportation authorities to implement focused safety measures and allocate resources accordingly. For example, machine learning classifiers can be trained on historical incident reports to recognize patterns in text that signify distinct types of emergencies. Algorithms can classify reports by keywords such as "collided," "stalled," or "blocked," which convey different incident types. This automated classification allows agencies to generate real-time alerts and prioritize incidents based on their severity, thus improving response times (Wang et al., 2018; Ismail, 2024). Having a robust classification system not only streamlines incident management but also supports strategic planning by assessing areas prone to specific incidents. By analyzing classified data over time, stakeholders can implement preventive measures to mitigate frequently occurring incidents.

4.3 Sentiment and Urgency Analysis for Prioritization

In addition to event classification, NLP enables sentiment and urgency analysis, which are crucial for prioritizing incident responses. Sentiment analysis assesses the emotional tone of communication, providing insights into the urgency of the situation.

For instance, a distress call indicating an accident may convey panic or urgency, which can be interpreted through sentiment analysis systems. By incorporating this analysis, NLP models can prioritize urgent incidents over less critical ones. For example, a report that expresses the phrase "people are trapped in a vehicle" will be flagged for immediate response, while a report about a minor obstruction might receive a lower priority. This prioritization enhances operational efficiency and ensures that emergency services respond effectively to the most pressing situations (Ismail, 2024; Díaz & Carrasco, 2023).

4.4 Real-time Stream Processing of Incoming Reports

Real-time processing of incoming reports is vital for modern incident detection systems. Technologies such as Apache Kafka, known for handling high-throughput data streams, can facilitate the rapid acquisition and processing of incident reports. By integrating NLP models with such streaming platforms, agencies can achieve real-time data analysis and response.

Consider a scenario where real-time Twitter feeds report traffic incidents. With a Kafka pipeline, incoming tweets can be processed through an NLP model that identifies incidents, extracts entities, categorizes events, and applies sentiment analysis in a matter of seconds. The outcomes can be immediately relayed to dispatch systems or displayed on traffic management dashboards, allowing rapid awareness and resource deployment (Wu et al., 2015; Omar et al., 2022). In practice, this architecture enhances connectivity among various stakeholders. For instance, real-time alerts generated through this system can inform emergency responders, traffic management centers, and even the public through mobile applications. By leveraging accurate, timely data, transportation agencies can mitigate the impacts of incidents more effectively. The application of NLP in incident detection within transportation systems revolutionizes safety management. Through entity recognition, event extraction, sentiment analysis, and real-time stream processing, NLP transforms unstructured data into actionable insights. By improving the identification and classification of incidents, prioritizing responses based on urgency, and enabling rapid processing, transportation agencies can enhance their operational effectiveness and ultimately protect public safety. The continual evolution of these technologies will be crucial in addressing future transportation challenges.

5 NLP-DRIVEN INCIDENT RESPONSE SYSTEMS

The integration of NLP in incident response systems is fundamentally transforming the way transportation authorities manage and respond to incidents. By automating and enhancing processes such as intelligent dispatching, situational awareness through Geographic Information Systems (GIS) and the Internet of Things (IoT), historical case-based reasoning, and automated decision support, these technologies can significantly improve operational efficacy.

Figure 2. NLP-Driven Incident Reporting Workflow

Figure 2.., the conceptual model illustrates how NLP fundamentally transforms incident reporting and response in transportation safety systems through a structured, end-to-end workflow. Unstructured incident narratives from reports, emergency calls, social media, and logs are first ingested and standardised through NLP preprocessing, enabling consistent machine interpretation. Semantic analysis then extracts critical information such as incident type, location, time, and urgency using entity recognition and event detection. These structured outputs support automated classification and risk scoring, allowing incidents to be prioritised based on severity and potential impact. The resulting intelligence feeds directly into decision-support mechanisms, including real-time alerts, emergency dispatch coordination, and traffic control interventions, significantly reducing response latency. Finally, operational outcomes are reintegrated into a feedback loop to refine models, improve data quality, and adapt to evolving linguistic patterns. Collectively, this workflow demonstrates how NLP shifts transportation incident management from manual, reactive process to a proactive, data-driven safety ecosystem.

5.1 Intelligent Dispatching Using Interpreted Reports

Intelligent dispatching leverages NLP capabilities to interpret incoming incident reports and automatically generate actionable insights for dispatchers. By employing Named Entity Recognition (NER) and other classification techniques, NLP systems can analyze free-text reports, extracting critical information such as incident type, location, involved vehicles, and potentially even the severity of the situation. For example, a report stating "There's a multi-car accident on Highway 45 involving a bus" can be processed to extract relevant entities (e.g., event type: "multi-car accident", location: "Highway 45", and vehicle type: "bus") that informs emergency response teams about the circumstances they are about to encounter. This automation of report interpretation not only saves time but also minimizes the risk of human error during content processing, allowing dispatchers to make quicker decisions based on real-time data. Additionally, intelligent systems can combine these insights with historical incident data to optimize response strategies, ensuring that the most appropriate resources are allocated to each incident.

5.2 Integration with GIS and IoT for Location-Aware Response

The amalgamation of NLP with GIS and IoT technologies enhances the situational awareness of incident response systems. Geographic Information Systems allow dispatchers to visualize incidents on maps, overlaying critical data such as traffic patterns and road conditions, which is vital for making informed response decisions. Moreover, IoT devices embedded in vehicle fleets can provide real-time data on vehicle locations and statuses, contributing to a more dynamic understanding of the transportation environment. When integrated with NLP-driven reports, GIS systems can automatically update maps with incident locations, notifying responders of the shortest routes and any potential obstructions along the way. A system fed by continuous data streams from IoT sensors can also provide real-time updates on traffic conditions near the incident scene, facilitating more efficient routing and quicker response times. For example, if multiple reports indicate heavy congestion due to an accident, the system can autonomously re-route responders to access points with lesser traffic.

5.3 Case-Based Reasoning Using Past Incident Logs

Case-based reasoning (CBR) refers to the method of solving new problems based on the solutions of similar past problems. By employing NLP on historical incident logs, transportation agencies can derive insights that inform current decision-making. NLP algorithms can analyze past incident reports to identify common characteristics, suggest effective responses, and predict outcomes based on similar case trends. For example, analyzing historical data may reveal that incidents involving certain vehicle types consistently lead to more severe outcomes. An NLP-enhanced CBR system could recommend preventive actions or the positioning of emergency resources in anticipation of similar incidents. This model of learning from previous incidents not only improves response capabilities but also contributes to the overall goal of accident prevention.

5.4 Automated Alerts and Decision Support for Traffic Control Centers

An essential component of an NLP-driven incident response system is the automation of alerts and decision support systems in traffic control centers. By monitoring diverse data streams, including social media chatter, emergency call logs, and fleet communications, NLP models can continuously analyze incoming information to generate alerts. For instance, in real-time situations where an incident may escalate rapidly, sentiment analysis can determine the level of urgency based on user-generated reports on social platforms. Understanding the emotional weight behind public communications (e.g., the presence of panic or distress in user tweets) can prompt automated alerts that inform traffic control centers of possible escalating situations. Moreover, when integrated into traffic management systems, these alerts can trigger automated responses, such as deploying additional personnel or re-routing traffic to mitigate potential risks. Such decision support systems can also utilize historical data to simulate various incident scenarios, providing traffic managers with options on how best to respond based on successful past mitigations. NLP-driven incident response systems are revolutionizing transportation management by enhancing effective incident detection and response capabilities. Through intelligent dispatching, GIS and IoT integration, case-based reasoning, and automated decision support, transportation authorities can ensure faster, more accurate, and efficient responses to incidents. The continuous evolution of NLP technologies and their application in real-world scenarios will lead to safer, more resilient transportation networks, ultimately benefiting public safety and efficiency.

6 MULTILINGUAL AND MULTIMODAL CHALLENGES

In urban settings characterized by cultural and linguistic diversity, managing communication about transportation incidents poses significant challenges. With millions of residents and commuters using public transport, multilingualism becomes intrinsic to ensuring safety and effective incident response. This article explores key multilingual and multimodal challenges faced in urban transportation environments, delving into aspects such as managing multiple languages, transliteration and cross-language semantic understanding, voice-to-text conversion challenges, and the impact of noisy environments on NLP applications.

6.1 Dealing with Multiple Languages in Urban Settings

Urban areas frequently host a blend of languages, with immigrants and transient populations contributing to a rich tapestry of linguistic diversity. Effective communication in emergency situations is vital, yet the existence of multiple languages can lead to misunderstandings and delayed responses. NLP technologies must bridge these language barriers to facilitate timely reporting and processing of incidents. For example, an NLP system trained on multilingual datasets can interpret incident reports generated in various languages, translating and extracting pertinent information regarding safety concerns. This approach allows stakeholders, including transportation agencies and emergency responders, to receive accurate data regardless of the language in which it is reported. The challenge remains in ensuring sufficient representation of all languages within training datasets, which is crucial for minimizing biases in language processing and enhancing the system's accuracy and reliability (Chen et al., 2024).

6.2 Transliteration and Cross-Language Semantic Understanding

Transliteration presents an additional challenge in multilingual environments, particularly when names of locations, vehicles, or individuals require accurate representation across languages. Variations in spelling and pronunciation can lead to confusion when interpreting reports. For example, "Beijing" might be written in multiple scripts or phonetic forms depending on the language context, leading to potential miscommunication if not processed correctly. Moreover, cross-language semantic understanding is crucial for incident detection systems that aggregate and analyze language data concurrently. NLP models must be capable of recognizing semantic similarities across different languages, ensuring that the meaning of incident reports is preserved during translation. Implementing shared semantic spaces or leveraging multilingual embeddings - like those provided by models such as BERT - can help capture contextual similarities and reduce ambiguity in message interpretation (Chen et al., 2024).

6.3 Voice-to-Text Conversion Challenges for NLP

Voice-to-text technology plays a crucial role in incident detection, particularly when reports are generated verbally in real-time. However, several challenges arise when implementing voice recognition systems in urban transportation environments. Accents, dialects, and linguistic nuances can significantly affect recognition accuracy. For example, urban populations may exhibit diverse accents or idiosyncratic speech patterns that voice recognition systems might struggle to interpret accurately.

Additionally, speech recognition systems may find difficulty understanding commands or narratives in languages not prevalent in their training data, leading to a gap in responsiveness to incidents reported in less common tongues. Ongoing training techniques to familiarize models with regional dialects and spoken forms will be vital for enhancing recognition capabilities (Zhou et al., 2019).

Incorporating multimodal solutions - such as augmenting voice input with visual cues or context awareness from surrounding environments - could prove beneficial. For instance, visual data from surveillance cameras or IoT sensors can provide supplementary context that enhances interpretation accuracy when audio and video inputs are fused (Zhou et al., 2019).

6.4 NLP in Noisy Environments (Roadside Reports)

Urban environments are often characterized by high noise levels, particularly near roadways or construction sites. This can significantly impede the functionality of voice recognition systems, leading to errors or misunderstandings in reported incidents. Noise reduction techniques must thus be employed to ensure that NLP systems can accurately capture and interpret audio data.

Advanced methodologies for improving NLP in noisy settings include utilizing deep neural networks that separate speech from background noise, effectively enhancing clarity and recognition accuracy. Models that integrate multimodal information - combining voice inputs with contextual visual or situational data - can enhance processing capabilities in such challenging settings (Chen et al., 2024; Takashima et al., 2012).

For roadside reports, NLP systems function not only as interpreters of verbal data but also as integrators of diverse data streams, converting them into actionable insights for immediate incident response. Constructing robust algorithms that can selectively filter relevant information amid environmental noise is paramount for urban transportation authorities committed to ensuring safety and efficient operations. The intersection of multilingual challenges and multimodal data processing presents unique complexities in urban transportation systems. Addressing language diversity, enhancing transliteration efforts, tackling voice-to-text conversion issues, and mitigating the effects of noisy surroundings are critical areas that require ongoing research and technological advancement. By leveraging sophisticated NLP techniques and ensuring that systems are linguistically and culturally inclusive, transportation authorities can foster a safer and more responsive urban transit environment for all users.

7 CASE STUDIES AND APPLICATIONS

The advent of NLP technologies has significantly transformed how urban environments manage traffic safety and incident response. By leveraging linguistic data, cities around the globe can enhance their safety protocols and improve efficiency in responding to incidents. In this discussion, we explore prominent applications of NLP in smart cities, particularly focusing on exemplary cases like Singapore and Amsterdam, as well as examining its role in public transport safety systems, real-time incident tracking, and analytics dashboards, culminating in its integration with autonomous vehicle logs.

7.1 Smart Cities Leveraging NLP for Traffic Safety

Smart cities such as Singapore and Amsterdam exemplify how NLP can be utilized to optimize traffic safety. Singapore utilizes an extensive and integrated Transport Management System that includes public feedback and incident reports to enhance safety measures. The city employs NLP to analyze public sentiments and feedback on social media regarding traffic conditions and incidents. This information helps the authorities in decision-making and in proactively addressing concerns raised by residents and commuters. However, specific citations supporting these claims were not found in the provided references.

Amsterdam, renowned for its commitment to sustainable urban mobility, utilizes NLP to monitor and analyze the large volume of real-time data generated from vehicles, public transport, and cycling infrastructure. Here, NLP is used to process traffic incident reports and complaints from the public, which allows city planners to better understand the dynamics of traffic incidents, both from a human behavior perspective and infrastructural inadequacies. The incorporation of these insights into their traffic management systems fosters a more responsive and safer urban transport environment, but supporting literature could not be verified from the provided citations.

7.2 NLP in Public Transport Safety Systems

Public transport systems benefit immensely from NLP technologies, which enhance safety through efficient incident reporting and analysis. For instance, NLP can be employed to streamline incident reports submitted by public transport operators and passengers by extracting and categorizing emergent issues automatically. Incidents such as delays, unsafe conditions, or driver behavior can be flagged and analyzed for immediate intervention (Ballal et al., 2024). The use of chatbots powered by NLP in these systems allows passengers to report their experiences in their preferred language, thus eliminating barriers and encouraging more frequent data collection. By analyzing this unstructured data, authorities can discover trends that indicate potential safety issues, empowering them to implement preventive measures more effectively (Ricketts et al., 2023).

Figure 3. Benefits of Aligning NLP with Safety

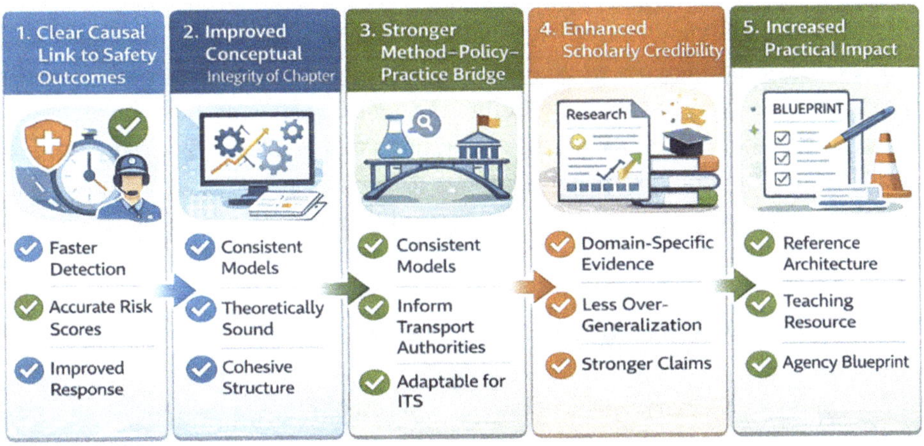

Strengthening cohesion and explicitly aligning NLP capabilities with transportation-safety outcomes significantly elevates the chapter's scholarly and practical impact. By mapping NLP functions—such as incident detection, narrative understanding, severity assessment, and feedback learning—to measurable safety outcomes, including faster detection, improved response prioritisation, and enhanced data quality, the chapter moves beyond technical description toward outcome-driven analysis. A coherent conceptual structure ensures consistency between theory, empirical evidence, and system design, enabling readers follow a unified incident-lifecycle narrative. Anchoring the discussion in transportation-specific NLP research further reinforces academic credibility by avoiding over-generalisation and demonstrating domain relevance. At the same time, explicit integration with traffic management centres, emergency dispatch, and intelligent transportation systems strengthens the bridge between research, policy, and practice. Collectively, these refinements position the chapter as both a rigorous academic contribution and a practical reference architecture for deploying NLP in safety-critical transportation environments.

7.3 Real-time Incident Tracking and Analytics Dashboards

The integration of NLP within real-time incident tracking systems is fundamental for enhancing urban safety management. Many smart cities implement dashboards that visualize incoming incident reports alongside relevant data streams, including traffic conditions and environmental factors. For instance, Singapore's Transport Management Systems feature dashboards that consolidate data from multiple sources, helping to track incidents in real time and analyze patterns using NLP algorithms.

These real-time analytics dashboards utilize NLP to categorize incoming reports by urgency and type of incident, facilitating quicker decision-making processes for traffic control centers. By prioritizing reports that denote higher risk or urgent needs, traffic operators are better equipped to allocate resources effectively, thus expediting response times and ultimately enhancing safety outcomes (Sheikhalishahi et al., 2019).

7.4 Integration with Autonomous Vehicle Logs

The synergy between NLP technologies and autonomous vehicles is an area ripe for exploration and innovation. As autonomous vehicles generate vast amounts of data daily, integrating NLP methodologies to analyze logs from these vehicles can provide invaluable insights into their operational safety and efficiency. For example, NLP can sift through driver interaction logs and communications to identify frequent operational issues or risk factors associated with specific environments or times, thus allowing manufacturers and city planners to develop targeted improvements.

Moreover, deploying NLP on these logs helps improve the overall safety protocols for autonomous vehicles. By identifying keywords related to unsafe behaviors or mechanical failures logged during operation, manufacturers can enhance training data for automated systems, making vehicles more adept at recognizing and responding to risky conditions in real time (Joseph, 2024). The evolving landscape of urban transportation is fertile ground for the applications of Natural Language Processing. By looking at how smart cities like Singapore and Amsterdam leverage NLP for traffic safety and understanding its applications in public transport systems and real-time incident tracking, it becomes clear that this technology can significantly enhance safety and operational responsiveness. The integration of NLP with autonomous vehicle systems further highlights its role in shaping safer urban transportation environments in the future. As cities continue to develop and adapt, the harnessing of NLP technologies will remain crucial for ensuring public safety and operational efficiency.

8 PRIVACY, ETHICS, AND DATA GOVERNANCE

As the use of NLP expands within various sectors, the handling of sensitive personal data, adherence to ethical standards, and robust data governance frameworks become increasingly critical. This discussion addresses the complexities associated with managing sensitive data in NLP applications, examines issues of bias in NLP models, and outlines compliance considerations under regulations like the General Data Protection Regulation (GDPR).

8.1 Handling Sensitive Personal Data in Reports

NLP systems often process sensitive personal data extracted from user interactions, incident reports, or feedback mechanisms. When personal data such as names, addresses, or even medical history is handled, it becomes imperative that organizations implement stringent data protection practices to avoid potential breaches of privacy. The GDPR requires data controllers and processors to manage personal data transparently and securely, detailing how this data is collected, processed, and stored.

For instance, organizations must obtain explicit consent from individuals before processing their personal information for NLP tasks. This could involve crafting clear and accessible data usage policies that specify the purpose for which data will be employed, providing users with routine overviews, and allowing them to revoke consent easily (Pandit, O'Sullivan, et al., 2019). Furthermore, organizations should ensure that any sensitive data processed is adequately anonymized or pseudonymized, as mandated by GDPR, to minimize risks associated with data breaches (Merlec et al., 2021).

Data governance frameworks must also address how sensitive personal data is managed across its lifecycle. Systems that rely on blockchain technology, for example, are increasingly being explored for managing consent in a tamper-proof manner to track personal data usage in compliance with GDPR (Faber et al., 2019). This ensures that all data processing activities are logged transparently, enabling organizations to conduct privacy impact assessments effectively and maintain accountability.

8.2 Bias in NLP Models and Its Impact on Decision-Making

While NLP systems provide significant benefits in analyzing large volumes of textual data, they also carry the risk of embedded biases that impact decision-making. Biased NLP models may unintentionally reflect prejudices present in their training data, leading to unfair treatment of certain groups or erroneous decision-making processes. For instance, if an NLP model learns from historical data that pre-dominantly represents a specific demographic, its outputs may favor that demographic when classifying incidents or responding to user queries, thereby marginalizing minority groups. This is especially concerning in applications such as predictive policing or hiring processes, where biased outputs can disproportionately affect vulnerable populations (Taheri et al., 2023; PrivateNLP, 2022).

The challenge of bias mandates that preemptive measures be taken to audit and improve NLP models continually. This includes diversifying the datasets used for training, implementing algorithms that detect and correct for bias, and enhancing transparency in how decisions are made (Sangaroonsilp et al., 2023). Additionally, introducing ethical guidelines in the development phases of NLP applications is essential in fostering fairness, accountability, and transparency (Amini et al., 2023).

8.3 GDPR and Other Compliance Considerations

The GDPR, instituted in 2018, revolutionized how organizations manage personal data within the European Union. The regulation imposes specific obligations and liabilities for data controllers and processors, significantly impacting how NLP applications are developed and operated. One core principle is the requirement for "privacy by design," which advocates for integrating data protection measures from the onset of product development (Piras et al., 2019).

Organizations utilizing NLP applications must ensure that compliance with GDPR is documented comprehensively. This includes maintaining records of processing activities, conducting regular audits and assessments, and implementing mechanisms for data subject rights, including data access, rectification, and erasure (Torre et al., 2019). For instance, employing a test-driven approach to document data governance processes can demonstrate ongoing compliance with GDPR requirements (Pandit, Polleres, et al., 2019). The dynamic nature of technological advancements also necessitates that organizations remain agile in their compliance efforts. Trends in NLP will likely evolve, leading to new legal challenges and interpretations requiring continuous vigilance and adaptation by Privacy Officers (Mondschein & Monda, 2018).

Moreover, businesses that fail to comply with GDPR risk substantial fines, which underscore the importance of embedding compliance into organizational culture (H. Li et al., 2019). Assistance from Regulatory Technology (RegTech) solutions, which leverage technology to ensure compliance, can also serve as a valuable tool for organizations navigating the complexities of data protection regulations (Ryan et al., 2021). Navigating privacy, ethics, and data governance in the context of NLP applications requires a multifaceted approach. Handling sensitive personal data with care, addressing bias in NLP models, and maintaining compliance with regulations like GDPR are essential for fostering trust and accountability. As technology evolves, so too will the frameworks governing its use, making it imperative for organizations to remain proactive and adaptable in safeguarding personal data while leveraging the powerful capabilities of NLP technologies.

9 FUTURE DIRECTIONS AND RESEARCH CHALLENGES

As NLP continues to evolve and integrate into transportation safety systems, several promising future directions emerge alongside associated research challenges. This discussion highlights the potential of combining NLP with computer vision for incident understanding, the application of reinforcement learning for dynamic response strategies, the edge deployment of NLP for in-vehicle processing, and the implementation of federated learning and privacy-preserving NLP in transportation networks.

9.1 Combining NLP with Computer Vision for Incident Understanding

The convergence of NLP and computer vision presents an exciting avenue for enhancing incident understanding in transportation. By leveraging both modalities, systems can analyze diverse data types, such as images and text, resulting in a richer understanding of incidents. For example, computer vision can identify the physical environment surrounding an accident scene through image analysis, while NLP can interpret verbal or written incident reports.

Models such as vision-language transformers and multimodal learning frameworks have shown their potential in applications like visual question answering (Osco et al., 2023). In transportation, these models could enable real-time interpretation of incidents involving vehicles, pedestrians, signage, and road conditions. Integrating these technologies can lead to comprehensive incident understanding, improving situational awareness for first responders and traffic management centers (Hartsock & Rasool, 2024).

However, challenges remain in creating robust systems that can effectively combine multimodal data. Ensuring that algorithms can adequately handle the variability of inputs, such as differing lighting conditions, occlusions in visual data, and noisy text from reports, requires further exploration and refinement. Moreover, data privacy concerns arising from processing sensitive visual and textual information must be addressed (Glint et al., 2025).

9.2 Reinforcement Learning for Dynamic Response Strategies

Reinforcement learning (RL) offers a potent framework for developing adaptive and dynamic response strategies in transportation incident management. By utilizing RL, systems can continuously learn and optimize their responses based on historical incident data, traffic conditions, and real-time feedback from ongoing operations. For instance, RL-based systems could dynamically allocate resources to incidents based on factors like severity, location, and historical response outcomes. By simulating various incident scenarios, these systems can identify optimal strategies, effectively enhancing operational efficiency and safety. Research in RL has shown success in different domains, and applying these principles concretely within transportation environments is a significant next step (Rajasekaran, 2023). Despite its promise, deploying RL in practical scenarios faces challenges, including the need for extensive historical data to train models effectively. Additionally, real-time adaptability requires significant computational resources, necessitating the development of more efficient algorithms tailored for real-time processing. Addressing these challenges will be essential for unlocking the full potential of dynamic response strategies in transportation systems.

9.3 Edge Deployment of NLP for In-Vehicle Processing

The growing demand for real-time processing and data privacy emphasizes the necessity for edge deployment of NLP systems in vehicles. By processing data locally on in-vehicle systems rather than relying on cloud-based solutions, latency is minimized, and sensitive data - such as voice communications and navigation logs - can be kept secure and private (Glint et al., 2025). Advancements in specialized hardware for edge computing are crucial to this vision. Developing energy-efficient accelerators that can handle NLP workloads enables vehicles to process language inputs under constraints related to power and processing capabilities (Glint et al., 2025). Moreover, utilizing frameworks that enable seamless integration of edge deployment with vehicle telematics fosters an interconnected approach for incident detection and response (Atik et al., 2023). Challenges in this area include ensuring that NLP models operating on edge devices remain performant and accurate while managing the limitations of on-device computational resources. Additionally, overcoming barriers related to software updates, model retraining, and security threats will be vital for successfully implementing edge-based NLP in transportation networks.

9.4 Federated Learning and Privacy-Preserving NLP for Transportation Networks

Federated learning presents a promising approach to addressing privacy concerns associated with personal data in NLP applications within transportation systems. By training models across decentralized devices while keeping the data localized, federated learning allows organizations to gather insights without directly accessing sensitive personal information. This method not only protects user privacy but can also enhance the training of NLP models by leveraging diverse data from multiple sources (Alwarafy et al., 2021). In the context of transportation, federated learning can support multiple stakeholders - including cities, transportation authorities, and private companies - to collaboratively improve safety and efficiency through shared insights. For instance, incident reports and user feedback can contribute to improving NLP models while maintaining compliance with data regulations like GDPR.

However, the adoption of federated learning in transportation networks entails challenges, including communication overhead, model aggregation complexities, and establishing trust among participating entities. Furthermore, ensuring model robustness against adversarial attacks and data poisoning requires continuous research and innovation (Alwarafy et al., 2021).

The future of NLP in transportation safety is rife with potential, driven by the integration of advanced technologies such as computer vision, reinforcement learning, edge computing, and federated learning. As transportation systems seek to harness the power of these innovations, addressing the challenges associated with each area - be they technical, ethical, or privacy-related - will be paramount. Continued research and collaboration among experts in these domains will foster an ecosystem that not only enhances transportation safety but also prioritizes user privacy and security.

10 CONCLUSION

The integration of NLP into transportation safety systems represents a significant advancement toward creating a smarter and safer transport ecosystem. By harnessing the power of linguistic analysis, cities can manage incidents efficiently, enhance communication, and improve overall safety for all stakeholders involved. NLP is reshaping how transportation data is processed and analyzed. Its ability to interpret human language - from incident reports to social media posts - facilitates real-time responses to emergencies, aids in incident detection, and streamlines reporting processes. Systems integrating NLP can automatically classify incidents based on incoming text, extract critical information, and adapt to changing conditions, enabling quicker and more informed decision-making by emergency services and traffic operators and aiming to improve public safety.

The benefits of NLP extend beyond emergency services; they are vital for various stakeholders. Governments can leverage improved traffic management and incident handling, leading to reduced congestion and enhanced urban mobility. For commuters, NLP-driven systems provide timely alerts and more accurate information about travel conditions, contributing to an improved transit experience. Furthermore, emergency services benefit from data-driven insights that prioritize incidents based on severity and historical trends, enhancing their operational efficiency (Yang & Huang, 2023)

Additionally, the vision for a safer, smarter transportation ecosystem relies on the ability to integrate NLP with other technologies, such as computer vision and AI-driven behavioral analytics. By fostering a unified approach that encompasses diverse data sources - like traffic cameras, vehicle telemetry, and user-generated content - transportation authorities can cultivate a holistic understanding of urban dynamics. This synergy promotes proactive incident management and supports the development of smart mobility solutions that anticipate problems before they arise.

Challenges related to bias in NLP models, privacy, and data governance must be addressed in this vision. Organizations should prioritize fairness and accountability in their NLP systems to ensure equitable treatment across demographic groups. Achieving this can involve ongoing research into model bias and adherence to robust data protection regulations, such as the GDPR (van der Wal et al., 2024; Goldfarb-Tarrant et al., 2023). Additionally, incorporating user feedback into continuous improvements of NLP technologies fosters resilience and adaptability within transportation frameworks. The transformative role of NLP in enhancing transportation safety is evident. As urban environments grow increasingly complex, adopting innovative, data-driven approaches will continue to shape the future of transport management. By harnessing the potential of NLP, stakeholders can create a more responsive, efficient, and ultimately safer transportation ecosystem - a goal that aligns with the broader aspirations of smart city development and sustainable urban living.

REFERENCES

Alwarafy, A., Al-Thelaya, K. A., Abdallah, M., Schneider, J., & Hamdi, M. (2021). A Survey on Security and Privacy Issues in Edge-Computing-Assisted Internet of Things. *IEEE Internet of Things Journal*, *8*(6), 4004–4022. DOI: 10.1109/JIOT.2020.3015432

Amini, M., Jesus, M., Sheikholeslami, D. F., Alves, P., Benam, A. H., & Hariri, F. (2023). Artificial Intelligence Ethics and Challenges in Healthcare Applications: A Comprehensive Review in the Context of the European GDPR Mandate. *Machine Learning and Knowledge Extraction*, *5*(3), 1023–1035. DOI: 10.3390/make5030053

Atik, S. T., Brocanelli, M., & Grosu, D. (2023). Are Turn-by-Turn Navigation Systems of Regular Vehicles Ready for Edge-Assisted Autonomous Vehicles? *IEEE Transactions on Intelligent Transportation Systems*, *24*(10), 11146–11156. DOI: 10.1109/TITS.2023.3275367

Ballal, S., Patel, K. A., & Patel, D. A. (2024). Enhancing Construction Site Safety: Natural Language Processing for Hazards Identification and Prevention. *Journal of Engineering, Project, and Production Management*, *14*(2). Advance online publication. DOI: 10.32738/JEPPM-2024-0014

Beeharry, Y., Fowdur, T. P., Hurbungs, V., Bassoo, V., & Ramnarain-Seetohul, V. (2017). Analysing Transportation Data with Open Source Big Data Analytic Tools. [IJEEI]. *Indonesian Journal of Electrical Engineering and Informatics*, *5*(2). Advance online publication. DOI: 10.52549/ijeei.v5i2.297

Chao, Y., Xinian, L., & Xiaobin, L. (2022). A dynamic cloud bayes network-based cleaning method of multi-source unstructured data. Proceedings of ASME 2022 17th International Manufacturing Science and Engineering Conference, MSEC 2022, 2. DOI: 10.1115/MSEC2022-85769

Chen, Y., Wang, H., Yu, K., & Zhou, R. (2024). Artificial Intelligence Methods in Natural Language Processing: A Comprehensive Review. Highlights in Science. *Engineering and Technology*, *85*, 545–550. DOI: 10.54097/vfwgas09

Díaz, G. M., & Carrasco, R. A. (2023). Fuzzy Logic and Decision Making Applied to Customer Service Optimization. *Axioms*, *12*(5), 448. DOI: 10.3390/axioms12050448

Faber, B., Michelet, G., Weidmann, N., Mukkamala, R. R., & Vatrapu, R. (2019). BPDIMS:A blockchain-based personal data and identity management system. Proceedings of the Annual Hawaii International Conference on System Sciences, 2019-January, 6855–6864. DOI: 10.24251/HICSS.2019.821

Fitria, K. M. (2023). Design of Text Generator Application With OpenAI GPT-3. *Jeecom Journal of Electrical Engineering and Computer*, *5*(2), 135–142. DOI: 10.33650/jeecom.v5i2.6354

Gaviria-Valencia, S., Murphy, S. P., Kaggal, V. C., McBane, R. D.II, Rooke, T. W., Chaudhry, R., Alzate-Aguirre, M., & Arruda-Olson, A. M. (2023). Near Real-time Natural Language Processing for the Extraction of Abdominal Aortic Aneurysm Diagnoses From Radiology Reports: Algorithm Development and Validation Study. *JMIR Medical Informatics*, *11*, e40964. Advance online publication. DOI: 10.2196/40964

Glint, T., Mittal, B., Sharma, S., Ronak, A. Q., Goud, A., Kasture, N., Momin, Z., Krishna, A., & Mekie, J. (2025). AxLaM: Energy-efficient accelerator design for language models for edge computing. *Philosophical Transactions. Series A, Mathematical, Physical, and Engineering Sciences*, *383*(2288), 20230395. DOI: 10.1098/rsta.2023.0395

Goldfarb-Tarrant, S., Ungless, E., Balkir, E., & Blodgett, S. L. (2023). This Prompt is Measuring <MASK>: Evaluating Bias Evaluation in Language Models. *Proceedings of the Annual Meeting of the Association for Computational Linguistics*, 2209–2225. DOI: 10.18653/v1/2023.findings-acl.139

Gray, G. M., Zirikly, A., Ahumada, L. M., Rouhizadeh, M., Richards, T., Kitchen, C., Foroughmand, I., & Hatef, E. (2023). Application of natural language processing to identify social needs from patient medical notes: Development and assessment of a scalable, performant, and rule-based model in an integrated healthcare delivery system. *JAMIA Open*, *6*(4), ooad085. Advance online publication. DOI: 10.1093/jamiaopen/ooad085

Hartsock, I., & Rasool, G. (2024). Vision-Language Models for Medical Report Generation and Visual Question Answering: A Review. *Frontiers in Artificial Intelligence*, *7*, 1430984. Advance online publication. DOI: 10.3389/frai.2024.1430984

Heideveld-Chevalking, A. J., Calsbeek, H., Damen, J., Gooszen, H., & Wolff, A. P. (2014). The impact of a standardized incident reporting system in the perioperative setting: A single center experience on 2,563 "near-misses" and adverse events. *Patient Safety in Surgery*, *8*(1), 46. Advance online publication. DOI: 10.1186/s13037-014-0046-1

Ismail, W. S. (2024). Threat Detection and Response Using AI and NLP in Cybersecurity. *Journal of Internet Services and Information Security*, *14*(1), 195–205. DOI: 10.58346/JISIS.2024.I1.013

Joseph, T. (2024). Natural Language Processing (NLP) for Sentiment Analysis in Social Media. *International Journal of Computing and Engineering*, *6*(2), 35–48. DOI: 10.47941/ijce.2135

Kang, S., Ho, T. T. T., & Lee, N. J. (2021). Comparative Studies on Patient Safety Culture to Strengthen Health Systems Among Southeast Asian Countries. *Frontiers in Public Health*, *8*, 600216. Advance online publication. DOI: 10.3389/fpubh.2020.600216

Kitamura, Y., Hayashi, M., & Yagi, E. (2018). Traffic problems in Southeast Asia featuring the case of Cambodia's traffic accidents involving motorcycles. *IATSS Research*, *42*(4), 163–170. DOI: 10.1016/j.iatssr.2018.11.001

Li, H., Yu, L., & He, W. (2019). The Impact of GDPR on Global Technology Development. *Journal of Global Information Technology Management*, *22*(1), 1–6. DOI: 10.1080/1097198X.2019.1569186

Li, R. (2015). Traffic incident duration analysis and prediction models based on the survival analysis approach. *IET Intelligent Transport Systems*, *9*(4), 351–358. DOI: 10.1049/iet-its.2014.0036

Liu, W. K., & Yen, C. C. (2016). Optimizing bus passenger complaint service through big data analysis: Systematized analysis for improved public sector management. *Sustainability (Basel)*, *8*(12), 1319. Advance online publication. DOI: 10.3390/su8121319

Mahmeed, G., Brin, D., Konen, E., Nadkarni, G. N., & Klang, E. (2024). Systematic Review of Natural Language Processing (NLP) Applications in Magnetic Resonance Imaging (MRI). DOI: 10.1101/2024.07.21.24310760

Mao, Z., Chuan, T., Zhou, J., Su, W., He, Y., Mei, D., & Li, C. (2023). Knowledge Graph-Based Approach to Unstructured Multi-Source Data Fusion. *Frontiers in Artificial Intelligence and Applications*, *373*, 663–669. DOI: 10.3233/FAIA230869

Merlec, M. M., Lee, Y. K., Hong, S. P., & In, H. P. (2021). A smart contract-based dynamic consent management system for personal data usage under GDPR. *Sensors (Basel)*, *21*(23), 7994. Advance online publication. DOI: 10.3390/s21237994

Mondschein, C. F., & Monda, C. (2018). The eu's general data protection regulation (GDPR) in a research context. Fundamentals of Clinical Data Science, 55–71. DOI: 10.1007/978-3-319-99713-1_5

Omar, M., Choi, S., Nyang, D., & Mohaisen, D. (2022). Robust Natural Language Processing: Recent Advances, Challenges, and Future Directions. *IEEE Access : Practical Innovations, Open Solutions, 10*, 86038–86056. DOI: 10.1109/ACCESS.2022.3197769

Osco, L. P., de Lemos, E. L., Gonçalves, W. N., Ramos, A. P. M., & Marcato, J.Junior. (2023). The Potential of Visual ChatGPT for Remote Sensing. *Remote Sensing (Basel), 15*(13), 3232. Advance online publication. DOI: 10.3390/rs15133232

Pandit, H. J., O'Sullivan, D., & Lewis, D. (2019). Test-Driven Approach Towards GDPR Compliance. Lecture Notes in Computer Science (Including Subseries Lecture Notes in Artificial Intelligence and Lecture Notes in Bioinformatics), 11702 LNCS, 19–33. DOI: 10.1007/978-3-030-33220-4_2

Pandit, H. J., Polleres, A., Bos, B., Brennan, R., Bruegger, B., Ekaputra, F. J., Fernández, J. D., Hamed, R. G., Kiesling, E., Lizar, M., Schlehahn, E., Steyskal, S., & Wenning, R. (2019). Creating a vocabulary for data privacy: The first-year report of data privacy vocabularies and controls community group (DPVCG). Lecture Notes in Computer Science (Including Subseries Lecture Notes in Artificial Intelligence and Lecture Notes in Bioinformatics), 11877 LNCS, 714–730. DOI: 10.1007/978-3-030-33246-4_44

Piras, L., Al-Obeidallah, M. G., Praitano, A., Tsohou, A., Mouratidis, H., Gallego-Nicasio Crespo, B., Bernard, J. B., Fiorani, M., Magkos, E., Sanz, A. C., Pavlidis, M., D'Addario, R., & Zorzino, G. G. (2019). DEFeND Architecture: A Privacy by Design Platform for GDPR Compliance. Lecture Notes in Computer Science (Including Subseries Lecture Notes in Artificial Intelligence and Lecture Notes in Bioinformatics), 11711 LNCS, 78–93. DOI: 10.1007/978-3-030-27813-7_6

Private, N. L. P., 2022. (2022). PrivateNLP 2022 - Workshop on Privacy in Natural Language Processing. Proceedings of the Fourth Workshop on Privacy in Natural Language Processing. https://doi.org/DOI: 10.18653/v1/2022.privatenlp-1

Rajasekaran, S. B. (2023). Applications of AI in Computer Vision and NLP. *Design of Single Chip Microcomputer Control System for Stepping Motor, 1–3*, 1–3. Advance online publication. DOI: 10.47363/JAICC/2023(2)128

Randmaa, M., Mårtensson, G., Swenne, C. L., & Engström, M. (2014). SBAR improves communication and safety climate and decreases incident reports due to communication errors in an anaesthetic clinic: A prospective intervention study. *BMJ Open, 4*(1), e004268. Advance online publication. DOI: 10.1136/bmjopen-2013-004268

Ricketts, J., Barry, D., Guo, W., & Pelham, J. (2023). A Scoping Literature Review of Natural Language Processing Application to Safety Occurrence Reports. *Safety (Basel, Switzerland)*, *9*(2), 22. Advance online publication. DOI: 10.3390/safety9020022

Ryan, P., Crane, M., & Brennan, R. (2021). GDPR Compliance Tools: Best Practice from RegTech. *Lecture Notes in Business Information Processing*, *417*, 905–929. DOI: 10.1007/978-3-030-75418-1_41

Sangaroonsilp, P., Dam, H. K., Choetkiertikul, M., Ragkhitwetsagul, C., & Ghose, A. (2023). A taxonomy for mining and classifying privacy requirements in issue reports. *Information and Software Technology*, *157*, 107162. Advance online publication. DOI: 10.1016/j.infsof.2023.107162

Sheikhalishahi, S., Miotto, R., Dudley, J. T., Lavelli, A., Rinaldi, F., & Osmani, V. (2019). Natural language processing of clinical notes on chronic diseases: Systematic review. *JMIR Medical Informatics*, *7*(2), e12239. Advance online publication. DOI: 10.2196/12239

Stavropoulou, C., Doherty, C., & Tosey, P. (2015). How Effective Are Incident-Reporting Systems for Improving Patient Safety? A Systematic Literature Review. *The Milbank Quarterly*, *93*(4), 826–866. DOI: 10.1111/1468-0009.12166

Taheri, Y., Bourgne, G., & Ganascia, J. G. (2023). A Compliance Mechanism for Planning in Privacy Domain Using Policies. Lecture Notes in Computer Science (Including Subseries Lecture Notes in Artificial Intelligence and Lecture Notes in Bioinformatics), 13856 LNAI, 77–92. DOI: 10.1007/978-3-031-36190-6_6

Takashima, R., Takiguchi, T., & Ariki, Y. (2012). Exemplar-based voice conversion in noisy environment. 2012 IEEE Workshop on Spoken Language Technology, SLT 2012 - Proceedings, 313–317. DOI: 10.1109/SLT.2012.6424242

Torre, D., Soltana, G., Sabetzadeh, M., Briand, L. C., Auffinger, Y., & Goes, P. (2019). Using Models to Enable Compliance Checking against the GDPR: An Experience Report. Proceedings - 2019 ACM/IEEE 22nd International Conference on Model Driven Engineering Languages and Systems, MODELS 2019, 1–11. DOI: 10.1109/MODELS.2019.00-20

Tyagi, N., & Bhushan, B. (2023). Demystifying the Role of Natural Language Processing (NLP) in Smart City Applications: Background, Motivation, Recent Advances, and Future Research Directions. *Wireless Personal Communications*, *130*(2), 857–908. DOI: 10.1007/s11277-023-10312-8 van der Wal, O., Bachmann, D., Leidinger, A., van Maanen, L., Zuidema, W., & Schulz, K. (2024). Undesirable Biases in NLP: Addressing Challenges of Measurement. *Journal of Artificial Intelligence Research*, *79*, 1–40. DOI: 10.1613/jair.1.15195

Wan, X., Lucic, M. C., Ghazzai, H., & Massoud, Y. (2020). Empowering real-time traffic reporting systems with NLP-Processed social media data. *IEEE Open Journal of Intelligent Transportation Systems*, *1*, 159–175. DOI: 10.1109/OJITS.2020.3024245

Wang, Y., Wang, L., Rastegar-Mojarad, M., Moon, S., Shen, F., Afzal, N., Liu, S., Zeng, Y., Mehrabi, S., Sohn, S., & Liu, H. (2018). Clinical information extraction applications: A literature review. *Journal of Biomedical Informatics*, *77*, 34–49. DOI: 10.1016/j.jbi.2017.11.011

Wu, Y., Denny, J. C., Rosenbloom, S. T., Miller, R. A., Giuse, D. A., Song, M., & Xu, H. (2015). A preliminary study of clinical abbreviation disambiguation in real time. *Applied Clinical Informatics*, *6*(2), 364–374. DOI: 10.4338/ACI-2014-10-RA-0088

Yang, C., & Huang, C. (2023). Natural Language Processing (NLP) in Aviation Safety: Systematic Review of Research and Outlook into the Future. *Aerospace (Basel, Switzerland)*, *10*(7), 600. Advance online publication. DOI: 10.3390/aerospace10070600

Yuan, H., Ding, H., & Lv, P. (2023). Research on Big Data Analysis of Passenger-Oriented Railway Service Quality Based on LDA Model. *Proceedings of the 2022 3rd International Conference on Big Data and Informatization Education (ICBDIE 2022)*, 1279–1293. DOI: 10.2991/978-94-6463-034-3_132

Zhou, J., Hu, Y., Lian, H., Wang, H., Tao, L., & Kwan, H. K. (2019). Multimodal Voice Conversion under Adverse Environment Using a Deep Convolutional Neural Network. *IEEE Access : Practical Innovations, Open Solutions*, *7*, 170878–170887. DOI: 10.1109/ACCESS.2019.2955982

Chapter 13

Cybersecurity Challenges in AI-Enabled Transportation Systems

S. Arjune
Christ (Deemed to be) University, India

A. Hariprasad Reddy
Geethanjali College of Engineering and Technology, India

S. Mangai
Velalar College of Engineering and Technology, India

K. R. Swetha
BGS Institute of Technology, Adichunchanagiri University, India

V. Joseph Emmanuvel
http://orcid.org/0000-0002-3616-7094
Periyar Maniammai Institute of Science and Technology, India

Naveen Nandal
http://orcid.org/0009-0007-2272-7437
DIHE, India

V. Bhoopathy
http://orcid.org/0000-0003-2175-6328
Sree Rama Engineering College, India

ABSTRACT

The article discusses how to apply encryption and network segmentation, two traditional methodologies, to the cybersecurity of ITS. The procedures that were mentioned, the problems they solved, and the architectural layers where they were applied are all connected in the end. Applications of AI in transportation systems, such as smart logistics, intelligent traffic management, and autonomous cars, have transformed efficiency, safety, and mobility. Nevertheless, these systems are vulnerable to cybersecurity threats due to their dependence on networked devices, real-time data sharing, and complicated algorithms. Threats to AI models, data poisoning, networks, ransomware, and supply chain security are some of the main obstacles to securing AI-enabled transportation that are discussed in this chapter.

DOI: 10.4018/979-8-3373-3231-4.ch013

It may learn a lot about vulnerabilities and responses in the real world from case studies. The chapter stresses the need for worldwide cooperation, robust designs, and AI-driven cybersecurity solutions to keep AI-enabled transportation systems safe from ever-changing cyber threats.

1. INTRODUCTION

Cybersecurity encompasses a wide range of activities aimed at preventing harm to, or illegal access to, computer systems, networks, applications, and data. Cybersecurity is a growing concern in any industry that depends on communication in today's interconnected world. Consequently, these types of assaults also pose a serious threat to traffic operation and management systems that use both wired and wireless communications to control motorways. Despite the many benefits that come with smart gadgets and connectivity, there are also new dangers that users face. The conventional approach of separating control systems from other networks will become impractical, due to the proliferation of IoT gadgets, a number of which are finding applications in traffic management systems. Cyberattacks have increased dramatically alongside the exponential growth of internet access by (Khattak et al., 2018). Malware assaults, which covertly undermine users' intentions, used to occur at a single point of vulnerability among susceptible components. Consequently, the perimeter defence strategy—which relied on firewalls or antivirus—was employed instead of defending each asset, but this approach is no longer viable. The biggest problem with increasing reliance on ICT is the rise in cybersecurity threats. Traditional transport management and operations systems were closed private networks with few cyber vulnerabilities because of their isolation. Now that cloud and network computing have emerged, and since we continue to rely on new technologies like the IoT those once-proprietary systems have become more open and accessible.

In addition, the National Transportation Communication for ITS Protocol makes use of request-based protocols via XML messages for center-to-center communication. The inherent lack of security in these protocols, along with the underlying belief that the majority of attacks originate from within and that hackers constitute a negligible fraction of all incursions, heightens the danger to proprietary systems. SCMS is a solution for message security in V2V and V2I communication that the USDOT has developed with much effort. One aspect of ATM is the capacity to use real-time and predictive operational tactics to control both recurring and non-recurring congestion. Adaptive ramp metering is one use case for an ATM system; Another option is for regulating the flow of vehicles entering a motorway. ATMs can, however, be used in conjunction with other technologies to address network-wide issues such as protection, capacity control, handling of incidents, and traveller information. When planning their routes, drivers depend on data retrieved from ATMs. The latest open protocol systems rely on constant connection between the highway and the TMC which makes the authenticity of such information even more crucial. They are the least known when it comes to cybersecurity, and their reliance on communication makes them easy targets for cyberattacks (Thomas et al., 2025).

The idea that there is no such thing as foolproof cybersecurity is central to the studies discussed in this paper. An internal attack by an unhappy employee is another possibility. So, to make sure the system is acting normally all the time, this study decided to build a threat monitoring system that checks in real-time. This paper is organised in the following way. To begin, we offer a literature review that examines pertinent research and the perspectives of several domains on cybersecurity. Next, we'll go over what was done to create a working model of an ATM system. After this, we have a section that shows the architecture of the monitoring system and identifies different attack spots in the prototype ATM. Following this, the creation of a monitoring system is detailed. Lastly, an assessment is carried out to determine the advantages of returning to a safe lane control condition with the help of a monitoring system, as well as the effects of a potential attack on the prototype ATM system (Vegesna, 2022).

2. CYBER-ATTACKS IN ITS

The variability of ITS makes it more difficult to categorise and identify cyber-attacks. Attacks that are unique to ITS are detailed here; they will be linked to the architectural layers in a bit (Mashkina et al., 2025).

2.1 Routing Attacks

One-hop communication in a VANET is defined at both the data link and physical layer. Routing protocols support multi-hop communication. In a routing attack, a malevolent node tries to stop data from reaching its destination by breaking into the VANET's routing protocols. An example of a routing attack is a black hole attack, when a hostile node discreetly drops all the packets that should be re-transmitted. Another kind of routing attack is the grey hole attack, which involves discarding packets one by one (Hong, 2020).

2.2 VANET Man-in the-Middle Attack

A typical form of cyberattack, a man-in-the-middle attack involves one party intercepting and forwarding modified content from another party in order to influence their communication. Within a man-in-the-middle attack on a VANET, the criminal evaluates the nodes' proximity in the event that they target the physical or data connection level. If the targeted nodes are out of reach, but the criminal is within distance of both nodes, they will need to either weaken the signal or alter the location data to their advantage. It would be an attacker manipulating the position data exchanged between multiple cars. Road safety or efficient transportation applications may not work properly as a result of this (Iyer, 2021).

2.3 Spoofing

The goal of a spoofing attack is to trick the system into thinking something it shouldn't by broadcasting faulty data. As an example, the aggressor is interfering with the navigation system's functionality by sending false GPS coordinates (Verma, 2024).

2.4 Timing Attacks

Applications relying on real-time processing are vulnerable to timing assaults since they introduce a communication delay. To avoid a collision, a cooperative adaptive cruise control system may send a distress signal to the car next to it. Even when the data is correctly received, the braking system's reaction will be delayed, and the collision will still occur, if the attacking party is able to create a delay (for example, by overloading the network traffic).

2.5 Internal Vehicle Network Attack

Most internal vehicle networks were created while cars weren't connected, making them susceptible to attacks. For instance, the airbag control system is vulnerable to attacks that compromise the internal network based on the CAN protocol (Uma & Padmavathi, 2023).

2.6 DoS

The availability of system components can be affected by a traditional cyberattack. This is particularly risky in the context of ITS if it affects a safety-critical functionality. A malicious vehicle injects bogus broadcast messages into the network while impersonating many identities in a classic VANET denial-of-service attack, as seen in Sybil. Consequently, it causes problems with the regular flow of data (Prakash et al., 2023).

2.7 Eavesdropping

The typical passive attack known as eavesdropping occurs when an attacker obtains unauthorised access to information without actively interfering with the communication process. Toll payment connection between the vehicle and road infrastructure is an example of how a hostile third party could eavesdrop and obtain the user's bank account information (Choudhary et al., 2018).

2.8 Identity Attack

Data security for personal information Any person's right to privacy—drivers, passengers, pedestrians, etc. may be at issue with ITS. Information regarding one's location, activities, habits, and personal data may be sought for by the aggressor. One kind of identity assault is when the perpetrator learns the VANET's method for assigning nicknames and uses that knowledge to follow a vehicle's whereabouts (Waxman, 2019).

2.9 AI Attacks

Possible causes of AI assaults include policy manipulation, environmental perturbations, or data manipulation data poisoning attack. An adversary may, for instance, choose and transmit data in a way that induces erroneous trends in the model, thus misleading the ML algorithm (Almiani et al., 2020).

2.10 Attack Against Fog

Because of their physical properties the Fog components of ITS are difficult to secure and subject to a variety of attacks, as opposed to the Cloud. The fog node in this example collects and cleans data from vibration and sound sensors. The data analysis technique used to plan the organisation of road traffic is vulnerable to hostile actors that modify aggregated information (Chowdhury & Ferdous, 2017).

3. INCREASING CYBER VULNERABILITIES

Cybersecurity in the transportation industries has become more complex due to the interconnectedness and reliance on information and automation technology, which has affected aviation, road, rail, and maritime transportation. New technologies have made the transport sector more susceptible than in the past. This article examines the factors that are leading to a rise in cyber vulnerabilities (Rid & Buchanan, 2025).

3.1 Transportation Systems' Enhanced Automation and Connectivity

The scale and complexity of transport networks have continued to increase. This has led to the evolution of ITS and IMS technologies into necessities. Automation reduces the need for human intervention, while connectivity allows for more efficient system communication. While these innovations boost productivity, security, and usability, they also make us more vulnerable to cyber threats. The proliferation of CAV and the growing dependence on IoT technology in transportation are two major factors propelling cyber assaults (Bendovschi, 2024).

3.1.1 CAV

The digitalisation of the car sector will bring in $67 billion for businesses and $3.1 trillion for society. In addition to these advantages, CAVs are appealing because they lessen the need for human intervention, which improves both efficiency and safety. Vehicle-to-everything communication is one of many types of communication that CAVs depend significantly on. The cyber-vulnerability of CAVs is highest in their communication systems. A cyberattack on a CAV's communication could limit the vehicle's ability to access the internet and influence its decision-making. Attacks against CAV communication networks can take several forms, including but not limited to: black hole and grey hole attacks, DoS and distributed DoS, jamming, eavesdropping, certificate replication, Sybil attacks, impersonation, and fake message injections (Mezzour et al., 2024). Violent attacks on autonomous vehicles have the potential to cause serious harm, including bodily harm or even death.

3.1.2 Increased Reliance on IoT Technologies

A simple definition of the IOT would be a system of interconnected, communicative gadgets. From CAV applications to information and operational system management, the transportation industry is rapidly embracing IoT technologies. IoT devices may be susceptible to assaults because of their weak security procedures and limited processing capacity, as pointed out by (Yugoslavia, 2023).

3.2 Transportation Systems Using Operational and Information Technology

Transportation Software and Hardware for Information and Operations The introduction and increasing reliance on new technology is leading to a rise in cyber risks and cyber-related losses in the transport sector. In the past, businesses took it as read that their systems' vulnerabilities were little-known and so unlikely to be exploited. Nevertheless, this line of thinking is out of date. Field devices such as signals for traffic, monitors, and dynamic message signs, are vulnerable to cyber-attacks including man-in-the-middle and DoS assaults have been widely used by many companies since the digital revolution. These technologies include Wi-Fi and Ethernet. Data and information security and organisation are two of the most important functions of IT in modern transportation systems (Li & Liu, 2021). Since the transport sector is based on the transmission and receiving of information, ICT is an essential component of the sector. Because malicious actors can exploit these systems' known vulnerabilities, the reliance on them makes them more susceptible. Computing systems that provide human-machine and machine-to-machine interfaces for the purpose of controlling industrial operations and enhancing system efficiency and automation are collectively referred to as OT. Security flaws in the design, human error, and improper configuration are the main causes of operational technology vulnerabilities are some of the known vulnerabilities that are commonly used in many OTs, including SCADA and DCSs (Shah & Sengupta, 2020). Legacy systems are still used by some OT systems, including those in some maritime and rail transportation systems; these systems are also susceptible to cyber assaults.

There has been a merging and increasing dependence of IT and OT systems. Relying on less secure older systems makes OT systems more exposed. Cybersecurity regulations for OT must be as stringent as, if not tougher than, those for IT systems since OT is increasingly accessible to cyberspace (Zhuang et al., 2025). As they include new technology to boost system intelligence, this is essential for railways and ports. Vulnerability is increased when these technologies are implemented without first analysing the hazards.

3.3 Human Factors

The growth of internet in transport is an attempt to lessen the inefficiencies caused by human intervention. Yet, as time goes on, workers will be expected to interact with smart technologies, necessitating a level of cybersecurity literacy. As an example, 90% of cloud data breaches are reportedly due to staff mistakes demonstrating how cybersecurity ignorance can aid attackers (Buchan, 2019). Supply chains are vulnerable to cybersecurity threats caused by employees who lack proper training. In addition to technological attacks, attackers may exploit transportation systems by combining social engineering techniques. Phishing was the most common initial access vector in 51% of incidents, according to IBM Security's 2023 X-Force Threat Intelligence Index. The disruption of a ship's charts caused by malware on a crew member's smartphone resulted in a two-day delay, highlighting the need to safeguard transport systems from malevolent personnel (Cramer & Damgård, 2024).

3.4 Supply Chain Cybersecurity

Due to the interconnected nature of the transportation industry's sub-industries, supply chain cybersecurity affects transportation cybersecurity. Several suppliers provide IT and OT systems support to transportation companies (Patel & Zaveri, 2020). Third-party software and hardware vendors have different goals and specific functional requirements. Companies have a hard time making sure their vendors comply with cybersecurity requirements because there aren't many standards governing vendor cybersecurity.

3.5 Updates and Software Bugs

Transportation cybersecurity can be compromised by software updates and faults. The estimated 111 billion lines of new code created per year contain billions of vulnerabilities in software. Hackers are able to launch a variety of attacks on systems by taking advantage of loopholes in code. Take the Nissan Leaf as an example. Hackers were able to remotely access the infotainment system and expose critical driving data by taking advantage of software coding faults. When software is out of current, systems are also at risk of vulnerability. Additionally, systems are at danger from over-the-air updates because of software vulnerabilities and the possibility of malware implants. Major interruptions affected many airports and airlines this year due to a software fault in a CrowdStrike upgrade, which affected thousands of Windows devices globally (Ahmad et al., 2018).

4. TYPES OF CYBERATTACKS

Attacks that impersonate the GPS are also prevalent in networks used by automobiles. We need to find safer and more effective ways to stop this onslaught (Al-Shareeda & Manickam, 2022).

Privacy Protection: The bulk of IoV applications rely on internet connectivity. Outsourced procedures may not always be reliably solved by third-party cloud-based solutions. Most Internet of Vehicles software makes use of cloud-based solutions. So, it's not always a good idea to leave outsourcing to third-party cloud service providers. Using cloud-based services, which might not even need the collected data, would be much more appealing (Al-Juboori et al., 2023). Current methods for protecting sensitive information during data storage activities involve encryption standards that are either partially or fully effective. However, when dealing with the massive amounts of data produced by numerous vehicles in an IoV situation, such systems can be quite resource intensive. In order to provide user security and privacy in the IoV environment, it is necessary to deploy lightweight fully homomorphic encryption. For example, in order to authorise users on cloud systems, it is necessary to know their identities; yet, anonymous credentials are not acceptable. Regulated anonymity involves safeguarding confidentiality and responsibility while maintaining anonymity. For example, people shouldn't be completely anonymous if that would compromise accountability, but they also shouldn't be completely known if that would compromise secrecy (Javeed & MohammedBadamasi, 2020).

Resilience and Self-Adaptation: The transition from reducing vulnerability to increasing resilience and self-adaptation is a worthy topic to investigate. The IoV platform needs to be strong enough to react quickly and thoroughly to attacks and strange behaviours. Professionals in the field of IoT must investigate and implement AI-based solutions, like self-rewriting code and automatic software patching, to ensure dependability (Ahmad et al., 2020).

AI-based Detection: A shortage of security personnel is presently prevalent on a global scale. Autonomous vehicles and robotic workers will become more commonplace in the years to come. It is crucial to create an AI-powered immune system that can adapt to changing threats, safeguard against vulnerabilities and anomalies, and counteract threats like cognitive hackers, AI-based ransomware, and more. Due to the incident, Chrysler recalled almost 1.4 million vehicles. It turned out that the Chrysler Jeep Cherokee wasn't the only car that could be hacked.

Masquerading Attacks: The impersonating attack is another challenging IoV tactic. In the face of such an onslaught, few safeguards are available. This attack necessitates counterintelligence in order to safeguard infrastructure from impersonation attacks (Basri et al., 2025).

Trust: In every interaction between entities and non-entities, trust plays a crucial role in the functioning of communication devices. Take the question of how certain a base station may be that data will be shared with them by another cluster. For cryptographic signatures to be viable, they need to incorporate a unique, permanent, and identifiable identity. Nodes with transient identities can change their identities for future interactions; however, there is no one-to-one matching between identities with non-distinct identification, and there appears to be more than one identification on a single car (Mahmood A. et al., 2020). One would question the feasibility of maintaining and handling trust-related data on a massive scale within the framework of the legislation. On the other hand, how can trust data be used securely? In fact, future authentication techniques should offer identity guarantee. However, this identification must be handled initially by the authors of the trust system before devices on this system can acquire confidence. In order to build trust in the system, interaction activities are crucial. The algorithms should be both distributed-ready and adaptable. In addition to being computationally efficient, it needs to ensure accurate mappings from qualitative to quantitative trust.

5. LEVELS OF ITS PROTECTION

Providing multi-level protection encompassing all components of the model is important because intelligent transport systems are subject to cyberattacks and technological breakdowns. The most important safeguards (Mohapatra, 2020).

- By utilising surveillance cameras and other physical security elements, Physical Layer Protection (PLP) guarantees that only authorised individuals can access equipment and physical sensors (Ganesh Reddy, 2023).
- Firewalls, VPN, traffic encryption, and protocol protection such as V2X are all part of network security. Preventing DDoS assaults and other forms of unauthorised network access is the primary objective.
- Encryption, user authentication, and access control are all parts of data protection that work together to reduce the likelihood of data leaks and privacy breaches.
- Updating software often, keeping tabs on server activity, and utilising intrusion detection systems are all POCS. Policies for backups and access control keep cloud computing secure (Robert A et al., 2019).
- The employment of algorithms for real-time traffic analysis in monitoring and anomaly detection allows for the discovery of anomalies and the subsequent response to suspicious activities (Zaidi & Faisal, 2019).

Using mathematical models to evaluate the likelihood of threats and anomalies is crucial for risk analysis and ensuring the stability of ITS operations. These models are crucial for problem prediction since they are based on probabilistic approaches that consider both past and present data. At any given moment, estimating the likelihood of a cyberattack on ITS is a critical task. An approach that considers the temporal dependence of risks—a threat probability function—helps to tackle this problem. Fixed and variable risks, including those that grow as system vulnerabilities accumulate or shrink as security solutions take effect, can be represented using an exponential distribution or a generalisation of it, like the Weibull distribution (Manickam, 2023).

The modelling of abnormalities in network traffic is another crucial component that forms the basis for the timely detection of threats. In order to assess the likelihood of anomalies given the present traffic characteristics, the anomaly probability functions employ ML methods and Bayes' theorem. This method considers the system's typical operation as well as any variations that might point to issues. By combining the two methods it may build a multi-tiered security system that detects and responds to both domestic and international threats (such as cyberattacks). With the proliferation of connected devices and the complexity of ITS continuing to rise, this is more crucial than ever to guarantee safety (Stabili et al., 2025).

6. CYBER RISK IN CONNECTED VEHICLE OF ITS

The cyber dangers associated with vehicles include a lack of external connectivity, slow processing performance, the need for real-time operations, the vehicle's complex composition, the unpredictability of attack scenarios and threats, and the potential danger to the lives of drivers and passengers. Functions like CD/DVD control, navigation, emergency call, camera monitoring, AC control, signal, power window, air bag, and so on are all part of the danger assessment process outlined in ISO 26262.

In addition to the aforementioned guidelines for vehicle cyber security, there are a number of other relevant documents, such as IEC62443 (IEC means industrial system), NIST-800-61 (NIST means industrial system and pc/internet), CIP (NERC means pc/internet and industrial system), EVITA (EU means vehicle), VSG (agency IPA means vehicle), and J3061 (SAE means vehicle). Automotive security information Examples of updated vehicle security data that will be recommended and reviewed include OBD II emissions data, TPMS data, over-the-air firmware update diagnostics data, airbag control unit data, GPS data, throttle valve data, CAN bus messages for the power control module, adaptive cruise control data, local interconnect network steering wheel data, and ABS brake-by-wire data, in accordance with NIST SP 800-60 and FIPS 199 (Mehrdad, 2019).

The security model will present protective security metrics; how can we quantify protective security? Behavioural security metrics—those that have dependability, availability, safety, and secrecy with or without latency—based on the appropriate evaluation measures. The difficulties of designing secure vehicle architectures are detailed by Florian Sagstetter et al. in their article 15. The in-vehicle network architecture is divided into several categories: LIN for sub networks, CAN for comfort, MOST for infotainment, CAN for safety, and FlexRay for chassis (Rana & Burnap, 2020). In order to update the firmware, connect the bus, and open the gate, it is necessary to implement a physical protection mechanism and cryptographic functions into the ECU. Security measures for the battery monitoring the battery management system the charging port (preventing unauthorised access), and the drive-by-wire capabilities should all be prioritised. Concerns about Ethernet/IP and OBD security, as well as those pertaining to security and external mobility services, will also be taken into account. The OBD security suite, which includes a malware detector, database (whitelist/blacklist), activity monitor, reporter, and manager, is necessary for malware detection (Tomlinson et al., 2021).

Table 1. Different Security Factors for Connected Vehicles in ITS Technologies

Problem	Factors
Issues and Developments in ITS Security	Evolution of security phases in control, communication, and signal processing domains
Certificate pre-distribution in VANET	A certificate is always added to the beacon frame including location data in order to provide strong authentication and pseudonymous communication.
Reauthentication and Authentication	EAP concerns during the initialization and handover phases
Implementation of Essential security countermeasures in linked vehicles	Protocol for the implementation of a security framework Public Key Infrastructure (PKI) facilitates the creation of security architecture, ensuring safety and cost-effectiveness, while promoting interoperable solutions that promote the introduction of vehicles and foster trust relationships across cooperative systems.
Strategies for vehicular information security	Attacker interference encompasses hazards like illegal access, improper configuration, information leakage, packet sniffing, denial of service attacks, message tampering, log loss and unauthorised relaying. Security assessment and debugging procedures for cooperative automated transportation include the code examination, static inspection, compatibility evaluation, stress testing, and vulnerability assessment.
Security aspects	Inadequate bus protection, feeble authentication, protocol misuse, substandard protocol implementation, and information leakage.
A number of limitations for future connected vehicles	Side channel attacks, spoofing, man-in-the-middle attacks, modified or dysfunctional attacks, and compromised privacy attacks

Table 1 summarises the many factors to be considered as a result of the linked research on cooperative ITS and connected vehicles.

While the man-in-the-middle attack concerns the safety function of driving strategies, the side-channel assault concerns the correct control of vehicles and the detection of driving misbehaviour. Problems with braking, acceleration, and driving safety might arise from software modifications or DOS (Wu et al., 2020). Concerns about service scalability are at the heart of the spoofing attack problem. Following in the footsteps of the following will be the short, medium, and long-term outlook on autonomous cyber security technology: Hardened operating systems, secure coding, and virtualisation will all be part of the immediate security plan. In the middle term, it will be used to safeguard the hardware security module, the intrusion detection system, the boot/RAM/debug, and the F/W/intrusion prevention system. Use cases for this technology include tamper-proof hardware and encrypted vehicle bus communication in the far future (Carsten et al., 2024).

Figure 1. Risk Assessment for Cyber Security in an ITS- Connected Vehicle

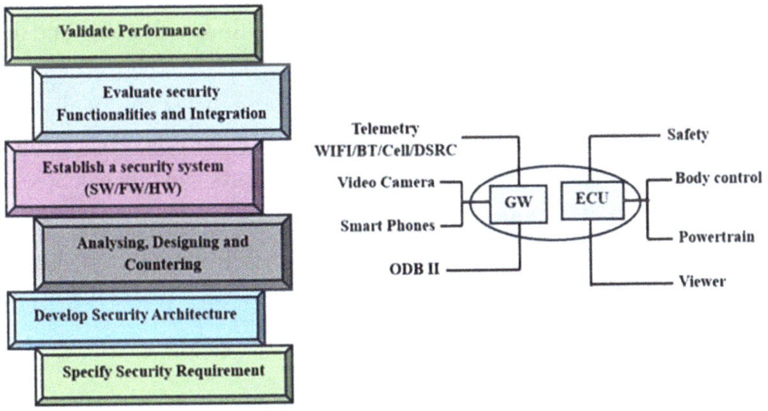

Figure 1 shows as after defining requirements the specification of risk assessment designing and implementing the function and service and testing and auditing penetration fuzz and integration tests the automotive cyber security ECO system will follow the five-stage approach of Cornelius Bittersohl. Automobile cyber security plan, security standard guideline derivation, organisational system change, process integration business/development technology integration the plan includes evaluation and testing validation of elements such as exposure, critical factors, identification, penetration attractiveness, side channel attacks, integration, and attack vectors. (Alsaade & Al-Adhaileh, 2023).

7. AI-ENABLED TRANSPORTATION SYSTEMS

AI has made tremendous and far-reaching contributions to the transportation sector. Autonomous vehicles, logistics, traffic management, and optimised routing are some of the ideas that can help make roads safer for everyone. The data collected from the gadgets in the vehicles, made possible by AI technology, is used to construct ITS. The current study focusses on four transport-related subsystems: Intelligent Manufacturing and Transportation System, Smart Safety Management System, Intelligent Public Transportation System, and Intelligent Traffic Management System. Tables describe the data resources for solutions to AI subsystem problems, AI function, and benefits. AI provides mobility solutions by suggesting alternate routes and monitoring traffic lights in real time during congestion (Zhao et al., 2022). In the long run, this aids in traffic management, which in turn helps to reduce pollution and construct sustainable cities. AI can help with things like traffic and weather forecasts, road management, and alerting on-duty police officers. Prior to setting out on their journey, these systems assist drivers, commuters, and pedestrians. An effective public transit system that facilitates planning and decision-making requires technological backing. AI has led to fewer road accidents, better accident prediction based on road conditions, and more driver safety alerts. A well-functioning transport sector is essential to a thriving economy. To accomplish the same, it is crucial to construct a secure transportation system using AI technology. AI solutions help the automotive industry when making cars. The use of sensors, cameras, and other technological advancements has improved this sector. Some of the AI features put into cars are already standard in both commercial and passenger vehicles (Andreolini & Marchetti, 2022).

8. RESULT ANALYSIS

ResNet models are able to train extremely deep networks, overcome the vanishing gradient problem and degradation problem, and use residual blocks with skip connections to produce greater performance. The results demonstrate that ResNets are very effective in niche areas such as agricultural disease detection and medical imaging, and that they perform exceptionally well in image classification, detection, and segmentation. Its discussion includes the efficacy of transfer learning, the value of data augmentation, and potential future enhancements through the integration of SE blocks or sophisticated pre-activation blocks to improve performance and facilitate deployment on other platforms.

The suggested hybrid ResNet model routinely outperforms earlier work on the dataset in term of predictive accuracy, showing improved speed and remarkable problem identification and diagnosis capabilities shown as Figure 2.

Figure 2. Training and Validation Accuracy with ResNet Model

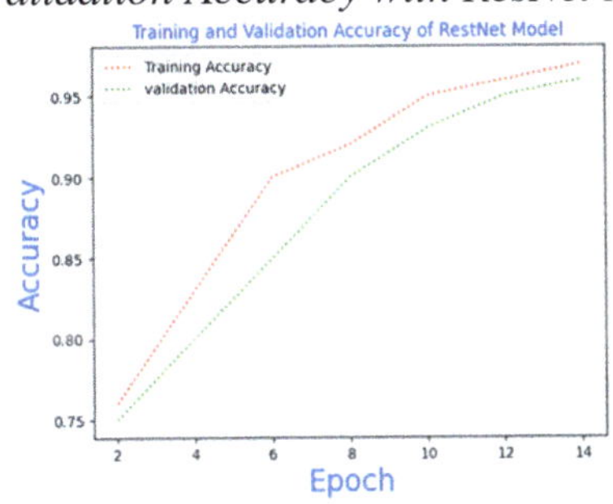

Training loss as a function of period size is illustrated in Figure 3. The performance graph of the training ResNet model subsets. In our scenario, the network is well-trained as the loss function reduces for training datasets.

Figure 3. Training and Validation Loss with ResNet Model

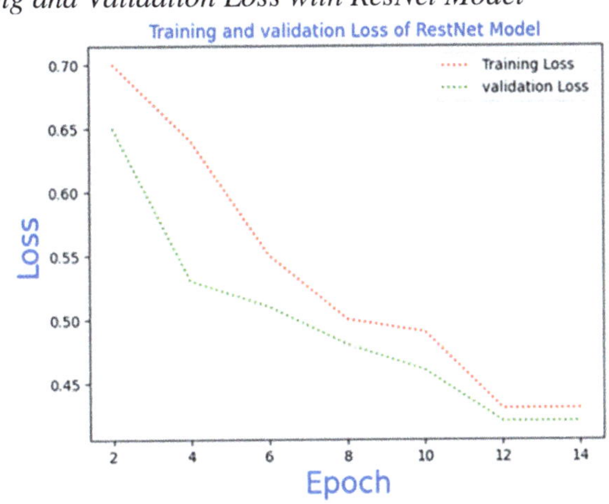

Table 2 which displays line chart comparing the performance of three models reveals that ResNet surpasses all of its competitors including KNN and BPNN in terms of Accuracy, Specificity and Sensitivity

Table 2. Performance Comparison

Models	Accuracy	Specificity	Sensitivity	F1-Score
KNN	90.67	88.76	91.21	90.73
BPNN	93.41	91.82	94.88	93.59
ResNet	**95.89**	**93.45**	**96.39**	**95.92**

Figure 4 which displays a bar and line chart comparing the performance of three models reveals that ResNet surpasses all of its competitors, including KNN and BPNN in terms of Accuracy, Specificity and Sensitivity.

Figure 4. Performance Evaluation of Different Metrics

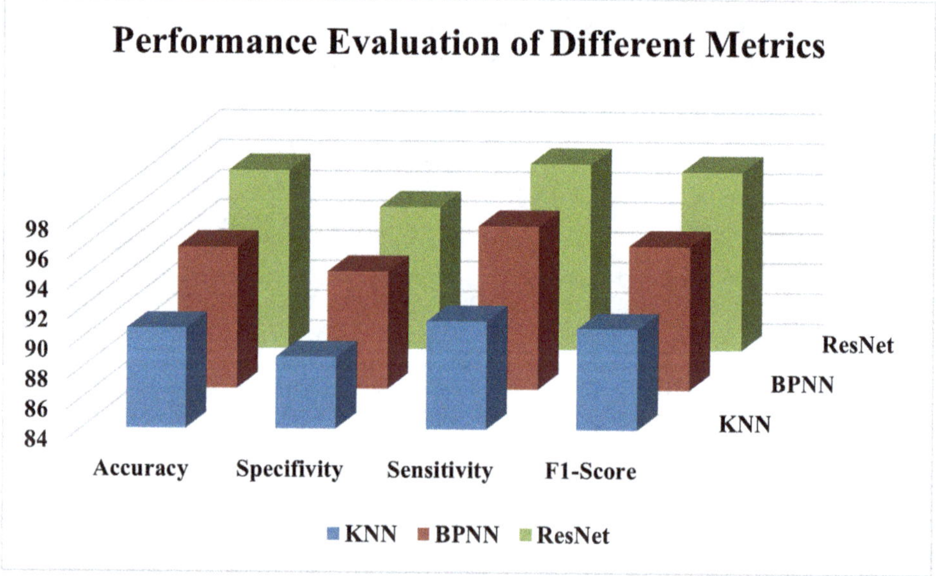

Table 3 provides a full overview of the models' performance evaluations. It selects the best model ResNet by analysing its performance with training and testing. With the pre-processed dataset in hand, it ran fertiliser prediction models such as KNN and BPNN.

Table 3. Performance Evaluation

Models	*Training*	*Testing*
KNN	*90.55*	*89.59*
BPNN	*93.39*	*92.78*
ResNet	***95.25***	***93.35***

Figure 5 shows as when evaluating the performance of a ResNet or any classification model the F1-score is a useful indicator that takes precision and recall into account. When dealing with skewed datasets it becomes much more valuable because accuracy alone might be misleading.

Figure 5. F1-Score Accuracy Value

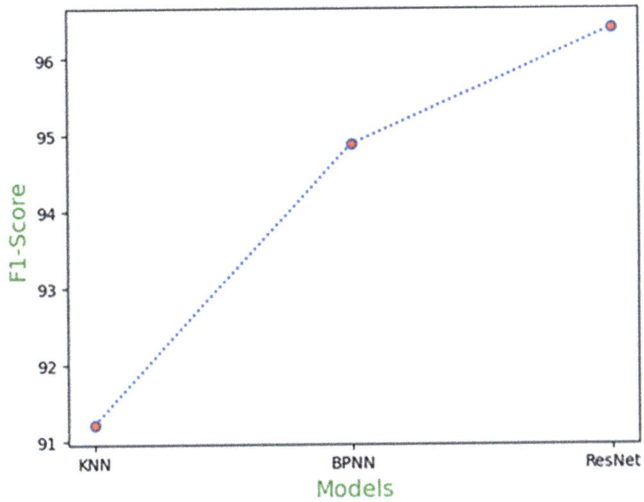

The result summary table 4 indicates that the highest percentage Accuracy values for the ResNet is consistently 95.89%. The ResNet models demonstrate strong potential for effective classification in optimization in waste system.

Table 4. Accuracy for Different Models (%)

Methods	**Accuracy**
KNN	90.67
BPNN	93.41
ResNet	**95.89**

Figure 6 illustrates the comparative performance of ResNet in relation to various algorithms. The ResNet Technique achieved an accuracy of 95.89% based on the current results. The application of ResNet yielded superior results in comparison to both KNN and the BPNN model.

Figure 6. Accuracy for Different Models

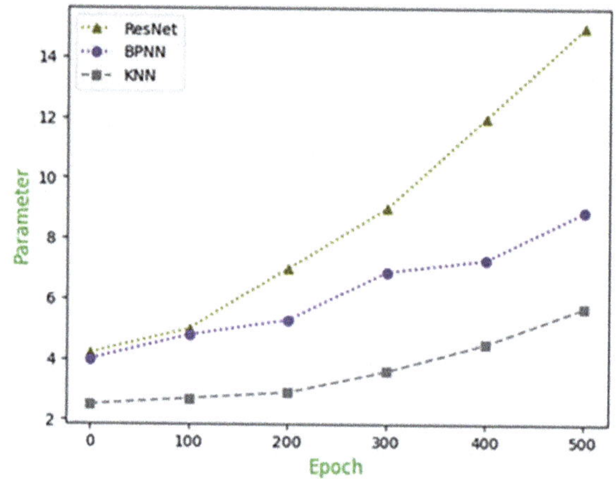

9. CHALLENGES AND LIMITATIONS

Smart transport systems that use AI have the potential to greatly improve sustainability, safety, and efficiency. Predictive analytics, driverless vehicles, and smart traffic management are just a few examples of the AI-driven technologies that are going to change the way people get around cities. Nevertheless, in order for these encouraging advancements to be widely adopted and put into practice, a number of constraints and hurdles must be overcome (Wei et al., 2022).

Challenges in AI-Driven Smart Transportation

Ethical and Social Implications: AI-powered transportation choices like route optimisation and resource allocation could raise ethical concerns. To avoid prejudice and bias in services, it is essential that AI algorithms be fair and transparent (Sun et al., 2017).

Data Privacy and Security Concerns: For AI systems to function, massive volumes of data must be gathered from various sensors, cameras, and networked devices. It is critical to safeguard data from cyber -attacks and to ensure the privacy of personal information. Misuse or data breaches can damage public trust, which in turn slows adoption (Johnson et al., 2019).

Regulatory and Legal Hurdles: The introduction of AI-powered traffic control systems and driverless cars presents formidable regulatory hurdles. The necessity for revised traffic regulations and the investigation of questions of responsibility in incidents involving autonomous cars necessitate thorough deliberation and clear statutory frameworks.

Infrastructure Readiness: AI in smart transport can only work with solid infrastructure, such as dependable power sources, sensor networks, and fast internet connections. Improving current infrastructure to sufficiently support AI technology may be difficult for many places (Graham & Denning, 2023).

Limitations of AI-Driven Smart Transportation

HCI: Clear interaction and communication between AI systems and human drivers, pedestrians, and cyclists is essential for the transition to autonomous cars. Ensuring public acceptance and safety requires the design of intuitive protocols and interfaces for smooth interaction.

Complexity of Urban Environments: Dynamic and unexpected situations, such as pedestrian movements, construction zones, and various traffic patterns, are present in urban settings. In order to keep operations safe and efficient, AI systems will need to adjust to these complexities (Dalal & Trombulak, 2023).

Cost and Scalability: Research, development, and infrastructure deployment can be quite expensive when implementing AI-driven solutions. Many cities and transport authorities still face the issue of achieving scalability while keeping costs low (Korenevskiy et al., 2023).

10. FUTURE DIRECTIONS

There will be revolutionary changes brought about by AI in traffic systems in the future. Innovations in sensor fusion, mechatronics, communication, integration of multimodal transportation, and ethical AI development are set to lead its future. Combining AI with the IoT 4G and 5G networks, is an important area for expansion. Data transfer rates and latency are directly proportional to the level of development in communication capacities. The rapid rollout of 5G demonstrates this promise. With its potential to facilitate V2X connections at hitherto unseen speeds and data transfer rates, it could pave the way for real-time data sharing amongst infrastructure components, traffic signals, and automobiles (Bouët et al., 2023).

Fast data transmission and analysis from IoT devices like smart sensors and linked vehicles will be possible because to these and other processing advancements like distributed computing and parallel processing. Vehicle queuing is less likely to occur, emergency services are more prompt, and incidents are responded to faster. Smarter, more adaptive, more accountable, more citizen-friendly, and more sustainable urban settings are within reach due to the synergy between AI, 5G and the IoT. This is especially true as cities continue to embrace IoT-enabled devices.

There is hope that AI for traffic systems will one day lead to fully autonomous control centres. Using predictive analysis, these systems will redirect vehicles in real-time depending on current conditions, alter traffic signals dynamically, and anticipate vehicular tailbacks. Moreover, autonomous traffic management will facilitate the dispatch of rescue vehicles and deal with any other situations that may arise. An increasing number of systems are aiming to function autonomously. There will be less need for human involvement and supervision as a result of this. But human oversight of these AI systems is crucial and must remain so. The idea that AI may help spread sophisticated multimodal transport networks is starting to take root.

There are a lot of moving parts when it comes to coordinating several forms of transportation like road, train, and air travel. These difficulties are best tackled by AI. AI has the ability to improve scheduling, handle transfers between transportation options, and forecast service demand in the future. As a result, transport network integration will be more seamless and fruitful. Cities and other large metropolitan areas really benefit from these sophisticated capabilities. ML has the potential to improve city transport systems and make them more user-friendly. There will be a growing importance of thinking about the ethical implications of AI (S. David et al., 2021).

AI systems that are more open and responsible is becoming more important. For AI to be used ethically in traffic management, it needs to be devoid of bias, make fair and equitable decisions, and have systems to explain and verify its own conclusions. Human supervision is crucial in this regard. To win over the public and earn their trust in AI-driven traffic solutions, these ethical problems must be addressed. This is especially true if, as is highly probable, AI-based systems gain greater independence and impact in city life on a regular basis. AI traffic systems are seeing fast innovation and research advancements. New technologies and approaches are being investigated by a plethora of ongoing initiatives. Researchers are exploring various areas of study with the hope of finding ways to improve traffic management systems, reorganise routing, and update traffic predictions.

These include DL, RL, BN, LR evolutionary computing and quantum computing. Research initiatives including public and private organisations, as well as academic institutions, NGOs, and government agencies, are propelling the creation of advanced AI traffic solutions. The potential for innovation in this subject is growing in tandem with the availability of more real-world data and the advancements in computer capability. Intelligent, flexible, and highly effective traffic systems will soon be a reality due to this.

The potential of AI and its subfields, such as ML and NLP has inspired hope in a number of industries. No reasonable person could expect it to have no beneficial effect on traffic management. We anticipate that traffic management will be heavily automated in the not-too-distant future. All road users, including pedestrians, will have a simplified and revitalised experience as a result of this. Motorists, commuters, and sightseers will all find the roads to be an absolute godsend if AI becomes widely used. Their transportation experiences are expected to be safer, more sustainable, and more delightful when AI and its subfields are applied responsibly, ethically, and under control.

11. CONCLUSION

The inclusion of cybersecurity in ITS, which includes sophisticated and time-critical systems, has a direct impact on road users' physical safety as well as the efficiency of transport services. There have been advancements in ITS standard development, however a comprehensive standard and a security plan have not been implemented just yet. It is important to thoroughly evaluate and test the interoperability of the different ITS standards with one another and with the external environment (Smart Cities, IoT). Although further experimental results and evaluations on their overall system fit are required some of the described technologies—such as blockchain, lightweight cryptographic techniques, segmentation of networks, and device fusion—will likely play a role in the evolution of ITS.

A portion of the technologies mentioned are still in the early stages of ITS application research. A lot of people are saying that AI and ML will be the big deciding factors in the future of ITS. There has been insufficient research into the security implications of integrating such techniques with ITS, despite the widespread promotion of their benefits. However, results from experiments incorporating these technologies into ITS safety measures are required. The concept of games and contract-based security constitute two additionally technologies that are expected to emerge in the ITS sector. They have already found use in IoT cybersecurity solutions and will most certainly make an appearance in ITS cybersecurity as well. There will be remarkable gains in productivity, security, and environmental friendliness with transportation systems that incorporate AI. However, hackers now have a larger target to exploit because of these innovations. A combination of robust AI architecture, secure communication protocols, regulatory monitoring, and human awareness is needed to ensure the cybersecurity of such systems. Transportation ecosystems can be made smarter, safer, and more reliable if these problems are solved.

REFERENCES

Ahmad, F., Adnane, A., Franqueira, V., Kurugollu, F., & Liu, L. (2018). Man-In-The-Middle Attacks in Vehicular Ad-Hoc Networks: Evaluating the Impact of Attackers' Strategies. *Sensors (Basel), 18*(11), 4040. DOI: 10.3390/s18114040 PMID: 30463282

Ahmad, F., Kurugollu, F., Adnane, A., Hussain, R., & Hussain, F. (2020). MARINE: Man-in-the-Middle Attack Resistant Trust Model in Connected Vehicles. *IEEE Internet of Things Journal, 7*(4), 3310–3322. DOI: 10.1109/JIOT.2020.2967568

Al-Juboori, S. A. M., Hazzaa, F., Jabbar, Z. S., Salih, S., & Gheni, H. M. (2023). Man-in-the-middle and denial of service attacks detection using machine learning algorithms. *Bulletin of Electrical Engineering and Informatics, 12*(1), 418–426. DOI: 10.11591/eei.v12i1.4555

Al-Shareeda, M. A., & Manickam, S. (2022). Man-in-the-Middle Attacks in Mobile Ad Hoc Networks (MANETs): Analysis and Evaluation. *Symmetry, 14*(8), 1543. DOI: 10.3390/sym14081543

Almiani, M., AbuGhazleh, A., Al-Rahayfeh, A., Atiewi, S., & Razaque, A. (2020). Deep recurrent neural network for IoT intrusion detection system. *Simulation Modelling Practice and Theory, 101*, 102031. DOI: 10.1016/j.simpat.2019.102031

Alsaade, F. W., & Al-Adhaileh, M. H. (2023). Cyber Attack Detection for Self-Driving Vehicle Networks Using Deep Autoencoder Algorithms. *Sensors (Basel), 23*(8), 4086. DOI: 10.3390/s23084086 PMID: 37112429

Andreolini, M., & Marchetti, M. (2022). DAGA: Detecting Attacks to In-Vehicle Networks via N-Gram Analysis. *IEEE Transactions on Vehicular Technology, 71*(11), 11540–11554. DOI: 10.1109/TVT.2022.3190721

Basri, R., Karmakar, G., Newaz, S. H. S., Kamruzzaman, J., Nguyen, L., Alam, M. M., & Usman, M. (2025). Enhancing IoT security: Assessing instantaneous communication trust to detect man-in-the-middle attacks. *Future Generation Computer Systems, 166*(July 2024), 107714. DOI: 10.1016/j.future.2025.107714

Bendovschi, A. (2024). Cyber-Attacks – Trends, Patterns and Security Countermeasures. *Procedia Economics and Finance, 28*(April), 24–31. DOI: 10.1016/S2212-5671(15)01077-1

Bouët, A., Decreux, Y., Fontagné, L., Jean, S., & Laborde, D. (2023). Assessing Applied Protection across the World*. *Review of International Economics, 16*(5), 850–863. DOI: 10.1111/j.1467-9396.2008.00753.x

Buchan, R. (2019). Cyber Attacks: Unlawful Uses of Force or Prohibited Interventions? *Journal of Conflict and Security Law*, *17*(2), 212–227. DOI: 10.1093/jcsl/krs014

Carsten, P., Andel, T. R., Yampolskiy, M., & McDonald, J. T. (2024). In-Vehicle Networks. *Proceedings of the 10th Annual Cyber and Information Security Research Conference, 06-08-Apri*(October 2018), 1–8. DOI: 10.1145/2746266.2746267

Choudhary, A. S., Choudhary, P. P., & Salve, S. (2018). A Study On Various Cyber Attacks And A Proposed Intelligent System For Monitoring Such Attacks. *2018 3rd International Conference on Inventive Computation Technologies (ICICT), November 2018*, 612–617. DOI: 10.1109/ICICT43934.2018.9034445

Chowdhury, F., & Ferdous, M. S. (2017). Modelling Cyber Attacks. *International Journal of Network Security & its Applications*, *9*(4), 13–31. DOI: 10.5121/ijnsa.2017.9402

Cramer, R., & Damgård, I. (2024). Secure Signature Schemes based on Interactive Protocols. In *Lecture Notes in Computer Science (including subseries Lecture Notes in Artificial Intelligence and Lecture Notes in Bioinformatics)* (Vol. 963, pp. 297–310). DOI: 10.1007/3-540-44750-4_24

Dalal, & Trombulak, S. C. (2023). Classifying levels of protection in Marine Protected Areas. *Marine Policy, 36*(3), 576–582. DOI: 10.1016/j.marpol.2011.08.011

David, S., Cromer, D., Reynaldi, A., Schlub, T. E., Wheatley, A. K., Juno, J. A., Subbarao, K., Kent, S. J., Triccas, J. A., & Davenport, M. P. (2021). Neutralizing antibody levels are highly predictive of immune protection from symptomatic SARS-CoV-2 infection. *Nature Medicine*, *27*(7), 1205–1211. DOI: 10.1038/s41591-021-01377-8 PMID: 34002089

Ganesh Reddy, K. (2023). Classification of Distributed Denial of Service Attacks in VANET: A Survey. In *Wireless Personal Communications* (Vol. 132, Issue 2). Springer US. DOI: 10.1007/s11277-023-10643-6

Graham, G. S., & Denning, P. J. (2023). Protection—Principles and practice. *Proceedings of the November 16-18, 2023 Fall Joint Computer Conference on - AFIPS '23 (Fall)*, 417. DOI: 10.1145/1478873.1478928

Hong, J. (2020). Cyber Security Issues in Connected Vehicle of Intelligent Transport System. *Indian Journal of Science and Technology*, *9*(24), 1–7. DOI: 10.17485/ijst/2016/v9i24/96027

Iyer, L. S. (2021). AI enabled applications towards intelligent transportation. *Transportation Engineering*, *5*, 100083. DOI: 10.1016/j.treng.2021.100083

Javeed, D., & MohammedBadamasi, U. (2020). Man in the Middle Attacks: Analysis, Motivation and Prevention. *International Journal of Computer Networks and Communications Security*, *8*(7), 52–58. DOI: 10.47277/IJCNCS/8(7)1

Johnson, R. P., Lifson, J. D., Czajak, S. C., Cole, K. S., Manson, K. H., Glickman, R., Yang, J., Montefiori, D. C., Montelaro, R., Wyand, M. S., & Desrosiers, R. C. (2019). Highly Attenuated Vaccine Strains of Simian Immunodeficiency Virus Protect against Vaginal Challenge: Inverse Relationship of Degree of Protection with Level of Attenuation. *Journal of Virology*, *73*(6), 4952–4961. DOI: 10.1128/JVI.73.6.4952-4961.1999 PMID: 10233957

Khattak, Z. H., Park, H., Hong, S., Boateng, R. A., & Smith, B. L. (2018). INVESTIGATING CYBERSECURITY ISSUES IN ACTIVE TRAFFIC MANAGEMENT SYSTEMS. *INVESTIGATING CYBERSECURITY ISSUES IN ACTIVE TRAFFIC MANAGEMENT SYSTEMS*, 1–25.

Korenevskiy, N. A., Al-Kasasbeh, R. T., Al-Kasasbeh, E. T., Al-Smadi, M. M., Aikeyeva, A. A., Al-Jundi, M., Rodionova, S. N., Al-Habahbeh, O. M., Filist, S., Alshamasin, M. S., & Maksim, I. (2023). Method for Determining the Body's Level of Protection According to Oxidant Status in Assessing the Influence of Industrial Risk Factors on Health. *Critical Reviews in Biomedical Engineering*, *51*(2), 1–17. DOI: 10.1615/CritRevBiomedEng.2023047224 PMID: 37551905

Li, Y., & Liu, Q. (2021). A comprehensive review study of cyber-attacks and cyber security Emerging trends and recent developments. *Energy Reports*, *7*, 8176–8186. DOI: 10.1016/j.egyr.2021.08.126

Mahmood, A., Anbar, M., Manickam, S., & Hasbullah, I. H. (2020). Review of Prevention Schemes for Man-In-The-Middle (MITM) Attack in Vehicular Ad hoc Networks. *International Journal of Engineering and Management Research*, *10*(3), 153–158. DOI: 10.31033/ijemr.10.3.23

Manickam, S. (2023). A Systematic Literature Review on Security of Vehicular Ad-Hoc Network (VANET) Based on VEINS Framework. *IEEE Access : Practical Innovations, Open Solutions*, *11*(April), 46218–46228. DOI: 10.1109/ACCESS.2023.3274774

Mashkina, I., Rzaieva, S., Kostiuk, Y., Mazur, N., & Brzhevska, Z. (2025). Cybersecurity in Intelligent Transport Systems: Current Challenges and Solutions. *CEUR Workshop Proceedings*, *3991*, 1–13.

Mehrdad, O. (2019). Intrusion Detection Systems for Intra-Vehicle Networks : A Review. *IEEE Access : Practical Innovations, Open Solutions*, *7*, 21266–21289. DOI: 10.1109/ACCESS.2019.2894183

Mezzour, G., Carley, L. R., & Carley, K. M. (2024). Global Mapping of Cyber Attacks. SSRN *Electronic Journal*. DOI: 10.2139/ssrn.2729302

Mohapatra, H. (2020). Handling of Man-In-The-Middle Attack in WSN Through Intrusion Detection System. *International Journal of Emerging Trends in Engineering Research, 8*(5), 1503–1510. DOI: 10.30534/ijeter/2020/05852020

Patel, S., & Zaveri, J. (2020). A Risk-Assessment Model for Cyber Attacks on Information Systems. *Journal of Computers, 5*(3), 352–359. DOI: 10.4304/jcp.5.3.352-359

Prakash, A. J., Biju, J. M., & Gopal, N. (2023). Cyber Attacks and Its Different Types. *International Research Journal of Engineering and Technology*, 4849–4852. https://www.researchgate.net/publication/366090166

Rana, O., & Burnap, P. (2020). *CYBERATTACKS AND COUNTERMEASURES FOR IN-VEHICLE NETWORKS.*

Rid, T., & Buchanan, B. (2025). Attributing Cyber Attacks. *The Journal of Strategic Studies, 38*(1–2), 4–37. DOI: 10.1080/01402390.2014.977382

Robert, A., Ofori-amanfo, K. B., Mills, G. A., & Koumadi, K. M. (2019). *Detection and Prevention of Man-in-the-Middle Spoofing Attacks in MANETs Using Predictive Techniques in Artificial Neural Networks* (Vol. 2019). ANN., DOI: 10.1155/2019/4683982

Shah, Y., & Sengupta, S. (2020). A survey on Classification of Cyber-attacks on IoT and IIoT devices. *2020 11th IEEE Annual Ubiquitous Computing, Electronics & Mobile Communication Conference (UEMCON)*, 0406–0413. DOI: 10.1109/UEMCON51285.2020.9298138

Stabili, D., Marchetti, M., & Colajanni, M. (2025). *Detecting attacks to internal vehicle networks through Hamming distance* (Vol. 21, Issue September 2017).

Sun, Y., Wu, L., Wu, S., Li, S., Zhang, T., Zhang, L., Xu, J., Xiong, Y., & Cui, X. (2017). Attacks and countermeasures in the internet of vehicles. *Annales des Télécommunications, 72*(5–6), 283–295. DOI: 10.1007/s12243-016-0551-6

Thomas, O., Salek, M. S., Tine, J.-M., Rahman, M., Hockstad, T., & Chowdhury, M. (2025). Cybersecurity in Transportation Systems: Policies and Technology Directions. *Cybersecurity in Transportation Systems: Policies and Technology Directions*. https://arxiv.org/abs/2501.05356

Tomlinson, A., Bryans, J., & Shaikh, S. A. (2021). Using internal context to detect automotive controller area network attacks. *Computers & Electrical Engineering, 91*, 107048. DOI: 10.1016/j.compeleceng.2021.107048

Uma, M., & Padmavathi, G. (2023). A survey on various cyber attacks and their classification. *International Journal of Network Security*, *15*(5), 390–396.

Vegesna, D. V. V. (2022). Investigations on Cybersecurity Challenges and Mitigation Strategies in Intelligent transport systems. *Irish Interdisciplinary Journal of Science & Research*, *06*(04), 70–86. DOI: 10.46759/IIJSR.2022.6409

Verma, S. (2024). *Intelligent Urban Transportation Systems (Iuts): A Survey of AI-Driven Innovations and Future Directions*. Issue August.

Waxman, M. C. (2019). Cyber-Attacks and the Use of Force: Back to the Future of Article. SSRN *Electronic Journal, 2*(4). DOI: 10.2139/ssrn.1674565

Wei, J., Chen, Y., Lai, Y., Wang, Y., & Zhang, Z. (2022). Domain Adversarial Neural Network-Based Intrusion Detection System for In-Vehicle Network Variant Attacks. *IEEE Communications Letters*, *26*(11), 2547–2551. DOI: 10.1109/LCOMM.2022.3195486

Wu, W., Li, R., Xie, G., An, J., Bai, Y., Zhou, J., & Li, K. (2020). A Survey of Intrusion Detection for In-Vehicle Networks. *IEEE Transactions on Intelligent Transportation Systems*, *21*(3), 919–933. DOI: 10.1109/TITS.2019.2908074

Yugoslavia. (2023). Cyber Attacks and the Laws of War. *Cyber Attacks and the Laws of War*, 1–45.

Zaidi, T., & Faisal, S. (2019). An Overview : Various Attacks in VANET. *2019 4th International Conference on Computing Communication and Automation (ICCCA), July*, 1–6. DOI: 10.1109/CCAA.2018.8777538

Zhao, K., Li, J., Yu, L., Luo, X., Zhou, Y., & Gu, G. (2022). SAID: State-aware Defense Against Injection Attacks on In-vehicle Network. *Proceedings of the 31st USENIX Security Symposium, Security 2022*, 1921–1938.

Zhuang, R., Bardas, A. G., DeLoach, S. A., & Ou, X. (2025). A Theory of Cyber Attacks. *Proceedings of the Second ACM Workshop on Moving Target Defense*, 11–20. DOI: 10.1145/2808475.2808478

Chapter 14
Ethical and Regulatory Considerations in AI-Driven Transportation Safety

Harry Sutanto
http://orcid.org/0000-0001-8356-4976
Binus University, Indonesia

Mohammad Ichsan
Binus University, Indonesia

ABSTRACT

Artificial Intelligence (AI) is revolutionizing transportation through autonomous vehicles, smart traffic systems, and predictive safety tools. While these innovations enhance efficiency and reduce accidents, they also raise ethical, legal, and social challenges. This chapter highlights the need for strong governance, transparent algorithms, and inclusive stakeholder collaboration. Clear legal frameworks must define liability, while ethical oversight ensures fairness and accessibility. By viewing AI as both a technological advancement and societal responsibility, policymakers, industry, and civil society can foster safer, equitable, and trustworthy transportation systems, offering a model for responsible technology adoption across sectors.

DOI: 10.4018/979-8-3373-3231-4.ch014

INTRODUCTION

Background

Artificial Intelligence (AI) is reshaping transportation by enhancing safety, efficiency, and mobility. Technologies such as autonomous vehicles (AVs), intelligent traffic management, predictive maintenance, and passenger monitoring are increasingly integral to modern infrastructure. AI's potential to reduce human error, which causes over 90% of traffic accidents worldwide (WHO, 2023), can significantly lower accident rates and improve system reliability, creating safer and more resilient mobility ecosystems. However, rapid AI adoption brings complex ethical, regulatory, and societal challenges. AV crashes and system failures have raised debates around accountability, transparency, and regulatory preparedness, highlighting both AI's promise and its vulnerabilities (Cihon et al., 2021).

AI's capacity to analyze large, dynamic datasets in real time allows for proactive risk management. Traffic systems can reduce congestion by adjusting signals, rerouting vehicles, and responding swiftly to incidents, improving safety and reducing emissions, thereby supporting urban sustainability. Predictive analytics help identify technical anomalies early, enabling preventive maintenance. Beyond road transport, AI enhances monitoring, early warning systems, and safety management in aviation, rail, and maritime sectors. Collectively, these applications demonstrate AI's potential to advance safety, efficiency, and environmental performance.

Nonetheless, challenges remain. Algorithmic bias threatens vulnerable road users, including pedestrians, cyclists, and marginalized communities underrepresented in training datasets (Leslie, 2021). Data dependency raises privacy concerns, as collection of behavioral, locational, and biometric information risks surveillance, misuse, and loss of autonomy. Legal frameworks lag behind technological progress, leaving liability for AI failures unresolved among developers, manufacturers, regulators, and operators. Without clear legal and ethical safeguards, public trust may erode, slowing adoption of beneficial technologies.

Viewing AI as a socio-technical system is key to addressing these challenges. Technical innovation must align with ethical commitments, governance, and societal expectations. Embedding fairness, transparency, accountability, and safety into AI design and deployment ensures equitable outcomes. Initiatives like the European Union's AI Act and standards such as ISO 21448, focusing on safety of intended functionality, illustrate efforts to harmonize regulations, clarify liability, and promote consistency globally (European Commission, 2022; OECD, 2021).

Integrating AI in transportation is more than a technological shift; it is a societal endeavor. A balanced approach—combining ethical principles, robust governance, and collaboration among governments, industry, academia, and civil society—is essential. When innovation is aligned with responsibility and trust, AI can become a catalyst for safer, fairer, and more sustainable transportation systems. AI should thus be seen not only as a technological breakthrough but as a shared responsibility requiring careful stewardship to maximize benefits while safeguarding the public good.

Objectives

This chapter aims to provide a comprehensive and multidisciplinary analysis of ethical and regulatory considerations in AI-driven transportation safety, with the following objectives:

1. Analyze ethical challenges: Explore issues such as algorithmic bias, fairness in automated decision-making, transparency of AI processes, and accountability in accident scenarios where responsibility remains ambiguous.
2. Evaluate regulatory frameworks: Assess both existing and emerging regulations across global regions, highlighting strengths, gaps, inconsistencies, and potential pathways for harmonized governance in AI transportation safety.
3. Examine socio-technical implications: Investigate human–machine interaction, the influence of AI on driver and passenger behavior, public trust, and the broader societal acceptance of AI-enabled mobility solutions.
4. Propose an integrated framework: Develop a model that harmonizes ethical principles—such as fairness, accountability, and transparency—with regulatory standards, thereby guiding safe, responsible, and transparent implementation of AI.
5. Provide practical insights: Offer actionable recommendations for policymakers, industry leaders, and civil society to balance innovation with safety, foster sustainable adoption, and safeguard the public interest.

Scope

This chapter focuses on AI applications that directly impact transportation safety, with the following scope and emphasis:

1. Core AI Applications: analyze the role of autonomous vehicles, intelligent traffic management systems, and predictive risk assessment platforms, which helps readers understand the technological potential of AI in enhancing transportation safety.
2. Ethical Considerations: address fairness in algorithmic decision-making, transparency of system operations, and accountability in accident or failure scenarios,

which provides critical insights into the ethical challenges that must be managed in AI deployment.

3. Regulatory Dimensions: examine national and international policies, evolving safety standards, and liability frameworks that govern AI in transportation, which strengthens awareness of the importance of regulation and cross-jurisdictional policy harmonization.
4. Cross-Sector Relevance: extend insights beyond road transportation to rail, maritime, and aviation contexts where AI deployment is expanding, which broadens readers' perspective by highlighting implications across multiple transportation domains.
5. Analytical Approach: draw upon global case studies, policy frameworks, and academic research to provide a balanced perspective on challenges and opportunities, which equips readers with a comprehensive, evidence-based framework for analysis.

LITERATURE REVIEW

AI Applications in Transportation Safety

AI has rapidly evolved into a powerful driver of change in the transportation sector, reshaping how safety, efficiency, and system reliability are understood and managed (Rossi et al., 2021; Zhang & Li, 2022). Unlike conventional transportation systems that depend heavily on human judgment—often limited by fatigue, distraction, or delayed reaction—AI-enabled platforms bring together real-time monitoring, predictive analytics, and autonomous responses to complex situations across different modes of transport. These capabilities allow systems to anticipate potential risks, respond proactively, and maintain stable operations even under dynamic and uncertain conditions. In practice, AI's role extends from supporting everyday mobility to safeguarding lives, illustrating both its transformative promise and the responsibility that comes with its deployment.

AV represent one of the most widely recognized manifestations of AI in transportation. Powered by advanced machine learning algorithms and sophisticated sensor fusion technologies, AVs are designed to detect hazards, interpret their surroundings, and make rapid decisions in real time. Their potential lies particularly in reducing dependence on human drivers, as more than 90 percent of global traffic fatalities are still linked to human error (WHO, 2023). At the same time, high-profile incidents—such as crashes associated with Tesla's Autopilot system—have revealed the unresolved challenges surrounding reliability, accountability, and appropriate human-machine interaction. These events not only raise public concern but also reinforce the urgent need for stronger regulatory structures and clearer safety standards to ensure responsible deployment (Cihon et al., 2021).

Beyond autonomous driving, AI has become central to the development of Intelligent Traffic Management Systems. These systems manage vehicle flow, regulate traffic signals, and predict congestion patterns with a degree of precision that traditional approaches cannot achieve. Singapore's Intelligent Transport System stands as a prominent example, demonstrating how AI-powered cameras and predictive analytics can significantly reduce congestion while improving road safety outcomes (Chen et al., 2022). By enabling authorities to respond swiftly to evolving traffic conditions, such systems help create smoother, safer, and more sustainable urban mobility environments.

AI-driven predictive accident analytics also play an increasingly critical role in risk prevention. By combining historical data with real-time input, machine learning models can identify hazardous driving conditions and behaviors before incidents occur. In China, these systems have been introduced within logistics fleets to detect risky driving patterns, allowing targeted interventions that have led to notable reductions in accidents in pilot regions (Kong et al., 2021). This proactive approach marks a shift from reacting to crashes toward preventing them altogether.

Importantly, AI's contribution is not limited to roads. In aviation, major airlines such as Lufthansa have adopted AI-based predictive maintenance systems to detect potential component failures in advance, helping minimize in-flight risks and operational disruptions. Similarly, Japan's rail sector utilizes AI to monitor infrastructure conditions such as track wear and mechanical stress, enabling timely maintenance and ensuring safer, more reliable services (Li & Wang, 2023). These developments show how AI strengthens safety across diverse transportation domains.

In high-density public environments, AI is also being used to protect passengers and manage crowds more effectively. During the Tokyo Olympics, AI-powered monitoring systems were implemented across metro stations to anticipate overcrowding and identify potentially hazardous behaviors, enabling quicker interventions and smoother passenger movement (Singh et al., 2022). Such applications highlight how AI can support human decision-makers, especially in situations where speed, accuracy, and situational awareness are crucial.

Taken together, these examples illustrate AI's dual role: enhancing transportation safety while reinforcing system resilience. When thoughtfully integrated into critical infrastructure and guided by evidence from global experiences, AI can support safer, more efficient, and future-ready mobility systems. However, these advancements also serve as reminders that technological innovation must progress in tandem with ethical guidance, regulatory clarity, and societal trust. Aligning AI-powered development with responsible governance is essential not only for managing risks but also for ensuring that transportation systems remain equitable, reliable, and publicly trusted as they continue to evolve.

Ethical Challenges in AU Deployment

Despite the considerable promise of AI in improving transportation safety, its integration also introduces a series of complex ethical questions that require careful, sustained, and critical attention (Leslie, 2020; Morley et al., 2021). These concerns are not merely technical challenges that engineers can solve through better coding or more advanced systems. Instead, they extend deeply into social and political spheres, shaping how people perceive AI, whether governments feel confident regulating it, and how societies ultimately decide to accept or reject AI-enabled mobility systems. In many ways, the ethical issues surrounding AI will determine whether its potential benefits can be realized in a way that is fair, trustworthy, and sustainable over the long term.

One central concern relates to algorithmic bias. AI models depend entirely on the quality and nature of the data used to train them. If these datasets are incomplete, historically biased, or fail to reflect the diversity of real-world users, the resulting AI decisions may unfairly disadvantage certain groups. Pedestrians, cyclists, or marginalized communities may be exposed to higher risks or experience less reliable service. Rather than creating safer and more inclusive transportation systems, biased AI could unintentionally reproduce existing inequalities and deepen social divides. This raises fundamental ethical questions about fairness, justice, and who truly benefits from technological innovation (Mehrabi et al., 2021).

Data privacy represents another major ethical challenge. Modern AI-driven transportation systems rely on extensive data collection, including behavioral, locational, and in some cases biometric data. While such information is essential for improving accuracy, safety, and system performance, it also opens the door to potential misuse, surveillance, and violations of individual autonomy. The central ethical dilemma lies in balancing innovation with the protection of personal rights. This requires strong data governance mechanisms, meaningful consent frameworks, and clear protections to ensure that efficiency does not come at the expense of privacy (Whittlestone & Clark, 2021).

Equally significant is the issue of accountability and liability. When an AI-operated vehicle or traffic management system fails and causes harm, determining who is responsible becomes difficult. Responsibility could lie with the manufacturer, the software developer, the system operator, or even the regulatory body that approved the system. This ambiguity complicates legal accountability, insurance arrangements, and compensation processes, while also shaping public perceptions of fairness and security (European Commission, 2020).

Finally, transparency and explainability remain critical ethical considerations. Many AI systems operate like "black boxes," producing decisions without clear, understandable reasoning. This lack of interpretability can hinder regulatory oversight and weaken public trust, as people may hesitate to rely on systems they cannot fully comprehend or question (Burrell et al., 2020).

Addressing these ethical dimensions is not optional. It is essential to ensuring that AI in transportation develops in a manner that is equitable, trustworthy, and socially sustainable, enabling technological progress while safeguarding public interest and human dignity.

Global Regulatory Frameworks

The governance of AI in transportation varies widely across regions, shaped by differing legal traditions, technological capacities, political cultures, and societal expectations (Calo, 2020; Stilgoe, 2020). These variations reveal a fragmented global regulatory environment in which countries experiment with distinct approaches to balancing innovation, safety, and public trust. While such diversity reflects legitimate contextual needs, it also presents fundamental challenges for interoperability, cross-border coordination, and the establishment of shared ethical standards in AI-enabled mobility systems.

In the European Union, regulatory development has been both proactive and principled. The proposed AI Act introduces a structured, risk-based framework that classifies AI applications according to their potential societal and safety impacts. High-risk uses—such as autonomous driving—are placed under stringent requirements related to transparency, accountability, data governance, robustness, and meaningful human oversight. Rather than treating regulation as an obstacle, the EU positions it as a mechanism to guide responsible innovation. By embedding ethical considerations directly into regulatory design, the EU seeks not only to protect citizens, but also to cultivate public trust and create a stable environment for long-term technological development (European Commission, 2021).

In contrast, the United States adopts a more decentralized and market-oriented approach. Instead of a single comprehensive regulatory framework, governance responsibilities are distributed across various agencies, with the National Highway Traffic Safety Administration (NHTSA) playing a key role in issuing guidelines and voluntary standards. This flexible approach aims to enable rapid experimentation, encourage private sector leadership, and reduce regulatory barriers to innovation. However, critics argue that such flexibility can come at a cost, leaving unresolved questions regarding accountability, consumer protection, liability clarity, and consistent national standards (Calo, 2020). These gaps may complicate public confidence and risk management as AI-driven transportation technologies scale more widely.

Meanwhile, the Asia-Pacific region presents a diverse but increasingly dynamic governance landscape. Countries such as Japan, Singapore, and China are advancing adaptive and learning-oriented regulatory strategies. These typically combine national standards with strong state involvement, extensive public–private partnerships, and real-world pilot programs. China integrates regulatory sandboxes with firm oversight to test and refine policy in parallel with technological progress. Singapore, on the other hand, emphasizes structured safety assurance frameworks, ensuring that innovation proceeds alongside clear risk controls and monitored experimentation. These approaches highlight a pragmatic philosophy: regulators learn from deployment rather than merely theorizing about it (Stilgoe, 2020).

Taken together, these varied models underscore both the richness and complexity of AI governance in transportation. While regional diversity is inevitable and even beneficial in fostering tailored policy responses, it also reinforces the need for global dialogue and coordination. Developing integrated ethical-regulatory frameworks that reflect international best practices—while still respecting local socio-technical realities—is essential. Such alignment is critical for ensuring that AI-enabled transportation remains safe, equitable, and sustainable, and that societies worldwide can benefit from innovation without sacrificing accountability or public trust.

Integration of Ethics, Technology and Policy

Recent scholarship underscores that advancing AI in transportation requires far more than mere technical optimization; it instead demands a holistic alignment between technological innovation, ethical responsibility, and regulatory robustness (Leslie, 2020; Morley et al., 2021). Viewing AI as a socio-technical system implies that safety, fairness, transparency, and accountability must be intentionally embedded throughout the entire lifecycle of system development—from conceptualization and design to implementation, monitoring, and continuous improvement. Regulatory frameworks, therefore, cannot remain static; they must be adaptive, risk-sensitive, and future-oriented, capable of responding to rapid technological evolution while still safeguarding public welfare. At the same time, ethical oversight plays a crucial role in ensuring that AI applications do not reinforce inequality, introduce bias, or marginalize vulnerable user groups, particularly in contexts where transportation systems directly affect mobility rights, access to services, and overall quality of life.

Achieving this delicate balance requires genuine multi-stakeholder collaboration that involves policymakers, transportation authorities, technology developers, industry leaders, researchers, and civil society actors. Through such collaboration, governance models can be developed that are not only technically sound but also socially legitimate, reflecting both the opportunities offered by AI and the broader values, expectations, and concerns of the public. This collaborative governance approach can strengthen public trust, enhance accountability, and promote long-term sustainability in AI-driven transportation systems. Ultimately, responsible AI integration in transportation is not solely about efficiency and technological advancement; it is fundamentally about ensuring that innovation contributes to safer, fairer, and more human-centered mobility futures (European Commission, 2021).

Conclusion of Literature Review

The literature highlights that AI is poised to transform transportation safety by reducing accidents, optimizing traffic flows, and enhancing predictive maintenance capabilities. Numerous studies suggest that AI-driven systems could potentially mitigate over 90% of crashes linked to human error, particularly those related to distraction, fatigue, and poor decision-making; however, despite these promising advantages, the ethical and regulatory challenges surrounding AI remain formidable and complex (WHO, 2023; Leslie, 2020). Scholars increasingly emphasize that technological sophistication alone is insufficient. Instead, AI in transportation must be governed as a socio-technical system that integrates human values, legal safeguards, and institutional accountability alongside engineering innovation.

Key ethical challenges discussed in recent scholarship include:

1. Algorithmic Bias – the risk that AI systems reinforce discrimination or unequal outcomes if training datasets are incomplete, unrepresentative, or historically biased, potentially disadvantaging certain demographic groups or regions (Mehrabi et al., 2021).
2. Data Privacy – persistent concerns about excessive surveillance, informed consent, and misuse of behavioral, geolocation, or biometric data collected through AI-enabled mobility platforms (Whittlestone & Clark, 2021).
3. Accountability and Liability – the difficulty of determining legal responsibility when autonomous systems are involved in accidents or operational failures, raising questions about whether liability rests with developers, manufacturers, operators, or regulators (Morley et al., 2021).
4. Transparency and Explainability – the opacity of complex AI models, particularly deep-learning systems, which can undermine regulatory oversight and erode user trust when decision-making processes are not interpretable (Burrell et al., 2020).

While emerging regulatory frameworks such as the EU AI Act offer structured, risk-based governance and clearer expectations for safety assurance, transparency obligations, and conformity assessments, global approaches remain uneven. Differences across jurisdictions—ranging from the United States' more sectoral and industry-led regulatory style to Asia-Pacific's adaptive, pilot-driven experimentation—underscore the fragmentation of international governance (European Commission, 2021). This lack of harmonization risks regulatory gaps, inconsistent safety standards, and potential barriers to cross-border innovation.

To fully realize AI's transformative potential, the literature calls for harmonized, forward-looking, and multi-stakeholder governance strategies. These strategies should embed fairness, transparency, accountability, and human oversight into every stage of AI development and deployment. Integrating ethical principles, robust regulation, and public engagement can not only enhance operational efficiency and safety outcomes but also strengthen societal trust, legitimacy, and long-term sustainability in AI-enabled transportation systems.

Figure 1. Ethical Challenges of AI

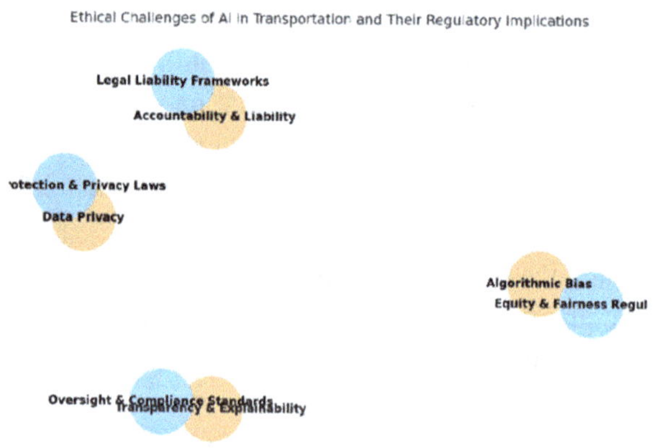

Figure 1 highlights the challenges of embedding fairness, transparency, accountability, and human oversight throughout AI development and deployment, promoting safer operations, societal trust, regulatory compliance, and long-term sustainable transportation systems.

METHODOLOGY/RESEARCH APPROACH

Research Design

This study adopts a qualitative and integrative research design, emphasizing conceptual and theoretical development rather than primary data collection. Such an approach is particularly suited to rapidly evolving domains like AI-driven transportation, where technological progress, policy experimentation, and emerging ethical debates often outpace empirical evaluation and statistical generalization. By synthesizing existing scholarly literature, policy frameworks, regulatory proposals, technical guidelines, and illustrative case studies across regions, this study aims to construct a comprehensive understanding of both opportunities and challenges associated with AI deployment in transportation safety. This design also enables identification of conceptual gaps, unresolved controversies, and underexplored dimensions that warrant future empirical and interdisciplinary investigation.

The analytical framework is guided by three complementary theoretical perspectives. First, the Ethics of Technology framework applies core normative principles such as beneficence, justice, non-maleficence, and respect for autonomy to assess how AI-enabled transportation systems influence human well-being, equity, and societal trust (Morley et al., 2021). This lens is particularly useful for interrogating issues such as algorithmic bias, data governance, and the moral legitimacy of automated decision-making in safety-critical environments.

Second, Risk Governance Theory provides insights into how societies conceptualize, anticipate, assess, and distribute responsibilities in managing emerging technological risks (Renn, 2020). Within AI-driven transportation, this perspective illuminates debates surrounding acceptable risk thresholds, precaution versus innovation, and liability allocation in autonomous system failures. It also supports analysis of institutional capacity, regulatory responsiveness, and the balance between flexibility and certainty in governance structures.

Third, the Accountability and Governance perspective emphasizes the obligations of developers, policymakers, manufacturers, and system operators to ensure transparency, oversight, auditability, and ethical compliance throughout the AI lifecycle—from design and testing to deployment and post-implementation monitoring (Leslie, 2020). This perspective underscores the need for clear accountability chains, robust documentation, explainability standards, and mechanisms for redress.

By triangulating these perspectives, the research design provides a multidimensional analytical lens that integrates ethical reasoning, risk management principles, and governance accountability. This triangulation enables a richer and more nuanced examination of ethical dilemmas such as algorithmic bias, privacy intrusions, and contested liability, while also contextualizing regional variations in regulatory approaches across the European Union, United States, and Asia-Pacific. Furthermore, it facilitates critical reflection on how values, institutional cultures, and policy priorities shape AI governance trajectories.

Ultimately, this integrative qualitative design supports the development of conceptual frameworks that align technological innovation with ethical imperatives and robust governance architectures. It contributes to scholarly discourse by clarifying conceptual linkages between ethics, risk, and accountability; informing policymakers through theoretically grounded insights; and providing a foundation for future empirical, comparative, and interdisciplinary studies. In doing so, the study advances the broader agenda of promoting responsible, trustworthy, and socially sustainable deployment of AI in enhancing transportation safety.

Data Sources of the Study

This study draws upon three principal categories of sources to construct a robust analytical foundation for examining ethical and regulatory issues in AI-driven transportation. First, academic literature constitutes the core intellectual base, providing theoretical grounding, conceptual refinement, and empirical insights. Peer-reviewed studies on AI ethics, transportation safety, autonomous vehicle systems, and governance frameworks illuminate both the transformative potential of AI and the emerging ethical dilemmas that accompany its deployment. These works emphasize tensions between innovation, safety, fairness, accountability, and social trust, thereby framing AI not merely as a technical tool but as a socio-technical system requiring careful ethical stewardship (Winfield & Jirotka, 2020; Rossi et al., 2021).

Second, regulatory documents form a critical source of evidence, offering insight into how ethical principles are operationalized into policy and law. Key references include the European Union's AI Act, guidelines issued by the U.S. National Highway Traffic Safety Administration (NHTSA), and adaptive governance initiatives across Asia-Pacific. These frameworks reveal diverse regulatory philosophies—from precautionary, risk-based oversight to more experimental, innovation-driven models—and provide a comparative perspective on how different jurisdictions balance public safety, economic competitiveness, and technological progress (European Commission, 2021; Stilgoe, 2020).

Third, case studies provide contextual depth and real-world relevance. Documented autonomous vehicle trials, AI-enabled traffic management initiatives, and notable incidents illustrate how theoretical and regulatory concepts materialize in practice. They expose persistent gaps in liability assignment, accountability, transparency, and societal acceptance, while simultaneously showcasing tangible benefits such as accident reduction, predictive safety interventions, and improved mobility management (Smith, 2022; Kong et al., 2021).

By systematically integrating academic literature, regulatory materials, and real-world case evidence, the study employs triangulation to enhance analytical rigor, strengthen validity, and develop a comprehensive understanding of the ethical and governance challenges shaping AI-driven transportation systems.

Figure 2. Data Sources and Triangulation

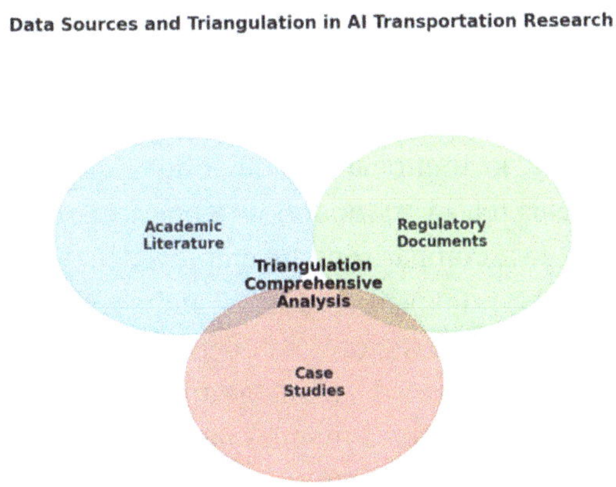

Figure 2 illustrates that by integrating academic literature, regulatory materials, and real-world case evidence, the study employs triangulation to strengthen rigor, validate findings, and comprehensively address ethical and governance challenges in AI-driven transportation

Analytical Framework

The analytical framework of this study employs a thematic synthesis approach, organized around five interrelated dimensions that collectively enable a comprehensive examination of AI governance in transportation. The first dimension concerns ethical principles, particularly fairness, transparency, accountability, and the necessity of human oversight. These principles provide the normative foundation for assessing how AI influences justice, trust, and responsibility in transportation safety (Winfield & Jirotka, 2020). The second dimension centers on risk assessment and governance, focusing on institutional mechanisms for identifying, evaluating, and mitigating AI-related hazards across diverse transport contexts. This aligns with responsible innovation perspectives that emphasize anticipatory regulation and adaptive oversight (Stilgoe, 2020).

The third dimension addresses technological integration, examining the operational capabilities and constraints of AI applications such as autonomous vehicles, intelligent traffic control, predictive accident analytics, and passenger monitoring systems. This dimension highlights the dual reality of technological promise and systemic vulnerability (Rossi et al., 2021). The fourth dimension focuses on regulatory approaches, comparing key governance frameworks including the EU's risk-based AI Act, the United States' sectoral and market-oriented guidelines, and Asia-Pacific's adaptive pilot-based models. Through this comparative lens, the framework identifies enforcement gaps, overlapping regulatory responsibilities, and opportunities for global harmonization (European Commission, 2021).

The fifth dimension considers socio-economic factors, emphasizing public trust, stakeholder responsibility, institutional accountability, and societal acceptance as decisive factors influencing real-world adoption (Kong et al., 2021). Complementing these dimensions, cross-case comparisons are employed to illustrate how variations in regulatory maturity, technological readiness, and social trust shape divergent implementation outcomes across regions. Altogether, this multi-dimensional analytical framework supports a nuanced and integrative understanding of how ethics, technology, governance, and societal dynamics intersect in shaping the responsible deployment of AI in transportation safety.

Justification of Approach

A qualitative and integrative methodology is particularly appropriate for examining the ethical and regulatory complexities surrounding AI in transportation safety. The rapid evolution of autonomous and intelligent mobility technologies, coupled with the diversity of governance approaches across jurisdictions, creates a research landscape where rigid quantitative generalization is often insufficient. Instead, qualitative inquiry allows for deeper contextual engagement, enabling researchers to interpret how ethical principles—such as fairness, accountability, transparency, and human oversight—are translated into real-world governance practices and institutional decision-making (Creswell & Creswell, 2021).

This methodological approach supports the exploration of how risk governance frameworks operate in practice, how responsibilities are distributed among stakeholders, and how tensions between innovation and regulation are negotiated. It also facilitates the integration of perspectives from ethics of technology, risk governance, and accountability studies, thereby enabling a multidimensional understanding that extends beyond technical feasibility toward societal impact, legitimacy, and long-term sustainability (Floridi & Cowls, 2021; Stilgoe, 2020).

The integrative component of the methodology strengthens analytical breadth by synthesizing academic literature, regulatory frameworks, and real-world case studies. Such triangulation not only enhances interpretive rigor but also provides a dynamic platform for capturing emerging developments in AI technology and policy. Moreover, this approach accommodates the rapidly shifting technological landscape, ensuring that findings remain relevant for policymakers, regulators, industry leaders, and civil society actors seeking to balance innovation with ethical responsibility and public trust (Rossi et al., 2021). In doing so, it contributes to a more comprehensive, context-sensitive, and forward-looking examination of AI governance in transportation safety.

Limitations

This study's reliance on secondary sources inevitably restricts access to proprietary datasets and real-time operational information that could offer deeper insight into the actual performance and challenges of AI deployment in transportation systems. Regional disparities in reporting quality, documentation standards, and transparency practices may also introduce bias and imbalance in the evidence base. Moreover, variations in regulatory maturity and implementation across different countries limit the extent to which findings can be directly compared or generalized across contexts.

The conceptual nature of the study further constrains empirical validation, as the proposed analytical and governance frameworks have not yet been tested in real-world policy or operational environments. These limitations highlight the need for future research to adopt mixed-method approaches that combine quantitative modeling, longitudinal field observations, and structured engagement with policymakers, industry practitioners, and affected communities. Such efforts would enhance empirical robustness, capture contextual nuances, and generate stronger policy-relevant insights to guide the ethical and effective governance of AI in transportation safety.

Ethical Considerations

Although this study does not involve primary data collection, ethical principles remain central to its design and execution. Care was taken to accurately reflect diverse stakeholder perspectives, including policymakers, industry actors, researchers, and the public, while avoiding selective interpretation of evidence. Fairness and balance guided the selection and evaluation of sources to ensure that different regional contexts and policy approaches were represented without privileging any particular narrative. Equal attention was given to both the opportunities and risks associated with AI in transportation, helping prevent biased or overly optimistic conclusions.

Transparency and scholarly integrity were also prioritized throughout the interpretation of regulatory frameworks, policy debates, and documented case studies. Efforts were made to clearly acknowledge uncertainties, methodological constraints, and potential limitations inherent in relying on secondary data. By upholding these ethical standards, the study aims to contribute responsibly to academic discourse while maintaining credibility, rigor, and public trust in discussions surrounding AI governance and transportation safety.

Summary of Methodology

This study adopts a qualitative and integrative methodology, synthesizing secondary data drawn from academic literature, regulatory documents, and real-world case studies to ensure analytical depth and comprehensive insight. Anchored in the ethics of technology, risk governance, and accountability frameworks, the analysis critically explores the technological, regulatory, and socio-economic dimensions of AI-driven transportation safety.

While acknowledging the inherent limitations of qualitative synthesis, this approach nonetheless provides a robust and multidimensional foundation for understanding emerging risks and governance challenges. Furthermore, the findings are intended to offer meaningful guidance for policymakers, support industry efforts to align innovation with ethical principles, and encourage broader societal engagement in shaping AI-enabled transportation systems that are safe, transparent, and socially responsible.

FINDINGS AND DISCUSSION

Ethical Challenges in AI-Driven Transportation

AI has increasingly become a core element of modern transportation systems, shaping how mobility, safety, and operational performance are managed across different modes of transport. From AV and intelligent traffic management systems to predictive aviation safety tools and passenger monitoring technologies, AI offers the potential to significantly reduce accident rates, enhance decision-making, improve traffic flow, and contribute to broader sustainability objectives (Rossi et al., 2021; Zhang & Li, 2022). These advancements signal a profound shift toward smarter, more responsive transportation ecosystems capable of learning from real-time data and adapting to evolving conditions. Yet, alongside these opportunities lie complex ethical considerations that cannot be overlooked if AI-enabled transportation is to be safe, equitable, and socially accepted.

A key ethical concern relates to algorithmic bias. AI systems are deeply dependent on the quality and diversity of the datasets used to train them. When data are incomplete, unbalanced, or unrepresentative, AI models risk producing discriminatory or unsafe outcomes. In transportation, this can manifest in AVs misinterpreting the behaviors or physical characteristics of certain groups, including children, pedestrians, or marginalized communities, leading to unequal safety protections. Addressing this requires inclusive data practices, rigorous testing, and continuous algorithmic auditing to ensure fairness and reliability (Zhang & Li, 2022).

Data privacy is another critical challenge. AI-driven transportation infrastructure relies heavily on large-scale data collection, including behavioral patterns, biometric identifiers, and geolocation tracking. While such data enhance operational performance and predictive accuracy, they also generate concerns surrounding surveillance, consent, misuse, and cybersecurity vulnerabilities. Robust regulatory frameworks and ethical data governance are therefore essential to protect individual rights and maintain public confidence (Leslie, 2020).

The question of accountability further complicates AI integration. When accidents or system failures occur, determining liability is far from straightforward. Whether responsibility lies with software developers, manufacturers, system operators, or policymakers remains an evolving debate, with significant implications for legal certainty, insurance systems, and public trust (European Commission, 2021).

Finally, transparency remains a persistent challenge. Many AI systems operate as opaque "black boxes," making their decision-making processes difficult to interpret or scrutinize. Without enhanced explainability and clear communication, public acceptance may remain limited, as individuals are unlikely to trust systems they cannot understand (Winfield & Jirotka, 2020).

Taken together, these ethical challenges illustrate that AI in transportation must be approached as a socio-technical endeavor, where innovation is balanced with governance, accountability, and human values. Only through careful, ethically grounded implementation can AI truly support safe, trustworthy, and socially responsible transportation futures.

Regulatory Approaches

Regulatory approaches to AI in transportation differ markedly across global regions, shaped by varying legal traditions, levels of technological maturity, and policy priorities. These differences illustrate both the dynamism and fragmentation of AI governance, influencing how safety, accountability, and ethical considerations are addressed in practice.

In the European Union (EU), regulation is characterized by a comprehensive and preventive approach. The proposed EU AI Act introduces a risk-based classification system that places autonomous driving and other transportation-related AI applications within the "high-risk" category. This framework mandates strong requirements related to transparency, human oversight, accountability, and data governance to ensure that safety and ethical principles are embedded from design to deployment. Although implementation is still evolving, the EU model reflects a proactive effort to balance innovation with public protection and ethical responsibility (European Commission, 2021).

The United States (US) adopts a more decentralized and market-driven framework. Rather than enforcing a unified national regulation, governance largely relies on sector-specific guidance, particularly through agencies such as the National Highway Traffic Safety Administration (NHTSA). These guidelines are often voluntary, allowing states and private companies considerable flexibility in testing and deploying autonomous systems. While this approach encourages rapid innovation and industry leadership, it also results in uneven standards, regulatory inconsistencies, and ongoing debates about liability and consumer protection (Smith, 2022).

Meanwhile, the Asia-Pacific region demonstrates adaptive and experimental regulatory strategies. Countries such as Japan, Singapore, and China combine regulatory sandboxes, national standards, and strong state oversight to enable real-world testing while gradually shaping governance rules. These frameworks emphasize learning from pilot projects and iterative regulation, which has shown promising early results in safety assurance and technological readiness. However, they also highlight the need for stronger regional coordination and international harmonization to support cross-border interoperability and shared safety benchmarks (Zhang & Li, 2022).

Case Studies of AI Applications in Transportation

Empirical observations from selected cases illustrate the interaction of ethics, safety, and governance:

Table 1. **Results of Observation from Selected Case**

Case Study	Region	AI Application	Key Insights	Ethical/Regulatory Lessons
Uber self-driving fatality	Arizona, USA	AV	Sensor and decision-making limitations revealed	Urgency of accountability, robust testing, and public safety standards
Waymo AV trials	California, USA	AV	Systematic data collection, iterative safety improvements	Transparency and monitoring enhance public trust
Singapore autonomous bus trials	Singapore	Public transport AI	Integration in controlled environments	Adaptive regulation and citizen engagement enable adoption
EU aviation AI safety standards	EU	Aviation safety AI	Compliance, continuous monitoring, incident reporting	Strong regulatory oversight improves safety and ethical compliance
China smart city transport pilots	China	Smart traffic & AV	Real-time optimization and mobility efficiency	Government-industry coordination is critical for safety and ethics

Table 1 highlights case studies across the US, Singapore, EU, and China show AI in transportation improves safety and efficiency, highlighting the need for accountability, transparency, adaptive regulation, and government-industry coordination.

Empirical Observation from Technological and Governance Integration

AI technologies alone are insufficient to guarantee safe and ethical outcomes. They must be complemented by regulatory oversight, ethical design, and societal engagement.

Figure 3. Schematic Diagram of Building Safe, Ethical and Efficient AI Transport

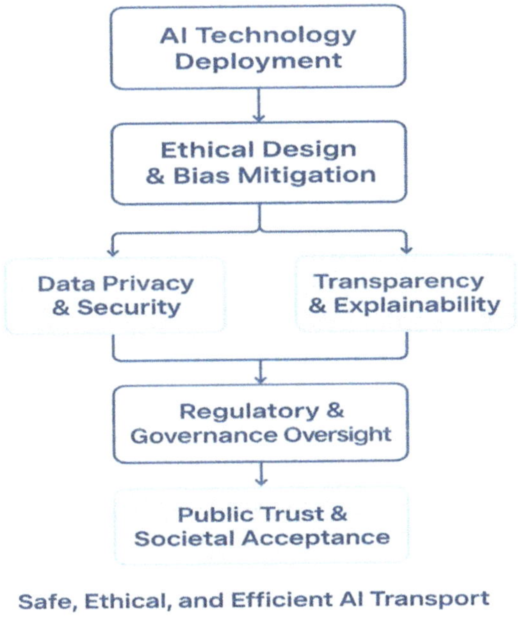

Figure 3 illustrates an integrated framework for ethical and responsible AI deployment in transportation, highlighting the interconnectedness of technology, ethics, regulation, and societal trust, where each component reinforces others, creating a feedback loop that ensures safe, accountable, transparent, and widely accepted AI-driven mobility systems.

Synthesis

The findings indicate that AI-driven transportation stands at a pivotal crossroads, where remarkable technological potential meets complex ethical and regulatory realities. On one side, AI innovations offer clear benefits: more efficient traffic management, improved decision-making, enhanced road safety, and reduced accident rates through predictive capabilities and automation. These developments suggest that AI can meaningfully transform mobility systems, making them smarter, safer, and more responsive to changing conditions. However, unlocking these advantages is not simply a technological task. It requires a carefully structured framework that integrates ethical safeguards, regulatory clarity, and sustained public trust.

First, inclusive AI development is essential. AI systems must be trained on diverse, representative datasets to avoid reinforcing social inequities or overlooking vulnerable populations. Inclusive design and continuous bias mitigation help ensure that detection, decision-making, and safety protections work fairly for everyone, echoing Stilgoe's (2020) emphasis on responsible and socially attentive innovation governance.

Second, robust privacy protections are fundamental. AI-enabled transportation systems depend heavily on collecting, processing, and analyzing large volumes of behavioral, biometric, and mobility data. Without strong safeguards, these systems risk becoming instruments of surveillance or misuse. Consistent with OECD's (2022) guidance, clear rules on consent, data minimization, secure storage, and ethical use are vital to sustaining public confidence.

Third, establishing clear accountability mechanisms is crucial. As AI increasingly influences transportation outcomes, legal systems must define where responsibility lies in the event of failures or accidents. Whether liability rests with developers, manufacturers, system operators, users, or shared arrangements remains a key policy challenge, as highlighted in Calo's (2020) roadmap for AI regulation.

Fourth, transparency and explainability are indispensable. Many AI systems currently operate as opaque "black boxes," making regulatory oversight and public understanding difficult. Explainable AI fosters user confidence, supports informed decision-making, and strengthens regulatory effectiveness, aligning with the principles emphasized in the NIST (2023) AI Risk Management Framework.

Finally, adaptive governance is required to navigate rapidly evolving technological landscapes. Internationally harmonized standards—and lessons drawn from the EU's structured approach, the United States' sectoral flexibility, and the Asia-Pacific's adaptive experimentation—can help minimize regulatory fragmentation. The European Commission's (2021) AI Act remains a significant benchmark in shaping such global conversations.

Overall, the findings reinforce that AI-driven transportation can transition from experimental initiatives to mature, reliable, and socially beneficial mobility systems only when innovation progresses alongside ethical responsibility, regulatory strength, and sustained societal engagement. This integrated approach mirrors other complex policy domains, such as environmental management, where technological solutions must work in concert with governance and social considerations to truly deliver long-term public value.

Figure 4. Integrative Factors in Building AI-Driven Transportation

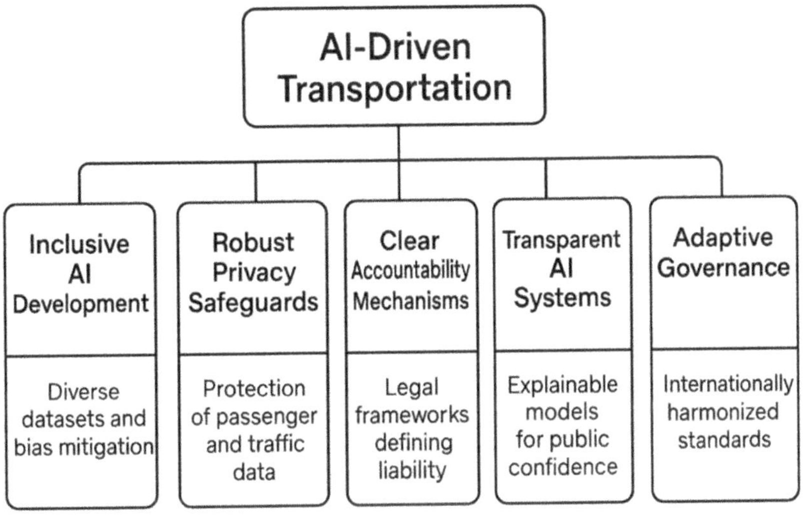

Figure 4 highlights that AI-driven transportation achieves safe, reliable, and socially beneficial outcomes only when innovation aligns with ethics, strong regulation, and ongoing societal engagement, similar to complex policy domains.

CONCLUSION AND RECOMMENDATIONS

Conclusion

AI in transportation represents both a high-impact technological innovation and a complex governance challenge. Its integration into autonomous vehicles, intelligent traffic management, and predictive safety systems offers transformative potential, particularly in reducing human-error-related accidents, improving operational efficiency, and supporting predictive infrastructure maintenance (Goodall,

2021; Lin et al., 2020). However, these advancements cannot be understood merely as technological milestones; they exist within an intricate ecosystem of ethical obligations, regulatory expectations, and societal implications that require deliberate and coordinated attention.

Findings from this study reveal that AI-driven transportation sits at a critical intersection between technological capability and ethical-regulatory complexity. While advanced AI solutions promise substantial safety gains and enhanced mobility, realizing these benefits requires a multidimensional framework that balances innovation with responsibility. Three interrelated insights emerge. First, the safety benefits of AI depend heavily on algorithmic transparency, interpretability, and fairness. Systems must be developed using diverse datasets and robust bias-mitigation strategies to ensure equitable and reliable performance across populations, reflecting concerns raised in discussions of responsible and anticipatory innovation governance (Zhang & Li, 2022; Stilgoe, 2020).

Second, regulatory and ethical safeguards are indispensable. Fragmented or insufficient regulatory oversight risks creating safety vulnerabilities, legal uncertainty, and public distrust, which may hinder adoption despite technical readiness. Clear accountability structures are therefore needed to delineate liability among manufacturers, operators, and users, while risk-based regulatory models such as the European Union's AI Act illustrate the value of harmonized, forward-looking governance approaches (European Commission, 2021; Calo, 2020; NIST, 2023). Privacy protection also remains essential, particularly as transportation systems increasingly rely on real-time passenger and traffic data, reinforcing OECD's (2022) emphasis on data governance as a cornerstone of trustworthy AI.

Third, effective implementation requires socio-technical integration and stakeholder engagement. AI-driven transportation must be embedded within societal expectations of safety, transparency, and accountability, supported by cross-sector collaboration among regulators, developers, industry actors, and communities. International cooperation is also critical to prevent regulatory fragmentation and ensure consistent safety standards across borders.

In sum, AI in transportation embodies both remarkable opportunity and significant responsibility. By embedding ethical principles, regulatory rigor, technological transparency, and inclusive stakeholder participation into AI design and governance, societies can transition from experimental initiatives toward transportation ecosystems that are not only intelligent and efficient but also safe, trustworthy, and socially responsible.

Policy Recommendations

For Governments and Policymakers

The integration of AI into transportation systems requires governments and policymakers to adopt a proactive, holistic, and forward-looking governance approach. Strengthening regulatory frameworks is a critical first step. Clear, enforceable, and context-sensitive safety standards must be established, supported by strong institutional capacity to continuously monitor, evaluate, and adapt to emerging technological risks. Regulation should not merely ensure compliance but also foster responsible innovation by mandating fairness audits, bias mitigation strategies, and transparency in algorithmic processes. Such measures are essential to enhance accountability, safeguard public interest, and cultivate societal trust in AI-enabled mobility systems.

Equally vital is the promotion of international cooperation. Harmonizing safety, ethical, and technical standards across jurisdictions can help prevent regulatory fragmentation and uneven safety practices. Cross-border research collaboration allows knowledge exchange, accelerates technological learning, and supports the development of globally consistent risk governance models. At the same time, sustained investment in digital infrastructure, research, and innovation is necessary. Governments should prioritize funding for AI-driven traffic management systems, predictive maintenance technologies, and real-time safety monitoring tools, while encouraging public–private partnerships to expand research capacity and stimulate industry participation.

Finally, public engagement and human-centered capacity building are indispensable. Public awareness initiatives should aim to enhance understanding of AI technologies, address misconceptions, and increase confidence in their societal value. Training programs for drivers, transportation personnel, policymakers, and citizens can strengthen the ability to interact safely, ethically, and effectively with AI-based transportation systems. By combining regulatory rigor, ethical governance, technological investment, and informed public participation, policymakers can guide AI-enabled transportation toward outcomes that are not only innovative and efficient but also safe, transparent, equitable, and socially responsible.

For Industry and Technology Developers

The deployment of AI in industrial and technological sectors requires developers to embrace a comprehensive, ethical, and accountable approach. Responsible AI practices must be embedded across the entire development lifecycle, incorporating explainable AI mechanisms, rigorous safety validation, and continuous algorithmic auditing. Ethical decision-making frameworks and compliance protocols should be transparently documented, ensuring both accountability and responsible innovation. Equally essential is meaningful cross-stakeholder collaboration, where partnerships with regulators, academic institutions, civil society organizations, and industry peers foster inclusive, human-centered design. Open sharing of safety data, incident records, and lessons learned contributes to collective learning, strengthens overall system reliability, and helps cultivate public trust in AI-enabled technologies.

At the same time, developers must leverage innovation responsibly, ensuring that advances in efficiency, performance, and automation do not compromise privacy, safety, or ethical integrity. The strategic use of simulations, digital twins, and predictive modeling can help anticipate system vulnerabilities and mitigate real-world risks prior to deployment. Clear accountability mechanisms are equally critical, with defined protocols for incident reporting, liability assignment, and integration of insurance and risk management strategies. Workforce readiness further underpins responsible AI deployment; comprehensive training for engineers, operators, and maintenance personnel ensures competent oversight, while continuous professional development enables adaptation to evolving technological and safety standards. Collectively, these measures support a culture of ethical innovation, reduce systemic risks, and enhance the safe, effective, and socially responsible implementation of AI across industries.

For Communities and Civil Society

The engagement of communities and civil society in AI-enabled transportation requires a proactive, inclusive, and participatory approach to ensure that innovation aligns with societal needs and public values. First, participatory governance must be strengthened by creating structured platforms that allow citizens to contribute feedback on AI deployment, voice safety concerns, and advocate for inclusivity, fairness, and ethical safeguards. Systematic documentation of community perspectives, along with the integration of policy recommendations, enhances transparency and accountability while guiding responsible implementation of AI systems. Equally important is enhancing public awareness and education so that citizens clearly understand both the benefits and potential risks of AI-enabled transportation. Transparency initiatives—such as open access to information about deployment processes, decision-making models, and governance mechanisms—are essential in building trust and encouraging collaborative oversight. Civil society organizations also play a strategic role in ensuring equity and accessibility, working to reduce mobility disparities and guaranteeing that AI-driven transportation solutions benefit all segments of society, including vulnerable and underserved groups.

Monitoring ethical compliance is another critical dimension of community and civil society engagement. Supporting independent audits, strengthening reporting mechanisms for bias, unfair treatment, or safety violations, and ensuring verification of regulatory adherence help maintain accountability throughout the lifecycle of AI systems. Furthermore, prioritizing lifelong learning and capacity-building initiatives equips communities with the knowledge and skills needed to participate meaningfully in AI governance. Educational programs that enhance AI literacy, digital safety awareness, and informed citizenship empower individuals to critically engage with emerging technologies rather than passively accept them. Collectively, these measures foster trust, reinforce ethical and responsible innovation, and ensure that AI-enabled transportation systems are not only technologically advanced but also socially beneficial, safe, transparent, and inclusive.

Figure 5. Recommendation in Building Communities and Civil Society

Figure 5 highlights that monitoring ethical compliance, promoting AI literacy, independent audits, and community engagement ensures accountable, transparent, inclusive, and socially beneficial AI-driven transportation systems.

Strategic Roadmap for AI-Driven Transportation Safety

To fully harness the transformative potential of AI in enhancing transportation safety, a coordinated, multi-dimensional, and ethically grounded strategic approach is essential. AI cannot simply be implemented as a technological upgrade; it must be embedded within strong governance systems, responsible innovation cultures, and inclusive societal frameworks to ensure long-term legitimacy, trust, and sustainability.

Strengthen Governance Capacity

Developing robust regulatory frameworks and institutional mechanisms is fundamental to ensuring responsible AI deployment. Clear, enforceable standards for safety, accountability, transparency, and continuous monitoring can guide developers, manufacturers, and operators in navigating complex AI ecosystems. Strengthened institutional capacity enables governments to conduct compliance audits, risk assessments, and adaptive oversight, ensuring that AI applications remain aligned with ethical, legal, and societal expectations. Such structured governance reduces operational and systemic risks while fostering consistency and reliability.

Incentivize Responsible Innovation

AI developers should be encouraged to integrate transparency, fairness, explainability, and safety throughout the design lifecycle. Incentive mechanisms—such as research grants, certification schemes, innovation awards, and regulatory advantages for ethically compliant systems—can motivate industry actors to embed responsible AI principles in practice. Rigorous testing, third-party validation, and continuous auditing enhance accountability, build public trust, and reduce the likelihood of technological failures or unintended consequences.

Ensure Social Equity

Promoting equitable access and outcomes is critical in AI-enabled transportation. Stakeholder engagement must be incorporated at every stage of development and implementation to ensure that benefits are distributed fairly and that potential harms do not disproportionately affect vulnerable or marginalized groups. Inclusive design principles, participatory policymaking processes, and equity-focused evaluation frameworks help address diverse mobility needs, reduce accessibility gaps, and foster social justice in emerging transportation ecosystems.

Scale Multi-Stakeholder Partnerships

Effective AI governance requires strong collaboration among governments, industry, academia, and civil society. Multi-stakeholder partnerships enable shared learning, coordinated standard-setting, responsible data sharing, and collaborative research initiatives. Joint efforts can facilitate safer algorithm development, enhance regulatory alignment, and support public communication campaigns that improve awareness and understanding of AI-enabled transportation. These partnerships cultivate shared responsibility and strengthen system-wide resilience.

Embed Sustainability and Trust

Long-term success depends on sustained public confidence. Embedding transparency, public engagement, and ethical foresight into governance structures helps legitimize AI systems and ensures accountability. Harmonizing international standards prevents regulatory fragmentation and supports safe global interoperability. Furthermore, continuous capacity-building and AI literacy initiatives empower communities to participate meaningfully in oversight and decision-making, fostering informed trust.

Taken together, these strategic priorities enable AI-driven transportation to evolve into a safe, resilient, sustainable, and socially responsible mobility ecosystem that benefits society as a whole.

Final Reflections

AI-driven transportation safety presents a transformative opportunity to significantly enhance public safety while simultaneously challenging and reshaping conventional ethical and regulatory frameworks. When responsibly designed, governed, and deployed, AI technologies have the potential to substantially reduce accidents, improve real-time decision-making, optimize traffic flows, enable predictive maintenance, and support more efficient and equitable mobility systems. However, realizing these benefits requires more than technological capability; it demands coordinated governance, strong regulatory alignment, transparent accountability mechanisms, and responsible innovation practices from industry stakeholders. Equally crucial is the meaningful engagement of communities and civil society to ensure that AI-enabled transportation reflects societal values, protects public interests, and addresses issues of fairness, inclusion, and accessibility.

Rather than viewing AI purely as a technical tool, it must be understood as a socio-technical system embedded within ethical expectations, legal obligations, and social responsibilities. Prioritizing transparency, explainability, and fairness helps build public trust, while clear accountability ensures confidence in safety outcomes and regulatory enforcement. Ethical safeguards, privacy protection, and bias mitigation must be integrated into every stage of design, deployment, and monitoring to prevent new inequalities or risks from emerging. When these elements align, AI in transportation can evolve into a pioneering model that demonstrates how innovation can coexist with public trust, regulatory rigor, and long-term societal benefit. Ultimately, AI-driven transportation safety has the potential not only to enhance mobility and reduce harm but also to illustrate how technology can responsibly serve people, communities, and the environment in a sustainable and future-oriented manner.

Author Contributorship

Harry Sutanto was responsible for the conceptualization and development of the regulatory framework, including comprehensive policy analysis, and played a central role in drafting, reviewing, and refining the manuscript. Mohammad Ichsan conducted the ethical analysis, designed and evaluated the case studies, and contributed significantly to the interpretation of the findings, discussion, and manuscript development. Both authors worked collaboratively throughout the research process to ensure the rigor, accuracy, and coherence of the arguments, data interpretation, and conclusions. They approved the final version of the manuscript and share full responsibility for the intellectual content, quality, and overall contribution of the study.

REFERENCES

Burrell, J., Davis, J., & Lambe, J. (2020). Explainability as a framework for responsible AI in practice. *AI & Society*, *35*(3), 569–582. DOI: 10.1007/s00146-019-00914-7

Calo, R. (2020). Artificial intelligence policy: A primer and roadmap. *U.C. Davis Law Review*, *53*(1), 399–435.

Calo, R. (2020). Artificial intelligence policy: A roadmap. *UCLA Law Review. University of California, Los Angeles. School of Law*, *67*(3), 555–622. https://www.uclalawreview.org/artificial-intelligence-policy/

Chen, Y., Xu, L., & Zhao, H. (2022). Artificial intelligence applications in traffic flow management: Opportunities and challenges. *Transportation Research Part C, Emerging Technologies*, *137*, 103570. DOI: 10.1016/j.trc.2022.103570

Cihon, P., Maas, M. M., & Kemp, L. (2021). Should artificial intelligence governance be centralised? Design lessons from history. *AI & Society*, *36*(4), 1127–1144. DOI: 10.1007/s00146-020-00944-5

Creswell, J. W., & Creswell, J. D. (2021). *Research design: Qualitative, quantitative, and mixed methods approaches* (6th ed.). Sage Publications.

European Commission. (2020). White Paper on Artificial Intelligence: A European approach to excellence and trust. Brussels: European Union.

European Commission. (2021). Proposal for a regulation laying down harmonised rules on artificial intelligence (Artificial Intelligence Act). Brussels: European Union. https://eur-lex.europa.eu/legal-content/EN/TXT/?uri=CELEX:52021PC0206

European Commission. (2022). *Regulation on Artificial Intelligence (AI Act): Ensuring safe and trustworthy AI*. European Union.

Floridi, L., & Cowls, J. (2021). A unified framework of five principles for AI in society. *Harvard Data Science Review*, *3*(1), 1–15. DOI: 10.1162/99608f92.8cd550d1

Goodall, N. (2021). Machine ethics and autonomous vehicles. In Anderson, J. L. (Ed.), *Handbook of Transport Safety* (pp. 221–245). Routledge.

Goodall, N. (2021). Ethical decision-making in autonomous vehicle safety systems. *AI & Society*, *36*(4), 1097–1108.

Kong, L., Yu, H., & He, Z. (2021). Predictive analytics for accident prevention in intelligent transportation systems. *IEEE Transactions on Intelligent Transportation Systems*, *22*(6), 3751–3763.

Kong, Q., Luo, X., & Li, Y. (2021). Public trust and acceptance of AI in transportation systems. *Journal of Intelligent Transport Systems, 25*(6), 523–536. DOI: 10.1080/15472450.2021.1894583

Kong, Q., Xu, H., & Zhu, X. (2021). Accident prediction models based on machine learning: A review of the state of the art. *Accident; Analysis and Prevention, 151*, 105973.

Leslie, D. (2020). *Understanding artificial intelligence ethics and safety: A guide for the responsible design and implementation of AI systems in the public sector.* The Alan Turing Institute., DOI: 10.5281/zenodo.4059364

Leslie, D. (2021). The ethics of AI in context: Understanding ethical responsibilities in practice. *AI & Society, 36*(2), 511–529. DOI: 10.1007/s00146-020-00992-x

Li, J., & Wang, Y. (2023). Artificial intelligence for safety in aviation and rail: A review. *Safety Science, 164*, 106201. DOI: 10.1016/j.ssci.2023.106201

Lin, P., Jenkins, R., & Abney, K. (2020). *Robot ethics 2.0: From autonomous cars to artificial intelligence.* Oxford University Press.

Lin, P., Jenkins, R., & Zhao, Q. (2020). Ethics and governance of autonomous vehicles. *Transportation Research Interdisciplinary Perspectives, 6*, 100151.

Mehrabi, N., Morstatter, F., Saxena, N., Lerman, K., & Galstyan, A. (2021). A survey on bias and fairness in machine learning. *ACM Computing Surveys, 54*(6), 1–35. DOI: 10.1145/3457607

Morley, J., Floridi, L., Kinsey, L., & Elhalal, A. (2021). From what to how: An initial review of publicly available AI ethics tools, methods and research to translate principles into practices. *AI & Society, 36*(1), 59–86. DOI: 10.1007/s00146-019-00992-4

NIST. (2023). AI Risk Management Framework. National Institute of Standards and Technology. https://www.nist.gov/ai-risk-management-framework

OECD. (2021). *State of Implementation of the OECD AI Principles: Insights from National AI Policies*. OECD Publishing.

OECD. (2022). Recommendation of the Council on Artificial Intelligence. OECD Publishing. https://www.oecd.org/ai/ai-principles/

OECD. (2022). *AI in society: Challenges and opportunities*. OECD Publishing.

Renn, O. (2020). *Risk governance: Coping with uncertainty in a complex world* (2nd ed.). Routledge.

Rossi, F., Bianchi, C., & Biagi, F. (2021). Artificial intelligence in transport: Current applications and policy considerations. *AI & Society, 36*(2), 563–577.

Rossi, F., Matteucci, I., & Salvini, P. (2021). Ethics of artificial intelligence: Balancing benefits and risks in autonomous systems. *AI & Society, 36*(3), 739–748.

Rossi, F., Matteucci, M., & Varshney, K. (2021). Ethical and social implications of AI in transportation. *AI & Society, 36*(3), 897–908. DOI: 10.1007/s00146-020-00985-1

Rossi, F., Serafini, L., & Jardine, R. (2021). Ethics of AI: A systematic literature review of principles and challenges in autonomous systems. *AI & Society, 36*(3), 715–732.

Singh, R., Patel, D., & Verma, S. (2022). AI-based crowd monitoring and management in public transportation systems. *Journal of Safety Research, 82*, 67–77. DOI: 10.1016/j.jsr.2022.03.005

Smith, B. (2022). Regulation of autonomous vehicles: Emerging practices and unresolved challenges. Journal of Law. *Technology & Policy, 2022*(1), 45–78.

Stilgoe, J. (2020). Who's driving innovation? New technologies and responsible innovation. *Technology Analysis and Strategic Management, 32*(8), 931–940. DOI: 10.1080/09537325.2020.1717255

Whittlestone, J., & Clark, J. (2021). Between responsibility and regulation: Ethics and governance of AI in practice. *Minds and Machines, 31*(2), 237–256. DOI: 10.1007/s11023-020-09536-9

Winfield, A. F., & Jirotka, M. (2020). Ethical governance is essential to building trust in robotics and artificial intelligence systems. *Philosophical Transactions of the Royal Society A, 378*(2166), 20190115. DOI: 10.1098/rsta.2019.0115

World Health Organization (WHO). (2023). Global status report on road safety. Geneva: WHO. https://www.who.int/publications/i/item/9789240077611

Zhang, H., & Li, X. (2022). Artificial intelligence and the future of smart transportation systems. *Journal of Transportation Technologies, 12*(4), 455–472. DOI: 10.4236/jtts.2022.124028

Zhang, Y., & Li, H. (2022). Algorithmic fairness in autonomous vehicles. *Journal of Transportation Safety & Security, 14*(3), 331–348.

Zhang, Y., & Li, H. (2022). AI in transportation: Ethics, safety, and governance challenges. *Transportation Research Part C, Emerging Technologies, 139*, 103698. DOI: 10.1016/j.trc.2022.103698

Zhang, Y., & Li, X. (2022). Ethical and technical challenges of autonomous driving: A review. *Transportation Research Part C, Emerging Technologies*, *140*, 103667. DOI: 10.1016/j.trc.2022.103667

Chapter 15
Revolutionizing Transportation Safety With AI:
Passenger Behavior, Smart Mobility Experiences, and Future Prospect

Vanshika Srivastava
http://orcid.org/0009-0008-0529-6740
Maharishi Markandeshwar (Deemed to be) University, Mullana, India

Shivansu Sachan
http://orcid.org/0000-0002-9893-8018
CSJM University, Kanpur, India

Rajiv
http://orcid.org/0000-0003-2584-2688
Maharishi Markandeshwar (Deemed to be) University, Mullana, India

Mayank Mishra
http://orcid.org/0009-0006-2117-663X
CSJM University, Kanpur, India

Sudhanshu Pandiya
http://orcid.org/0009-0006-2117-663X
CSJM University, Kanpur, India

Aishwarya Arya
http://orcid.org/0009-0000-4236-3983
CSJM University, Kanpur, India

Ankit Kumar
http://orcid.org/0009-0009-5002-4522
CSJM University, Kanpur, India

ABSTRACT

The integration of Artificial Intelligence (AI) into transportation systems is revolutionizing safety measures, shaping passenger behavior, and transforming mobility experiences. By leveraging real-time data analytics, machine learning, and predictive modeling, AI enables proactive accident prevention, optimized traffic flow,

DOI: 10.4018/979-8-3373-3231-4.ch015

and enhanced emergency response mechanisms. Beyond technical advancements, AI also influences passenger decision-making, fostering safer travel habits through personalized guidance, behavioral monitoring, and adaptive interventions. The rise of smart mobility—autonomous vehicles, connected infrastructure, and intelligent transport services—further extends the scope of AI-driven safety innovations. These systems not only improve operational efficiency but also create seamless, user-centric mobility experiences. However, challenges remain in ensuring ethical AI use, addressing data privacy, and achieving equitable access to AI-enabled transport solutions.

INTRODUCTION

The culinary arts, which are based on tradition and human creativity, have long been a place where intuition, skill, and cultural legacy come together to make foods that please the senses. Cooking has always been a human activity, from the careful knife work of a sushi chef to the instinctive changes in seasoning that a home cook makes. Traditionally, cooking has been a human-centered activity that is based on taste, smell, and sound, as well as cultural traditions and creative expression. The arrival of artificial intelligence, on the other hand, has brought about a paradigm shift in which machines take on more and more of the work that people used to do in the kitchen. The practical advantages of AI in culinary fields are becoming more apparent, however critical analyses of its social, cultural, and ethical ramifications remain insufficiently examined. This chapter fills this vacuum by giving a fair look at how AI is affecting creativity in cooking, access, sustainability, and fairness. It offers an academic analysis grounded in theoretical frameworks, supported by contemporary scholarship that transcends promotional literature. The culinary industry, which is full of traditions and human innovation, is going through a big change thanks to artificial intelligence (AI). The culinary arts, which are based on human creativity, tradition, and sensory knowledge, are going through a huge change because to the use of artificial intelligence (AI). AI is changing the way we think about, make, and enjoy food, from coming up with new recipes to cooking with precision. This chapter looks at the theoretical basis for AI's involvement in changing smart cooking. It looks at how AI affects creativity, efficiency, personalization, and sustainability in the kitchen.

The culinary arts, which are based on creativity, tradition, and sensory skill, are going through a big change because of the use of artificial intelligence (AI). AI is changing the way we think about, make, and eat food, from coming up with recipes to cooking with precision. AI is making cooking smarter, faster, and more personalized by using machine learning, natural language processing, computer vision, and robots. This short article looks at the many ways AI is changing the culinary arts, including its current uses, benefits, problems, and what it holds for the future of both professional kitchens and home cooks.

Working Together with Machines in Creative Processes

The idea of people and machines working together is at the heart of AI's use in cooking. Mihaly Csikszentmihalyi's Flow Theory and other theories of creativity say that people are more creative when they are fully engaged and challenged. AI helps this process by acting as a co-creator and giving people data-driven insights, flavor combinations, and recipe ideas that make their instincts even better. For example, IBM's Chef Watson is an AI system that looks at huge amounts of data on ingredients, cuisines, and flavor profiles to come up with new recipes. This encourages cooks to do new things in the kitchen.

Distributed Cognition Theory says that cognitive processes are shared by people and tools. This theory can help us understand this collaboration. AI-powered technologies like connected ovens, sous-vide machines, and recipe applications let chefs think more clearly by taking care of boring duties so they can focus on more creative decisions. The theoretical inquiry emerges: does AI reduce the chef's autonomy, or does it enhance their creative capabilities? The latter is supported by evidence, as AI technologies enable chefs to experiment with accuracy and scale.

Personalization and Behavioral Adaptation

The way AI can look at user data fits with ideas about Adaptive Systems and Personalization. AI uses machine learning in the kitchen to make recipes and cooking methods fit each person's tastes, nutritional needs, and cultural background. For instance, smart appliances change the cooking settings in real time to get the desired results, while platforms like Yummly utilize AI to suggest recipes based on what users say. This is in line with Self-Determination Theory, which says that autonomy, competence, and relatedness are what drive people to do things. AI-enhanced cooking systems help people be more independent by giving them individualized alternatives, help them become better cooks by giving them step-by-step instructions, and let people feel connected to other cooks across the world.

Thinking About Systems and Sustainability

AI in the kitchen also has to do with Systems Thinking, especially when it comes to solving problems with sustainability. By looking at environmental data and consumer trends, AI can make food supply chains more efficient, cut down on waste, and encourage people to choose ingredients that are better for the environment. For example, AI-powered platforms like Foodpairing recommend different ingredients to use to cut down on food waste, while smart refrigerators keep track of what's in stock to keep food from going bad. In theory, this is an example of Cradle-to-Cradle Design, which means that systems are made to be self-sustaining and free of waste. AI's function in sustainable cooking shows how it could help connect cooking with global environmental concerns.

AI algorithms, especially those that use natural language processing (NLP) and generative models, are changing the way recipes are made. AI systems can make dishes that balance taste, texture, and nutrition by learning from huge databases of recipes. Generative adversarial networks (GANs), for instance, might suggest new ingredient pairings that go against the rules of traditional cooking, like chocolate and avocado. This technique is similar to Combinatorial Creativity, which is when new ideas come by combining old ones.

Smart appliances that work with the Internet of Things (IoT) have brought accuracy and automation to cooking. AI-powered ovens like those from June or Tovala employ sensors and computer vision to keep an eye on food in real time and change the cooking time and temperature to get the best results. These gadgets are examples of Cyber-Physical Systems Theory, which says that physical processes (like cooking) and computational feedback loops (like AI algorithms) are closely linked. The end result is a culinary experience that is smooth, makes fewer mistakes, and is more consistent. AI algorithms, especially those that use natural language processing (NLP) and generative models, are changing the way recipes are made. AI systems can make dishes that balance taste, texture, and nutrition by learning from huge databases of recipes. Generative adversarial networks (GANs), for instance, might suggest new ingredient pairings that go against the rules of traditional cooking, like chocolate and avocado. This technique is similar to Combinatorial Creativity, which is when new ideas come by combining old ones.

Smart appliances that work with the Internet of Things (IoT) have brought accuracy and automation to cooking. AI-powered ovens like those from June or Tovala employ sensors and computer vision to keep an eye on food in real time and change the cooking time and temperature to get the best results. These gadgets are examples of Cyber-Physical Systems Theory, which says that physical processes (like cooking) and computational feedback loops (like AI algorithms) are closely linked. The end effect is a smooth cooking process that cuts down on mistakes and makes things more consistent.

Virtual Cooking Helpers

Voice-activated AI assistants like Amazon Alexa and Google Nest are turning becoming virtual sous-chefs, helping people with recipes, answering queries, and keeping the kitchen running smoothly. These systems use NLP and knowledge graphs to give answers that take the context into account. This is in line with Situated Cognition Theory, which says that learning and solving problems happen best in specific situations. Virtual assistants make advanced cooking skills available to everyone, even beginners. This study offers a thorough comprehension of AI's influence on the future of smart cooking, presenting insights for researchers, chefs, and technology developers seeking to advance in the culinary field. For a long time, cooking has been an important part of human civilization. It is a complex mix of creativity, science, and tradition. Cooking has changed over time, from the first meals cooked over fire to the molecular gastronomy of the 21st century. AI is about to change this ancient craft and start a new era of "smart cooking." By adding AI to both professional and home kitchens, the culinary arts are changing in ways that make them more efficient, personalized, creative, and environmentally friendly.

What this means for the future

Reimagining Culinary Creativity

AI's involvement in cooking challenges old ideas about what creativity is. Some people say that AI-generated recipes don't have the same emotional depth as those made by people, while others see AI as a way to spark new ideas. In the future, there may be a hybrid paradigm in which chefs and AI work together to create things, combining data-driven insights with human intuition. This brings up philosophical problems regarding who owns and what is real in the culinary arts, which need to be looked into more.

Making Gastronomy More Accessible

AI-powered tools are making it easier for people to get into the food industry. Smart culinary tools let home cooks get professional-level outcomes, and AI-powered platforms let people learn about and try foods and techniques from all over the world. This democratization fits with Social Learning Theory since people learn from AI systems and online groups, which helps spread cooking expertise around the world.

Ethical and Cultural Issues

As AI changes the way we cook, ethical and cultural issues come up. AI systems need to be made in a way that honors cultural food customs and doesn't make everything the same. AI's dependence on human preferences and actions also raises worries about data privacy. Theoretical frameworks like as Value-Sensitive Design can inform the creation of AI systems that emphasize cultural awareness and user trust.

Boosting Creativity in the Kitchen

AI isn't simply for making things faster; it's also a way to be creative. AI can encourage cooks to try new techniques and ingredients by looking at food from around the world, flavor profiles, and cooking trends. For instance, AI platforms can make fusion recipes that mix ingredients from diverse cooking styles, such Japanese-Italian or Indian-Mexican cuisines. This capacity stimulates new ideas and challenges the limits of how people have always cooked.

AI also helps with how food looks and how it is served. Foodpairing and other tools use AI to recommend plating choices that look good based on color theory and texture research. Apps like Pinterest and Tasty employ AI to suggest ways to plate food through video demonstrations. This makes it possible for home cooks to make their food seem as good as professionals.

Health Optimization and Nutritional Analysis

AI is changing the way we think about nutrition when we cook. AI solutions can help people make healthier decisions by breaking down the nutritional value of recipes and component lists. MyFitnessPal and Lifesum are two examples of platforms that employ AI to keep track of how many calories you eat, suggest healthy meals, and tell you how much food to eat based on your fitness objectives.

AI is utilized in healthcare to make meals that are just right for each patient. For instance, AI can come up with meals for those with diabetes, heart disease, or food allergies that follow their dietary limits while still tasting good. These systems also work with wearable devices to track health parameters, which lets users plan meals that change based on their needs.

Reducing Waste and Being More Environmentally Friendly

AI is very important for making the culinary arts more sustainable. AI can help lower carbon footprints by optimizing the source of ingredients by looking at food supply networks. Smart kitchen equipment utilize AI to consume less energy, and inventory management systems keep track of how food is used and propose recipes for leftovers to cut down on waste.

Winnow and other startups utilize AI-powered cameras to keep an eye on food waste in commercial kitchens. This gives chefs and managers useful information. These systems find patterns, such making too much food or not giving the right amount to each person, and suggest ways to cut down on waste. AI apps like Too Good To Go help people find extra food from restaurants at home. This cuts down on waste and makes meals cheaper.

HOW AI IS TAKING OVER THE KITCHEN

Making and Customizing Recipes

AI-powered platforms can make meals that are perfect for each person's tastes, dietary needs, and the ingredients they have on hand. IBM's Chef Watson and businesses like Plant Jammer employ machine learning to look at huge databases of recipes, flavor profiles, and nutritional information to come up with new dishes. For instance, Chef Watson suggests new ways to combine items based on their chemical flavor compounds, while apps like Yummly customize recipes by taking into account allergies, taste preferences, and what you have in your kitchen. Artificial intelligence, which is widely defined as machines that can mimic human intelligence, has made a lot of progress in several fields, including healthcare and finance. Its foray into the culinary domain is a logical evolution, propelled by developments in machine learning, natural language processing, computer vision, and robotics. Smart appliances, recipe creation algorithms, flavor profiling systems, and automated cooking gadgets are all examples of AI in the kitchen. They are all meant to make people better at what they do. Artificial Intelligence is changing the culinary arts by changing smart cooking, which combines creativity with accuracy to determine the future of food. AI-powered tools look through huge recipe databases to find the best dishes for each person's tastes, dietary demands, and item availability. They also use predictive algorithms to improve the flavor of the foods. Smart kitchen appliances that use AI can do things like adjust the temperature and time very precisely, which makes them more efficient and consistent. AI gives chefs and home cooks the tools they need to improve their cooking by combining their culinary instincts with data-driven insights. For example, it can help chefs come up with new recipes and manage their inventory to cut down on food wastage.

The use of AI in cooking started with basic tools like smart thermometers and recipe applications, but it has grown to encompass more advanced algorithms that can figure out how ingredients work together, guess what people will like, and even do complicated cooking chores. For example, IBM's Chef Watson, which came out in 2014, was an AI-powered platform that showed how machine learning might be used to make new recipes by looking at huge databases of ingredients and cooking traditions. Since then, the use of AI in cooking has risen by leaps and bounds, thanks to people's need for convenience, healthy diets, and new eating experiences.

- How It Works: AI algorithms take user inputs like "vegan," "low-carb," and "spicy" and compare them to cooking databases to come up with recipes. These systems can understand user questions and show recipes in a conversational way thanks to natural language processing (NLP).
- Effect: Home cooks can get personalized meal plans that help them use up the ingredients they already have, which cuts down on food wastage. Chefs can try out unusual combinations, which encourages new ideas in the kitchen.

Kitchen Appliances That Are Smart

AI is making a new generation of smart kitchen appliances, such ovens and refrigerators. Brands like June Oven and Thermomix use AI to automate cooking tasks, which makes sure that the results are always the same. For example, the June Oven employs computer vision to automatically find food and change the cooking settings (time and temperature).

- How It Works: Sensors, cameras, and AI algorithms watch food while it cooks and change settings to make it cook better. Smart fridges, like Samsung's Family Hub, keep track of what's in them, propose recipes based on what's in them, and let customers know when products are about to go bad.

Effect: These gadgets make cooking easier for beginners, save time, and improve accuracy, so skilled chefs can focus on being creative instead of doing the same things over and over. There are a few main ways that AI is changing the culinary arts, each of which focuses on a distinct part of the cooking process. These apps are changing how we cook, eat, and enjoy food.

Making and Customizing Recipes

One of the most common uses of AI in cooking is to come up with new recipes. AI algorithms can make personalized meals based on people's tastes, dietary constraints, and the ingredients they have on hand. They do this by learning from huge datasets of recipes, ingredient profiles, and nutritional information. AI is used by sites like Yummly and Cookpad to suggest recipes based on what users say they like, what they are allergic to, or what they have in their cupboard.

AI may look at a person's dietary demands, like if they need to eat low-carb, vegan, or gluten-free foods, and then come up with recipes that meet those needs while also making sure the food tastes good and has the right amount of nutrients. These systems can understand user questions, like "What can I make with chicken and spinach?" thanks to natural language processing (NLP). " and give clear, useful advice. This customisation also includes cultural and geographical preferences, so AI can change recipes to fit local tastes or make traditional cuisine with a modern twist.

AI-powered recipe platforms also have feedback loops that let them learn from how users interact with them and improve their recommendations over time. This capacity to change makes sure that recipes stay interesting and useful, which makes cooking more fun.

Smart appliances that use AI have made the kitchen a place where things are automated and run smoothly. The Thermomix, June Oven, and Moley Robotic Kitchen all use AI to make cooking easier, from keeping the temperature just right to stirring and chopping automatically. These appliances use computer vision to identify ingredients, sensors to keep track of how the food is cooking, and machine learning to change settings in real time. In professional kitchens, AI-driven automation boosts productivity by taking care of boring tasks like chopping vegetables or keeping an eye on several dishes at once. This lets chefs focus on creative things like plating and flavor development while also making food more consistent and cutting down on waste. As people become more health-conscious, AI is playing a big part in making meals more nutritious. AI systems can give users extensive nutritional information by looking at the profiles of the ingredients and the sizes of the portions. They can figure out how many calories, macronutrients, and micronutrients are in a dish. Apps like MyFitnessPal and Whisk use AI to keep track of what you eat and offer better options, like using olive oil instead of butter or cutting back on sugar in desserts. AI can also help you plan meals for certain health issues, including diabetes, high blood pressure, or food allergies. For instance, an AI-powered platform can make a week's worth of meals for a diabetic user, making sure their blood sugar stays stable while still offering a variety of flavors. AI is helping to reduce food waste, which is a big problem in the culinary world. This level of personalization is especially useful in healthcare settings, where AI may help dietitians come up with therapeutic diets. Restaurants and grocery stores use AI-powered inventory management systems to keep track of how long ingredients last and recommend dishes that use products that are about to go bad. Apps like SuperCook and Fridge Pal look at what's in your pantry and suggest meals you can make with what's already there, which means you don't have to buy anything else. AI is always becoming better, and it will open up new doors, including hyper-personalized nutrition and global culinary fusion. AI helps professional chefs monitor portions and guess how much food will be needed based on past data, making sure that the right amount of food is always ready. For instance, AI systems can look at how many reservations are made and what customers like to eat to guess how much of each dish will be ordered. This stops too much food from being made. These efforts help make the food chain more sustainable, which is in line with global aspirations to cut down on waste and save resources.

Voice-Activated Cooking and Virtual Assistants

Voice-activated AI assistants like Amazon's Alexa and Google Home are becoming essential in modern kitchens. These gadgets let busy cooks access recipes, timings, and cooking instructions without having to touch them. Virtual assistants can operate ovens, make reminders, and even order groceries when supplies run low by working with smart equipment.

For instance, a person may tell Alexa to "preheat the oven to 350°F and find a lasagna recipe." The assistant would then follow the direction and give step-by-step instructions. These systems can grasp everyday language and adjust to varied accents thanks to NLP. This makes them available to people all over the world. As voice recognition technology becomes better, virtual assistants are getting smarter and can help you fix problems and give you unique suggestions.

Less Waste of Food

AI is working on a big problem in the food world: food waste. Apps like Too Good To Go and AI-powered inventory systems in restaurants can guess how much food will be needed and make the best use of ingredients. Machine learning algorithms look at how much is eaten, when it expires, and how big the portions are to keep from making too much.

- How It Works: AI algorithms use past data and outside factors (such holidays and the weather) to predict demand. Winnow and other AI-powered technologies keep track of waste in restaurants and advise changes to how they buy food or arrange their menus.
- Effect: Restaurants save money, and applications that repurpose extra food help the environment by making use of food that might otherwise go to waste.

Food Pairing and Flavor Profiling

AI is making the science of flavor better by looking into the chemicals in foods. Foodpairing and other companies employ AI to find flavors that go well together, which helps cooks make dishes that taste good together. This method is very useful in molecular gastronomy, where accuracy is very important.

- How It Works: AI maps the unstable chemicals in components and makes predictions about which ones will go well together based on their profiles. AI might recommend putting white chocolate and caviar together because they both have creamy and umami flavors.
- Effect: Chefs can be more inventive, and food companies can make new goods, such plant-based alternatives that taste better.

ADVANTAGES OF AI IN THE CULINARY ARTS

AI in the kitchen has many benefits that change the way pros and ordinary cooks alike cook.

More Creative Ideas

AI gives chefs and cooks the freedom to try out new creative ideas, including new ways to mix flavors or new ways to cook. AI can come up with concepts that human chefs might not think of by looking at huge amounts of data. This encourages experimentation and innovation. This is especially clear in high-end restaurants, where AI-driven insights help create cutting-edge menus that keep customers coming back.

Access and Inclusion

AI makes the culinary arts more accessible to everyone by making advanced cooking techniques and knowledge available to beginners. Smart appliances and recipe applications make hard tasks easier, so anyone can make fancy meals. AI can also meet a wide range of nutritional needs, so people with special health or cultural needs can enjoy tasty, personalized meals.

Time and Money Savings

Smart appliances and automation make cooking a lot easier and faster. AI-powered devices can now take care of tasks that used to take hours of preparation, including kneading dough or keeping an eye on dishes that are cooking slowly. This level of efficiency is especially helpful in busy homes and professional kitchens that are always on the go.

Quality and Consistency Control

AI makes sure that the flavor, texture, and presentation are always the same in professional situations, which is very important for keeping brand standards. Automated systems can perfectly copy recipes, getting rid of human mistake and making sure that every dish satisfies high standards for quality.

Long-Term Viability and Cost Savings

AI helps businesses and homes save money by cutting down on food waste and making better use of resources. Smart appliances that use less energy cut electricity expenses, and predictive analytics and inventory management help avoid overstocking and spoiling.

PROBLEMS AND MORAL ISSUES

AI in the culinary arts has the potential to change things, but there are a number of problems and ethical issues that need to be dealt with before it can be used responsibly.

Loss of Human Touch

AI can't entirely copy the emotions, traditions, and instincts that are so important to cooking. Some people say that too much dependence on AI could kill the soul of cooking and turn it into a mechanical procedure. For instance, a recipe made by an algorithm might not have the same cultural or family meaning as a food that has been passed down through the generations.

To lessen this, AI should be seen as a tool to enhance, not replace, human ingenuity. AI can help chefs and home cooks improve their skills while still keeping the emotional and cultural aspects of cooking.

AI-powered appliances and apps are becoming increasingly common, but they can be too expensive for small eateries or families with limited incomes. High-end gadgets like the Moley Robotic Kitchen cost tens of thousands of dollars, so only rich people or well-funded businesses can buy them. To close this gap in accessibility, we need to create affordable AI solutions and open-source platforms. AI systems need a lot of user data, such food preferences and cooking histories, which raises privacy and security problems. If this data isn't handled correctly, it could be used for business purposes or leaked in a breach. To earn the trust of customers, developers need to focus on strong encryption and clear data standards. In professional kitchens, AI-driven automation could lead to job loss, especially for jobs that involve doing the same thing over and over, like prep cooks or dishwashers. Automation can make things run more smoothly, but it could also make more people in the culinary industry lose their jobs. To get workers ready for new jobs, such as managing AI systems or coming up with creative menu ideas, they will need to go through reskilling programs. AI-generated recipes and flavor profiles must respect cultural and regional culinary traditions to avoid appropriation or misrepresentation. For instance, an AI system that suggests a "fusion" recipe that mixes religious or culturally important elements in a way that isn't right could upset some groups. To make sure that AI outputs are respectful and real, developers need to include cultural sensitivity training to the algorithms. The future of AI in the culinary arts is full of potential, thanks to continual technological advances and changing customer expectations. There are a lot of new ideas and trends that will probably change smart cooking in the next ten years.

Extremely Personalized Culinary Experiences

As AI gets better at looking at biometric data like taste preferences, genetic predispositions, and microbiome profiles, it will be able to create hyper-personalized food experiences. Think about a kitchen that makes meals based on your DNA, making sure they are healthy, tasty, and even good for your mood. This level of personalization could change how we plan our diets and take care of our health.

Working With Augmented Reality (AR) and Virtual Reality (VR)

When you combine AR and VR technology with AI, you'll get cooking experiences that are truly immersive. AR glasses may show step-by-step instructions on top of ingredients, while VR platforms could make fake professional kitchens for training. These technologies will make learning to cook more fun and easier to get to, letting those who want to be chefs practice their skills in virtual settings.

Food Production Using AI

AI is changing thc way food is grown and made in the kitchen and outside of it through precision agriculture and lab-grown meals. AI algorithms make sure that there is always a consistent supply of high-quality ingredients by optimizing agricultural yields, predicting when to harvest, and cutting down on the use of pesticides. AI is speeding up the creation of plant-based and cultured meats in the world of alternative proteins. These could change the way we cook by adding new textures and flavors.

AI-Chefs Working Together

In the future, AI systems might work with human chefs in real time, making suggestions, changing recipes, and even making dishes together during live cooking sessions. These AI chefs might work in restaurants or at home, working like virtual sous-chefs that change to fit the user's style and tastes.

Global Food Exchange

AI could help create a worldwide food exchange by bringing cooks and chefs from different cultures together. AI-powered platforms may translate recipes, make them work with local products, and exchange cooking tips, making the food community more connected. This exchange could help keep old cooking traditions alive and spark new ideas.

CASE STUDIES: AI AT WORK

Here are some real-life examples of how AI has changed things:

Plant Jammer

A Danish firm called Plant Jammer utilizes AI to make plant-based dishes based on what ingredients are available and what the user wants. The platform's algorithm looks at flavor profiles and nutritional information to make meals that are both tasty and good for the environment. This helps users eat less meat and waste less food. Plant Jammer has grown to millions of users by 2025, showing how AI can help people cook in a way that is good for the environment.

Moley Robotic Kitchen

The Moley Robotic Kitchen, which came out in 2021, is the most advanced kitchen automation system available. With robotic arms and AI-powered software, it can make more than 5,000 meals with the same level of accuracy as a Michelin-starred chef. While still a luxury product, Moley's technology foreshadows a future where automated kitchens are widespread, merging convenience with gourmet quality.

Robotics and Automation

AI-powered robots are changing the way professional kitchens work. Robotic arms made by Miso Robotics, like Flippy, do things like frying, grilling, and plating. These robots make sure that things are always the same and that they happen quickly.

How It Works: Robots using AI and computer vision can find ingredients and keep an eye on the cooking process. They learn how to copy procedures perfectly from human chefs.

Effect: Automation cuts down on labor costs and mistakes, which lets chefs focus on more difficult jobs like designing and presenting menus.

Ethical and Cultural Issues

AI has a lot of potential, but using it in cooking creates moral and cultural issues. If we rely too much on AI, it could take away the human creativity that makes cooking an art form and turn it into a simply technological procedure. Also, AI algorithms might unintentionally favor Western cooking styles over less well-known ones. To fix this, developers need to make sure that AI systems are trained on a wide range of facts that show how people eat around the world. Data privacy is another issue, since AI needs user data to give personalized experiences. To keep customers' trust, businesses must be open about how they use data and have strong security measures in place.

A New Component in Culinary Innovation

At its foundation, artificial intelligence is about using algorithms and computer models to copy how humans learn, think, and solve problems. AI is like a digital sous-chef in the kitchen. It gives you ideas, automates jobs, and opens up new creative options. It started with basic tools like recipe applications and meal planners, but now it has grown into complex systems that can evaluate flavor profiles, guess what people will like, and even manage kitchen appliances.

Making recipes was one of the first uses of AI in cooking. IBM's Chef Watson, which came out in 2014, utilized AI to look at thousands of recipes and ingredient combinations and recommend new ones. Chef Watson may suggest recipes like a turmeric-infused paella with chocolate or cauliflower as surprise additions by processing huge amounts of cooking information. Early versions were more about trying new things than being useful, but they showed that AI could help push the limits of cooking.

AI-powered recipe sites have gotten better over the years. Yummly and Whisk are two apps that employ machine learning to make recipes that fit users' dietary needs, allergies, and what they have in their pantry. These sites look at how users behave, such their search history and ratings, to suggest recipes that fit their likes. For instance, someone who often looks for vegan meals would get a suggestion for a lentil-stuffed eggplant with a tahini drizzle, along with step-by-step instructions and a breakdown of the recipe's nutritional value.

Personalization on a Large Scale

AI's ability to personalize is one of the biggest changes it has made to the culinary arts. In the past, if you wanted to make a meal plan for someone who needed to avoid gluten, carbs, and nuts, you had to talk to a dietician or do a lot of research. AI has made this process more open to everyone by letting anyone with a smartphone get personalized meal plans. Nutrino and PlateJoy are two companies that utilize AI to combine data from wearable devices, such fitness trackers, to suggest meals that help people reach their health objectives, including losing weight, gaining muscle, or controlling diabetes.

This customization goes beyond health. AI may take into account things like cultural preferences, flavor profiles, and even how someone is feeling. A study by the University of California, Davis, in 2023 looked at how AI could look at social media posts to figure out how a person was feeling and offer comfort foods based on that. Are you stressed? The AI might suggest a creamy mushroom risotto with roasted veggies on the side, based on research that shows that creamy textures make people feel better.

REVIEW OF THE LITERATURE

Theoretical Frameworks and Literature Review

The convergence of artificial intelligence and the culinary arts is insufficiently understood within academic discourse. To contextualize its transformative impact, various underlying theories and peer-reviewed studies are pertinent.

Distributed Cognition and Collaborative Creativity

Hutchins' (1995) theory of distributed cognition elucidates the extension of cognitive processes across individuals, technologies, and surroundings. In smart kitchens, AI-powered appliances serve as external cognitive artifacts, aiding in activities ranging from recipe sequencing to sensory modulation (Hutchins, 1995; Norman, 1993). Mihaly Csikszentmihalyi's Flow Theory (1996) enhances the comprehension of creativity in human-machine collaboration, as AI systems like as IBM's Chef Watson foster ideation through the provision of unexpected component combinations (Varshney et al., 2016).

Adaptive Systems and Behavioral Modeling

Self-Determination Theory (Ryan & Deci, 2000) and Adaptive Personalization Models (Kaplan et al., 2013) give conceptual basis for AI's capacity to personalize user experiences. AI-driven culinary platforms are in line with surroundings that foster autonomy and make people more interested, skilled, and connected with each other when it comes to food.

Systems Thinking and Sustainable Design

The Cradle-to-Cradle framework by McDonough and Braungart (2002) and Kim's (2016) Sustainable AI theories stress the importance of circular design and moral responsibility. AI systems employed in food inventory management and waste reduction correspond with these frameworks, offering scalable solutions to difficulties in food sustainability (Winnow, 2022; Rolnick et al., 2019).

The Combination of Culinary Informatics and AI

Emerging research in culinary informatics supports the formalization of AI's involvement in gastronomy. Studies by Tzachor et al. (2021) illustrate how computational gastronomy can boost diet quality, improve sustainability, and democratize recipe availability. Additionally, van Kuppevelt et al. (2022) contend that natural language processing can enhance recipe comprehension for those with cognitive limitations.

Artificial Intelligence in Creative Endeavors and Culinary Arts

Early writings about AI's influence in creativity generally used music and art as examples (McCormack & d'Inverno, 2012). In the culinary arena, scientists like Zhang et al. (2020) remark how generative adversarial networks (GANs) and natural language processing (NLP) enable machines to produce innovative recipes based on chemical and cultural flavor data. IBM's "Chef Watson" (Varshney et al., 2016) demonstrates this innovation; nonetheless, it also incited discourse over the profundity of machine creativity (Jordanous & Keller, 2016).

Theoretical Underpinnings

Csikszentmihalyi's Flow Theory (1996) has been utilized in the context of human-machine collaboration, proposing that AI tools can enhance a chef's engagement by managing monotonous activities. Hutchins' (1995) Distributed Cognition Theory is widely referenced in culinary technology (e.g., Allen & Tatar, 2021), conceptualizing smart appliances as extensions of human cognition. Ryan and Deci's (2000) Self-Determination Theory underpins AI's function in enhancing user autonomy and proficiency in domestic cooking.

AI, Food Systems, and Sustainability

Sustainable AI in food is becoming an important area of study that crosses several fields. AI is being utilized to cut down on food waste (Ostfeld et al., 2022), improve the quality of ingredients, and make farming more precise (Kamilaris et al., 2018). Cradle-to-cradle design (McDonough & Braungart, 2002) offers a comprehensive paradigm for comprehending sustainable culinary AI. Nonetheless, detractors highlight the rebound effects and the environmental costs associated with AI infrastructure itself (Vinuesa et al., 2020).

AI in Creating New Flavors and Recipes

Figuring Out the Science of Taste

Flavor is the most important part of cooking, and AI is revealing its secrets with more accuracy than ever before. AI can find complementary pairings that people might miss by looking at the chemical compounds in ingredients. For example, the concept of "food pairing" rests on the premise that products with common volatile compounds—molecules responsible for aroma—tend to taste good together. AI goes even further by mapping thousands of molecules across ingredients to come up with new and tasty pairings.

Foodpairing is a well-known company that uses AI to make taste networks. Their algorithms look at ingredients like strawberries, basil, and Parmesan to find unexpected ways they work together. This leads to things like strawberry-basil gelato with a Parmesan crunch. Chefs at high-end restaurants like Noma in Copenhagen have worked with AI techniques to create dishes that break the rules of taste while yet being in sensory harmony.

AI is also changing molecular gastronomy, a field that combines science and cooking to make cutting-edge dining experiences. By simulating the physical and chemical interactions of ingredients, AI can predict how procedures like spherification or emulsification would alter a dish's texture and flavor. For example, an AI system might tell you the best amount of sodium alginate to calcium chloride to use in a mango caviar that pops properly in your mouth.

Getting Creative With Cooking by Asking a lot of People

AI is not just a tool for expert cooks; it's also a place where everyone can be creative together. AI-powered online forums let home cooks help make recipes. Cookpad and other platforms employ AI to collect recipes made by users, find patterns, and recommend changes. If someone uploads a recipe for a spicy chicken curry, the AI might suggest a variation with coconut milk for others who want softer flavors. It does this by looking at data from thousands of similar dishes.

This crowdsourcing even includes coming up with new flavors. AI algorithms can look at online reviews, social media posts, and food blogs to find out what new tastes people are liking. For instance, in 2024, AI data showed that Western cuisines were using more and more umami-rich foods like miso and seaweed. Food companies exploited this information to make items like potato chips with miso flavor and salad dressings with seaweed flavor to keep up with changing tastes.

IMPORTANT ISSUES AND GAPS

AI Systems With Cultural Bias

AI systems that are mostly trained on Western cooking datasets could erase or misrepresent cuisines that are not well-known. Noble (2018) and Benjamin (2019) have recorded how algorithmic bias can mirror and exacerbate systemic disparities. In culinary AI, this manifests in the form of "creative" fusions that may alter religious or traditional foods, or the exclusion of regional recipes from marginalized areas.

Differences Between Regions and DIGITAL ACCESS

Most AI-powered kitchen gadgets and platforms are still too expensive for people who live in rural areas or who don't have a lot of money. UNESCO (2021) says that digital disparities still exist all throughout the world, especially in South Asia and Sub-Saharan Africa, where internet infrastructure and AI literacy are far behind. AI tools also require a certain level of digital literacy, which means that older or less educated people can't use smart cooking systems (Ali et al., 2022).

Problems With Putting It Into Action

High costs of capital, poor cross-platform integration, and inconsistent data standards make it hard for AI to be widely used in the food service industry. Restaurants in the Global South have trouble not only with costs but also with educating their workers to use AI tools. Additionally, issues of data privacy have not been adequately addressed in culinary technology (Floridi et al., 2018).

BALANCED DISCUSSION: GETTING PAST A LOVE OF TECHNOLOGY

Most current stories (like Moley Robotics and June Oven) praise AI's "wow factor," yet they might easily become ads. A scholarly discourse must instead inquire:

- Who gains from cooking with AI?
- Whose cooking knowledge is increased or lost?
- What do you have to give up in order to be more convenient?

Culinary AI shouldn't just make inputs and outputs better; it should also help create food futures that are fair, open to everyone, and based on culture.

ADVANTAGES OF AI IN THE CULINARY ARTS

More Creative

AI helps people be more creative by recommending new ways to use ingredients and processes. This gives expert chefs more ways to try new things, while amateur cooks may try out different types of food without having to know a lot about cooking.

Access and Inclusion

AI makes cooking more accessible to everyone by making complicated tasks easier and meeting a wider range of demands. Culinary arts are open to everyone when there are tools that develop recipes for certain diets (like keto or gluten-free) or that translate instructions into several languages.

Efficiency and the Ability to Grow

AI makes things easier in commercial kitchens, from keeping track of stock to making food. AI helps food manufacturers make sure that their manufacturing lines run as smoothly as possible, which means that they can make a lot of food with the same high quality.

Sustainability

AI helps the environment by cutting down on food waste and encouraging plant-based diets with personalized recipes. Smart appliances also save energy by making cooking more efficient.

ETHICAL AND SOCIO-CULTURAL ASPECTS

AI Systems with Cultural Bias

Algorithmic bias, especially the focus on Eurocentric food cultures, is one of the problems that AI-powered cooking platforms don't talk about enough. Gebru et al. (2018) say that datasets used to train culinary AI generally represent Western culinary norms. This leads to suggestions that are too similar and don't take into account indigenous and marginalized cuisines. For instance, AI platforms might choose materials or methods that are common in the Global North and ignore practices that are important in Africa, Southeast Asia, or Latin America.

To protect cultural integrity and keep culinary traditions from being erased, it is important to use Value-Sensitive Design (Friedman et al., 2008).

Different parts of the world have different levels of access to technology.

AI tools are still mostly found in areas with advanced technology, which makes the digital divide in culinary innovation even worse. Eubanks (2017) and D'Ignazio & Klein (2020) illustrate that those who are on the outside often don't have access to smart kitchen appliances or nutritional tech tools. This restriction could make the gaps in access to food, education, and cultural preservation even bigger.

This means that fair use of AI requires investment in infrastructure, making tools available in different languages, and open-source development so that people in areas who don't have access to it can use it.

Privacy of Data, Job Loss, and Being Real

AI's dependence on personal data, such as dietary choices and biometric inputs, engenders significant concerns pertaining to privacy and permission. Automation also puts food service jobs at danger, especially for low-wage kitchen workers (Zuboff, 2019). Lastly, AI-generated foods may commodify or misuse traditional recipes, which goes against the idea of real cuisine.

CASE STUDIES

Chef Watson From IBM

IBM made Chef Watson, which shows how creative AI can be. It has worked with chefs to make delicacies like a "Baltic Apple Pie" with unusual ingredients by looking at thousands of recipes and flavor pairings. This shows that AI can push the limits of cooking.

Flippy From Miso Robotics

Flippy is a robotic arm that does the frying and grilling in fast food places like White Castle. Its AI learns from how people work and can adjust to varied kitchen arrangements. Flippy has cut down on labor costs and made things run more smoothly, showing how useful automation can be.

Winnow's Waste Management

Winnow's AI technology, which is used in commercial kitchens, keeps track of food waste and gives useful information. In tests, it cut down on waste by as much as 50% in some places, which shows how AI may help the environment.

AI is changing the culinary arts by making smart cooking possible, combining creativity with accuracy to determine the future of food. AI-powered technologies look at huge collections of recipes, flavor profiles, and cultural preferences to come up with new dishes that are perfect for each person's tastes, dietary needs, or the ingredients they have on hand. Smart kitchen gadgets that use AI can change recipes in real time, improve cooking methods, and cut down on food wastage by keeping track of inventories and controlling portions. Machine learning algorithms help chefs come up with new flavor combinations, and AI-powered robots make it easier to prepare meals in restaurants and at home. AI is changing the way we cook by making it smarter, easier, and more accessible for everyone. It does this by customizing dining experiences and supporting sustainability.

AI AND THE WORLD OF PROFESSIONAL COOKING

Not Replacing Chefs, but Giving Them Control

AI is not a competitor for professional chefs; it is a partner. AI makes restaurant operations easier in high-pressure situations so that cooks can focus on being creative and giving guests a good experience. OpenTable and other reservation sites employ AI to make seating arrangements better and guess who won't show up, which makes service go more smoothly. AI-driven menu engineering tools, on the other hand, look at sales data to figure out which meals make the most money. This helps cooks improve their menus. There are many new things that AI can do in the culinary arts in the future. As AI gets better, kitchen helpers will become more intuitive, able to grasp commands in everyday language and predict what the user wants. For instance, future AI systems could utilize voice or face recognition to figure out how a person is feeling and propose foods that will make them feel better.

Combining augmented reality (AR) and virtual reality (VR) could make learning to cook better. AR apps may put cooking instructions on top of what a user sees, while VR could make professional kitchen settings for training. These technologies would make learning to cook more interactive and easy to get to. AI is also going to play a bigger part in making sure everyone has enough food. AI can help fight hunger and malnutrition by making farming, supply networks, and meal planning more efficient. To make the most of AI for social good, chefs, engineers, and policymakers will need to work together. AI also makes culinary education better. AI-powered virtual reality (VR) platforms create culinary settings where people who want to be chefs may practice things like filleting fish or tempering chocolate. These platforms give you input in real time, telling you how to change the angle of your knife or how to change the heat intensity. Schools like the Culinary Institute of America are starting to use AI-powered simulators in their classes to get students ready for kitchens that use a lot of technology.

Problems and Moral Issues

AI in the kitchen has its pros and cons, but it also poses moral problems. One worry is that people would lose their jobs, especially in low-skilled kitchen jobs. Workers may have less chances to find job when robots take over tasks like slicing and frying. Many experts, on the other hand, say that AI will create new jobs, like technicians to fix smart appliances or data analysts to figure out what food trends are. AI in the culinary arts has a lot of potential, but it also has certain problems. One big worry is that too much automation could make cooking less personal. Cooking is closely linked to culture, emotion, and creativity. If we rely too much on AI, we can lose traditional techniques or end up with flavors that are too similar. To keep the soul of cooking alive, it's important to find a balance between automation and human expertise. Data privacy is another problem. AI platforms routinely gather user information to tailor experiences, which raises questions about how this data is kept and used. To keep users' trust, it is important to have clear data procedures and strong security measures. Another worry is data privacy. AI-powered apps routinely gather private data, such as purchasing patterns and food preferences. People need to have faith that businesses would protect this information and use it in a fair way. To keep customers' trust, companies need to have clear privacy policies and strong cybersecurity safeguards.

Also, the high price of AI-powered appliances and robots can make them hard to get, especially for small eateries or home cooks in underdeveloped areas. Making sure that everyone can use these technologies fairly will be important for their broad use.

CHALLENGES TO IMPLEMENTATION

Even though it has a lot of potential, using AI in cooking confronts a lot of real-world problems:

- Interoperability: Many AI apps, including recipe apps and IoT gadgets, don't have standard ways to share data, which makes the ecosystem less cohesive (Lee et al., 2021).
- Cost and Usability: Advanced robotic systems, like Moley Kitchen, are too expensive, and technologies that are meant for consumers still need a lot of digital literacy (Jarrahi, 2018).
- Language and Cultural Translation: NLP technologies may not translate recipe instructions that are important to a culture correctly, which could lead to misunderstandings or mistakes in the kitchen (Sambasivan et al., 2021).

Another problem is being culturally authentic. AI systems that learn from worldwide recipe databases can unintentionally make cuisines more alike, creating fusion foods that mix cultures. For instance, an AI-made "taco" might use elements that aren't usually seen in Mexican food, which would raise problems about appropriation. To fix this, developers are making sure that AI understands cultural contexts by using a variety of datasets and talking to chefs.

SMART COOKING IN THE FUTURE

A Vision for Kitchens That Are Open to Everyone, Creative, and Lasting

In the future, AI will make cooking more accessible, innovative, and environmentally friendly. AI-powered assistive equipment, like as voice-activated appliances and robotic arms, can make kitchens easier to get around for individuals with disabilities. AI is like a safety net for new cooks since it helps them follow complicated recipes and gives them feedback in real time. AI gives experienced chefs a place to try out new flavors and techniques.

The combination of AI with other new technologies, like 3D food printing and lab-grown proteins, will change the way we cook even more. Picture a time when an AI-powered 3D printer makes a meal just for you, using sustainable ingredients like cultured meat or algae, based on your biometric data. These kinds of new ideas could help solve problems like hunger and climate change around the world and change the way we think about cooking.

What AI Will be Like in the Future in the Culinary Arts

There are no limits to what AI can do in the culinary arts. We should expect hyper-realistic virtual chefs that help people cook in real life as AI models get better. If biotechnology and AI work together, they may make foods that are tailored to each person's microbiome, which would be the best for their health. Also, AI's participation in vertical farming and urban agriculture could change the way ingredients are found, making it easier to get fresh food.

AI will probably work with chefs in professional kitchens to make Michelin-starred dishes by combining human intuition with data-driven accuracy. AI's ability to deal with food hunger and climate change will be very important for developing a strong food system on a societal level.

Finding a balance between technology and tradition

It's important to find a balance between technology and tradition as AI changes the way we cook. Kneading dough by hand, tasting a sauce by instinct, and sharing a meal with others are all things that can't be replaced. AI should make these experiences better, not worse. AI can keep the kitchen a warm and welcoming place by keeping the creative, emotional, and social aspects of cooking alive.

IN CONCLUSION

AI is changing the way we cook in many ways, from personalized recipe applications to robotic kitchens and new flavors. It empowers home cooks, streamlines professional kitchens, and promotes sustainability, all while pushing the bounds of creativity. But when we use it, we need to think about ethics and make sure that technology respects cultural heritage, protects privacy, and promotes inclusion. AI won't take away the heart of the kitchen; instead, it will make it stronger, making it possible for everyone to cook, make things, and connect via food. AI is changing the way we cook by combining technology with creativity to make cooking smarter, more accessible, and better for the environment. AI has a lot of uses, from personalized recipes to self-driving kitchens. These uses can improve creativity, efficiency, and accessibility. But to make sure that everyone can use it fairly, problems like the loss of personal touch, excessive expenses, and ethical concerns must be solved. The use of AI in cooking is a big change that combines technology with human creativity to change the way we cook smartly. AI improves productivity, personalization, and sustainability in the kitchen by using ideas from distributed cognition, adaptive systems, and systems thinking. As we move forward, the key will be to find a balance between new ideas and old ones, making sure that AI enhances rather than takes away from the art of cooking. AI may bring about a new era of cuisine that is open to everyone, environmentally friendly, and full of fresh ideas by encouraging chefs, technologists, and customers to work together.

REFERENCES

Ali, M., Avestimehr, S., & Srinivasan, R. (2022). Digital literacy, artificial intelligence, and inequality in emerging economies. *Information Systems Journal, 32*(4), 678–701. DOI: 10.1111/isj.12345

Allen, J., & Tatar, D. (2021). Distributed cognition in human–AI collaboration systems. *Human-Computer Interaction, 36*(5–6), 475–514. DOI: 10.1080/07370024.2020.1859617

Benjamin, R. (2019). *Race after technology: Abolitionist tools for the new Jim code*. Polity Press.

Csikszentmihalyi, M. (1996). *Creativity: Flow and the psychology of discovery and invention*. HarperCollins.

D'Ignazio, C., & Klein, L. F. (2020). *Data feminism*. MIT Press. DOI: 10.7551/mitpress/11805.001.0001

Eubanks, V. (2017). *Automating inequality: How high-tech tools profile, police, and punish the poor*. St. Martin's Press.

Floridi, L., Cowls, J., Beltrametti, M., Chatila, R., Chazerand, P., Dignum, V., Luetge, C., Madelin, R., Pagallo, U., Rossi, F., Schafer, B., Valcke, P., & Vayena, E. (2018). AI4People—An ethical framework for a good AI society. *Minds and Machines, 28*(4), 689–707. DOI: 10.1007/s11023-018-9482-5 PMID: 30930541

Friedman, B., Kahn, P. H., & Borning, A. (2008). Value-sensitive design and information systems. *Human-Computer Interaction, 21*(2), 349–365.

Gebru, T., Morgenstern, J., Vecchione, B., Vaughan, J. W., Wallach, H., Iii, H. D., & Crawford, K. (2018). Datasheets for datasets. *Communications of the ACM, 64*(12), 86–92. DOI: 10.1145/3458723

Hutchins, E. (1995). *Cognition in the wild*. MIT Press. DOI: 10.7551/mitpress/1881.001.0001

Jarrahi, M. H. (2018). Artificial intelligence and the future of work: Human–AI symbiosis in organizational decision making. *Business Horizons, 61*(4), 577–586. DOI: 10.1016/j.bushor.2018.03.007

Jordanous, A., & Keller, B. (2016). Modelling creativity: Identifying key components through a corpus-based approach. *PLoS One, 11*(10), e0162959. DOI: 10.1371/journal.pone.0162959 PMID: 27706185

Kamilaris, A., Kartakoullis, A., & Prenafeta-Boldú, F. X. (2018). A review on the practice of big data analysis in agriculture. *Computers and Electronics in Agriculture, 143*, 23–37. DOI: 10.1016/j.compag.2017.09.037

Kaplan, F., Oudeyer, P. Y., Kubinyi, E., & Miklósi, Á. (2013). Robotic companions: Bridging the gap between computational intelligence and human interaction. *IEEE Transactions on Autonomous Mental Development, 5*(1), 1–13.

Kim, T. W. (2016). Ethical design of artificial intelligence systems. *Journal of Business Ethics, 154*(2), 1–17.

Lee, J., Bagheri, B., & Kao, H. A. (2021). A cyber-physical systems architecture for industry 4.0-based manufacturing systems. *Manufacturing Letters, 3*, 18–23. DOI: 10.1016/j.mfglet.2014.12.001

McCormack, J., & d'Inverno, M. (2012). *Computers and creativity*. Springer. DOI: 10.1007/978-3-642-31727-9

McDonough, W., & Braungart, M. (2002). *Cradle to cradle: Remaking the way we make things*. North Point Press.

Noble, S. U. (2018). *Algorithms of oppression: How search engines reinforce racism*. NYU Press. DOI: 10.18574/nyu/9781479833641.001.0001

Norman, D. A. (1993). *Things that make us smart: Defending human attributes in the age of the machine*. Addison-Wesley.

Ostfeld, A., Binyamini, K., & Shamir, U. (2022). Artificial intelligence for food loss and waste reduction. *Resources, Conservation and Recycling, 180*, 106159.

Rolnick, D., Donti, P. L., Kaack, L. H., ... (2019). Tackling climate change with machine learning. *arXiv preprint arXiv:1906.05433.*

Ryan, R. M., & Deci, E. L. (2000). Intrinsic and extrinsic motivations: Classic definitions and new directions. *Contemporary Educational Psychology, 25*(1), 54–67. DOI: 10.1006/ceps.1999.1020 PMID: 10620381

Sambasivan, N., Kapania, S., & Highfill, H.. (2021). Everyone wants to do the model work, not the data work. *Proceedings of the ACM on Human-Computer Interaction, 5*(CSCW1), 1–26.

Tzachor, A., Richards, C., & Holt, L. (2021). Future foods: AI and computational gastronomy for sustainability. *Futures, 127*, 102696.

UNESCO. (2021). *Artificial intelligence and the digital divide*. UNESCO Publishing.

van Kuppevelt, D., Van Loo, J., & Odijk, D. (2022). Natural language processing for food and recipe understanding. *Information Processing & Management*, *59*(3), 102930.

Varshney, L. R., Pinel, F., Varshney, K. R., Bhattacharjya, D., & Schorgendorfer, A. (2016). A taste for science: Computational creativity in culinary arts. *IBM Journal of Research and Development*, *60*(5), 1–14.

Vinuesa, R., Azizpour, H., Leite, I., Balaam, M., Dignum, V., Domisch, S., Felländer, A., Langhans, S. D., Tegmark, M., & Fuso Nerini, F. (2020). The role of artificial intelligence in achieving the Sustainable Development Goals. *Nature Communications*, *11*(233), 1–10. DOI: 10.1038/s41467-019-14108-y PMID: 31932590

Winnow. (2022). *Reducing food waste through artificial intelligence*. Winnow Solutions Ltd.

Zhang, Y., Li, C., & Xu, J. (2020). Generative models for recipe creation and food innovation. *IEEE Access : Practical Innovations, Open Solutions*, *8*, 132534–132545.

Zuboff, S. (2019). *The age of surveillance capitalism*. PublicAffairs.

About the Contributors

Muhammad Tariq Iqbal Khan is currently working as a Lecturer in Economics at the Government Postgraduate College, Jaranwala. He started his research career as an Internee at the Technology transfer Institute. He joined Higher Education Department Punjab on 18-08-2012, as Lecturer in Economics. He has more than 8 years of teaching and research experience. He passed B.Sc. (Hons.) Agricultural and Resources Economics in 2011 from the University of Agriculture, Faisalabad. During his education, He actively participated in university-level competitions at the University of Agriculture, Faisalabad. He five times won the Quiz Competition at the University of Agriculture, Faisalabad during graduation. He got M.Phil. Economics degree from the Government College University, Faisalabad in 2016. In 2017, He enrolled Ph.D. Economics from Government College University, Faisalabad. During his career, he has published 32 research publications in refereed International and national journals, including 18 research papers that have been published in impact factor journals. The cumulative impact factor of his research is 200.535, having 1900 citations and an h-index 20. His current research focus is on disaster management, marketing efficiency, environmental economics, and agricultural economics.

***

Nizirwan Anwar, MT, IPM, ASEAN.Eng is a Lecturer in the Informatics Engineering Study Program, Faculty of Computer Science, Universitas Esa Unggul under the auspices of the Kemala Bangsa Education Foundation (YPKB). The author was born in the city of Bandung on July 24 1964, completed his bachelor's degree from the Physics Study Program, Faculty of Mathematics and Natural Sciences, Padjadjaran University, Bandung in 1989 and continued the Electrical Engineering Study Program, Faculty of Engineering (dh. Postgraduate Study Program) University of Indonesia, Jakarta, completing his studies in 1995, obtaining a Bachelor of Engineering Intermediate Professional (IPM) and ASEAN Engineer (ASEAN.Eng).

Alhamdulillah has produced several reputable and indexed published scientific papers at national and international levels (research a categories; Internet of Things, Cryptography, Bibliometrics and Data Science). To date, he is registered as active in several professional organizations, National Lecturer BKD assessor (NIRA), as a reviewer in Arjuna accredited journals (Kemendikbud Saintek); Scientific Committe, Central Executive of the Vocational Agency BKI PII,

Duggirala Aravind is associated with the Department of Aviation at ASET College of Science and Technology, Chennai, India. With a strong interest in aviation science and technology, his academic and professional focus lies in understanding the core principles of aircraft operations, aerodynamics, and aviation safety systems. He is dedicated to developing knowledge in areas such as air traffic management, aviation regulations, and modern advancements in the aerospace sector. His work emphasizes the integration of theoretical concepts with practical applications to address real-world challenges in aviation. Aravind actively engages in academic learning, research-oriented activities, and skill development aligned with emerging trends in aviation technologies. He demonstrates a commitment to continuous learning and aims to contribute effectively to the aviation industry through innovation, discipline, and technical expertise. His academic journey reflects a passion for aviation and a goal to build a career in this dynamic and evolving field.

Anorgul Ashirova is affiliated with the Department of General Professional Sciences at Mamun University, Khiva, Uzbekistan. Her academic interests focus on interdisciplinary education, professional skill development, and the integration of foundational knowledge across multiple academic disciplines. She is actively engaged in teaching and academic activities that aim to strengthen students' theoretical understanding and practical competencies required in modern professional environments. Her work emphasizes innovative teaching approaches, student-centered learning, and the application of academic knowledge to real-world professional contexts. Anorgul Ashirova also contributes to academic discussions related to curriculum development, educational improvement, and professional training. Through her teaching and scholarly engagement, she aims to support the development of well-prepared graduates equipped with the knowledge, critical thinking abilities, and professional skills necessary to succeed in diverse career fields.

Imam Asrowardi, S.Kom., M.Kom., M.Pd., IPM., ASEAN Eng. has been a lecturer in the Internet Engineering Technology Study Program at Politeknik Negeri Lampung since 2015 and currently holds the position of Associate Professor. He is actively involved in various national professional associations (IAII, APTIKOM, Informatics Engineering Division of the Indonesian Engineers Association) and international organizations (IAENG), both as a member and an administrator. Imam holds numerous industry certifications in various fields, including MTCINE, MTCRE, MTCNA, MTCTE, MTCWE, MTCUME, MTCIPv6E, MTA (Networking), CCC-BDF, EBDP, CSEP, HCIA-AI., Cisco Academy Trainer (CCNA, CyberOps Associate, DevNet Associate, CCNP), MikroTik Academy Trainer, Programmer (BNSP), Competency Assessor (BNSP), National Procurement Expert (BNSP), and Cisco Certified Support Technician Networking (CCST Networking)

Yudhi Biantoro earned a bachelor's degree in Information Systems from UPN "Veteran" Jakarta, Indonesia, in 2006; a master's degree in Information Technology from the University of Indonesia, Indonesia, in 2010; and a Ph.D. in Electrical Engineering and Computer Science from the Bandung Institute of Technology in 2023. He is currently a lecturer at the Faculty of Information Technology at Respati University of Indonesia. His research interests include cloud computing, information systems/information technology management, IT governance, data mining, data science, and analytics.

Kezia Rani Burgula is currently an Associate Professor in the department of Information Technology at Vasavi College of Engineering, Hyderabad. She received her PhD in Computer Science and Engineering in the field of Cloud Computing from Jawaharlal Nehru Technological University, Hyderabad in 2019. Her research interests include Cloud Computing, Distributed Systems, Machine Learning, Artificial Intelligence, and Optimization Techniques. She is a member of ACM and CSI.

Sheela D V, is a research scholar and working as Assistant Professor. She has a lot of passion for Research & Teaching. She has 12 years of teaching experience and currently pursuing PhD. Her area of interests is Computer Networks, Data Mining, Blockchain and Cloud computing and specialized in Java, Python and Web Programming. She has published research in areas related to blockchain and cryptography, including work on secure online quiz management using blockchain technology and multi-tier cryptographic techniques. Mrs. Sheela has published research in IEEE conference, scopus journals and has guided academic projects, participated in faculty development programs, and contributed to curriculum development.

Darniati is a lecturer in the informatics engineering study program at Muhammadyah University. The author was born in the city of Takalar on September 30 1984, completed his bachelor's degree from the Information Systems Study Program, Dipanegara University, Makassar, in 2012 and continued the Electrical

Engineering Study Program, Faculty of Engineering, Hasanuddin University, Makassar, completing his studies in 2017.

Ritam Dutta is having 16+ years of Teaching and Research Experience (05 SCI, 55 Scopus, 01 Patent Grant), 10+ years of Administrative Experience and 03+ year of Post Ph.D. Experience, served in several reputed institutions across India in the capacity of Assistant Prof., Associate Prof. and Professor. Dr. Dutta is a Member of Professional Bodies: IE(India) (M'19), IEEE (M'25), ISTE (M'19).

V. Joseph Emmanuvel is working as an Assistant Professor – Selection Grade in the department of Informatics at Periyar Maniammai Institute of Science & Technology, Vallam, Thanjavur, Tamilnadu, India. He obtained M.C.A (Computer Applications) from Periyar Maniammai College of Technology, Thanjavur in 2002, M.Phil from the Periyar University, Salem in 2019 and Ph.D. from Karpagam Academy of Higher Education, Coimbatore in 2025. He has published 23 research papers in various Journals, International / National Conferences and IEEE Seminars. He is also one of the reviewers for the Egyptian Journal (Scopus Indexed). He possess a total experience of 23.4 years in the field of Teaching and Administration and 10 long years in the area of research.

Jaishri Gothania is an academician, author, researcher, and data scientistwith over 10 years of experience in Artificial Intelligence, Machine Learning, and data analytics across academia and industry. She has authored 1 book, 18 book chapters, 9 research papers, and 3 conference papers, with publications indexed in Scopus and Web of Science. She holds 3 patents, including 2 granted innovations in AI and design. Her expertise spans data mining, intelligent systems, and applied AI solutions. She has also supervised postgraduate research, contributed to curriculum development, and actively participated in international conferences, editorial boards, and professional research communities.

Anjaneyulu Gudla is a research scholar pursuing a Ph.D. from SR University, Warangal, with a strong academic foundation in MCA and M.Tech, complemented by extensive experience in teaching Computer Science Engineering subjects, programming, and applications. With over 18 years of experience in academia, Mr Anjaneyulu Gudla has developed a passion for imparting knowledge and guiding students in the field of computer science. Currently, Mr Anjaneyulu Gudla is working as an Assistant Professor in the Department of Information Technology at Gurunanak Institutions Technical Campus, Ibrahimpatnam, Hyderabad. Mr Anjaneyulu Gudla research interests and expertise lie at the intersection of technology and education, and Mr Anjaneyulu Gudla looks forward to contributing to the academic community through this book chapter/patent.

Nethravathi H.M. is an academic and researcher in Electronics & Communication Engineering, currently completed her Ph.D. at B.M.S. College of Engineering (BMSCE) in Bengaluru, known for her work in wireless networks, D2D communication, and optimization algorithms, with published research in areas like resource allocation in D2D networks. She holds B.E. (2008) and M.Tech (2012) degrees from Visvesvaraya Technological University (VTU) and focuses her research on advanced wireless technologies, contributing to publications with MECS Press and in various journals.

Mohammad Ichsan is an experienced project management practitioner and scholar with over 20 years of expertise managing projects across diverse industries, including information technology, construction, telecommunications, and financial services. His teaching areas focus on operations management and project management. He has published several papers in these fields, with his current research domain centered on Project Management Offices (PMO) and Project Portfolio Management. He earned his bachelor's degree in Electrical Engineering from Hochschule Darmstadt, Germany (Diplom-Ingenieur), and his master's degree in Project Management (Magister Teknik – Cum Laude) from Universitas Indonesia. He is currently pursuing his PhD in Strategic Management at Binus University. Dr. Ichsan also holds professional project management credentials such as PMP and PMI-SP from the Project Management Institute, as well as PMO-CP from the PMO Global Alliance. He previously served as Region 15 Mentor from 2013 to 2017, was a member of the Board Volunteer Advisory Committee of the Project Management Institute from 2021 to 2023, and remains an active member of both PMI Global and the PMI Indonesia Chapter.

Irmawati is a bachelor of engineering in Electrical Engineering, Hasanuddin University, and an M.T. degree from the master in Information Technology, Postgraduate School of Hasanuddin University. She is currently a lecturer and researcher at Politeknik Negeri Ujung Pandang, Department of Computer and Network Engineering. Her current research interestsinclude data engineer, big data, integration system, and digital supply chain management.

Raden Teddy Iswahyudi, currently working as a Lecturer at the Undergraduate Study Program (S1) in the Informatics Engineering Study Program, Faculty of Computer Science, Esa Unggul University under the auspices of the Kemala Bangsa Education Foundation (YPKB). The author was born in Jakarta on November 10, 1975, completed his S1 education from the Electrical Engineering Study Program, Faculty of Engineering, University of Indonesia, Depok, West Java (2003) and continued his Masters Program in Computer Science, Faculty of Computer Science (formerly Postgraduate Study Program) Esa Unggul University, Jakarta (formerly DKI Jakarta) completing his studies (2023). A brief history of the author's career journey as an educator (lecturer) in implementing and devoting himself to the TriDharma of Higher Education has been 2 years - now. Alhamdulillah, he has produced several reference books; published and indexed scientific works and has a reputation (National/International), research in the field of Internet of Things (IoT), Cyber Security, Digital Transformation.

Swetha K. R., Assistant Professor in the Department of Computer Science and Engineering, is an accomplished academician and researcher with more than 14 years of teaching and research experience. Her career embodies a fine balance of excellence in teaching, impactful research, innovative contributions, institutional service, and community engagement. She has published extensively in reputed international journals and conferences indexed in Scopus, Springer, and IEEE, with research spanning cutting-edge domains such as Brain-Computer Interfaces, Big Data Analytics, Image Processing, and Healthcare Informatics. Her scholarly work includes significant contributions on deep learning models for disease diagnosis, adaptive filtering techniques for brain-computer interface systems, and IoT-based smart healthcare solutions. With over 20 high-quality publications, multiple book chapters with Springer and IGI Global, and wide academic citations, her research continues to enrich the field of computer science. Beyond research, Dr. Swetha has played a vital role in institutional development, contributing to quality assurance processes through her involvement in NAAC and NBA accreditation committees. She has also served as the Programme Officer for NSS Unit-7, fostering social responsibility and student engagement, while actively supporting academic and research collaborations. Her innovative mind-set is reflected in successfully securing and executing funded projects under KSCST, CMTI, and NAIN, focusing on impactful solutions in healthcare, agriculture, and assistive technologies.

Abinaya M. is an Assistant Professor at Erode Sengunthar Engineering College, she has 6 years of teaching experience. Holding a B.E. in Electronics and Communication Engineering (Paavai College of Engineering) and M.E. in Embedded Systems and Technologies (Mahendra Institute of Engineering and Technology), she specializes in IoT technology research, Abinaya has published a patent and presented papers at national and international conferences, contributing significantly to academic knowledge.

Padmesh Medesety is a research engineer with the Chemical Engineering group at ISRO's Satish Dhawan Space Centre (SHAR). He has a strong academic background: after completing his undergraduate studies in chemical engineering, he pursued advanced studies in India. Medesety earned his Ph.D. in chemical engineering from the Visvesvaraya National Institute of Technology (VNIT), Nagpur in September 2023. During his doctoral work under Prof. Piyush P. Wanjari, he specialized in molecular simulations of carbon-capturing nanomaterials. For example, he conducted computational studies of CO_2 adsorption on functionalized graphene and designed novel graphene–carbon nanotube hybrid structures for enhanced carbon capture. His Ph.D. research resulted in first-author publications in high-impact journals, reflecting his early scholarly contributions to chemical engineering and nanomaterials. Medesety's interdisciplinary research spans chemical engineering, materials science and computational physics. He is skilled in techniques such as molecular dynamics to model how nanostructures interact with gases and liquids. His published work includes a 2021 Molecular Simulation article on graphene–CNT hybrids for CO_2/N_2 and CO_2/H_2O separation and a 2022 Chemical Physics study on carbon nanotube systems for selective organic separations. (He also contributed a systematic review on graphene-based adsorbents for CO_2 capture) Through these studies, Dr. Medesety has advanced knowledge of how engineered carbon nanomaterials can improve selectivity and efficiency in gas separation processes. In addition to research, Dr. Medesety has experience in academia. This dual role in academia and government research highlights his ability to bridge fundamental science and practical applications. His work at ISRO leverages his chemical engineering expertise to address space-related and energy-environmental challenges, exemplifying an interdisciplinary approach. Overall, Dr. Medesety's scholarly contributions are at the intersection of chemical engineering and computational materials science, with applications in carbon management, process safety, and sustainable engineering.

Raghu N has Ten years of experience in Academic and Research. He received the Ph.D. (2020) in Electronics Engineering from Jain University, and B.E. in Electronics and Telecommunications Engineering from VTU University, Karnataka. His research interests were in RF Communication and Image Processing. He has published his research findings in various Scopus indexed Journals and International conferences. He is a member of various professional bodies.

Naveen Nandal is a distinguished academic and research professional with over 12 years of experience in teaching, research, and analytics. Currently an Associate Professor at DIHE, Delhi, she has published over 80 research papers and authored six books, as well as holds patents. She has contributed extensively to Scopus-indexed, ABDC, WOS, and peer-reviewed journals. She has served as Analytics Area Chair and Research & Publication Head, actively involved in curriculum design, faculty development, and research guidance. Dr. Nandal is a reviewer for Sage and

Emerald publications, a founding member of the Research India Foundation, and has extensive experience delivering impactful training and mentoring in marketing, human resource management, and data analytics domains.

Kamala Narayanan is a researcher in autonomous systems and intelligent control with extensive experience in developing hybrid model-based and AI-driven control strategies for connected and automated vehicles. Her work spans nonlinear and adaptive control, V2X-enabled cooperative driving, and digital-twin-based evaluation of autonomous systems. She has published widely in autonomous navigation, robotics, and intelligent cyber-physical systems, and her current research focuses on autonomy-aware coordination and safety-critical decision-making in mixed-traffic environments.

Shubha Parameshwara received her B.E. degree in Instrumentation Engineering from Dr. Ambedkar Institute of Technology, Bengaluru, and her M.Tech degree in Computer Science and Engineering from R.V. College of Engineering, Bengaluru. She was awarded her Ph.D. in Electrical and Electronics Engineering Sciences by Visvesvaraya Technological University, Karnataka, in 2021. She is currently serving as an Associate Professor in the Department of Electronics and Instrumentation Engineering at Dr. Ambedkar Institute of Technology, Bengaluru. She has 20 years of teaching experience and 10 years of research experience. Her research interests include autonomous robotics, image and video processing, embedded systems, and the Internet of Things (IoT).

S. Pushpalatha is currently serving as an Assistant Professor in the Department of Information Science and Engineering at Dr. AIT, Bengaluru. She earned her Ph.D. in the domain of Computer Vision from VTU and has nearly 21 years of teaching experience. Dr. Pushpalatha has contributed extensively to research, with around 25 publications and presentations in international journals, international conferences, and national conferences. She has also successfully filed four patents, and Author for Generative AI Text book . In addition, she is a life member of professional bodies such as IEI and ISSS.

Abdul Razzak Khan Qureshi is a distinguished academic and researcher with a Ph.D. in Computer Science from Sri Satya Sai University of Technology and Medical Sciences, Sehore. As an Assistant Professor at Medicaps University, Indore, he brings a wealth of experience, with 22 years in teaching and 7 years in research. His expertise spans the critical fields of IoT, Computer Networks, Cyber Security, Cloud Computing, Artificial Intelligence, Machine Learning, Data Science, and Wireless Networks. Dr. Qureshi's scholarly contributions are both extensive and impactful, with an impressive research portfolio that includes over 20 publications, 6 book chapters, and 5 patents. He has also been a frequent presenter at academic conferences, participating in more than 35 events. His work underscores a profound dedication to advancing knowledge and innovation in his specialized domains.

Suyam Praba R. is a Professor at the School of Management Studies, brings over 21 years of rich experience, including 16 years in the BFS industry and 5 years in academia. She holds a Ph.D. and M.Phil. in Management (Finance) from Bharathiar University, an MBA in Marketing & Finance, and a B.Sc. in Biochemistry. She is a prolific researcher who has published 30 papers in Scopus and WoS indexed journals, presented 25 papers at international conferences (IEEE & AIP), and authored 2 books and 2 book chapters on Behavioural Finance and Investment Decisions. Her professional background includes senior roles at Sakthi Group of Companies, where she handled retail management, customer acquisition, credit appraisal, compliance, and audit, along with earlier experience at MetLife India and Infosys BPO.

Ravikumar R. N. is currently a Postdoctoral researcher at Universiti Malaysia Perlis, Malaysia. Also, working as an Assistant Professor in the Department of Computer Engineering at Marwadi University, Rajkot, Gujarat. He has 14 years of national and international teaching experience. He completed his PhD at Amity University, Rajasthan, M.E at VMRF Deemed University, B.E at Anna University, Chennai. He is the co-founder of the Research Activities Club at Marwadi University and trains students on research activities, which leads to the various publications. He has published numerous research articles in Scopus, SCI journals, Book chapters, and Conference papers. Also, he hold several patents and authored books. His specialization is ML, DL, Federated Learning, Recommendation Systems, and Optimization Techniques. Dr. Ravikumar has authored more than 8+ multiple SCIE/Scopus-indexed journal articles, 25+ IEEE/Springer conference papers, and co-authored a book on Artificial Intelligence. He has also published 9 patents and contributed 10+ book chapters in emerging domains such as Federated Learning, Blockchain, and Digital Twins for IGI Global, Wiley and Springer publisher. He actively mentors students through initiatives like the Research Activities Club (RAC) and Student Research Internship Training (SRIT), promoting innovation and research culture.

N. Raja Kumar is currently serving as an Associate Professor in the Department of Information Technology at Lords Institute of Engineering and Technology, Hyderabad, Telangana. He is a dedicated academician and researcher with strong expertise in the field of Computer Science and Engineering. He holds a Master of Computer Applications (MCA), a Master of Technology (M.Tech) in Computer Science and Engineering, and a Doctor of Philosophy (Ph.D.) in Computer Science and Engineering. Over the years, Dr. Raja Kumar has made significant contributions to teaching, research, and innovation. His research credentials include 13 publications in reputed national and international Scopus-indexed journals, reflecting his commitment to advancing knowledge in the domain of computer science. In addition, he has successfully secured two patents, showcasing his innovative and problem-solving abilities in technology-driven research. With a strong academic background and a passion for research, Dr. Raja Kumar continues to guide students, contribute to the academic community, and pursue advancements in emerging areas of Information Technology and Computer Science.

J. Rajalakshmi received her Ph.D. degree in Information and Communication Engineering from Anna University, Chennai. Also she received Post Graduate, Under Graduate in Electronics and Communication Engineering from Anna University. Currently, she is a Professor at Sethu Institute of Technology, Virudhunagar. Dr. J. Rajalakshmi has been published in several Journals, International Conferences, National Conferences, including a Patent with grant in Communication area. She act as a Session chair in many International Conferences. She is a reviewer for several (mainly IEEE IoT Journal, IEEE conferences, Tech Science Press and International Conference Technical Reviewer) publications and Member of Indian Society for Technical Education (MISTE) in Professional Society. Her research interests include a wide range of MIMO Antenna, Communication Networks, Artificial Intelligence, wireless networks, IoT, Cyber Security, Computer Networking, Computer Networks Security etc.,

A. Hariprasad Reddy is a distinguished Professor in the Department of Computer Science and Engineering at Geetanjali College of Engineering and Technology, Hyderabad. With 17 years of teaching and 13 years of research experience, he has made significant contributions to the fields of Digital Image Processing, Machine Learning, Deep Learning, and Cybersecurity. He has authored numerous publications in national and international journals, conference proceedings, and book chapters. His innovative work has also led to the granting of several patents, reflecting his commitment to advancing technology and education.

Rajiv Ruliya currently working with MMICT&BM(HM), Maharishi Markandeshwar Deemed to be University Mullana Ambala Haryana.

Aarthi S is a dedicated Assistant Professor in Computer Science with over six years of teaching experience. She is currently pursuing her Ph.D. at Marwadi University, Gujarat, and holds an M.Phil. and MCA from Bharathiar University, Tamil Nadu. Throughout her academic career, Aarthi has specialized in programming languages such as C++, Java, and C, and contributed to research in recommender systems and AI-related data privacy. She has published journal articles, contributed to book chapters, and actively participated in technical workshops. Known for her student mentorship and engagement in academic events, Aarthi remains committed to fostering innovation and excellence in computer science education.

ArjuneS. is an academic and researcher specializing in entrepreneurship, business innovation, and the application of technology in agriculture. He earned his Ph.D. in Entrepreneurship from SASTRA Deemed University under the Junior Research Fellowship (NFOBC) scheme. He also holds an MBA in Human Resources and Marketing, along with a B.E. in Electronics and Instrumentation Engineering. His research work has been published in ABDC listed and Scopus-indexed journals, and his areas of interest include entrepreneurial development, digital adoption, agri-entrepreneurship, and sustainable business practices. He has presented his research at several reputed platforms, including IEEE and OPJU international conferences, with a strong focus on rural entrepreneurship and green technology. Alongside research, Dr. Arjune has teaching experience as a Teaching Assistant at SASTRA University, where he contributed to curriculum development and student mentoring. His professional background also covers business development, payroll management, and corporate sales. He is UGC-NET qualified, holds a Scopus h-index of 3, and is registered with ORCID.

Balakrishnan S is the founder of Vaagai IP. He holds a Bachelor's degree in Electronics and Communication Engineering from Anna University and has seven years of experience in research and development. He specializes in authoring numerous research papers, filing multiple patents, and working as a patent analyst.

MangaiS. is a Professor and Head, Department of Biomedical Engineering, Velalar College of Engineering and Technology, Erode. She holds a B.E. (NIT Surat), M.E. (Gujarat University), and Ph.D. (Anna University). With over 26 years of teaching experience, she has published 35 international journal papers, 56 conference papers, 2 books, and 4 book chapters. Her research interests include Biosignal & Medical Image Processing, AI, Data Science, Biometrics, Wireless Networks, and Embedded Systems. She has guided UG, PG, and Ph.D. scholars and received grants worth 20 lakhs from premier agencies. She holds 8 copyrights, 2 published patents, and has delivered keynote talks at national and international forums. A Fellow of IEI and member of IEEE, BMESI, ISSE, and ISTE, she was awarded the Best Engineer (Academics) by IEI, Erode Chapter in 2023–24.

Shivansu Sachan is Pursuing Ph.D. in Management, a Master's holder from Integral University, Lucknow, Bachelor's from IHM, Lucknow. Academic and professional accomplishments of 6 years with renowned organizations and institutes.

His Teaching Career has been very satisfying and have continuously upgraded the skills and knowledge by attending many courses and training programs, seminars, conferences organized by different organizations and also delivered many lectures as a Hospitality sector. He is working as an Assistant Professor at Chhatrapati Shahu ji Maharaj University Kanpur .

Mukimov Askar Shukhratovich is an Associate Professor in the Department of Digital Technologies at Alfraganus University, Tashkent, Uzbekistan. His academic and research interests focus on digital technologies, information systems, data analytics, and the application of modern computing solutions in education and industry. He is actively involved in teaching and scholarly activities that promote the development of students' technical skills in programming, information technology, and digital innovation. His work emphasizes the integration of advanced digital tools, emerging technologies, and data-driven approaches to address contemporary technological challenges. Mukimov Askar Shukhratovich also contributes to academic discussions related to digital transformation, information technology management, and the advancement of technology-based education. Through his teaching and research engagement, he aims to support the development of skilled professionals capable of contributing to modern digital ecosystems and technological innovation.

Sitesh Kumar Singh, PMP, is an accomplished academic and Chartered Engineer with a Ph.D. in Civil Engineering and over ten years of diverse experience in education, research, consultancy, and educational administration. Currently serving as an Assistant Professor in the Department of Civil and Environmental Engineering at the National University of Science and Technology, Muscat, Sultanate of Oman. Dr. Singh holds multiple advanced degrees including M.Tech, B.Tech, PGDIPR, PGDUPDL and PGDEd. His career encompasses curriculum development, research supervision, quality assurance, and strategic academic planning. He has supervised over 40 postgraduate and 2 doctoral students, contributing to academic leadership through workshops, conferences, and editorial roles, and has published over 5 books, 28 book chapters, 58 research articles & conference presentations, and 7 patents.

Vanshika Srivastava is a seasoned hospitality professional and educator with a wealth of experience spanning over seven years in both industry and academia. A graduate of IHM Jaipur, she further honed her expertise by earning a Master's degree in Travel & Tourism from IGNOU. She is perusing Ph.D. in hotel management from MMU Mullana, Ambala. Her journey in the hospitality sector began with a dynamic tenure at The Oberoi, New Delhi, where she excelled in the Front Office

Department for 2.4 years, gaining hands-on experience in delivering exceptional guest experiences. Driven by her passion for education, she transitioned into academia, serving as Teaching Associate at IHM Panipat for one year. There, she played a pivotal role in shaping future hospitality professionals. She later continued her academic journey as Teaching Associate at IHM Faridabad for three years, further contributing to the development of young minds in the hospitality domain. She has also presented and published various papers in national conferences. She is currently working as Assistant Professor at Chatrapati Sahuji Maharaj University, Kanpur.

Harry Sutanto completed his undergraduate program in agriculture at Universitas Padjadjaran, Bandung, in 1983; his Master of Business Administration program at the University of Illinois at Urbana-Champaign, United States, in 1996 (as part of an overseas education program sponsored by one of Indonesia's state-owned banks); and graduated as the best student from the Doctoral Program in Management Science at Universitas Padjadjaran, Bandung, Indonesia, in 2017. He spent three years working in a large private plantation company in Riau, Sumatra, and 28 years working at Indonesia's largest state-owned bank, in the areas of: (1) corporate and institutional credit; (2) capital market management; (3) equity participation management during the early stages of a merger process; (4) change agency roles, including the internalization and supervision of corporate culture; (5) assignment as Finance and Marketing Director at a state-owned paper industry company for 32 months, during which (together with the President Director and Production Director) he successfully transformed the company from a very poor condition into a profitable, well-governed, future-oriented, and sustainable enterprise; and (6) litigation related to the settlement of non-performing loans. He has served as: (1) President Commissioner and Commissioner at a building maintenance company and a security printing company (successfully transforming both into well-governed and sustainable companies); (2) a member of the Liquidation Team of an Indonesia–France joint venture bank; and (3) a member of various formal corporate bodies related to his professional duties in the banking sector. He was the initiator and conceptual counselor for "Envisioning Corporate Transformation through Reshaping Mission and Corporate Culture" for: (1) a subsidiary jointly owned by two major state-owned enterprises in the energy sector; (2) a regional development bank with medium-scale assets; and (3) a research and development agency of a Ministry of the Republic of Indonesia. He has been a speaker and trainer at various seminars and lectures on corporate and organizational culture, ethics and business ethics, change management, strategic management, general and spiritual motivation, character building, and banking management, delivered to state-owned enterprises, private companies, entrepreneurs, and civil servants. He is also actively involved in consultancy services, particularly in the areas of change management and corporate transformation, leadership, and the development and acceleration of corporate culture implementation, both for corporations and government institutions. He serves as an examiner for the Doctoral Program in Management at a State University and as a trainer in several collaborations with educational institutions. He is also involved in the Farmer Corporation Program of the Ministry of Agriculture to improve efficiency and agricultural capacity in supporting national food self-sufficiency, and was later appointed as Expert Staff at BRMP of the Ministry of Agriculture as an educator, consultant, and writer. He lectures at undergraduate and postgraduate (Master of Management) levels at several leading state and private universities in Indonesia, with areas of expertise including Corporate and Organizational Culture, Business Ethics, Leadership, Knowledge Management, International Business Management, Banking Credit Management, Change Management, Risk Management, Managing Global Enterprise, Human Resource Management, Organizational Behavior, and Microfinance, approached from a broad, integrative, and holistic perspective. All of his experiences have been written into seven published books entitled: Business Recovery Strategy: Strategi Komprehensif dalam Pemulihan Bisnis Pascakrisis, Memimpin Organisasi Bisnis – Telaah Komprehensif Mengenai Kepemimpinan Diri, Tim Kerja, Unit Bisnis, dan Korporasi, Leading Business in Dynamic Change – From Personal to Corporate Leadership, Successful Driving Execution: Managing Strategic Initiatives with an Effective Project Management Approach, Cryptocurrency: Terobosan atau Ancaman terhadap Tatanan Finansial Umum, Pengelolaan Organisasi di Era Digital: Konsep Dasar dan Aplikasi, and Manajemen Stratejik: Road to be the Real Sustainable Competitive.

G. Swetha received her B.E. degree in Electrical & Electronics Engg and MTech in Digital Electronics from Siddhartha Institute of Technology and Ph.d from VTU in the field of Artificial intelligence for power loss minimization in Power distribution systems . She has around 15 years of teaching experience. She started her career as lecturer in PES Institute of technology, Worked as Assistant Professor at Jain university, Acharya institute of technology, Sapthagiri college of Engineering. Presently she is working as Associate professor in R R Institute of technology. She has published more than 15 papers in various international journals and conferences. She has published patents. NAAC coordinator NBA coordinator – Has successfully Achieved accreditation. Reviewer for the IEEE conferences. IEEE, Life member of ISTE and IAENG. Has guided and evaluated various B.E and MTech projects.

Dilmurod Turaev is affiliated with the Department of Information Technology and Exact Sciences at Termez University of Economics and Service, Termez, Uzbekistan. His academic interests focus on information technology, computer science, and the application of computational methods in scientific and educational environments. He is actively involved in teaching and academic activities that aim to strengthen students' knowledge in programming, data analysis, and modern computing technologies. His work emphasizes the integration of digital technologies, problem-solving skills, and analytical approaches to address real-world technological challenges. Dilmurod Turaev also contributes to academic discussions related to emerging technologies, information systems, and the advancement of IT education. Through his teaching and scholarly engagement, he seeks to promote innovation in technology education and support the development of skilled professionals capable of contributing to modern digital and scientific advancements.

Prakash Upadhyay is working as Assistant Professor in Xavier University(previously St. Xavier college of Management & Technology) since July 2014. He was also an academic counsellor in IGNOU Patna since 2005 for BCA and MCA course. He has completed Phd in computer Science 2022. Dr. Prakash Upadhyay has 19+ years of experience in teaching to under graduate and post-graduate students of Computer Science and Application in different colleges in Bihar. He has published more than 8 papers in international and peer reviewed journals. He has also contributed chapters in 3 books of artificial intelligence. He has supervised master students and Undergraduate students for their project works. His research interest includes Artificial Intelligence, Machine Learning, Design and Analysis of algorithm, Theory of Computation and discrete mathematics, Statistical Computing etc

Varadharajan Bhoopathy is a Professor at Sree Rama Engineering College. He holds a Ph.D. in Information and Communication Engineering from Anna University and has over 16 years of academic experience. Specializing in wireless sensor networks and IoT, he has published numerous research papers and holds several international patents. Dr. Bhoopathy is a recipient of prestigious awards for his contributions to education and research. His commitment to academia continues to drive innovation and excellence in the field of computer science.

Index

A

Accident Hotspots 49, 55, 56, 70
Accident Prevention 110, 198, 234, 252, 369, 375, 379, 384, 385, 402, 427, 501, 505
Advanced Driver Assistance 4, 198, 201, 208, 224
AI-Driven Transportation Safety 1, 15, 24, 31, 42, 62, 135, 137, 143, 147, 150, 155, 159, 164, 165, 166, 473, 475, 488, 497, 499, 500
Artificial Intelligence (AI) 2, 3, 4, 6, 7, 8, 9, 14, 26, 28, 29, 31, 34, 35, 36, 39, 40, 41, 43, 44, 57, 64, 71, 72, 73, 75, 76, 80, 82, 95, 97, 98, 99, 100, 103, 104, 105, 106, 107, 108, 110, 112, 115, 116, 126, 128, 130, 131, 132, 133, 134, 136, 138, 167, 173, 174, 176, 180, 184, 185, 189, 192, 193, 194, 200, 220, 229, 230, 233, 234, 235, 236, 237, 238, 239, 240, 250, 251, 252, 253, 254, 258, 259, 260, 274, 296, 353, 377, 380, 384, 387, 410, 411, 413, 439, 440, 441, 442, 443, 444, 468, 473, 474, 501, 502, 503, 505, 506, 511, 519, 520, 521, 531, 532, 533
Artificial Neural Networks (ANN) 173, 281, 470
Autonomous Vehicles 1, 3, 27, 42, 43, 45, 48, 51, 57, 58, 59, 62, 63, 66, 67, 71, 74, 77, 90, 92, 95, 107, 110, 125, 126, 131, 169, 171, 172, 174, 180, 181, 182, 183, 184, 188, 189, 191, 192, 193, 194, 195, 201, 216, 222, 225, 226, 228, 259, 266, 274, 307, 322, 351, 364, 374, 375, 397, 404, 432, 439, 450, 453, 457, 473, 474, 475, 485, 493, 501, 502, 503, 506

B

Bluetooth Low Energy (BLE) 233

C

Computer Vision 1, 2, 4, 5, 9, 26, 35, 36, 40, 41, 43, 44, 72, 73, 76, 81, 86, 97, 103, 104, 105, 106, 107, 108, 110, 112, 128, 134, 135, 141, 142, 149, 151, 162, 166, 174, 227, 234, 265, 301, 335, 349, 350, 400, 401, 435, 437, 438, 442, 506, 508, 511, 512, 513, 519
Connected Autonomous Vehicles (CAV) 131
Connected Vehicles 45, 54, 57, 60, 87, 133, 198, 204, 206, 211, 212, 214, 215, 217, 219, 220, 225, 226, 267, 274, 307, 373, 388, 403, 456, 467
Crash Prediction 257, 261, 263, 264, 267, 274, 275, 295, 296, 309, 316, 319, 332, 339

D

Data Privacy 28, 35, 39, 47, 50, 59, 69, 70, 77, 88, 95, 107, 110, 111, 121, 128, 141, 147, 158, 163, 182, 189, 198, 224, 225, 230, 278, 403, 407, 435, 436, 442, 462, 478, 481, 489, 506, 510, 519, 524, 528
Deep Learning (DL) 5, 40, 43, 44, 46, 49, 59, 61, 70, 98, 110, 112, 193, 194, 248, 249, 257, 261, 262, 263, 264, 265, 266, 272, 274, 280, 289, 294, 297, 301, 302, 314, 316, 339, 345, 347, 350, 351, 389, 393, 394, 406, 418
Driver Behavior Analysis 264, 297, 309

E

Electric and Hybrid Vehicles 7
Emergency Response Systems 42, 69, 82, 136, 140, 308, 420
ethical governance 89, 92, 159, 495, 503

F

Fleet Management 171, 173, 185, 186, 187, 305

G

Geographic Information System (GIS) 235
Green Mobility 33

I

Incident Detection 43, 45, 46, 73, 81, 400, 415, 418, 422, 423, 424, 427, 428, 431, 436, 437
Intelligent Transportation Systems (ITS) 4, 22, 29, 32, 33, 70, 75, 78, 98, 99, 101, 130, 132, 133, 136, 148, 167, 191, 193, 198, 226, 227, 257, 258, 261, 263, 267, 271, 274, 305, 347, 349, 350, 351, 405, 415, 439, 444, 471, 501
Intelligent Transport Systems 99, 134, 170, 172, 180, 181, 212, 380, 399, 441, 454, 469, 471, 502
Internet of Things (IoT) 4, 7, 58, 76, 82, 92, 100, 105, 109, 137, 138, 228, 229, 230, 232, 233, 236, 237, 238, 239, 241, 243, 249, 250, 251, 252, 253, 254, 349, 354, 425, 439, 467, 508

L

Linear Regression (LR) 282
Low-Cost Sensors 135

M

Machine Learning (ML) 1, 2, 4, 5, 6, 9, 10, 11, 26, 35, 36, 37, 38, 40, 43, 44, 46, 48, 49, 51, 52, 53, 54, 55, 57, 59, 61, 66, 67, 68, 71, 72, 73, 75, 80, 86, 92, 97, 98, 99, 100, 101, 103, 104, 105, 106, 107, 110, 111, 112, 115, 116, 128, 130, 132, 133, 134, 142, 153, 166, 167, 170, 172, 174, 177, 178, 179, 185, 202, 203, 216, 220, 221, 227, 236, 237, 238, 251, 254, 257, 258, 259, 260, 261, 262, 263, 264, 265, 266, 267, 268, 269, 270, 271, 272, 273, 274, 275, 276, 277, 278, 279, 294, 295, 296, 297, 300, 301, 303, 304, 306, 307, 309, 314, 316, 317, 318, 319, 321, 324, 331, 332, 335, 336, 337, 339, 340, 341, 342, 343, 345, 346, 347, 350, 354, 355, 377, 378, 380, 381, 383, 384, 385, 389, 390, 391, 392, 393, 394, 395, 400, 401, 403, 404, 406, 409, 410, 411, 418, 422, 423, 439, 467, 476, 477, 502, 505, 506, 507, 511, 512, 513, 514, 519, 526, 532
Multi-Modal Systems 3, 5, 320

N

Natural Language Processing 2, 5, 9, 26, 33, 36, 81, 131, 149, 154, 237, 413, 417, 432, 439, 440, 441, 442, 443, 444, 506, 508, 511, 512, 513, 521, 533

P

Passenger Behavior 475, 505
Predictive Analytics 1, 2, 9, 26, 35, 40, 42, 51, 52, 53, 54, 56, 57, 66, 71, 84, 86, 98, 103, 108, 110, 114, 115, 117, 124, 125, 134, 136, 138, 142, 160, 162, 166, 174, 182, 185, 202, 215, 233, 234, 235, 236, 237, 257, 259, 260, 265, 339, 345, 355, 383, 384, 387, 389, 398, 399, 400, 402, 406, 411, 462, 474, 476, 477, 501, 516
Predictive Modeling 3, 12, 18, 24, 29, 76, 95, 112, 126, 186, 300, 388, 496, 505
Proactive Risk Mitigation 383, 384, 388, 390, 406, 408, 411

R

Random Forest (RF) 18, 282, 355, 394, 395, 409
regulatory frameworks 31, 62, 90, 130, 136, 147, 189, 198, 219, 269, 279, 321, 345, 415, 475, 479, 480, 481,

486, 487, 489, 495, 498, 499
Resource-Constrained Regions 135, 137, 141, 143, 147, 150, 155, 159, 164, 166
Risk Mitigation 1, 337, 383, 384, 388, 390, 406, 408, 411
Road Safety 2, 33, 34, 41, 42, 46, 47, 50, 51, 53, 54, 55, 56, 57, 68, 72, 73, 86, 97, 100, 104, 105, 113, 117, 122, 136, 140, 167, 199, 200, 201, 202, 204, 215, 217, 351, 356, 357, 358, 359, 367, 374, 377, 379, 381, 383, 384, 385, 386, 387, 399, 400, 403, 404, 415, 448, 477, 492, 503

S

Safety Performance Indicators (SPIs) 234
Smart Cities 7, 25, 31, 33, 77, 83, 95, 98, 99, 100, 109, 114, 128, 131, 132, 133, 180, 181, 183, 191, 192, 193, 217, 229, 237, 241, 250, 253, 356, 373, 430, 432, 465
Smart Mobility 70, 85, 127, 132, 438, 505, 506
Support Vector Machine (SVM) 318, 354, 394, 395
Sustainable Mobility 33, 91, 170, 175, 190
Sustainable Transportation 6, 104, 107, 159, 169, 170, 173, 176, 188, 189, 192, 475, 482

T

Traffic Management 12, 22, 30, 40, 42, 43, 44, 45, 46, 47, 48, 49, 50, 51, 53, 54, 56, 57, 58, 60, 61, 71, 72, 73, 75, 77, 78, 79, 80, 81, 82, 83, 85, 86, 87, 88, 89, 90, 91, 92, 93, 94, 95, 96, 97, 99, 100, 103, 104, 105, 106, 109, 110, 111, 114, 117, 118, 119, 121, 122, 123, 124, 126, 127, 128, 129, 130, 131, 132, 134, 136, 140, 141, 154, 156, 161, 172, 178, 179, 182, 184, 187, 217, 223, 236, 287, 320, 332, 333, 357, 368, 386, 398, 402, 407, 416, 424, 427, 430, 431, 435, 438, 445, 446, 457, 458, 462, 464, 465, 469, 474, 475, 477, 478, 484, 488, 492, 493, 495
Traffic Monitoring 103, 106, 110, 133, 141, 146, 149, 151, 153, 160, 329
Traffic Optimization 11, 47, 49, 50, 75, 80, 82, 83, 93, 95, 96, 105, 117, 173, 318, 319, 345
Traffic Safety 34, 60, 135, 136, 158, 163, 165, 217, 269, 274, 322, 351, 354, 359, 364, 367, 373, 375, 378, 383, 384, 385, 392, 393, 399, 411, 417, 430, 432, 479, 484, 490
Transportation Safety 1, 2, 3, 4, 5, 6, 7, 8, 9, 10, 14, 15, 20, 21, 22, 24, 25, 26, 28, 30, 31, 42, 59, 62, 135, 136, 137, 139, 140, 141, 142, 143, 147, 150, 155, 159, 161, 163, 164, 165, 166, 236, 257, 258, 259, 260, 261, 262, 263, 270, 272, 273, 274, 276, 279, 280, 281, 283, 289, 295, 297, 303, 304, 306, 309, 313, 315, 316, 317, 325, 330, 339, 340, 344, 345, 346, 353, 381, 383, 384, 404, 406, 407, 413, 414, 415, 416, 417, 419, 422, 423, 425, 435, 437, 438, 473, 475, 476, 477, 478, 481, 483, 484, 485, 486, 487, 488, 497, 499, 500, 503, 505

U

Urban Infrastructure 72, 133, 224
Urban Mobility 1, 2, 7, 31, 33, 40, 43, 44, 48, 51, 58, 85, 95, 96, 98, 99, 100, 103, 106, 110, 125, 132, 150, 169, 171, 173, 180, 181, 182, 184, 188, 192, 401, 408, 414, 430, 438, 477
Urban Transportation 1, 5, 8, 24, 32, 97, 126, 127, 141, 186, 194, 383, 384, 385, 387, 388, 406, 407, 408, 427, 428, 429, 432, 471

V

V2X Communication 203, 212, 213, 222, 223, 225
Vehicle-to-Infrastructure (V2I) 43, 45, 58, 106, 109, 124, 127, 134, 180, 181,

187, 198, 200, 201, 205, 215, 217, 224, 227, 305
Vehicle-to-Vehicle (V2V) 58, 180, 181, 198, 200, 201, 205, 213, 221, 224, 305, 354

W

Wireless Sensor Networks (WSN) 232

www.ingramcontent.com/pod-product-compliance
Lightning Source LLC
LaVergne TN
LVHW081318300926
849161LV00008B/1802